BIOLOGY AND CONSERVATION OF RIDLEY SEA TURTLES

BIOLOGY AND CONSERVATION OF RIDLEY SEA TURTLES

Edited by PAMELA T. PLOTKIN

THE JOHNS HOPKINS UNIVERSITY PRESS
Baltimore

Printed in the United States of America on acid-free paper

9 8 7 6 5 4 3 2 1

The Johns Hopkins University Press
2715 North Charles Street
Baltimore, Maryland 21218-4363
www.press.jhu.edu

Library of Congress Cataloging-in-Publication Data

Biology and conservation of ridley sea turtles /
edited by Pamela T. Plotkin.
p. cm.
Includes bibliographical references and index.
ISBN-13: 978-0-8018-8611-9 (hardcover : alk. paper)
ISBN-10: 0-8018-8611-2 (hardcover : alk. paper)
1. Lepidochelys. 2. Wildlife conservation.
I. Plotkin, Pamela T., 1962–
QL666.C536B55 2007
597.92′8—dc22 2006027168

A catalog record for this book is available from the British Library.

Photo credits: Chapter opening art courtesy of SeaPics.com. All photographs are by Doug Perrine except those for Chapters 4 (Carlos Villoch / V & W), 10 (Robert L. Pitman), 12 (Peter C. H. Pritchard), and 13 (Lisa Ballance).

CONTENTS

Preface vii
Abbreviations, Acronyms, and Conventions ix

Introduction
Pamela T. Plotkin 3

1 *Arribadas* I Have Known
Peter C. H. Pritchard 7

2 Understanding Human Use of Olive Ridleys: Implications for Conservation
Lisa Campbell 23

3 Evolutionary Relationships, Osteology, Morphology, and Zoogeography of Ridley Sea Turtles
Peter C. H. Pritchard 45

4 An Evolutionary Perspective on the *Arribada* Phenomenon and Reproductive Behavioral Polymorphism of Olive Ridley Sea Turtles (*Lepidochelys olivacea*)
Joseph Bernardo and Pamela T. Plotkin 59

5 Age and Growth in Kemp's Ridley Sea Turtles: Evidence from Mark–Recapture and Skeletochronology
Melissa L. Snover, Aleta A. Hohn, Larry B. Crowder, and Selina S. Heppell 89

6 Phylogeography and Population Genetics
Kristina Kichler Holder and Mark T. Holder 107

7 Respiratory and Endocrine Physiology
Roldán A. Valverde, Erich K. Stabenau, and Duncan S. MacKenzie 119

8 Reproductive Physiology of the Ridley Sea Turtle
David C. Rostal 151

9 Sex Determination and Sex Ratios in Ridley Turtles
Thane Wibbels 167

10 Population Sex Ratio and Its Impact on Population Models
Michael Coyne and Andre M. Landry, Jr. 191

11 Adult Migration and Habitat Utilization: Ridley Turtles in Their Element
Stephen J. Morreale, Pamela T. Plotkin, Donna J. Shaver, and Heather J. Kalb 213

12 Effect of Land-Based Harvest of *Lepidochelys*
Stephen E. Cornelius, Randall Arauz, Jacques Fretey, Matthew H. Godfrey, Rene Márquez-M., and Kartik Shanker 231

13 Human–Turtle Interactions at Sea
Jack Frazier, Randall Arauz, Johan Chevalier, Angela Formia, Jacques Fretey, Matthew H. Godfrey, Rene Márquez-M., Bivash Pandav, and Kartik Shanker 253

14 Head-Starting the Kemp's Ridley Sea Turtle
Donna J. Shaver and Thane Wibbels 297

15 Kemp's Ridley Recovery: How Far Have We Come, and Where Are We Headed?
Selina S. Heppell, Patrick M. Burchfield, and Luis Jaime Peña 325

Near Extinction and Recovery
Pamela T. Plotkin 337

Contributors 341
Index 345

PREFACE

THIS BOOK WOULD NOT HAVE BEEN POSSIBLE without the valuable contributions, knowledge, energy, and enthusiasm of the chapter authors. Collectively they represent the world's experts on ridley sea turtles. Individually their diverse interests and expertise have contributed to furthering our understanding of the biology and conservation of ridley sea turtles. Their willingness to contribute, coupled with their patience and flexibility, made this book a truly collaborative and rewarding effort.

Each chapter of this book was reviewed by two or three external reviewers. Charles Caillouet offered critical reviews of almost every chapter in the book and greatly enhanced them. The other reviewers whose comments improved the book include George Balazs, Christopher Binckley, Marydele Donnelly, Karen Eckert, Nancy FitzSimmons, Nat Frazer, Jack Frazier, Matthew Godfrey, Cynthia Lagueux, Valentine Lance, Duncan MacKenzie, J. Nichols, Frank Paladino, Kartik Shanker, Anthony Steyermark, Manjula Tiwari, and Tony Tucker.

Finally, I thank several individuals who worked behind the scenes to inspire and encourage this project. I acknowledge my husband, Joseph Bernardo, for his encouragement and support. I also thank my editor, Vincent Burke, without whom this book would not have been possible. Numerous delays held up this project, and his patience, guidance, and encouragement were unfailing. I thank James Spotila for his generosity and friendship. I thank the Johns Hopkins University Press for publishing this book and recognizing the value of all the world's sea turtles.

Support for this project came from a grant awarded by the East Tennessee State University Research Development Committee. Publication of this book was supported by the Betz Chair of Environmental Science at Drexel University.

ABBREVIATIONS, ACRONYMS, AND CONVENTIONS

ACTH	adrenocorticotropic hormone
ADIO	Asociación de Desarrollo Integral de Ostional
AMH	antimullerian hormone
ASEAN	Association of Southeast Asian Nations
ASM	age at sexual maturation
AVT	arginine vasotocin
BPH	body proportional hypothesis
CCL	curved carapace length
CNS	central nervous system
CRH	corticotropin-releasing hormone
CTF	Cayman Turtle Farm
CWT	coded wire tag
DNA	deoxyribonucleic acid
E_2	estradiol
EEZ	exclusive economic zone
FAD	fish-aggregating device
FSH	follicle-stimulating hormone
GAM	generalized additive models
GH	growth hormone
HDT	high disturbance threshold
HEART	Help Endangered Animals—Ridley Turtle
HPA	hypothalamo-pituitary-adrenal
IAC	Inter-American Convention for the Protection and Conservation of Sea Turtles
IATTC	Inter-American Tropical Tuna Commission

INP	Instituto Nacional de la Pesca
IOSEA	Indian Ocean and Southeast Asia
IRD	inner ring deiodinase
IRENA	Instituto Nicaragüense de Recursos Naturales y del Ambiente
IUCN	International Union for the Conservation of Nature
LAG	line of arrested growth
LH	luteinizing hormone
MARENA	Ministerio de Agricultura y Reforma Agraria
mtDNA	mitochondrial deoxyribonucleic acid
MY	million years
MY BP	million years before present
N_e	effective population size
nDNA	nuclear deoxyribonucleic acid
NGO	nongovernmental organization
NMFS	National Marine Fisheries Service
NP	neurophysin
ORD	outer ring deiodinases
P	probability
PINS	Padre Island National Seashore
PIOSA	Pesquera Industria de Oaxaca
PIT	passive integrated transponder
PRL	prolactin
PROFEPA	Procuraduria Federal de Protección al Ambiente
PROPEMEX	Productos Pesqueros Mexicanos
RFLP	restriction fragment length polymorphism
RIA	radioimmunoassay
SCL	straight carapace length
SEAFDEC	Southeast Asian Fisheries Development Center
SEMARNAP	Secretaría del Medio Ambiente Recursos Naturales y Pesca
T	testosterone
T_3	triiodothyronine
T_4	tetraiodothyronine, thyroxin
TED	turtle excluder device
TEWG	Turtle Expert Working Group
TPWD	Texas Parks and Wildlife Department
TRH	thyrotropin-releasing hormone
TRT	transitional range of temperatures
TSD	temperature-dependent sex determination
TSH	thyrotropin
USFWS	United States Fish and Wildlife Service
VBGF	Von Bertalanffy growth function

BIOLOGY AND CONSERVATION OF RIDLEY SEA TURTLES

PAMELA T. PLOTKIN

Introduction

RIDLEYS ARE ETHOLOGICALLY the most enigmatic of all sea turtles. Unlike other species that emerge individually from the sea to lay their eggs in the sand, ridleys come ashore en masse to reprise the cycle of life. These mass emergences are not mere coincidence; rather, they are notably premeditated. Throngs of turtles migrate from distant reaches of open ocean to the nearshore waters of specific beaches, gathering there for days or weeks, and then suddenly—much like the start of a race when a whistle blows, gates burst open, and the race is on—turtles synchronously emerge. Waves of turtles crawl ashore for many days to sometimes weeks and vie for limited space to lay their eggs. These events, or *arribadas,* occur at only a few particular beaches worldwide. The chosen beaches are usually adjacent to other, larger stretches of sand and do not appear to be radically different. The sheer number of turtles nesting in spatially limited areas (1,000–500,000 turtles) results in moderately high levels of density-dependent mortality of eggs during a single arribada. The turtles return approximately every month during a discrete nesting season (3–6 months), and the eggs that survived intact during the previous month are at risk again when new waves of turtles crawl ashore.

Surprisingly, the behavior that characterizes *Lepidochelys* is neither well studied nor understood. There exists a fundamental gap in our understanding of the proximate and ultimate cues that regulate this behavior and its impact on the population biology and ecology of the species. Why did this behavior evolve in ridleys but not in other sea turtles? What is the adaptive significance of this behavior? What cues regulate the synchronicity of the aggregations and emergences? Why do arribadas occur where they do? A detailed general description of the behavior is lacking as well. For example, scientists studying olive ridley sea turtles are aware that some of them nest during arribadas, some nest individually, and yet others exhibit a mixed strategy, switching between behaviors. This behavioral polymorphism is one of the most intriguing aspects surrounding the arribada phenomenon, and it has yet to be described in the literature.

There are multiple reasons that the trait unique to *Lepidochelys* is so poorly studied. First, the nesting behavior and beaches for both the Kemp's ridley (*Lepidochelys kempii*) and the olive ridley (*Lepidochelys olivacea*) species of sea turtle were discovered late in the game: the sole mass nesting site for the Kemp's ridley was revealed to the scientific community in the early 1960s and, as a result of years of overexploitation of eggs, was a relict of its former state by the time scientists learned about its existence. In the late 1960s several mass nesting sites were discovered for olive ridleys, many of which were similarly overexploited and nearly extinct by the time they became widely known. Amazingly, some mass nesting sites have gone undetected for decades despite efforts to find and study these animals. For example, several olive ridley mass nesting sites were just discovered in the 1990s along the coast of India.

Second, historically, great emphasis has been placed on studying Kemp's ridley because the species was on the brink of extinction. Saving it from extinction greatly overshadowed the need to study and understand its unique nesting behavior. From the mid-1960s well into the 1990s, research conducted on Kemp's ridley focused on applied rather than basic science. During this time, multilateral, last-ditch efforts were implemented to save Kemp's ridley from extinction. Although a great deal was learned about Kemp's ridley basic biology, it was ancillary to the required focus on critical management needs. The First International Symposium on Kemp's Ridley Sea Turtle Biology, Conservation and Management was held in Galveston, Texas, October 1–4, 1985. The *Proceedings,* published in 1989 (Caillouet and Landry, 1989), was the first thorough assemblage and review of research conducted on Kemp's ridleys during the prior two decades. It also served as an important springboard for identifying gaps in our knowledge of Kemp's ridley biology, conservation, and management. Several other important publications followed, the most notable being Rene Márquez-M.'s (1994) *"Sinopsis de datos biologicos sobre la tortuga lora."* He summarized taxonomy, biogeography, life history, and ecology and included a great deal of data collected from his own studies. A special session on the biology of the Kemp's ridley was held during the Nineteenth Annual Symposium on Sea Turtle Biology and Conservation, South Padre Island, Texas (March 2–6, 1999), and these papers were recently published in a special issue of the journal *Chelonian Conservation and Biology.*

Finally, the behavior that defines *Lepidochelys* is so poorly understood in part because of the distribution of olive ridleys. Their arribadas tend to occur in remote areas not easily accessible by land, and they frequent deep oceanic waters where they are also generally inaccessible. Furthermore, because olive ridleys are the most abundant sea turtle in the world, they have not been a "conservation priority," and funding to study them has been less available. As a result, comparatively little research has been conducted and published on this species. There have been no symposia focused on the species, only a few hard-to-obtain and now out-of-print books exist about it, and most olive ridley studies have been published in the gray literature or in obscure journals. One such book is *The Turtle Paradise Gahirmatha (An Ecological Analysis and Conservation Strategy)* by M. C. Dash and Chandra Sekhar Kar (1990). This book is based on the dissertation work of Kar and is a vast reservoir of data on olive ridleys in general and, more specifically, on the natural history, ecology, and conservation threats to olive ridleys at their most prominent arribada nesting beach, Gahirmatha, located in the northern Indian Ocean. Another important resource is Stephen Cornelius' *The Sea Turtles of Santa Rosa.* This book, written for a lay audience, provides information about all of the sea turtle species nesting at Santa Rosa National Park,

Costa Rica (now part of the Area de Conservación Guanacaste) and contains specific information about the biology and ecology of the olive ridleys that nest at Nancite Beach, the only arribada beach that has been completely protected and remains in an intact wild ecosystem to this day.

Much of what we do know about ridleys is summarized in the chapters herein. It is my intent and hope that this book provides a scholarly and synthetic review of the ridley literature and serves to educate students, scientists, resource managers, and conservation policymakers about ridleys. More importantly, I hope that this work will stimulate further research, not only in the areas represented in this book but in other areas as well, to provide better understanding of both species and the roles they play in their marine and terrestrial ecosystems, and to provide the best available science on which global and local conservation policy decisions must be based.

LITERATURE CITED

Caillouet, C. W., Jr., and Landry, A. M., Jr. 1989. *Proceedings of the First International Symposium on Kemp's Ridley Sea Turtle Biology, Conservation and Management*. Texas A&M University, Sea Grant College Program, TAMU-SG-89-105.

Dash, M. C., and Kar, C. S. 1990. *The Turtle Paradise Gahirmatha (An Ecological Analysis and Conservation Strategy)*. New Dehli: Interprint Press.

Márquez-M., R. 1994. *Sinopsis de datos biologicos sobre la tortuga lora*. FAO Sinopsis sobre la Pesca, No. 152. Mexico: Instituto Nacional de La Pesca.

PETER C. H. PRITCHARD

1

Arribadas I Have Known

THINGS HAVE CHANGED IN THE Galápagos Islands since I started field work there over 30 years ago, when there was neither telephone nor radio, and precious little mail. Today I sit, pen in hand, in one of the many internet cafes in Puerto Ayora during a 3-day hiatus between field trips in the Islands, having just heard via e-mail that Henry Hildebrand has died. My path intersected only occasionally with that of this quiet Texan, but those encounters were always both enlightening and pleasurable. I see vividly in my mind's eye the broad-cheeked face with eyes almost closed against the harsh Texas sun, and I hear again the slow drawl as Henry talks about ridley turtles and about *arribadas.* I have long considered arribadas to be the most spectacular manifestations of reptile life anywhere, and to aficionados they are a genuine and absolute wonder of the world. It was Henry, in 1961, who introduced arribadas to the global herpetological audience.

The warm-up act to Henry's revelation was Archie Carr himself, who had wandered the shores of Florida, the Bahamas, and the Caribbean for more than two decades, pondering the weird mystery of Kemp's ridley. This was a familiar Florida turtle, by no means rare, that had never been known to breed, not in Florida and not anywhere else. To this day, the Germans call it *Bastardschildkröte,* an allusion to the innocent turtle having something profoundly wrong with its parentage or its sexuality or just its origins in general. Yet the other kind of ridley, *Lepidochelys olivacea,* not all that different from *kempii* in appearance and of nearly circumglobal distribution, nested "normally," or so it was thought.

Archie included the chapter "The Riddle of the Ridley" in his 1957 book *The Windward Road* (Carr, 1957). Subsequently, Archie started to search the Mexican coast and eventually found three adult Kemp's ridley carapaces nailed to the wall of a small restaurant south of Alvarado in Veracruz. These were the first adult ridley shells he had ever seen (their adulthood was an assumption, but he was assured that the bearers of the carapaces had nested locally), and he pondered the possibility that the ridley mystery would end not with a bang but a whimper of isolated stories and rumors and individual nesting records. Later, Archie's attention was drawn to J. E. Werler's account, published many years earlier, in 1951, of two ridleys nesting on the coast of Padre Island, Texas, in 1948 and 1950, by daylight at that, which was a further hint that the western Gulf was a crucial area for ridley investigations.

But Henry Hildebrand's independent lines of investigation, which finally led him to the office of the Mexican rancher and engineer Andres Herrera, dispelled any possibility of an anticlimax. Herrera's famous film of an estimated 40,000 turtles nesting in the midday sun on a Tamaulipas Beach in June 1947 may not have taken Hollywood by storm, but it surely stunned the sea turtle world. Archie Carr (1963) and Hildebrand (1963) later jointly made the formal announcement to the scientific world, a couple of years after the Herrera film was shown at the Austin, Texas, meeting of the American Society of Ichthyologists and Herpetologists in 1961 under circumstances eloquently described by Carr (1967).

Archie obtained a copy of the Herrera film, and he showed it to me several times. Although magenta with age and somewhat scratched, it was well shot and well focused. And just as American households today always have to ask their youngest member to handle computer tasks for them, it seemed that Archie's youngest son, David, was the only one who could work the projector. As the old film clattered through the complex sprockets and gates of the 16-mm machine, Archie would point out the highlights of the epic production with unfailing enthusiasm, never hiding his disgust when some of the Mexicans in the film actually stood on the backs of the already overburdened little ridleys. For sheer demonstrative excitement, the frenetic perennial fans of the *Rocky Horror Picture Show* had nothing on Archie.

The film inspired me too, and in the subsequent years and decades, arribadas have provided me with a host of lifetime memories. For several years (1968, 1970, 1973) I drove my Land Rover to Rancho Nuevo, Mexico, each season to help with the ridley patrols. I used this mobility to deploy the Mexican marines assigned to turtle patrol to far-flung locations up and down the beach. Later we flew to Rancho Nuevo in Florida Audubon's little Cessna 182, and we not only did radiotracking with Mary Mendonça of offshore ridleys from this lofty base (in contrast to *olivacea,* their surface time was absolutely minimal) but also "buzzed" egg-laden poachers leaving the beach by secret trails that were hidden from the ground but totally visible from above. Other Rancho Nuevo adventures included creating the first airstrip at the site and testing it with an aircraft much larger and faster than our informal specifications would have permitted.

We always landed on the highway at Barra del Tordo, sometimes even taxiing up to the motel and parking for the night while we slept; and once we almost hit a school bus as we came in to land. Who could forget, too, the night that a screeching hurricane tore the roof off our camp and, 12 hours later, brought in the biggest arribada of the season. Or the year when an outside company poured a concrete slab just behind the dune to build a slaughterhouse so they could slaughter several hundred adult *kempii*—with full permission from the Mexican authorities. Or the time when a huge pig lived in camp for months as a pet; when it was finally scheduled for slaughter, tender-hearted Laura Tangley gave it soothing backrubs all night; and the next day it was dispatched and barbecued—and charred almost beyond edibility.

Pacific Arribadas

Great adventures lay with Pacific arribadas too. For several years we conducted intense negotiations and discussions with Antonio Suarez, who managed trading in ridleys in the State of Oaxaca, about the merits of his "sustained complete utilization" policy for Oaxaca ridleys. We despised his arguments but nonetheless could not categorically deny their reasoning, and in some ways he turned out to be right (see below). Later, I invited him to explain his theories of ridley population dynamics to the assembled turtle biologists of the world in a meeting at the U.S. State Department in Washington, D.C. He reluctantly agreed to come but was arrested by the feds as he finished his speech, and he was convinced that I had put them up to it. I hadn't, but I was the party responsible for producing a sample of certified fresh ridley meat from Mexico for the feds investigating the case. There had been no time to issue a federal import permit, so I had asked to be arrested on my arrival in Miami, a cloak-and-dagger operation that ended up with representatives of three federal departments meeting me at the steps of the aircraft.

We once built an elaborate camp at Piedra de Tlacoyunque in which to await the local arribada, only to have the thatched roof blown away the first night by hurricane-force winds and driving rain, and the next day the entire party became deathly sick from drinking the tainted waters of a local river inaptly known as Purificación.

And never to be forgotten is the time we set off to Playa Nancite by boat from Playas del Coco with a small group of rather elderly tourists, only to have the boat start to fill with water to the point where we had to head for a deserted beach near Punta Mala, where we lived like Robinson Crusoe for 24 hours before being rescued. On the return trip, we experienced the loss of our engine (literally: it dropped off in over 100 feet of water), but we were rescued by a lobster fishing boat and laboriously towed into port. This adventure was written up and published by Fred Caporaso (1987) in a florid composition entitled "Indiana Pritchard and the Dinghy of Doom." And despite all, we didn't miss the arribada: when we arrived a day late from Punta Mala, we landed in the thick of one comprising at least 30,000 turtles.

There were numerous other adventures during the many weeks that I have waited at Nancite for arribadas to arrive. I recall a tranquil month when we camped together with the Japanese film crew making a documentary for *Waku Waku Animal World*. The arribada was weeks late, and we all missed it; but the evenings spent singing songs with people with whom we had absolutely no language in common but a great deal of mutual fondness was memorable indeed. Another time at Nancite, I shared a crowded tent with a highly pregnant Peruvian turtle person, who insisted that I place my hand on her belly during each of her numerous fetal gyrations, and who protected the virtues of the youthful Costa Rican kitchen staff by constantly spraying me with an imaginary spray can of "repellente sexual." Other events had an almost surrealistic quality: the time we had to transport a 500-lb water-cooled generator to the beach so that we could illuminate the arribada for filming purposes; we ended up borrowing the helicopter from the Office of the President and slinging the massive machine below it while all other parties hiked in.

Perhaps the most dangerous Nancite incident was when we spotted a small crocodile in clear water a dozen or two yards from shore, and I returned to camp to enlist a volunteer to catch it. It was a very dark night, but the eye-shine was still visible in exactly the same place. We waded out ready to seize it with a coordinated grab at the tail-base and the base of the neck when a small wave lifted it up slightly, and only then did we realize that the meter-long animal had been replaced in my brief absence by a monster three times that length.

Western Atlantic Arribadas

I never got to Gahirmatha. We became hopelessly lost in the mangrove forests as we tried to break through to the coast and eventually had to turn back. But having serendipitously discovered the first recorded nesting by olive ridleys in the Western Atlantic, in Guyana in 1964, I selected the genus *Lepidochelys* as the subject of my doctoral dissertation, and this happy development led to many other adventures. I continued my Guianas programs each summer, progressively shifting operations eastward, until my travels took me to the little beach of Eilanti, Suriname.

Serendipity struck again. When, after a long sea voyage from Bigisanti, we got to this 900-m scrap of sand, day was almost done, and dozens of Carib men were milling about, seemingly with high expectations of something. Many were sharpening sticks, as if for war. It was going to be a Night of the Warana: an arribada, the first I have ever heard of for the olive ridley, was expected, and the sticks were for the perfectly unwarlike purpose of marking turtle nests.

And the turtles came. Not by day, as the Kemp's ridleys did, but by night, with the rising tide, and with a spanking breeze from the sea. Over several nights, hundreds of ridleys came ashore, all with the agitated, hurrying look of the *kempii* in the Herrera film, and always with the same thump-thump-thump as the turtle, with limbs fixed and a look of unfrolicsome resignation or determination on its face, rock-and-rolled from side to side as it compressed and packed the loose sand that it had excavated as it dug its nest. We were seeing the "dance of the ridley," peculiar to the genus *Lepidochelys*. In Costa Rica, the olive ridley is called "carpintera" because the sand-thumping operation sounds like a carpenter hammering. The arribada was not as massive as that of the *kempii* in Archie's favorite movie, but it was constrained into a small space by the limited size of the beach, and it was good enough for me.

It also turned out that a small, compact arribada was much easier to study than a huge one. By enlisting the help of the Caribs, it was possible to get a remarkably exact count of the nesters, and even though we used all the 130 tags we had brought in a few hours, this was enough to establish that the majority of the females returned the following year to renest. And allowing for mortality and tag loss, perhaps all the survivors returned. Nesting in successive years was at least the norm for the population, but this had not yet been demonstrated for any other sea turtle population in the world.

I returned to Eilanti the next two seasons, 1967 and 1968, not just to gather scientific data but also to carry out a conservation project. It was clear that, even though the Caribs did not kill any of the turtles, the 100% egg harvest was not a good idea. They did not normally eat the eggs themselves but rather took or sent them to Paramaribo, where they were purchased largely by the Javanese community, still mindful of a crucial dietary priority in their ancestral homeland.

Our strategy was a simple, unsubtle one: we would buy the eggs. With a small grant from the Word Wildlife Fund, we paid one Suriname cent per egg and had 300,000 in the hatchery by the end of the season. It seemed to work, and everyone cooperated. The next year we gambled that we would have more hatchlings if we never moved the eggs but rather marked each nest site with a code number and rewarded the man who showed us the nest. This required more trust than simple purchase, but again the system worked. After that season, the beach was declared a nature reserve, and egg collecting was prohibited. The economic impact was reduced somewhat by hiring as many of the erstwhile egg collectors as possible to work in the reserve.

Despite our success in converting the Eilanti ridley colony from one with no reproductive success to one with remarkably high hatchling production, the number of turtles themselves gradually diminished in the following years. Possibly the steady trawler mortality was the main cause; we learned of this problem from the trawlermen themselves, who regularly sent in tags of turtles that they had caught in the waters of all three of the Guianas in the "old days" when the trawler by-catch problem had not become politically sensitized. And obviously the many years of total egg harvest must also have been a problem. But the beach itself had changed. The last time I saw it, in 2000, it was unrecognizable, with a thick mud flat with a dense stand of young mangroves fronting the whole beach, which itself had become very muddy; and a new sand ridge had built up in front of the whole of Eilanti Beach, where numerous leatherbacks now nested! The new beach already had a name, Samsambo. Meanwhile, those ridleys that were still around had shifted a couple of hundred kilometers to the east, to the beaches around Cayenne itself.

One of the identified problems in marine turtle conservation, and in particular population monitoring, is the issue of the "shifting baseline." Turtle populations may become depleted progressively and gradually, and the current generation of beach patrollers may have a relatively short-term personal experience of how things were in the past. They may be tempted to reckon that all is well if they see 100 turtles nesting if they never saw that many before in the few years they have studied the colony; yet there may have been 1,000 turtles each season a century ago. Another problem is that truly antique "baseline" data are entirely anecdotal, reflecting methodologies no more advanced than casual or incidental observation and guesswork, and the written word, sometimes committed to paper (or parchment) years later, may incorporate the tricks of imperfect memory or even gross exaggeration to make a wonderful discovery even more "wonderful."

The Shifting Baseline

With ridleys, the "shifting baseline" takes on a new twist, for the historical baseline is nonexistent! Somehow, even though nesting sea turtles, especially the highly edible green turtles, feature time and time again and in considerable detail in the accounts of early explorers, especially of the New World, the extraordinary phenomenon and spectacle of ridley arribadas escaped them completely. Even in Orissa, India, where currently very large arribadas occur, it seems bizarre that such occurrences could have completely escaped the attention of local people until a couple of decades ago. Mohanty-Hejmadi (2000) did unearth an interesting eighteenth-century report (Hamilton, 1727), in which the author visited northeastern India (Ganjam to Ballasore) in 1708 and commented on a sandy bay "where prodigious Number of Sea Tortoises resort to lay their Eggs" between Cunnaca (Maipura) and Balasore (Budha Balanga), but whether this is the precursor of the Gahirmatha arribada remains unknown. Nowadays, despite the enormous numbers of turtles nesting at Gahirmatha, the arribada is concentrated in a few kilometers near the Maipura River. Perhaps extensive diffuse nesting becomes concentrated into arribadas that persist for some decades and then rediffuse when the arribada becomes self-limiting.

The explanation for the recent discovery of all arribadas may lie partially in the remote locations of these events. Local people may indeed be present and indeed look on the ridleys as the bearers of a great seasonal bounty of highly edible eggs, but communications, literacy, and especially the habit of publishing one's observations may simply be so limited that the outside world never got the word.

But this may not be the whole story. Explorers and travelers surely reached almost all tropical and subtropical coastal areas of the world at least briefly or occasionally—hence the abundance of accounts of green turtles, so excellently summarized by Parsons (1957) in his book *The Green Turtle and Man*. Nevertheless, one difference may be that green turtles often nest on islands, and even very remote, isolated ones were not beyond the reach of some of the intrepid eighteenth- and nineteenth-century voyagers. Ridleys, on the other hand, are primarily—indeed, almost exclusively—mainland nesters, and the interests of explorers and travelers in continental areas may have been mostly inland.

Another part of the answer may lie in the nature of ridleys themselves and the peculiar survival advantages of the arribada habit. A nesting ridley is a small, wide animal that shuffles and slides over the sand of the beach, leaving a very shallow track. Also, ridleys often nest during high winds, which may quickly obliterate these shallow tracks and also, possibly, the characteristic scent of a fresh nest, leaving little or no indication that anything has nested. Furthermore, if these highly ephemeral signs are combined with the habit of the majority of the population nesting in a sudden, vast assemblage, emerging unpredictably, nesting rapidly, and retreating promptly, a casual visit may give no indication that the beach is one that is used by turtles at all. Only if you live there will you know that turtles ever come.

In view of all of these observations, with arribadas in the eastern Pacific and northern Indian oceans as well as in the western Atlantic, one wonders whether it is only a matter of time before a West African arribada is discovered. In a major new monograph on sea turtles of the western coast of Africa, Fretey (2001) documents nesting of *L. olivacea* on mainland shores from much of tropical West Africa and also on the islands of Bioko and São Thomé, with "strong" nesting in both Guinea Bissau and Angola. My personal conviction is that, somewhere in this region, there must be arribadas. After all, the major nesting colony of green turtles in the Bijagos Archipelago and the huge Gabon leatherback colony, perhaps the largest in the world, were unknown until recently, even though green turtles and leatherbacks leave very persistent, deep tracks and, in contrast to ridleys, do not concentrate their nesting into just a few nights of the year.

And finally, in view of the fact that the olive ridley must be the most numerous sea turtle in the world, it is amazing that it was confused with the loggerhead for most of its known history. It was not just arribadas that were overlooked but the species itself. Although olive ridleys were described as *Chelonia olivacea* by Eschscholtz (1829) and renamed *Lepidochelys olivacea* by Fitzinger (1843), the name and to a large extent the actual species to which the name referred were ignored (with rare exceptions, such as Baur [1890] and Deraniyagala [1933]) until about a century later, when Carr (1942) demonstrated the close relationship between *kempii* and *olivacea* and their distinctness from *Caretta*.

Significance of Arribada Behavior

The question also arises of what is the point, the advantage, of arribada nesting (see Bernardo and Plotkin, Chapter 4). Examination of hatching success at Nancite and Ostional, Costa Rica, indicates that participation in an arribada is a singularly poor way of ensuring that one's genes are well represented in the next generation, and the late Douglas Robinson hypothesized that the future of the species statistically lay with the turtles that nested singly rather than those in the arribadas. Yet if this were true, there would have been great selective pressure against aggregated nesting ever occurring unless there were many turtles and only a few beaches that were physically suitable, and even then, "abundant nesting" every night of the season is not the same thing as nesting in arribadas.

Probably, an arribada is highly advantageous only in its early stages. Turtles that nest singly may be subject to devastating, one-by-one predation on their nests, especially by mammals (including mankind), whereas an unexpected large group of turtles nesting and then quickly returning to the sea may saturate the scattered local predators with a very ephemeral overabundance of food, and the surplus may be left to hatch in safety.

The former situation is exemplified by the nesting colony at Isla Ratones, in the Golfo de Fonseca, a picturesque Pacific bay shared by El Salvador, Honduras, and Nicaragua. Carr (1948) visited the beach in 1947 and reported intensive exploitation of olive ridley eggs by *hueveros* who encamped on the beach for the duration of the nesting season each year. I spent two months there myself 20 years later, in 1967. Egg collection had intensified to the extent that several hueveros competed for each emerging turtle, rushing to the surf line, and the winner carried the surprised turtle on his head and dumped it heavily into a preexisting body pit. (To my amazement, about half of the turtles so treated actually proceeded to nest.) And after 20 more years, in 1987, Cynthia Lagueux was there and reported similar exploitation, but the turtles were still nesting.

How on earth could the turtles have maintained their nesting colony in the face of half a century of total egg exploitation? The answer must lie with the then-undiscovered arribadas farther south, at Chacocente and La Flor and in Costa Rica too. On these beaches, there may have been millions of hatchlings produced each year, and when the survivors matured, they may well have fanned out over a considerable distance to nest singly on hundreds of kilometers of coast (including Isla de Ratones), even though the majority, the philopatric ones, returned to their natal beaches.

The irony is that these philopatric arribada turtles, or at least their eggs and hatchlings, will become victims of their own abundance as the arribada grows, and the effective replacement or recruitment rate will then drop, ultimately to a very low figure. Clearly, an arribada, to maintain effectiveness as a reproductive strategy, needs a switch-off mechanism, but nature seems unable to provide one. Turtle nesting cues and behavior are extremely "hard-wired," and they show remarkably little flexibility. (Large aggregations of human beings may also show no great group wisdom or centralized planning ability!) So an arribada will eventually die out, and some genetically fortunate solitary nester will become the mother of a new or recrudescent arribada elsewhere.

So the final consideration is this: although the casual observer may assume that an arribada is a ritual of nature that has persisted from ancient times in each of a few, secret places in the world, the reality is that, even within a single human lifetime, new arribadas may form and others shift or disappear. Older inhabitants of Ostional, Costa Rica, indicate that the olive ridley arribadas for which their community is now famous may not even have occurred until about 1961 (although others remember arribadas in the 1940s; Campbell, 1998). Others in the Playa Grande area, to the north near Tamarindo, suggest that this beach used to be a ridley beach but is now almost entirely leatherbacks, a nesting population that, in turn, may have peaked a few decades ago and is now in serious decline. Perhaps none of the major arribadas discovered during the decade of the 1960s had existed, at least at those precise locations, for very long. Others peak, decline, and disappear. The Eilanti arribada now no longer takes place, the Kemp's ridley arribada in Tamaulipas, although recovering, nearly disappeared a couple of decades ago, and arribadas no longer occur at Piedra de Tlacoyunque (Guerrero), Playon de Mismaloya (Jalisco), and Bahia Chacagua on the Mexican Pacific coast.

This will-o'-the-wisp, ephemeral nature of arribadas presents unique challenges for conservation, a sort of counterpoint to the problems encountered in conservation of the hawksbill, whose nesting emergences are rarely aggregated and may occur singly on any of thousands of beaches, large and small, so that it is very difficult to plan useful, cost-effective patrols. The usual rule of thumb for any marine turtle conservation program is to make sure the nesting females survive and to ensure also that as many hatchlings reach the sea as possible. "The more turtles the better," is the unspoken but unquestioned philosophy of most marine turtle conservationists, although others condemn this approach as ecologically distorted "favorite species" conservation, such as is practiced by game departments to maximize deer populations, with inadequate regard for the predators and competitors of the turtles and their "equal rights" within their ecosystem. Much depends, of course, on whether sea turtles are to be regarded as endangered species in need of protection and recovery action or as management species of commercial and subsistence interest; opinions on this issue remain divided and partisan.

It is also widely suggested that egg harvests are relatively tolerable, whereas take of near-adult and adult turtles is unacceptable (Crouse et al., 1987). Yet some turtle populations (loggerheads in Florida, green turtles at Tortuguero) appear to have resisted decline in recent decades despite extensive loss of both subadult and adult loggerheads in shrimp trawls and heavy subsistence take of Tortuguero-hatched green turtles in half a dozen western Caribbean nations. However, other sea turtle populations, including leatherbacks and green turtles in Malaysia and possibly Kemp's ridleys in Mexico and olive ridleys in Suriname, have collapsed as a result primarily of egg harvest, although combined with some trawler capture.

Arribada Management

So what is the goal in arribada management, and how do we achieve that goal? Perhaps we should abandon the usual, unstated goal of "as many turtles as possible," not only because natural population constraints will eventually be felt on the feeding grounds but also because there is almost certainly some level of density of an arribada at which the sheer number of turtles is counterproductive, leading to degradation of the beach and massive, although accidental, destruction of eggs laid by previous nesters. Enormous, maximum-density arribadas are powerful spectacles and potentially could generate significant ecotourism revenues, but they are probably not sustainable phenomena.

One might hypothesize several alternative desired outcomes of an arribada management strategy:

1. To preserve the natural rhythms and fluctuations of the nesting population.
2. To provide the maximum sustainable take of turtles or turtle products (i.e., eggs) for human utilization.
3. To stabilize the arribada close to its present level or around some arbitrarily chosen different level.
4. To recover a (presumably depleted) population to a semblance of primordial abundance.
5. To achieve a population level higher than that which might have existed in an entirely natural state.

Criteria for success may be based primarily on numbers of turtles or on numbers of arribadas. It is often assumed that these criteria would have a direct relationship, but this may not be justified. From the evidence and accounts of the Herrera observations and film of the Kemp's ridley arribada, it would seem that only one enormous arribada appeared in the month of June 1947, without even scattered nesting on other dates. However, during the 1970s and 1980s, when the ridley population was depleted to 10% or less of the 1947 level, arribadas (now numbering only a few dozens or at the most some hundreds of turtles) would be seen many times during the season, or even each month, although the individual turtles in these events were now widely spaced so that only one or two could be seen at a time. The hypothesis, still unproven, governing these events is that the offshore aggregation of the turtles is a response to the pheromonal attraction of the secretion of the Rathke's glands, developed to a unique degree in the ridleys, and when there are many turtles in the area, a random concentration of animals will quickly grow in geometric fashion as each turtle adds to the pheromonal concentration in the water around the group. If only small numbers of turtles are present, they will form small clumps, but these may never achieve critical mass to coalesce into a single, tight aggregation.

An interesting additional point is that each arribada has its own distinct culture, a sort of group idiosyncrasy that may correspond to local geographic and ecological considerations but that, at least to some degree, may be independent of genetic programming. Thus, the Kemp's ridley arribada has the unique feature of always being diurnal, whereas olive ridley arribadas are almost always nocturnal, although they may carry over for an hour or two into early morning daylight. This difference may be so ingrained that it is reflected in the clear-cut difference in the diameter of the orbit in the two species, the orbit of the diurnal *kempii* being much smaller (Pritchard, 1969). The Suriname arribada had a reasonably fixed cycle with one or two precursive, small arribadas in May and early June, followed by three full arribadas at intervals of about 15 days in June and July. Each of these was characterized by a single, peak night of maximal nesting, with the fine details of timing controlled by meteorological conditions: specifically, a strong wind from the sea was necessary for many turtles to nest. Many turtles participated in two of the three major arribadas, but very few participated in all three (Pritchard, 1969). My own experience indicates that both the Rancho Nuevo and the Eilanti arribadas occurred during periods of strong onshore winds and that an arribada, even though expected in terms of normal interemergence intervals, would be delayed until the wind picked up. However, Cornelius (1991) reported that, in Costa Rica, there was no link between arribada timing and wind speed or temperature.

In Costa Rica, the two very large arribadas show some interesting differences. The one at Nancite is spatially confined between rocky headlands and thus does not shift, whereas the Ostional turtles emerge on a lengthy series of beaches with occasional small rocky outcrops, and successive arribadas may come ashore on slightly different sections. The Ostional arribadas, during the later months of the year, tend to follow a predictable rhythm, whereas that at Nancite is virtually unpredictable, at least in recent years, even though there is significant exchange between individuals nesting at the two sites. Furthermore, in certain years (but not all) there may be arribadas at Ostional in all months of the year, although those early in the year are relatively small (20,000–60,000 nests per month compared with 90,000–180,000 during the wet season; Richard and Hughes, 1972; Ballestero, 1994).

A striking peculiarity of the Orissa (India) arribadas is that, although they are very large, they completely fail in certain years. The reasons are unknown but may appertain to irregularities in the nutritional base of the population, causing individual turtles to switch from the typical annual nesting cycle to one of two years (or more).

Each one of the above-mentioned options has been the implicit goal in the management of one or more of the world's arribadas. It is worth reviewing each case to gain clues for the development of an enlightened management strategy that will associate specific manipulations with real-world, observed results. The known arribadas are described in the following sections.

Kemp's Ridley

The single known site of aggregated nesting of this species is the southern part of the coast of Tamaulipas, Mexico. Once thought to be solely near Rancho Nuevo in the Municipio de Aldama, nesting is now known to occur over several hundred miles of almost continuous undeveloped beach from southern Texas (Padre Island) to the northern and central Veracruz coast, but aggregated nesting is usually still in the vicinity of Rancho Nuevo.

Olive Ridley

Atlantic Ocean

The only known site of aggregated nesting in the Atlantic is the now-dispersed (or disappeared) colony at Eilanti, eastern Suriname, formerly with scattered nesting in Guyana as well. Today nesting has shifted to eastern French Guiana and Brazil, but arribadas have not been reported. It is now known that olive ridley nesting is widespread in West Africa as well, but arribadas have not yet been reported.

Indian Ocean

Almost all the olive ridleys in the Indian Ocean nest in two or three large aggregations near Gahirmatha in the State of Orissa, India. Scattered nesting occurs elsewhere on the Coromandel Coast and in Sri Lanka, but in most areas of the Indian Ocean ridleys are a rarity.

Pacific Ocean

Ridleys are rare in the western and central Pacific, and the known arribadas are confined to the tropical eastern Pacific, in Mexico and Central America. In Costa Rica, large arribadas occur at Nancite and Ostional Beaches. There are two in Nicaragua (Chacocente and La Flor) and a small one in Pacific Panama. Formerly, there were arribadas in Mexico at Playon de Mismaloya, Jalisco; Piedra de Tlacoyunque (San Luis la Loma), Guerrero; and Playa Escobilla in Oaxaca. The last of these still exists.

Management History

Arribada management historically has been based on little more than guesswork, common sense, or economic and cultural pressures rather than on scientific theory or population modeling. Nevertheless, the diversity of management strategies that have been undertaken and the results revealed in the ensuing decades are very instructive. Case histories are described below, under Options 1–4.

Option 1: Preservation of Natural Rhythms and Demographic Changes

This option usually represents the "total protection" or "no intervention" policy, although conceivably, with enough resources and information, natural population rhythms could be duplicated by intensive management of some kind. Natural progressive changes in a turtle population may be assumed to proceed if human intervention, both negative and positive, is avoided or prevented. This is not always easy, and some human impacts may be unnoticed or difficult to investigate. Nevertheless, the inclusion of Playa Nancite, Costa Rica, in the Guanacaste Conservation Area (Santa Rosa National Park) comes close to the hands-off ideal. The nesting beach is remote and is only reached by a small number of Park visitors. The beach itself is less than 1 km in length and is constrained between high rocky cliffs and headlands.

The outcome of total protection at Playa Nancite has been a peaking of the population (an arribada of nearly 150,000 in 1980) followed by a progressive but somewhat erratic decline, although with an isolated large arribada (80,000 turtles) in 1995 (Valverde et al., 1998). There also has been a drop in the numbers of arribadas, with four per season in the early 1980s and only one or two in recent years. In the years of major arribadas, Cornelius and Robinson (1983, 1985) found that only about 1–4% of the eggs laid, on average, yielded viable hatchlings as a result of both egg destruction by later-nesting turtles and frequent complete failure of nests that were not physically destroyed. In the absence of evidence to the contrary, it may be assumed that the changes are natural and that the very large arribadas of the 1960s and 1970s were ultimately self-limiting. The alternative explanation might be that the population was undergoing decimation by human take during the oceanic phase of their life cycle (with heavy take of olive ridleys in Ecuadorian waters until 1981 and an estimated 100,000 taken there in 1979 alone), although if this were the case, it would have been expected that the neighboring arribada at Playa Ostional would show a similar decline.

Management of the Orissa arribadas has taken the form of theoretical legal protection, with closing down of the markets in Calcutta where the turtles used to be sold, but with episodes of intensive incidental mortality in the trawls of the local fishing fleets. At times, the beaches are littered with turtle carcasses, and recent studies have revealed that the vast majority of these deaths take place in nearshore waters. The arribadas follow no obvious pattern, and at times shifts occur, with abandonment of an earlier arribada site and development of a new arribada elsewhere; furthermore, there are certain years in which no arribada appears. Thus, one is faced with a choice between two extreme interpretations of the significance of the mass mortalities: either the population is being severely stressed by this loss of breeding adults and will soon show significant decline, or the numerous mortalities are an inevitable result of huge numbers of adult turtles converging for an aggregated reproductive effort, especially when extensive trawling occurs in the same area, and the loss of these animals may actually serve to increase recruitment by reducing overcrowding and nest destruction on the nesting beach. Every turtle has to die some time, and the reproductive period may be the most dangerous, apart from the neonatal stage.

Option 2: Management for Maximum Sustainable Yield of Turtles or Eggs

In those cases where the utilization is combined with a conservation program, the (usually unstated) additional objective is to stabilize the population at an approximation of that which prevailed at the onset of the sustained yield program.

This approach has been followed in several diverse cases. A highly controversial, centrally structured program was followed at Playa Escobilla (Oaxaca, Mexico) in the 1960s and 1970s, in the course of which olive ridleys were captured in large numbers (hundreds per day) by turtle fishermen in 6-m outboard-powered fiberglass boats operating offshore from the nesting beach, one man operating the motor in each boat and the other leaping on turtles resting at the surface and manhandling them into the boat. Forty could be caught in a few hours, at which point the heavily laden boat would return to San Agustinillo to discharge, and then to home port at Puerto Angel. The proprietor of this operation, Antonio Suarez, had established a monopoly throughout the State of Oaxaca whereby all turtles caught had to be sold to his company (PIOSA), and part of the profits from the sale of turtle leather and meat was utilized to operate a hatchery and pay for reasonably effective protection of the nesting beach. The operation was controversial not only among turtle conservationists but also among local residents, who were deprived of access to the egg resource. For the last decade of operation, the business was owned and operated by the Mexican government. On June 1, 1990, the turtle fishery and slaughterhouse were closed by presidential decree, and all sea turtles enjoyed permanent legal protection throughout Mexico. Of course, some illegal take of both eggs and turtles continues (Guerrero and Flores, 2004; Taniguchi, 2004).

The outcome of this operation, which flew in the face of all conventional wisdom about the importance of protecting adult turtles, was surprising. Unlike the other Mexican Pacific arribadas, this one did not disappear. In 1968, turtle harvests were excessive throughout the Mexican Pacific, ultimately leading to a 2-year closed season (1972–1973). From 1973 to 1990, the population was statistically stable, with around 200,000 nests per year at La Escobilla (Márquez-M. et al., 2002), with one outstanding year of 400,000 in 1975 and low years of about 100,000 in 1979, 1983, and 1984. Following protection in 1990, the annual nest count immediately surged, from 200,000 in 1991 to nearly 400,000 in 1992–1993, over 800,000 in 1996 and 1997, and about 1,100,000 in 2000 and 2001. Furthermore, although the Jalisco and Guerrero arribadas disappeared some decades ago, a new one has developed at Morro Ayuta, and the beginnings of one have been reported at Playa Ixtapilla, Michoacan, with more than 10,000 nests per year.

By contrast, at Ostional, Costa Rica, the large olive ridley population nesting there—previously subject to uncontrolled exploitation, mainly of eggs, for an uncertain period—has been subject to a management program including a controlled harvest of eggs since 1987, with essentially complete protection of the turtles themselves. The program is somewhat controversial, mainly because the presence of some legal turtle eggs in national markets can complicate the enforcement of the law protecting eggs of all other species as well as eggs of olive ridleys from other beaches. The program has also given rise to a variety of human social interactions and problems bordering on the lurid at times, as well as legal ambiguities, the area of egg collection being within a designated Wildlife Refuge that also includes a thriving village. Nevertheless, the turtle population continues to be strong, and the members of the local egg-collecting cooperative are highly protective of the turtles themselves; they also conduct various programs to protect eggs and hatchlings. A vast arribada that lasted 11 nights and was informally estimated to number a million turtles, or at least a million nesting attempts, reportedly occurred in late 2000. Many of the turtles may have been displaced from Nancite, where a pattern of erosion resulting in formation of vertical cliffs in the beach may have made nesting very difficult.

The future of the Ostional arribada is clearly a matter for speculation. Douglas Robinson, as long ago as 1977, proposed a "rational" egg harvest to replace the irrational and wasteful operation that existed at the time (Campbell, 1998), and the idea has been widely circulated that removal of a relatively significant number of eggs laid by the earlier nesters could actually promote a higher production of hatchlings than would occur with total protection. Furthermore, Nancite, totally protected and located deep within a national park, provided an excellent control site for any management experiments at Ostional. And indeed, the available data support this theory. Typically, 20–30% of the dry season eggs and 3–15% of the wet season production have been harvested (Ballestero and Ordonez, 1991), and during the August and September arribadas over a 5-year period, Cornelius and Robinson (1983, 1985) found the hatching success rate at Ostional to be no less than 17 times that at Nancite.

More recent data on hatching rates at Ostional (Mehta et al., 2000) are very interesting. It was found that, in general, the harvested areas of the beach were much more productive of hatchlings than the "protected" ones. Nonharvested beach sections always showed high percentages of rotten nests (45.5–67.8%), whereas in the harvested areas the percentage of rotten nests was very low until late in the year, when it increased steadily from October to December (less than 1% to 25.1–69%); these data are strongly suggestive of rotten nests being in some way a result of high nest density. By November, the percentage of developing nests in harvested areas was 24.9%, and in the nonharvested areas 6.1%, but confusingly, by the end of the season, the relationship was reversed (7.1% in the harvested areas, 30% in the nonharvested areas). The simplest explanation is that, even with the harvest, nest density (including accumulated rotten nests) became dangerously high in the harvested area by the close of the wet season and its associated large arribadas.

Option 3: Recovery of a Depleted Arribada to a State of Primordial Abundance

The most high-profile example of this approach has been the management of the unique arribada of Kemp's ridley in Tamaulipas, Mexico. Early "management" of this population probably consisted simply of large-scale collection of eggs, possibly approaching 100%, although details are uncertain. Hildebrand (1963) alludes to the annual spectacle of long mule trains loaded down with eggs being taken to market during turtle season in Tamaulipas, and although this may be optimistically classified as "Type 2 management," in reality there would have been little concept of sustainability in theory or in practice. Data for the 1950s are totally lacking, but when field protection was initiated in the 1960s, the ridley population was severely depleted, and local people (as well the local coyotes) still continued with traditional egg collection and were only gradually dissuaded by the presence of fisheries inspectors, research biologists, and armed Mexican marines.

The subsequent management of the population reflected a total reverse, with a progressive sequence of protective laws and regulations and strong enforcement action at the field level. Eggs were transferred to guarded hatcheries at Barra Coma; the United States joined Mexico in legal protection of the species; a binational beach protection and hatchery management program was instituted; and two U.S. agencies, the National Marine Fisheries Service (NMFS) and the U.S. Fish and Wildlife Service (USFWS), cooperated in a massive head-start program, based in Galveston, Texas. NMFS and the Mexican Departamento de Pesca also initiated turtle excluder device (TED) deployment programs to reduce or eliminate the incidental catch problem.

These multiple conservation steps were apparently successful, and the nesting population now seems to have embarked on a progressive increase since 1987. The nesting number trend lines for Kemp's ridley, which is an annual nester, do not show the wild alternations typical of *Chelonia mydas* and some other species, and trends are thus easier to detect and are evident within a smaller number of successive seasons. Nevertheless, the sheer multiplicity of steps taken to protect and proliferate the species make it hard to tease apart the causes from the results and identify the crucial lessons for management that may lie hidden. However, an approach I took earlier (Pritchard, 1997) is worth considering. This analysis noted the staggered years of initiation of the various conservation measures and correlated the timing of initiation of each supplementary technique with outcomes 10 years later (the estimated time to maturity of the species [Márquez-M., 1972]; known-age head-started nesters at Rancho Nuevo reported by Shaver and Caillouet [1998] were 11, 12, and 14 years old). Because the pattern of exploitation of the resource took the form of intensive egg collection for many years preceding introduction of field protection of eggs and turtles in 1965, the passage of at least 10 years would be expected before the operation of the hatchery would have a positive impact on the wild population, and certainly the numbers of nesting adults continued to decline during these years.

The head-start experiment began in 1978, 13 years after the instigation of the hatchery, with an annual quota of about 2,000 eggs set aside for the Kemp's Ridley Recovery Program. The adult population decline continued until 1988, 10 years after the onset of head-starting and a full 23 years after the initiation of the hatchery, at which point the nesting population curve bottomed out and started its steady rise (Márquez-M. et al., 2001). Subsequently, new initiatives including mandatory use of TEDs in the Gulf of Mexico were instigated, but these occurred too late to have been responsible for the population curve reaching a point of inflection and starting to rise in 1988.

Parallel evidence has come forth in recent years indicating that head-started turtles can grow to maturity in the wild following release at about 1 year of age and may nest with the truly wild adults at Rancho Nuevo (see Shaver and Wibbels, Chapter 14). It remains undetermined whether they return in higher numbers than if they had been released as hatchlings, but if it is possible for at least some head-started turtles to complete their life cycle alongside truly wild turtles, this is an important point. Although the initial phase of the Head Start Experiment came to a dramatic conclusion with one federal service (USFWS) denying an import permit for ridley eggs destined for head-starting to another federal service (NMFS), my own conclusion is that hasty negative judgments (e.g., by Woody, 1990, 1991; but see Allen, 1990, for a good counterpoint) were made on an experimental procedure for which at least a decade would be necessary for concrete results to be obtainable. Nevertheless, bringing Phase 1 (collection of eggs and captive rearing hatchlings) to a close did not mean that the whole operation was a failure, simply that the plan was entering Phase 2 (seeking evidence of long-term survival of the released hatchlings) and Phase 3 (successful nesting of head-started animals, either at Rancho Nuevo or elsewhere).

Head-starting is an expensive and controversial procedure, certainly, and is not to be recommended except in extremely rare, desperation cases, but, when carried out on an adequate scale and with appropriate safeguards, it may work. A less controlled but quite large-scale project has been undertaken in recent years at Cayman Turtle Farm, utilizing farm-raised juvenile green turtles released in the waters of the Cayman Islands. Many appear to be surviving and growing, and some have remained in Cayman waters, although others have been caught in Cuba. Recently several mature individuals of both sexes have been encountered; several were only 14 years of age (Bell and Parsons, 2002), and one was encountered while nesting at only 17 years of age. These data may require a recalibration of the customarily accepted growth curve for free-ranging green turtles.

A much smaller-scale effort to combine approaches 3 and 4 took place at Eilanti, Suriname. The small but intense arribada there was not known to have declined, but common sense indicated that collapse would be inevitable because virtually all eggs had been collected for many years by the local Carib people. The program initiated by Landsbosbeheer (Suriname) and me in 1966 involved complete protection of both the turtles and their eggs at Eilanti, and this was continued in succeeding years. Presumably the population would have recovered, although trawler captures were frequent, and before long the beach itself changed in character and progressively became unsuitable for turtle nesting of any kind.

Option 4: Management for a Higher than Natural Adult Population

This outcome corresponds to situations in which turtles in an arribada that is still at preexploitation level not only receive legal protection of both turtles and eggs but in which additional positive management occurs. The management may take the form of assisting oviposition in turtles with deformities or injuries that prevent nest excavation; operation of hatcheries for eggs that might have been lost to erosion, predation, or destruction by later-nesting turtles; and head-starting. Standard ecological theory would suggest that the number of turtles in the adult population is governed by the available ecological space or the carrying capacity of the environment, a level that would not be raised by artificially increasing recruitment at the hatchling stage. A slight complication in sea turtle population-modeling theory is that it is not just the carrying capacity of the feeding grounds that is critical but also the availability of nesting space where eggs have a good chance of hatching, something that may indeed be limiting in a strongly philopatric species with "panspecific reproductive convergence," in the elegant words of Archie Carr.

It probably will always be difficult to demonstrate that these population-boosting techniques are successful because they are undertaken on populations that are already large and undiminished. An observed subsequent increase could equally be the result of natural fluctuation rather than successful artificial enhancement.

Conclusions

1. Management for the simple goal of maximal numbers of turtles, although a realistic approach for marine turtles in most places, is untenable or inappropriate for arribada-nesting species unless it is recognized that the resulting abundance of turtles will be unsustainable.
2. It is possible to restore highly depleted nesting populations if the nesting habitat is intact by means of a multiapproach program that may include TED usage, bans on commercial take of turtles, hatchery operation, and beach patrols. It is also possible that head-starting, with appropriate safeguards, can be a valuable component of a restoration program.
3. Under undisturbed conditions, a given arribada may not be a permanent phenomenon but rather may ultimately fail to replace itself because of deterioration of the nesting substrate and egg failure or destruction.
4. An arribada colony may demonstrate surprising stability in the face of substantial exploitation of turtles themselves (Escobilla), incidental capture and mortality of turtles (Gahirmatha), or controlled egg harvest (Ostional), provided that provision is made for the protection of the greater part of the egg production. However, excessive egg collection can result in a population collapse that will require decades of protection of

both the surviving turtles and their eggs if it is to be reversed.

5. There is no objection at all to simply studying and monitoring an arribada and avoiding both active population enhancement and consumptive exploitation of any kind, if these approaches are politically feasible, and only in this way will we gain real-world insight into the fundamental cycles of the arribada phenomenon.
6. In a changing and modernizing world, the most profitable management scheme for a major arribada may be ecotourism. "Primitive" (tent-based) ecotourism may be flexible enough to shift geographically as and when turtle nesting shifts, but investment in permanent tourist facilities should be accompanied by management calculated to stabilize the turtle population. This may involve controlled egg collection by local people, as has been undertaken at Ostional, Costa Rica. However, excessive egg collection, such as was practiced for many years with the leatherback turtles at Terengganu, Malaysia, may result in inexorable decline and total loss of the nesting colony, with severe economic impact on an ecotourism enterprise with heavy investment in infrastructure in the nesting zone.

LITERATURE CITED

Allen, C. 1990. Guest editorial: give headstarting a chance. *Marine Turtle Newsletter* 51:12–16.

Ballestero, J. L. 1994. Plan de manejo para los huevos de la tortuga marina lora (*Lepidochelys olivacea*) en el Refugio Nacional de Vida Silvestre de Ostional, Santa Cruz, Guanacaste, Costa Rica. Ostional, Costa Rica.

Ballestero, J. L., and Ordonez, G. 1991. Untitled, unpublished paper. Ostional Wildlife Refuge.

Baur, G. 1890. The genera of the Cheloniidae. *American Naturalist* 24:486–487.

Bell, C. D. L., and Parsons, J. 2002. Cayman turtle farm head-starting project yields tangible success. *Marine Turtle Newsletter* 98:5–6.

Campbell, L. M. 1998. Use them or lose them? Conservation and consumptive use of marine turtle eggs at Ostional, Costa Rica. *Environmental Conservation* 25:305–319.

Caporaso, F. 1987. Indiana Pritchard and the dinghy of doom! *Tortuga Gazette* 1987 (Feb.):5–7.

Carr, A. F. 1942. Notes on sea turtles. *Proceedings of the New England Zoology Club* 21:1–16.

Carr, A. F. 1948. Sea turtles on a tropical island. *Fauna* 10:50–55.

Carr, A. F. 1957. *The Windward Road.* New York: Alfred A. Knopf.

Carr, A. F. 1963. Panspecific reproductive convergence in *Lepidochelys kempii. Ergebnisse der Biologie* 26:298–303.

Carr, A. F. 1967. *So Excellent a Fishe.* New York: Natural History Press.

Cornelius, S. 1991. *Lepidochelys olivacea.* In Janzen, D. H. (Ed.). *Historia Natural de Costa Rica.* San José: University of Costa Rica, pp. 407–410.

Cornelius, S., and Robinson, D. 1983. *Costa Rica: Project 3085. Olive Ridley Sea Turtles.* Cambridge: World Conservation Monitoring Centre.

Cornelius, S. E., and Robinson, D. 1985. *Counting Turtles in Costa Rica.* WWF Monthly Report for August (Project 3085), Cambridge: World Conservation Monitoring Centre.

Crouse, D., Crowder, L. B., and Caswell, H. 1987. A stage-based population model for loggerhead sea turtles and implications for conservation. *Ecology* 68:1412–1423.

Deraniyagala, P. E. P. 1933. The loggerhead turtles (Carettidae) of Ceylon. *Ceylon Journal of Science* 18(B):61–72.

Eschscholtz, J. F. 1829. *Zoologischer Atlas enthaltend Abbildungen und Beschreibungen neuer Thierarten, während des Flottcapitains von Kotzebue zweiter Reise um die Welt, auf der Russisch-Kaiserlichen Kriegsschlupp Predpriaetië, in den Jahren 1823–26 beobachtet,* Heft 1. Berlin: G. Reimer.

Fitzinger, L. J. 1843. *Systema Reptilium.* Vienna: Vindobonae, Braunmüller et Seidel, Bibliopolas.

Fretey, J. 2001. *Biogeography and Conservation of Marine Turtles of the Atlantic Coast of Africa.* CMS Technical Series No. 6. Bonn: UNEP/CMS Secretariat.

Guerrero, H., and Flores, S. 2004. Hay 22 playas de riesgo para tortugas marinas. *Reforma,* Mexico, January 13.

Hamilton, A. 1727. *A New Account of the East Indies.* Vol. 1. Edinburgh, Printed by John Mosman One of His Majesty's Printers, and sold at the King's Printing House in Craig's Clof.

Hildebrand, H. H. 1963. Hallazgo del area de anidacion de la tortuga marina "lora," *Lepidochelys kempii* (Garman), en la costa occidental del Golfo de Mexico (Rept., Chel.). *Ciencia* 22:105–112.

Márquez-M., R. 1972. Resultados preliminares sobre edad y crecimiento de la tortuga lora, *Lepidochelys kempii* (Garman). In *Memorias IV Congreso Nacional de Oceanografia,* Mexico DF, pp. 419–427.

Márquez-M., R., Burchfield, P., Carrasco, M. A., Jiménez, C., Diaz, J., Garduno, M., Leo, A., Pena, J., Bravo, R., and Gonzalez, E. 2001. Update on the Kemp's ridley turtle nesting in Mexico. *Marine Turtle Newsletter* 92:2–4.

Márquez-M., R., Carrasco, M. A., and Jiménez, C. 2002. The marine turtles of Mexico: an update. In I. Kinan (Ed.). *Proceedings of the Western Pacific Sea Turtle Cooperative Research and Management Workshop.* Honolulu: Western Pacific Regional Fishery Management Council, pp. 281–285.

Mehta, S., Russell, A., and Arauz, R. 2000. Solitary nesting activity in the Ostional Refuge, Costa Rica, and the impact of poaching (during solitary nesting) and harvesting (during arribadas) on olive ridley nests (*Lepidochelys olivacea*). In Kalb, H. J., and Wibbels, T. (Compilers). *Proceedings of the Nineteenth Annual Symposium on Sea Turtle Biology and Conservation.* NOAA Technical Memorandum NMFS-SEFSC-443, pp. 69–71.

Mohanty-Hejmadi, P. 2000. Earliest record of Gahirmatha turtles. *Marine Turtle Newsletter* 88:11–12.

Parsons, J. J. 1957. *The Green Turtle and Man.* Gainesville: University of Florida Press.

Pritchard, P. C. H. 1969. Studies of the systematics and reproductive cycles of the genus *Lepidochelys.* Ph.D. diss., University of Florida, Gainesville.

Pritchard, P. C. H. 1997. A new interpretation of Mexican ridley population trends. *Marine Turtle Newsletter* 76:14–17.

Richard, J. D., and Hughes, D. A. 1972. Some observations of sea turtle nesting activity in Costa Rica. *Marine Biology* 16:297–309.

Shaver, D., and Caillouet, C. W., Jr. 1998. More Kemp's ridley turtles return to South Texas to nest. *Marine Turtle Newsletter* 82:1–5.

Taniguchi, H. 2004. Identifica profepa rutas de productos de tortuga. *Reforma,* Mexico, January 15.

Valverde, R. A., Cornelius, S. E., and Mo, C. 1998. Decline of olive ridley sea turtles (*Lepidochelys olivacea*) nesting assemblage at Nancite beach, Santa Rosa National Park, Costa Rica. *Chelonian Biology and Conservation* 3:58–63.

Werler, J. 1951. Miscellaneous notes on the eggs and young of Texan and Mexican reptiles. *Zoologica* 36:37–48.

Woody, J. B. 1990. Guest editorial: Is headstarting a reasonable conservation measure? On the surface, yes; in reality, no. *Marine Turtle Newsletter* 50:8–11.

Woody, J. B. 1991. Guest editorial: it's time to stop headstarting Kemp's ridley. *Marine Turtle Newsletter* 55:7–8.

NCAA

LISA CAMPBELL

2

Understanding Human Use of Olive Ridleys

Implications for Conservation

OLIVE RIDLEY SEA TURTLES (*Lepidochelys olivacea*) are used widely by humans, on land and at sea, and the implications for olive ridley populations are considered in Chapters 12 and 13 of this book. Here I consider the context of olive ridley use, that is, its economic, political, social, and cultural aspects, and is based on published and unpublished results by researchers working in a variety of disciplines. Although understanding the context of use is increasingly recognized as important, there is relatively little published research related to sea turtles (Campbell, 2003). Here research is distinguished from project descriptions or reports that reflect on the context of sea turtle use rather than the study of the subject.

The purpose of this chapter is to (1) highlight some of the socioeconomic research on olive ridley use and, in so doing, draw attention to the contributions from a variety of social science disciplines and (2) look for common themes arising from what is most often site-specific research, which may inform attempts to conserve olive ridleys and sea turtles more generally. The chapter does not attempt to assess whether olive ridley use is biologically sustainable but rather considers economic, political, social, and cultural aspects of use that influence human response to conservation programs, be they designed to ensure use is sustainable or to eliminate use altogether.

The definition of conservation adopted in this chapter is that of the World Conservation Union: "the management of human use of organisms or ecosystems to ensure such use is sustainable. Besides sustainable use, conservation includes protection, maintenance, rehabilitation, restoration, and enhancement of populations and ecosystems" (IUCN, 1980). In this definition, sustainable use is considered a legitimate component of a conservation strategy, and an assumption in this chapter is that if use is to be sustainable, the human context of it must be fully understood. In only one of the case studies discussed is there some agreement that use might be biologically sustainable (egg collecting at Ostional, Costa Rica), and biological sustainability will ultimately influence whether humans can or will continue to derive benefits from such use. The issues are clearly linked, however, and should biologically sustainable extraction rates be identified, then social, economic, political, and cultural issues will determine whether these rates will be respected by resources users. Thus, although this chapter considers only part of the use equation, it is an important part.

The IUCN definition of conservation is not shared by all sea turtle biologists, and the idea that turtles can be used consumptively in a sustainable manner, as opposed to nonconsumptively—for example, via tourism—has been a controversial one in the IUCN Marine Turtle Specialist Group (Campbell, 2002). The possibilities for use are different for olive ridleys than for other sea turtle species for two reasons. First, the olive ridley is the most abundant of all sea turtle species (Pritchard, 1997a). Second, some olive ridley populations have mass nesting behavior and gather to nest in the thousands in what are referred to as *arribadas* (see Bernardo and Plotkin, Chapter 4). On arribada nesting beaches, high nesting densities and related high levels of egg loss can justify egg collections; if eggs are not harvested, they are likely to be destroyed by later nesters. At sea, turtles aggregating for arribadas off the Pacific coast of Mexico were legally fished until 1990, with their dense concentrations making fishing relatively efficient (high catch rates per unit effort). Although few biologists would consider the capture of reproductively active females to be consistent with conservation, egg collecting on arribada beaches has received some support (Mrosovsky, 1983; Pritchard, 1984; Cornelius et al., 1991; Mrosovsky, 1997, 2001).

This chapter focuses on olive ridleys nesting in or aggregating offshore for arribadas (with one exception) at various locales in Latin America (including Suriname). The justification for this focus is twofold. First, although a second ridley species, Kemp's ridley (*Lepidochelys kempii*), is also known to nest in arribadas, the small numbers of Kemp's ridleys and their critically endangered status make it inappropriate to compare conservation options for Kemp's and olive ridleys. Furthermore, whereas Kemp's ridleys were exploited historically, existing nesters are protected, and human use of them is prohibited. Second, although olive ridleys nest in arribadas in other regions (and a particularly large aggregation nests in Orissa, India), the Latin American focus reflects the interests and expertise of the author and the geographic focus of the majority of published research on the human context of use.

The chapter is structured around case studies, the approach taken in most existing research. As a result, the descriptions below sometimes rely heavily on one piece of work, and basic information on methods used in the work is provided to the reader. There is an imbalance in the treatment of specific cases, based on the extent of the research conducted and its accessibility. For example, research published in graduate theses is described in more detail than that available in more readily accessible publications. The final section of the chapter outlines some of the common themes or lessons that emerge from the case studies.

Mexico: Economics of Turtle Fishing and Egg Collecting

Historically, olive ridley turtles nested in arribadas at several beaches along the Pacific coast of Mexico. Egg use has been illegal in Mexico since 1927, but use of olive ridley turtles was legal until 1990 (Trinidad and Wilson, 2000). Turtles were taken for both meat and leather, and there has been research on the impacts of harvesting on population numbers and responses since turtle capture was banned in 1990 (Márquez-M. et al., 1996; Ross, 1996; Godfrey, 1997; Pritchard, 1997b). Although many of the beaches no longer host arribada nesting, the phenomenon continues at Escobilla, Oaxaca, and illegal use of both eggs and turtles is believed to be widespread. From 1995 to 1998, the Mexican enforcement agency, Procuraduria Federal de Protección al Ambiente (PROFEPA), seized approximately 1,000–8,000 kg of turtle meat, 100–1,800 units of turtle leather, and several hundred dead and live whole turtles each year in Oaxaca (species not specified). In the same period, approximately 300,000–600,000 turtle eggs were seized each year (Trinidad and Wilson, 2000). It can be assumed, given enforcement constraints described below, that these figures reflected only a portion of the actual take.

Trinidad and Wilson (2000) considered the economics of egg and turtle use in Mexico and its legislative context. They argued that traditional economic models for understanding illegal activities, both outside the legal fishery when it existed and generally since the ban, are insufficient. Such models are based on theories of crime and punishment and assume fishermen are "rational, amoral, and apolitical profit maximizers." Rather, Trinidad and Wilson (2000) posed illegal behavior as a result of decoupling the political process of sea turtle conservation from the resource users themselves. To this end, they reviewed changes in legislation over time (for both conservation and industrial development) "to understand the economic history of the management failure" (Trinidad and Wilson, 2000). They also considered the economic context of egg harvesting and constraints on enforcement activities. Trinidad and Wilson (2000) relied on interviews conducted in the summer of 2000 with fishermen, egg collectors, government employees, egg and turtle consumers, and researchers (although details on methods—e.g., number of interviews, their structure, methods of data analysis—were not provided). In addition, they reproduced data published by Mexican researchers in technical reports to support their argument.

Turtle Fishing

Industrial fishing for turtles on the Pacific coast of Mexico began in the 1960s and focused on the olive ridley. Unlike commercial fishing, which refers to the sale of the products of fishing for cash, industrial fishing refers to large-scale, mechanized operations with centralized slaughterhouses and processing plants. These fishermen worked in cooperatives, and the processing industry was centralized at a slaughterhouse in San Agustinillo. The valued commercial product was olive ridley leather, taken primarily from the flippers. There was considerable waste of other turtle parts, including meat, because of lack of developed markets, and Trinidad and Wilson (2000) suggested that waste was a result of underselling the full value of the resource to original concession holders. With only a few bidders for concessions, lack of competition meant there were no incentives for efficiency. By 1969, Mexican law stipulated that the entire turtle was to be used in exploitation (Cliffton et al., 1995).

By 1968, many Pacific olive ridley fisheries had dwindled, and Oaxaca remained predominant. Annual take was also declining in Oaxaca, however, which led to the total ban of 1971–1972 and to industry restructuring. From 1973 to 1980 a private firm, Pesquera Industria de Oaxaca (PIOSA), controlled the reopened fishery. The shift to the private sector may have been designed to encourage more entrepreneurial management of the resource, as a long-term concession held as a monopoly can encourage more rational use of the resource because the exploiter has exclusive rights to it (Trinidad and Wilson, 2000). PIOSA did make fuller use of the olive ridley (Cliffton et al., 1995), selling meat for food; bone, blood, shell, and entrails for meal; and calipee for soup, thus decreasing waste. It also protected nesting beaches, an action Cliffton et al. (1995) cited as central to postponing the collapse of the turtle population. However, PIOSA also pressed the government for increased quotas, following initial acceptance of decreases, and olive ridley takes continued to dwindle. Trinidad and Wilson (2000) suggested that the terms of the original concession to PIOSA did not take into account the depreciation of the resource. Thus, PIOSA could overexploit the resource, recover its initial investment, and sell when the industry no longer looked profitable. Indeed, the government purchased three PIOSA processing plants in 1980. Trinidad and Wilson (2000) pointed to failing profitability (because of decreased export markets and growing international pressure to stop the turtle fishery) to explain PIOSA's departure from the industry. However, the legal difficulties faced by PIOSA owner Antonio Suárez for exporting olive ridley meat disguised as river turtle to the United States may also have played a role in the transition. This deception was first exposed by Tim Cahill, a journalist, in his article "The Shame of Escobilla" in *Outside Magazine* in 1978. The article was reprinted and updated in his 1987 book *Jaguars Ate My Flesh* (Cahill, 1987).

When the government acquired PIOSA's processing plants, it created an agency, Productos Pesqueros Mexicanos (PROPEMEX), to be responsible for the turtle fishery and sold the fishing cooperatives 45% ownership. The cooperatives' share was to be paid for with turtles sold exclusively to PROPEMEX. The turtle take continued to decline, and, in 1986, the government attempted to sell the San Agustinillo slaughterhouse and an additional processing plant to the cooperatives. Again, debt was to be paid with product (67% of the price of each turtle), and cooperatives were obliged to sell to PROPEMEX. Five cooperatives agreed to this arrangement, but four did not (Trinidad and Wilson, 2000).

Based on interviews with members of one fishing cooperative, and using price data from 1989–1990, Trinidad and Wilson (2000) described the conundrum faced by fishermen. Membership in the cooperative went from 80 fishermen at its founding in 1975, to 250 during the peak harvesting years, to 35 by 2000. Although it is not surprising that there are few cooperative members now that the fishery is closed, decreases in membership began before the total ban was introduced in 1990. Fishing quotas were so low that profits did not cover fishing costs, which forced many members to move from the legal cooperative-based fishery to the illegal one. Fishermen working legally earned approximately 14% of what they could earn illegally in the black market.

Eggs

Turtle eggs have long been an important source of food for coastal peoples in Mexico, including some indigenous groups. Trinidad and Wilson (2000) collected basic socioeconomic data and conducted interviews in Escobilla, where illegal egg collecting was evident. For example, in 1988 an estimated 3 million eggs were collected illegally, with a value to collectors of US$64,430 (and an estimated final market value of US $1,962,922) (Aridjis, 1990, cited in Trinidad and Wilson, 2000). (All values are in U.S. currency, converted from local currencies by the individual authors at the time of their research.) Trinidad and Wilson argued that the importance of egg collecting could be understood in the overall economic context of the region, where subsistence agriculture was important, there were few cash-earning jobs, and migration for employment was common. Although the income earned in egg collecting was significant, *acaparadores* received the most benefits. (*Acaparadores* derives from the verb *acaparar*, meaning "to monopolize." In this context, it refers to middlemen who buy eggs directly from collectors and resell them to distributors. In fact, the existence of many collectors and fewer buyers implies a monopsony rather than a monopoly.) These intermediaries paid egg collectors only after eggs were sold, and if the eggs were confiscated, collectors were unpaid. Villagers in Escobilla coordinated themselves to collect and sell eggs and took their own risks of injury and arrest.

Cahill (1987) relays a local man's description of the egg harvest: local people gather nightly to decide who will collect eggs, and 10 people are chosen to work in pairs. His informant was chosen to work four times that year and made about US$300 to supplement his main income from growing corn (US$500), all of which supports a family of 10. At the time Cahill was there (late 1970s), a driver transporting eggs could make up to US$4000 per shipment. The differences in Cahill's and Trinidad and Wilson's (2000) descriptions of the harvest may be accounted for by the time difference, who they collected the information from, or site-specific differences in collection practices. Cahill, a journalist, acknowledges that his description is based on a conversation with one egg collector, whereas Trinidad and Wilson interviewed an unspecified number of people in Escobilla.

Because egg collection was illegal, it was difficult for collectors to organize to demand better prices and treatment from acaparadores. Trinidad and Wilson (2000) found that egg collecting was practiced mainly by women and children, although youths were often involved before the school year to earn money for fees and supplies. To sell the eggs sometimes required transportation, which increased chances of discovery and decreased individual profit. If alternatives were available, villagers said they would harvest eggs only for home consumption.

In considering why turtle fishing and egg collecting have continued in spite of extensive legislative commitments, Trinidad and Wilson (2000) drew a number of conclusions, some relevant to turtles or eggs and others to both. All concerned the incentives for individuals to disregard the laws surrounding turtle use and the inability of the government to enforce such laws.

First, economic incentives for illegal harvesting were strong in both cases. In terms of turtle fishing, however, the contradictions between government policies for conservation and development had the consequence of increasing incentives to act illegally. As the olive ridley fishery appeared destined to fail because of falling catch rates and increased external opposition to the harvest, ownership was increasingly transferred to fishermen. Rather than reduce dependence on the turtle fishery, the government increased the cooperatives' stakes in it. Thus, when the fishery closed in 1990, the cooperatives had much to lose (the private sector having recouped its investment and disappeared). In this context, it is less surprising that the ban was resisted.

Second, alternatives to both egg collecting and turtle fishing did not materialize. Both fishermen and egg collectors cited the lack of alternative economic activities as a reason they continued to operate illegally, and egg collectors specifically said they would harvest only for household consumption if viable alternatives existed. One attempt at diversification was through ecotourism. In 1990, the World Bank supported a "campsite" program to develop basic services at arribada beaches with the intention of attracting ecotourists. A loan was provided with the aim of seeing campsites become self-sufficient in 7 years, a goal that was not reached. Trinidad and Wilson (2000) suggested further investments in exploring ecotourism as an alternative development strategy.

Third, the overall management of the fishing industry has been centralized, and existing social and cultural institutions have been ignored. Centralization has impacts on fishermen. The small number of concessions in the early years and the monopsony held by PIOSA in the 1970s meant buyers held market power over fishermen. When the government took over, market power was still centralized, and PROPEMEX was designated the only buyer to ensure that cooperatives would repay government loans (Trinidad and Wilson, 2000). Fishermen have been external to most decisionmaking, and they were not consulted on the development and implementation of the moratorium. Nor, in 1999, did fishermen or egg collectors attend meetings held in Escobilla to discuss the olive ridley, specifically enforcement, alternative economic development, and the possibility of reintroducing quotas (Trinidad and Wilson, 2000).

Having failed to facilitate alternative economic activities or to engage resource users in decisionmaking, the government has had to rely on regulatory incentives to pursue conservation. Such centralized enforcement policies may work when costs of enforcement are low and the likelihood of compliance is reasonable (Trinidad and Wilson, 2000). However, these criteria do not exist in the case of olive ridley use in Mexico. A PROFEPA inspector cited lack of personnel as a key constraint to effective enforcement (six inspectors, whose concerns are not restricted to turtles, for all of Oaxaca), along with other administrative constraints. Furthermore, corruption among government personnel was evident (Trinidad and Wilson, 2000).

In 2000, the Secretaría del Medio Ambiente Recursos Naturales y Pesca (SEMARNAP) pursued a legal initiative that would have changed the status of olive ridley protection. The initiative encouraged the sustainable exploitation of animals whose life cycles depend on water and provided the means to amend and repeal laws and agreements that prohibit sustainable exploitation. Also in 2000, amendments to the penal code that would remove prison terms for persons engaged in egg collection for subsistence, or for satisfying basic needs, were proposed. The proposed changes were "an attempt to provide some relief to members of the coastal communities already hard hit by the bans on turtle captures and egg collection" (Trinidad and Wilson, 2000). Although the proposed changes were defeated, and opposition was based on a number of concerns including definitions of subsistence and individual eligibility, the increase in olive ridley turtles nesting at Escobilla ensures that pressure to at least partially lift the ban will continue. Increased nesting will also further affect illegal activities, as they make laws protecting eggs and turtles appear unjust.

Honduras: Economics of Olive Ridley Egg Collecting and Selling

Unlike other case studies considered in this chapter, olive ridley nesting around the Gulf of Fonseca is not in arribada concentrations. Lagueux (1989) estimated that, in 1987, 2,022 nests were laid at 46 beaches around the Gulf and cited earlier studies and interviews with long-time residents to suggest that nesting has never reached arribada levels. In 1987, Lagueux assessed the economic value of olive ridley turtle egg collecting by communities around the Gulf of Fonseca (Lagueux, 1989) and evaluated the contribution of egg harvesting to average cost of living in one village, Punta Ratón (Lagueux, 1991). She used a variety of data collection tools to measure value, including household surveys, interviews with collectors, cost-of-living surveys, and key informant interviews. Lagueux (1989) also assessed commerce in olive ridley eggs, focusing on relationships between egg collectors and sellers, and income earned along the market channel. Methods of data collection included interviews with egg sellers and buyers, regular surveys of primary egg buyers, market surveys of egg availability and pricing, and interviews with egg vendors.

Lagueux's (1989, 1991) approach was empirical, focused on describing economic value, with little reference to economic or other theory related to use and conservation of resources. This empirical approach contrasts with that of Trinidad and Wilson (2000), which relied heavily on existing data and applied economic theory to it. Both approaches offer insight into the respective case studies.

Egg Collecting

Based on a sample of households in seven communities, Lagueux (1989) found that the number of households involved in egg harvesting varied from a low of 20% of households to a high of 95%. When interviewees were asked about economic activities, fishing was the most frequently identified activity for both the rainy and dry season, and egg harvesting was the second most frequently identified rainy season activity. However, even during the rainy season, only 7 of 71 household interviewees identified it as the most important economic activity, compared to 35 interviewees identifying fishing. When asked about the perceived benefits of sea turtle eggs, the most frequent response by household interviewees was that there were no known benefits (28 of 71 interviewees), and 35 interviewees said they would be not be affected by the loss of the sea turtle resource.

In her case study of Punta Ratón, Lagueux (1991) found that 88% of egg clutches laid at Punta Ratón were collected (total eggs collected in 1987 = 63,798), worth the equivalent of approximately US$10,000 to collectors. Collectors from Punta Ratón households earned US$7,680 of the total. The remaining US$2,320 was earned by collectors from outside of Punta Ratón, the majority (41%) of whom came from a town 10 km inland. Income earned by collectors from Punta Ratón was unevenly distributed. Of the 82 households (total households = 93) participating in egg collection, four households earned 23% of all income, and 15 households earned nothing. The average earned was US$93.66 per household (range, US$0 to US$684.56), and 80% of households earned US$160 per year or less. Because the focus of Lagueux's (1989) study was on the Punta Ratón households, the value of egg collecting to external collectors was not investigated. This value could have been higher because while in Punta Ratón, external collectors were dedicated to this activity.

Lagueux (1991) calculated basic cost of living expenses for Punta Ratón households as US$1 per day per person and the average household size as six people. Based on these figures, and assuming net income equals net expenditure, she calculated that egg collecting contributed to 4.3% of the yearly expenditure of the average household. Given the unequal distribution of benefits, however, egg collection was clearly much more valuable to some households than to others, and the seasonality of egg collecting suggests the 4.3% contribution to expenses was concentrated during the collecting season. Lagueux (1989) identified cash earned by egg collecting as important, given that most households purchase the majority of staple and other foods and that alternative income-earning activities were limited by geographic isolation and by lack of education and other social services. She also reflected on the social benefits associated with egg harvesting, which provided an opportunity to exchange information and news with friends and family.

Using past observations made by Carr (1948) and Pritchard (1979) and her own observations in 1987, Lagueux suggested that 100% of olive ridley eggs have been harvested from the beaches in the Gulf of Fonseca since at least 1940 and possibly since 1920. She predicted a steep decline in the population and argued that "unless an improved conservation effort is made . . . both the olive ridley sea turtle . . . and the economic benefit that human populations derive from collecting eggs will be known as a historical occurrence" (Lagueux, 1991). Because turtles are valuable, Lagueux argued, local people should be encouraged to conserve them. However, alternative interpretations of the results are possible; with an average of 4.3% of cost of living expenses generated through egg collecting, this activity could be characterized as relatively unimportant. If so, total protection might be pursued with small social and economic costs borne by collectors (although the uneven distribution of benefits suggests costs would be high for some households, and seasonality concentrates contributions to particular times of year). Or the small percentage of household income earned and the perception that loss of the turtle resource would have "no effect" for most households might mean that incentives for conservation would be lacking. Further research may be required to evaluate these potential outcomes, but the baseline economic data provided by Lagueux (1989, 1991) would be critical to any conservation planning for the region.

Egg Commerce

Lagueux (1989) found a variety of market channels in operation in coastal Honduras, composed of anywhere from two to five buyers and sellers. Most primary egg buyers (86%), those who purchased directly from the egg collector, were residents of the coastal communities, and the relationship between collector and primary buyers was constant (i.e., the same collectors sold to the same buyers, even if other buyers were offering higher prices). This relationship provided security to collector and buyer, and collectors also accessed loans from buyers based on this relationship. A small number of primary buyers owned restaurants or other retail outlets and sold directly to consumers, but most sold to secondary buyers.

Secondary egg buyers came to the coastal communities to purchase eggs, and some bought other products at the same time (with the importance of turtle eggs varying across buyers). No buyers were dependent exclusively on turtle eggs. Lagueux calculated the average income earned by primary egg buyers in six communities, which ranged from US$9.72 to US$186.31 across communities. She also calculated price inflation from egg collector to final vendor and showed that price inflation varied from 62% to 262%, depending on the time of year. In her survey of markets, Lagueux (1989) found high availability of eggs in markets, though sometimes by few vendors. She concluded that because egg buying and selling is relatively easy with few entry costs, movement in and out of the business was fluid, and egg commerce potentially generated income for a large number of people. She argued that the varied form of the market channels and numbers of people involved in egg commerce would make management of such commerce difficult.

Lagueux's (1989, 1991) study showed how the context for egg use in Honduras must be considered at various scales, from the household to the village to the regional level. Although insights gained at any one level are useful, results are most interesting when combined. For example, access to income from egg selling might not be as important to egg collectors as access to loans from primary egg buyers. The importance of egg collector and primary buyer relationships is an issue Lagueux (1989) may not have uncovered had she focused only on household activities and ignored market channels and mechanisms.

Nicaragua: Politics and Economics of Egg Collecting at Chacocente and La Flor

Olive ridleys nest in arribadas at two beaches on the Pacific Coast of Nicaragua, Playa La Flor and

Chacocente, and managed egg collection projects exist at both. Chacocente has received some attention by analysts interested in environmentalism in Nicaragua, particularly as it relates to the country's political history (Faber, 1993). Research by Stewart (2001) and Hope (2000, 2002) is reviewed here; both conducted archival research in Nicaragua and undertook field site visits, interviews, and, in Stewart's case, participant observation. These methods were employed in very different theoretical contexts. Stewart was interested in the geopolitical context of conservation at Chacocente, whereas Hope attempted to assess the sustainability of egg collection regimens by applying an "Arribada Sustainability Framework" to four arribada beaches (Playa La Flor and Chacocente in Nicaragua and Ostional and Nancite in Costa Rica).

Chacocente

Faber (1993) and Stewart (2001) linked conservation activities in Nicaragua to the country's political history, and they traced prerevolutionary (pre-1979), revolutionary (1979–1990), and postrevolutionary (post-1990) views on environment. The history of olive ridley use at Chacocente is linked to this history, and Chacocente is an important case study for understanding it. As one of the protected areas established by the revolutionary Sandinista government, Chacocente has symbolic value. The Nicaraguan revolution was, in some ways, an environmental revolution, and environmentalists were key supporters of it (Weinberg, 1991; Faber, 1993). As a result of environmental degradation and control of resources by the ruling elite under Somoza, environmentalists "saw that only a fundamental transformation in the country's power structure could open the door to ecologically sound and socially beneficial development." So central was environment and control over natural resources to the aims of the Sandinista government that their revolution has been labeled an experiment in "ecological socialism" or "revolutionary ecology" (Faber, 1993).

Within weeks of the revolution, the Sandinista government established the Instituto Nicaragüense de Recursos Naturales y del Ambiente (IRENA). Under its initial leader, Jorge Jenkins, IRENA pursued productive conservation, that is, conservation to benefit people. A plan was developed to create 36 national reserves, covering 17% of the country, where resources would be managed in a productive manner. As part of this plan, the 4,800-ha Chacocente–Rio Escalente Wildlife Refuge was established in 1983 (Stewart, 2001). Its symbolic value is derived from three sources. First, according to Faber (1993), Chacocente was one of the only protected areas implemented of the intended 36. Once civil war between the Sandinistas and the Contras began, funds were scarce, and conservation activities were scaled back. Furthermore, the guerilla war tactics of the Contras were based in wilderness zones, many of which were intended for protection. Even if they were declared protected areas, operationalizing protection was infeasible. The 1993 UN List of Parks and Protected Areas shows 17 protected areas established in 1983, but few received adequate funding or support (IUCN, 1992). Chacocente, however, was geographically removed from most war activities, and it could feasibly be protected. Second, the Sandinista government turned its attention to sea turtles in general and, according to Faber (1993), the "most exemplary of IRENA's wildlife initiatives was the Sea Turtle Conservation Campaign." The campaign sought to reduce harvesting of green turtles on the Miskito Coast and to regulate egg collecting at Chacocente, and a national educational program was a critical component of the campaign. Third, as an arribada beach, Chacocente presented a good opportunity to implement the principles of productive conservation. As discussed below, the symbolic value of Chacocente has been an important factor influencing its management.

Each year between July and January, turtles nest in four or more arribadas at Chacocente (Stewart, 2001; Hope, 2002). There are four villages within the bounds of the refuge and 13 villages surrounding it. In the early 1980s, 350 families lived in these 17 villages. The area was one of extreme poverty, with high illiteracy rates, few educational or health services, limited access to potable water and sanitation, high fertility, and no electricity. Of the 350 families, 317 practiced slash and burn agriculture and cattle raising as their primary economic activity, 27 fished, and 19 worked in semiskilled labor (AID, 1991, cited in Stewart, 2001). Subsistence farmers pursued wage labor on larger cooperative farms and migrated seasonally for agricultural work or worked as domestic servants in cities. During Stewart's research, wage labor paid a maximum of the equivalent of US$2 per day but was rare, and unemployment was high. Communities within the refuge were some of the poorest in the province, relying on basic grain crops, fuelwood collection, and citrus cultivation (FUNDENIC-SOS, 1999, cited in Hope, 2002). Although this regional picture may be dated, Nicaraguan standards of living remain some of the poorest in the world. Nicaragua ranked 121 out of 175 countries on the 2003 UNDP Human Development Index and had an estimated GDP per capita of US$2,450 (www.undp.org/hdr2003). The only country in the region to rank lower was Haiti (151).

Stewart (2001) described the communities around Chacocente as having their backs to the sea. With the exception of one village fishing cooperative, there were few sea-based economic activities. None of the villages were immediately proximate to the nesting beach; villagers (and others) traveled (sometimes short distances) to Chacocente to collect turtle eggs. The authorized egg harvest was theoretically for consumption only by the collector and immediate family, but eggs were almost always sold, as the market value (US$0.60–0.90 per dozen: Stewart, 2001; Hope, 2002) outweighed their attraction as food, and profits from two nests exceeded what could be earned in a week by unskilled labor (Stewart, 2001). Hope (2002) suggests that families make decisions on whether to sell or consume eggs based on price, alternative food sources available, and size of egg allocation. Once collected, eggs were sold to traders, and collectors and traders may have long-term relationships. Traders transported eggs to urban markets, where they were resold to vendors who resold them to consumers and restaurants. A few extended families traditionally controlled much of the egg trade, and some traders interviewed by Stewart (2001) had been working as such for over 20 years. Stewart concluded that "the egg trade is deeply imbedded in the economic and social life of the area" and examined the political context of this trade and what it implied for conservation in prerevolutionary, revolutionary, and postrevolutionary time periods.

PREREVOLUTIONARY CHACOCENTE UNDER SOMOZA. Before the Sandinista revolution, egg collecting at Chacocente was unregulated. Eggs were an open-access resource that was important to the destitute majority in this rural hinterland, where the ruling elites controlled most land and resources. Although there were some attempts to control egg harvesting at Chacocente in the last years of Somoza's regime (e.g., a 2-month ban on harvesting was introduced in the late 1970s), rules were generally flouted by both local people and Somoza's National Guard (Stewart, 2001).

CHACOCENTE DURING THE REVOLUTIONARY SANDINISTA PERIOD. At the time of the revolution, egg collectors and sellers at Chacocente numbered in the thousands. The revolutionary government sought to "change the social ecology of egg harvesting, distribution, and consumption in ways consonant with the larger national transition toward socialism" (Stewart, 2001). Sandinista activities to achieve these goals at Chacocente can be divided into three phases.

Hope (2002) suggests that, before 1993, no access to the beach at Chacocente was permitted. The level of detail provided by Stewart (2001) on egg harvesting under the Sandinista government and reference to Chacocente's turtle program by Faber (1993), however, clearly indicate that egg harvesting was ongoing throughout the Sandinista period.

Following the revolution of 1979, and until 1982 (phase 1), IRENA, under the leadership of biologist Magali Ubina, faced egg collectors and sellers "involved in a sort of combined squatter's movement and nonagricultural land invasion laying claim to beaches, turtles, and nests in the absence of the controls exerted by . . . local manifestation of the Somoza dictatorship." People came from distant cities and neighboring countries and set up a temporary beachfront shantytown housing as many as 3,000 people (Stewart, 2001). An impromptu market was established to serve the needs of collectors, and competition in harvesting, an outsider-versus-insider divide, and alcohol all combined for a turbulent and sometimes violent collection (for both people and turtles).

Ubina and her students first set out to study the turtles and the human community. On the human side, they found that declining social and economic conditions had changed collection activities; collecting had gone from being an extension of women's and children's domestic duties to a cash-generating activity undertaken by men. These men, many of them farmers, neglected farms during egg collection, which undermined production of staple foods. Collectors were "victimized" by the egg traders, intermediaries who transported the eggs from beach to market. Egg collectors received only 14–27% of final value (Stewart, 2001). Furthermore, profits were unevenly distributed among families. Based on Ubina's initial assessment of the economic and social situation, a scheme for productive conservation was developed and implemented in the second phase.

Productive conservation was operationalized from 1982 to 1987 (phase 2). IRENA established itself as the egg buyer and paid collectors twice what they had received previously. IRENA then set the sale price of eggs to be competitive with chicken eggs and used the profits from the sale to fund conservation at Chacocente. The number of people participating in the harvest was reduced, and 350 families with historical links to the egg harvest from the 17 surrounding villages were given collection permits and some role in decisionmaking. Egg collecting was restricted to the first three arribadas, and all nests laid in the final arribada were protected. Traders were not totally eliminated under IRENA's plan but rather had to organize in cooperatives of six or more members. Cooperatives would deposit money to buy eggs with IRENA, and IRENA would then buy eggs from collectors and transport them to market, where traders could sell to vendors (Stewart, 2001).

In his archival research, Stewart (2001) found only one remaining management plan for Chacocente during this period, written in 1985. The listed objectives included avoiding exploitation of collectors, restricting price speculation, increasing standards of living, keeping children in school, keeping collectors from abandoning their farms, and restricting collecting to the poorest women and the elderly. The wider and explicit revolutionary agenda in this plan is evident. It "envisioned far-reaching transformation of local communities in line with national scale revolutionary goals. The aim was not simply to control access to turtle eggs but to dampen the destabilizing effects of unrestrained egg trade on the national society and to promote social order of a particular kind." Whether many of these objectives were achieved or even pursued is uncertain, but some activities did reflect revolutionary socialist thinking. First, IRENA operated a store where families licensed to collect eggs could buy goods at subsidized prices and avoid trips to distant markets. Second, it attempted to ensure that collectors received a higher proportion of the profit from the collection by restricting the role of intermediaries.

Revenues skimmed off egg sales by IRENA were used to support park guards, technicians, and staff (Stewart, 2001). The environmental claims of the program included reducing the total number of eggs collected by restricting access to nearby communities only (Faber, 1993), reducing the consequences of the collection itself by controlling activities on the beach, and investing in infrastructure, with a research station constructed in 1982 to house IRENA employees and serve as a base of operations during collections (Stewart, 2001). IRENA sponsored environmental education about turtles, focusing on the conservation of turtles in line with human interests (Faber, 1993; Stewart, 2001). A seasonal ban on commercial sale of eggs was implemented and supported by military patrols and roadblocks as well as inspectors in markets, restaurants, and bars (Stewart, 2001). Because of this combination of a charismatic wildlife phenomenon, strong state action, and local defense of livelihoods, Stewart (2001) labeled Chacocente a high-profile exercise in productive conservation during this period. However, financial support for Chacocente was under pressure. By 1985, IRENA's budget had been cut by 40%, and in 1986 it was reduced a further 10% (Faber, 1993).

Between 1988 and 1990 (phase 3), state funding was diverted increasingly to the war effort. In 1988, the program was shut down temporarily, and large-scale invasions of the beach by egg collectors recommenced (Stewart, 2001). By 1989, the government could not pay to guard Chacocente (Faber, 1993). Opposition from traders and collectors mounted during this period. Traders, unhappy that IRENA had taken a major role in egg selling, discouraged the idea of egg-collecting cooperatives by convincing collectors that cooperatives were a ruse to conscript people to fight the Contras at the border. Traders also argued that IRENA's role as egg-selling intermediary "violated the traders' right under revolutionary ideology to a livelihood and to access to commonly held resources." This phase was marked by a change (and reduction) in personnel, the withdrawal of IRENA from egg selling, and the end of self-financing for Chacocente through the egg trade (Stewart, 2001). IRENA as a whole saw an 85% cut in personnel and was demoted to a subunit of the Ministerio de Agricultura y Reforma Agraria (MARENA) (IUCN, 1992).

POSTREVOLUTIONARY CHACOCENTE: CHAMORRO AND BEYOND. Violetta Chamorro's United States–supported coalition government defeated the Sandinistas in national elections in 1990. Following this, Stewart (2001) describes Chacocente as a "poster child of the vicious cycle of human desperation and environmental degradation," as clashes over access to the egg resource increased among squatters, park guards, police, and the army. A less powerful and autonomous IRENA, working under MARENA, at one point tried to reduce the number of communities that were licensed to take eggs from 17 to 9 (a decision soon reversed). MARENA was also criticized for failing to cooperate with student volunteers and nongovernment organizations (NGOs) interested in sea turtle conservation.

By 1993, some control at Chacocente was reestablished, and Chacocente was staffed seasonally by soldiers and permanently by a rotating MARENA staff of five (Stewart, 2001). The community representatives currently meet monthly with MARENA and army staff to discuss refuge operations (Hope, 2002). Any attempt to understand MARENA's management plan for turtle eggs from Chacocente (or La Flor) must be made in the wider national context. There is a national ban on collecting and selling eggs from all beaches between October 1 and January 31 of each year, but eggs from beaches outside of protected areas can be collected and sold at other times of the year. In Chacocente, the ban on egg collecting applies from July 1 to January 31, and sale of eggs from the Refuge is not permitted. In spite of the July 1 ban, a managed collection of eggs occurs from July to October, but theoretically for consumptive purposes only.

The strategy for egg collection during this third phase has changed over time. An initial harvesting strategy allowed egg collecting in Chacocente (and Playa La Flor) from February 1 to June 30 (dry season) because of the high sand temperatures and related low hatching rates. In 1995–1996, the seasonal strategy was abandoned because of concerns about temperature-dependant sex determination in hatchlings and because the seasonal harvesting strategy may have biased the sex ratio. Production and reproduction zones were established on the beach; eggs were collected from the production zone (the lower half of the beach, where eggs were more likely to be washed out by tides and surf) and protected in the reproduction zone (upper half of the beach). Spatially, the 800 m of beach where most turtles converge to nest was divided into 17 sections (one for each village), and, during arribadas, communities sent a representative to help collect eggs and guard the beach in their section. Eggs were then distributed to families in 17 surrounding communities according to a quota of several dozen per family per season. Also in the 1995–1996 season, a Christmas quota of five dozen eggs was distributed to each child under 9 years of age from families involved in the collection (involving over 1,000 children) (Stewart, 2001).

In 1998, reproduction and production zones were abandoned. Instead, MARENA allowed unlimited collection during the first night of the arribada and protected the turtles and nests during subsequent nights. Unlimited collection was allowed on the outer perimeter of the refuge because of enforcement limitations, and on portions of beach likely to be washed out. This policy produced more than the number of eggs required to meet the quota of 10 dozen per family for that year. Excess eggs were used to pay designated communities' members and additional "hired" men who helped collect eggs. Women and children from nearby villages also showed up during collections and were given leftovers. Stewart (2001) calculated that the total of these payments and free eggs was about one-fifth of the harvest, representing "a significant siphoning-off of eggs from the official (and more equitable) quota system of egg distribution to families."

Hope (2002) and Stewart (2001) both refer to a "rotation" system in which different communities participate in turn in different arribadas. Neither provided details on how the rotation works. Does the rotation involve participating in collecting activities only while still receiving eggs for each arribada? Or do families receive eggs only during the arribada their community participates in? But both imply that it results in an equitable distribution of eggs.

The quota of eggs per family and the number of families involved in the harvest have also changed, with the quota deceasing as the number of families has increased. The quota fell from 15 dozen eggs per family in 1995 to 11 in 1997, 10 in 1998, and 2 or 3 in 1999–2000 (Hope, 2002). The number of families involved has increased from 350 families in the 1980s to 800 in 1995 and 1,036 in 1997 (Stewart, 2001). Hope (2002) identified 5,754 individuals with harvesting rights in 1999. Neither Stewart (2001) nor Hope (2002) discussed the implications of this change (decreased benefits to individual families but distribution to more families), but Hope did suggest that communities living closer to the nesting beach felt their use rights should take precedence of those living farther away. Stewart suggested that the social unrest resulting from large-scale invasions was often blamed on outsiders living far from the refuge.

In spite of the law that prohibits the sale of eggs from the refuge, eggs are quickly sold to traders and taken to market, where, according to Stewart (2001), they were easily found, even during the period of the national ban. Based on 1996–1997 figures, Stewart calculated that earnings from the egg trade were distributed as follows: US$4,600 for collecting families, US$11,600 for market vendors, and US$46,000 for food vendors and restaurants. Hope (2002) reported a 32–33% price spread for eggs; that is, collectors received approximately one-third of the final sale price of eggs. Both Stewart (2001) and Hope (2000) saw problems arising from the large-scale collection of eggs from Chacocente, prohibition on all legal commerce of these eggs, but legal commerce from July to October in eggs from elsewhere. Given the longstanding commercial nature of egg collecting, high levels of poverty in the region, and a powerful group of egg traders, prohibition of egg commerce would require considerable enforcement if implemented. From July to October, when commerce in eggs from other beaches is legal, the inability to determine the source of eggs is problematic (Stewart, 2001). Hope (2000) argued that this legal dichotomy leads to high price spreads; prices are inflated in the black market for eggs, but egg collectors, numbering in the thousands and acting illegally and as individuals, have little bargaining power with egg traders, who are fewer in number (a situation of monopsony).

Playa La Flor

Located farther south on the Pacific coast of Nicaragua, Playa La Flor is a 1.6-km arribada nesting beach (five to seven arribadas per year: Hope, 2002), and, as in Chacocente, limited egg collection is permitted. Hope (2002) included La Flor in his study of the economics of arribada nesting beaches, but there is little else published. Reports written by the Sea Turtle Restoration Project (Arauz, 1996) and Fundación Cocibolca (Cocibolca, 1997) provide background detail to supplement Hope's (2002) analysis. Original research by Nicaraguan NGOs and researchers was cited in the Cocibolca report and is referred to in this chapter where appropriate.

Playa La Flor was declared a wildlife refuge in 1996 under General Environmental Law No. 216 (Hope, 2002). Prior protection had been facilitated by a private land owner (the Sequeria family) in cooperation with MARENA. Their agreement included a permanent military presence at La Flor from July to January, beginning in 1992. In 1993, this presence was supplemented with nine rangers contracted by MARENA from July to January, hired to conduct research, control the newly instituted egg-collecting program, and protect nests. A dormitory was built to house staff (Arauz, 1996). Currently, the refuge is co-managed by Fundación Cocibolca and MARENA, in consultation with communities and the army (Hope, 2002).

Arauz (1996) stated that egg collecting at La Flor was minimal before 1983. However, with the creation of Chacocente Wildlife Refuge in that year, demand increased when people excluded from egg harvesting at Chacocente turned their attention to La Flor. In 1993, a collection program was initiated, and three communities participated. This number has increased; Hope (2002) suggested that the current egg harvesting benefits eight communities consisting of 598 families and 2,618 people, figures in line with the eight communities and 576 families identified by Cocibolca (1997).

The legal framework for egg collecting at Playa La Flor is the same as that described above for Chacocente. Approximately 4% of eggs laid at Playa La Flor are collected under the approved program (Arauz, 1996; Hope, 2002) from areas of the beach where highest nest loss is anticipated. Each community involved in the collection elects a Community Commission to organize the distribution of eggs. Commissions are made up of six or seven community members, and 98% of commissioners are male (Cocibolca, 1997). Commissioners participate in extracting the eggs and distributing eggs among families (seven to eight dozen per family per season: Arauz, 1996). Commissioners also participate in meetings with MARENA to discuss problems and to receive information. Commissioners receive two extra nests per season for their work, as do individuals who help with routine monitoring activities in the refuge (Cocibolca, 1997).

As at Chacocente, many families at La Flor sell their eggs to earn cash to purchase other products and services. However, the situation for collectors at La Flor is superior to that at Chacocente, as the proximity of the tourist town of San Juan del Sur and its transportation links to and from Managua promote easier access to markets. Outside the national ban on egg sales, collectors retained 37.5% of final price (Hope, 2002). There is a discrepancy between Hope's and Cocibolca's estimated value of egg collecting to the community. Whereas Hope suggested collectors make the equivalent of US$38–63 per capita income per month, Cocibolca (1997) referred to recent (unspecified) studies showing that egg collection did not generate significant income and had a negligible effect on the precarious economy. However, Cocibolca did recognize that, given the ease of earning income via eggs, the community was unwilling to give it up.

Cocibolca (1997) reported "reproachable" acts by MARENA within the refuge, including alcohol abuse, illegal egg extraction, sexual violence against women, and excessive violence against apprehended poachers. The NGO Nixtayolero conducted community research at La Flor and found several points of local dissatisfaction and social unrest. Communication, availability of information, and function with process were all identified as problems. Local people complained about irregularities in the project's functioning and lack of transparency in decisionmaking. At the same time, Nixtayolero found that a high level of dependence on eggs, in addition to economic poverty, increased demands for eggs (Nixtayolero, 1997, cited in Cocibolca, 1997). Hope (2002), Arauz (1996), and Cocibolca (1997) all called for greater community participation at La Flor. Cocibolca saw this as essential for resolving problems experienced with MARENA staff. For example, if local people replaced MARENA staff as rangers (two local people were hired as rangers in 1996), social problems might decrease.

Few economic activities exist in the La Flor area. Arauz (1996) and Cocibolca (1997) both identified the need to develop alternatives and agreed that ecotourism was one option, calling for development of tourist services and training local people as tourism guides. They differed in their views of egg harvesting. Arauz argued that high quotas of eggs could be allowed if scientific research on productivity on the beach supported it. Some research has been done on this issue (von Mutius and van den Berghe, 2002). Cocibolca, however, aimed to convert egg collectors into ecotourism guides and to reduce extraction from the refuge, with a goal to eliminate egg collecting in 5 years. Their rationale: "As communities become more involved and sources of employment are created through ecotourism, they will see for themselves that this option is better than harvesting turtle eggs." Neither Arauz nor Cocibolca considered possible links between tourism and egg consumption, which may be indicated by the existence of markets for eggs in San Juan del Sur.

Conservation is always a political activity, involving decisions about who is allowed access to specific resources. The use of sea turtle eggs in Nicaragua, as described above, exemplifies this reality. In the case of Chacocente, conservation was an overtly political activity tied to revolutionary ideology in the Sandinista era. There were, and still remain, implications of such overt linkages. Ideology dictated that price inflation through intermediaries should be avoided, and the Sandinistas ignored the long history of egg traders and their political power when they tried to eliminate them from the process. Not only did egg traders work to undermine the government's cooperative program as a result, but they waged a rhetorical battle, arguing that their rights to a livelihood were being denied, counter to revolutionary doctrine. Such problems show the difficulties of putting ideology into practice.

The political legacy of Chacocente will continue to dictate how it evolves. Although it has been more than 10 years since the Sandinistas were defeated, they remain a force in Nicaraguan politics, and local people will not easily be deprived of the sea turtle resource. Stewart (2001) commented that even as the collection program in the 1990s continued to regulate social access, seek an equitable distribution of eggs, and provide opportunities for local people to stay informed about management and influence it to some degree, egg collectors and traders continued to subvert the program through poaching, illegal trade, and occasional mass invasions of the nesting beach. "This combination of "participation" via cooperation with the Chacocente program and direct action via poaching and beach invasions has maintained pressure on state authorities to stick to a model of sea turtle conservation that is oriented toward serving the immediate economic needs of surrounding communities through egg harvesting" (Stewart, 2001). The political parties have changed, but the people's focus on productive conservation has remained.

The links between La Flor and Chacocente illustrate the need to consider the repercussions of local activities. The thousands of people engaged in egg collecting in Chacocente at the beginning of Sandinista rule were reduced to several hundred, and both Faber (1993) and Stewart (2001) identified this as a "success." Neither considered where the excluded collectors went: if Arauz (1996) was correct, protection at Chacocente accelerated egg harvesting at La Flor.

In both Playa La Flor and Chacocente, the absolute amounts earned by individual egg collectors may seem small, but these earnings need to be viewed in the overall context of high economic stress and few cash-earning opportunities. Combined with the political atmosphere and beliefs about rights to harvest, the absolute value of egg collecting is perhaps the least important measure for consideration in decisionmaking. Hope (2002) also argued that the key issue is who captures "rents" from egg collecting, that is, how profits are distributed. When a significant proportion of rents are captured by legitimate harvesters, they may be more likely to take a long-term view and work to ensure the sustainability of egg collecting, particularly if their rights are clearly outlined. At both Chacocente and La Flor, too much rent is siphoned off by intermediaries. This imbalance is supported by the illegal nature of the sale, as collectors cannot easily organize to challenge the monopsony power of traders, and the history of Chacocente shows that the traders, who earn more and are politically powerful, can effectively undermine organized egg collecting. Any effort to redistribute income that ignores traders will do so at its own peril.

The legal context of egg selling in Nicaragua is problematic. Stewart (2001) cites failure to address legal / illegal contradictions as illustrating a "lack of follow-through on the institutionalization of Chacocente program and the principle of productive conservation the program has historically embodied." If eggs sales were legal, the "productive" element of the program would be official. With eggs collected only for household consumption, the program looks like welfare. The distinction is not simply ideological. Stewart argued that this failure leaves the area vulnerable to land grabbing and the new "parquismo" movement. MARENA, for example, includes in its aims reducing social pressures for egg collecting through developing alternatives in the buffer zone, regardless of whether egg collecting is sustainable.

Both Stewart (2001) and Hope (2002) criticized the top-down nature of egg collecting and the failure to involve local people adequately in decisionmaking. Such externally enforced compliance decreases incentives for ownership or stewardship, and the occasional invasions at Chacocente can be interpreted as the collectors' assertions of rights in a system that affords them few.

Costa Rica: Egg Collecting at Ostional

Ostional has received considerable attention over the years as the site of a widely publicized legal egg collection project (Pritchard, 1984) and because of conflicts that have plagued the project since the mid-1990s (Valverde, 1999). The discussion of Ostional is based on the author's research (Campbell, 1997, 1998, 1999), other published work (Arauz Almengor et al., 2001; Hope, 2002), and unpublished documents from the Asociación de Desarrollo Integral de Ostional (ADIO) and the University of Costa Rica. The egg harvest itself and its effects on national egg marketing are discussed below.

The Egg Project

Campbell's (1998) study was based on 8 months of residence in Ostional (11 months in Costa Rica) over the course of 1994–1995. She conducted in-depth interviews with community members, government employees, and biologists associated with the project, surveyed 91% of households, and analyzed unpublished management and consulting reports. Hope's (2002) methods are described above (Nicaragua section).

The arribada nesting beach at Ostional is protected in the Ostional Wildlife Refuge (established in 1985), where olive ridleys nest almost every month throughout the year. Arribada nesting varies from month to month, but in the rainy season (peak nesting season) of 2001, estimated numbers of turtles ranged from 20,000 to 130,000 turtles per arribada (Chaves, 2002). Eggs are collected by members of ADIO during the first 36 hours of each arribada, packaged in plastic bags stamped with the Ostional insignia, and transported throughout the country. One objective of the project is to saturate the national market for eggs, and the price of Ostional eggs is kept low to discourage illegal collection of eggs from other beaches. All aspects of the collection and distribution are managed by ADIO, and although various government agencies have responsibilities for the refuge and egg collection, their involvement is often minimal. Profits are distributed among salaries paid to cooperative members (70%) and community development projects and expenses of the association (including the biologist's salary) (30%) (Campbell, 1998).

Campbell (1998) evaluated the social and economic aspects of egg collecting that contributed to its socioeconomic sustainability. She determined that 70% of households identified egg harvesting as their most important activity and that salaries earned in egg harvesting were superior to those earned in other available employment, with the exception of tourism (see below). Benefits of the project were well distributed in the village, although some large families with many associates earned more than the average household, and individuals holding egg-selling and distribution contracts (which theoretically rotate among community members) earned extra income. Households that did not participate in egg harvesting were also supportive of the project, as they received secondary economic benefits by selling goods and services to cooperative members. Social and environmental impacts of the project were believed to be primarily positive, although there was recognition of the need to diversify the economy and reduce dependence on egg harvesting, partly because egg harvesting, although lucrative, occurs for only several days each month.

Received and perceived economic benefits were critical to community support for the project. However, there were other important factors, including the legality of the project (thus legitimizing local livelihoods and removing fear of arrest), community management of the project by ADIO (rather than by an outside agency), and the establishment of the community's management role in law; Wildlife Conservation Law 6919, which allows for the egg collection, stipulates that a community development association be formed to manage the project. The combination of these factors provided security and encouraged wide distribution of benefits to both cooperative members and some nonmembers, reinvestment of profits into local development projects (lessening individual profit but increasing community benefits), investment in conservation through both paid and unpaid activities (beach guarding, escorting hatchlings to the surf), and adoption of voluntary development restrictions to minimize interference with marine turtle nesting and habitat. Hope (2002) calculated that the Ostional project retained more profits locally than did egg harvesting projects in Nicaragua through control over marketing and distribution and linked this to the overall lower price spreads experienced in Ostional. Hope (2000) cited price spreads of 75% for Ostional but also noted the existence of regional price spreads: the farther eggs moved away from Ostional, the greater the price difference.

Campbell's (1998) conclusions regarding the importance of the legal nature of the project to ensuring community support for conservation activities have been tested over the past three years, as several legal challenges to the project have been instigated by biologist Anny Chaves, formerly director of marine turtle research at the University of Costa Rica. The many *recurso de amparo* (petitions) registered with Constitutional Court by Chaves have ranged in scope from challenging the definition of the dry season (and related nest location techniques) adopted in ADIO's management plan and questioning the responsibilities of the government to provide a clean environment to demanding the cancellation of permission for commercial sale of eggs from Ostional. The details of these legal debates are described by Monge Artavia and Jiménez Gómez (2001), and although the project has continued to function, the sense of security that Campbell (1998) deemed crucial to sustainability undoubtedly has been threatened. However, no research on the effects of this latest stage in Ostional's history has been conducted.

Campbell (1999) also evaluated the effects of tourism in Ostional and its potential to coexist or compete with the egg-harvesting project. At the time of the study, levels of tourism in Ostional were low but increasing. In 1995, there were an estimated 852 overnight stays in Ostional, generating approximately US$6,500 for the two *cabinas* (small motels, four to eight rooms) operators in that year. For those two households, tourism earned them four and seven times more than the egg-harvesting project, respectively. In-depth interviews revealed that people of Ostional perceived economic benefits of tourism but saw these as concentrated among a few families. Furthermore, there was a strong belief that tourism development could have negative impacts on the nesting turtles. Some saw tour guiding for turtle watching as a possible way to overcome these potential negative effects, but others saw limitations on guiding in Ostional. For example, there are many access points to the beach, it is easy to observe turtles during arribadas, and there are few English-speaking residents to serve as tour guides. There was also an awareness of existing and potential social impacts, including increased land speculation and ownership by foreigners. In contrast, household surveys showed wide support for tourism and beliefs in the financial benefits. However, respondents had trouble identifying specific economic benefits or further economic opportunities to invest in tourism, and approximately half the surveyed respondents did not want to work in tourism. Thus, the contradiction between interview and survey responses may be linked to survey respondents' expressed desire for any type of additional development.

Campbell (1999) identified constraints to the community's ability to profit further from tourism. First, the existing accommodation sector was underutilized with occupancy rates at 12%, and the expense of investing in accommodations coupled with increasing land prices meant that investment in infrastructure (e.g., cabinas) would be out of reach for many community members. Foreign ownership was already evident in the small tourism economy of 1994–1995. The opportunities for tour guiding were limited, although laws requiring that tourists be accompanied by a guide while on the beach (in operation at Tortuguero, Costa Rica) might change this if introduced. Most importantly, tourism, rather than merely offsetting dependence on the egg project, could conflict with it, as tourist and local uses of turtles (as spectacle and as food) may be incompatible. With a large number of people depending on the egg harvest and a small but relatively wealthy number on tourists, the potential for conflict is high and is complicated by the interests of foreign land owners investing in tourism. Hope (2002) noted that tensions existed around building restrictions, road maintenance, and lighting, and at least part of this was linked to the desire by some residents for further tourism development (Ostional resident, personal communication).

National Markets

Because one of the objectives for legalizing the Ostional project was to flood the market with a legal supply of cheap eggs, thereby decreasing illegal collection elsewhere, the project's success in this area should be evaluated (Valverde, 1999). A study by Arauz Almengor et al. (2001) sought to "determine the demands and characteristics of turtle-egg marketing from Ostional Wildlife Refuge." The results are dated, as the study was conducted in 1991 when ADIO contracted the distribution of eggs to an outside party (associates of ADIO currently distribute eggs along a number of national routes). Nevertheless, their results are reviewed here. Arauz Almengor et al. interviewed ADIO executive members, two egg distributors, and two egg retailers in San José. In addition, they visited 64 establishments where eggs were sold. Although they did not say so explicitly, their data suggest they interviewed at least one person in these establishments.

Under the old distribution system, eggs were delivered to four distributors to cover six national routes, and 81 resellers (contractors, distributors, fish markets, street vendors, bar restaurants) were involved in resale (Arauz Almengor et al., 2001). When Campbell was conducting her research (1994–1995), eggs were distributed directly by ADIO, on nine national routes (Campbell, 1998). Under both systems, the Central Valley was well supplied with eggs, but the southern zone and Limón were inadequately supplied (Campbell, 1997; Arauz Almengor et al., 2001). Arauz Almengor et al. (2001) found major price variations in San José, and Hope (2002) found the geographic price spread described above.

The main market for eggs is bars, where approximately 90% of eggs were prepared raw in a red sauce as an appetizer. Although Arauz Almengor et al. (2001) reported that eggs were believed to hold aphrodisiac qualities, they also found that only 22% of persons interviewed favored eggs over other appetizers.

One conclusion of Arauz Almengor et al. (2001) was that ADIO should consider taking over the distribution of eggs to minimize the number of intermediaries and resulting price inflation. ADIO did so (whether linked to this study or not is unknown), and although there are still issues of national coverage (Campbell, 1998; Hope, 2002), profits retained by egg selling and distribution were significant to project member households (Campbell, 1998).

Suriname: Culture and Resistance with Egg Collecting

Schulz's (1975) overview of use of sea turtles in Suriname included all species of turtles nesting in the country, but there was specific reference to the use and conservation of olive ridleys. He combined a review of historical documentation and anthropological research undertaken by Kloos (1971, cited in Schulz, 1975) with a description of management efforts. Published in 1975, it is an early example of attention to the human context of use and to a related anthropological study. It is worth noting that the management plan, one that allowed for continued exploitation of sea turtle eggs in Suriname, was devised before the contemporary sustainable use dialogue began; it was based both on a belief in the potential to rationally exploit sea turtles and on the recognition of the infeasibility of a total ban, given human demand.

Schulz (1975) cited records of turtle use in historical documents dated as early as the late 1600s. The people involved in the harvest of turtles and eggs in Suriname were primarily the Kali'na, indigenous peoples residing in coastal areas. In Schulz's work, the Kali'na are referred to as Carib. These Amerindians, who live on both sides of the Marowijne / Maroni river that separates Suriname and French Guiana, have been given various names over time (Carib, Galibi, and Kaliña). Kali'na is the name they currently adopt for themselves (Collomb and Tiouka, 2000). Although both turtles (primarily greens, *Chelonia mydas*) and eggs (greens, leatherbacks [*Dermochelys coriacea*], and olive ridleys) were used historically, meat was not consumed by the Kali'nas (but was traded nationally and internationally). Green turtles were the targeted species for meat export, but olive ridleys were also included (Schulz, 1975). Before the 1900s, it is not clear whether eggs were used only by Kali'nas or traded with other peoples, but by 1945, Geijskes (1945, cited in Schulz, 1975) reported that egg harvesting had increased to meet non-Kali'na demand. At the time of World War II, most olive ridley (and green turtle) eggs laid in Suriname were collected and distributed to national markets and to French Guiana (Schulz, 1975).

Following World War II, meat exports stopped. Because there was little internal demand for meat, attention turned to egg collecting. In 1954, the Game Ordinance and the Nature Preservation Ordinance came into force, and some of the nesting beaches were protected. Because of high levels of beach erosion, however, some of Suriname's initial attempts to protect nesting beaches were undermined. The first nesting turtle population protected in 1954 had shifted outside the protected zone by 1973. The main nesting beaches for olive ridleys were originally outside the areas subject to the wildlife laws and outside the limits of protected areas. In 1964, about 750,000 green and olive ridley eggs were harvested from around the Marowijne Estuary and sold in national markets. This represented about 90% of the production from the beach. When Eilanti Beach was identified as the most important olive ridley nesting beach on the Atlantic, efforts to stop egg collecting began. In 1967 and 1969, money from the World Wildlife Fund was used to purchase eggs from collectors for reburial in hatcheries. In 1969, the area was declared a nature reserve, and by 1970, a complete ban on collecting olive ridley eggs was enforced. Some compensation was provided through a limited collection of green turtle eggs established and run by the government. Kali'nas would eventually be paid approximately double the amount they previously had received from intermediaries, and yet the project initially met with resistance and became highly politicized. "The 'turtle affair' was even made an issue in electoral tactics in 1969–1970" (Schulz, 1975).

The basis of Kali'na disagreement included their challenge to the notion that turtle populations were decreasing, that compensation was too low, and that they did not want interference in their territory. Anthropologist Kloos (1971, cited in Schulz, 1975), however, found that more was at stake and that the issue of money was not as important as the issue of freedom. Only a small number of individuals participated in olive ridley egg collecting and were to be affected by the ban, but the issue became important for all Kali'nas. Meetings held around the creation of the nature reserve became hostile, and threats of violence were made, in spite of Kali'na cultural preferences to avoid open disagreement. Although the Kali'nas eventually yielded to official regulations, Kloos believed the process was misunderstood. The conservation project remained incompatible—and ultimately irreconcilable—with Kali'na visions of freedom, but as the Kali'nas and their leaders continued to yield to official pressure, officials failed to understand this problem. This yielding was linked to the cultural preference to avoid open disagreement, a preference that also meant Kali'na dissatisfaction with their leaders was hidden from officials.

Schulz (1975) concluded that, regardless of their reasons, the Kali'nas cooperated with the conservation program beginning in 1970. His account of this process highlights two important issues. First, a more complete understanding of the culture of the Kali'na people from the outset could have avoided some of the problems experienced in the initial stages of the project and may have changed the entire approach of the government in creating the project. Kali'na values of freedom might have been better accommodated if Kali'na leadership had been involved initially and consulted in the establishment of the plan. Second, Schulz's summary highlighted the limited use to which the results of social science research are sometimes put. Schulz was interested in explanations of Kali'na dissatisfaction, but as the ultimate goal of ensuring Kali'na cooperation with the ban on olive ridley egg collecting was eventually secured, his interest waned. What Schulz did not discuss were the ramifications of these events for the Kali'na people. For example, the effects of the challenge to Kali'na leadership and the erosion of freedom were unexplored. These issues are both of general interest to the anthropologist but also could have had further long-term implications for conservation efforts.

Discussion and Conclusions

Given the limited research on the human aspects of use of sea turtles in general and olive ridleys specifically, it is not surprising that the research described above is in the form of case studies. Case study research is common in social sciences, particularly when one is trying to understand contextualized problems about which little is known. The case studies discussed here provide data that can be used to understand specific scenarios and inform the design of conservation or development interventions. For example, further research similar to that of Arauz Almengor et al. (2001) could identify remaining problems with Ostional egg distribution and pricing and shape policies designed to address these. Case-specific information can advise against the application of "blind" theory. For example, eliminating intermediaries is theoretically a good way to increase rents retained by local people (Hope, 2002), but in the case of Nicaragua, it would be politically difficult if not inappropriate (Stewart, 2001).

One of the weaknesses of case studies is that they provide a snapshot in time, and the specific details can change (e.g., numbers of people involved in harvesting, the laws guiding harvesting). Another shortcoming relates to the site specificity of findings. For example, Hope (2002) compared egg harvesting in Costa Rica favorably with that in Nicaragua, but it is clear that the Costa Rica model is not directly transferable. For one thing, the number of people involved varies; approximately 220 people from one village have rights to harvest in Ostional, versus 600 families from 8 villages in Playa La Flor and 1,070 people from 17 villages in Chacocente. Although it is not impossible that 17 communities could cooperatively manage a commercial egg harvest, the structure of such management would undoubtedly look different from what is in place at Ostional. Nevertheless, there are lessons to be learned in contrasting case studies. For example, Hope argued that one of the reasons Ostional fared better than Nicaragua is the larger proportion of rents retained by collectors. The objective of increasing rents for collectors (while keeping the powerful traders in mind) could be pursued in Nicaragua.

With the strengths and weaknesses of case study research kept in mind, findings from the site-specific research contain several general themes:

- Legal frameworks were important in several case studies, some of them supporting conservation (Costa Rica) and some confusing the situation (Nicaragua, Mexico). Campbell (1998), Hope (2002), and Trinidad and Wilson (2000) all promote implementation of legal frameworks to guide use and establish community participation in decisionmaking in order to maximize incentives for conservation and socioeconomic benefits.
- The economic value of olive ridleys to communities was a focus in many of the case studies (Lagueux, 1991; Campbell, 1998; Trinidad and Wilson, 2000; Hope, 2002). However, absolute values are only one piece of information needed to understand incentives for use, and perhaps not the most important. Relative monetary value compared to other livelihood activities, contributions to cash versus subsistence income, economic return on effort expended, and seasonality of livelihoods are other important issues to assess. If absolute monetary value of use is the only information collected, the overall value of use might be underestimated with distorted implications for both conservation and development. Furthermore, many of the case studies highlighted explicitly or implicitly the need to look beyond economic assessment. Political (Stewart, 2001), social (Campbell, 1998), cultural (Schulz, 1975), and legal (Campbell, 1998; Trinidad and Wilson, 2000) structures provide an important context for understanding use and conservation. Hope's (2002) economic analysis recognized how legal and social structures influence economic incentives for conservation and act as incentives or disincentives on their own.
- Lack of alternative development opportunities was stressed in several studies (Campbell, 1998; Trinidad and Wilson, 2000; Stewart, 2001; Hope, 2002), and conservationists interested in reducing stress on resources are often told to look for, promote, and even provide such alternatives. However, in almost all case studies, communities involved in the use of olive ridleys were marginalized economically, and in such scenarios, opportunities for income substitution through alternative activities may be limited. If people are living in poverty, the point at which income can be effectively substituted for (rather than added to) may be distant. If so, the expressed willingness of people to reduce use if other opportunities become available (e.g., Trini-

dad and Wilson, 2000) should be treated with caution. Ecotourism in particular was promoted (Arauz, 1996; Cocibolca, 1997; Trinidad and Wilson, 2000; Hope, 2002) or existed (Campbell, 1999) as an alternative economic development strategy in several cases. Further research on ecotourism, the structure of the industry, the potential for local people to participate in it, and its compatibility with other economic activities is needed.

- Participation of local people and communities influences conservation outcomes. With the exception of Ostional, Costa Rica, levels of participation by resource users in decision-making about conservation are deemed inadequate in the case studies (Schulz, 1975; Arauz, 1996; Cocibolca, 1997; Stewart, 2001; Hope, 2002). Repercussions of this linkage varied; in Nicaragua, Stewart (2001) categorized beach invasions at Chacocente as an expression of rights that are mostly denied in the management regime, and Cocibolca (1997) suggested that increased participation would reduce specific conflicts related to the presence of MARENA staff members, who were considered outsiders. In Ostional, Campbell (1998) concluded that community participation was a critical component of socioeconomic sustainability in the egg-collecting project. These calls for participation reflect wider thinking about conservation (Western and Wright, 1994; Ghimire and Pimbert, 1997; Agrawal and Gibson, 2001; Hulme and Murphree, 2001).

Ultimately, if lessons learned regarding the context of use are to contribute to conservation programs, they will have to be combined with biological research. When the goal is sustainable use, there is limited utility in ensuring that profits from olive ridley use are equitably distributed if that use undermines the survival of the turtle population. Nevertheless, use, whether it contributes to conservation or threatens it, is ongoing in many parts of the world, and understanding the context of use is critical to managing it.

ACKNOWLEDGMENTS

The author's research was supported by the Canadian Social Sciences and Humanities Research Council. I thank Matthew Godfrey, Randall Arauz, and Rob Hope for sharing documentation.

LITERATURE CITED

Agrawal, A., and Gibson, C. C. (Eds.). 2001. *Communities and the Environment: Ethnicity, Gender, and the State in Community-Based Conservation.* New Brunswick, NJ: Rutgers University Press.

Arauz, R. 1996. *La Flor Wildlife Refuge, Rivas, Nicaragua: Action Plan.* San José: Sea Turtle Restoration Project.

Arauz Almengor, M., Mo, C. L., and Vargas M., E. 2001. Preliminary evaluation of olive ridley egg commerce from Ostional Wildlife Refuge, Costa Rica. *Marine Turtle Newsletter* 63:10–13.

Cahill, T. 1987. The shame of Escobilla. In *Jaguars Ripped My Flesh.* New York: Random House, pp. 144–173.

Campbell, L. M. 1997. International conservation and local development: the sustainable use of marine turtles in Costa Rica. Ph.D. diss. University of Cambridge, Cambridge.

Campbell, L. M. 1998. Use them or lose them? The consumptive use of marine turtle eggs at Ostional, Costa Rica. *Environmental Conservation* 24:305–319.

Campbell, L. M. 1999. Ecotourism in rural developing communities. *Annals of Tourism Research* 26:534–553.

Campbell, L. M. 2002. Science and sustainable use: views of conservation experts. *Ecological Applications* 12:1229–1246.

Campbell, L. M. 2003. Contemporary culture, use, and conservation of sea turtles. In Lutz, P. L., Musick, J. A., and Wyneken, J. W. (Eds.). *The Biology of Sea Turtles,* Volume 2. Boca Raton, FL: CRC Press, pp. 307–338.

Carr, A. F., Jr. 1948. Sea turtles on a tropical island. *Fauna* 10:50–55.

Chaves C., G. 2002. *Plan de aprovechamiento para la utlización racional, manejo y conservacion de los huevos de la tortuga marina lora,* Lepidochelys olivacea, *en el Refugio de Vida Silvestre de Ostional, Santa Cruz, Guanacaste, Costa Rica.* San José: University of Costa Rica.

Cliffton, K., Cornejo, D. O., and Felger, R. S. 1995. Sea turtles on the Pacific coast of Mexico. In Bjorndal, K. (Ed.). *Biology and Conservation of Sea Turtles, revised edition.* Washington, DC: Smithsonian Institution Press, pp. 199–209.

Cocibolca. 1997. Management plan for La Flor wildlife refuge, Rivas, Nicaragua. Managua: Cocibolca.

Collomb, G., and Tiouka, F. 2000. *Na'na Kali'na. Une Histoire des Kali'na en Guyane.* Guadeloupe: Ibis Rouge.

Cornelius, S. E., Alvarado Ulloa, M. A., Castro, J. C., Malta De Valle, M., and Robinson, D. C. 1991. Management of olive ridley sea turtles (*Lepidochelys olivacea*) nesting at playas Nancite and Ostional, Costa Rica. In Robinson, J. G., and Redford, K. H. (Eds.). *Neotropical Wildlife Use and Conservation.* Chicago: University of Chicago Press, pp. 111–135.

Faber, D. 1993. *Environment under Fire: Imperialism and the Ecological Crisis in Central America.* New York: Monthly Review Press.

Ghimire, K. B., and Pimbert, M. P. (Eds.). 1997. *Social Change and Conservation.* London: Earthscan.

Godfrey, M. H. 1997. Further scrutiny of Mexican ridley population trends. *Marine Turtle Newsletter* 76:17–18.

Hope, R. 2000. Egg harvesting of the olive ridley marine turtle (*Lepidochelys olivacea*) along the Pacific coast of Nicaragua and Costa Rica: an arribada sustainability analysis. MA thesis, University of Manchester, Manchester.

Hope, R. A. 2002. Wildlife harvesting, conservation and poverty: the economics of olive ridley egg exploitation. *Environmental Conservation* 29:375–384.

Hulme, D., and Murphree, M. 2001. *African Wildlife and Livelihoods: the Promise and Performance of Community Conservation.* Oxford: James Currey.

IUCN. 1980. *The World Conservation Strategy.* Gland: IUCN.

IUCN. 1992. *Protected Areas of the World: a Review of National Systems,* Volume 4: *Nearctic and Neotropical.* UNEP World Conservation Monitoring Centre. Gland: IUCN.

Lagueux, C. 1989. Olive ridley (*Lepidochelys olivacea*) nesting in the Gulf of Fonseca and the commercialization of its eggs in Honduras. M.A. thesis, University of Florida, Gainesville.

Lagueux, C. 1991. Economic analysis of sea turtle eggs in a coastal community on the Pacific coast of Honduras. In Robinson, J. G., and Redford, K. H. (Eds.). *Neotropical Wildlife Use and Conservation.* Chicago: University of Chicago Press, pp. 136–144.

Márquez-M., R., Peñaflores, C., and Vasconcelos, J. C. 1996. Olive ridley turtles (*Lepidochelys olivacea*) show signs of recovery at La Escobilla, Oaxaca. *Marine Turtle Newsletter* 73:5–7.

Monge Artavia, K., and Jiménez Gómez, G. 2001. *Protección y conservación de las tortugas marinas a la luz de derecho internacional y nacional ambiental. Análisis de casos en Costa Rica.* San José: University of Costa Rica, pp. 95–114.

Mrosovsky, N. 1983. *Conserving Sea Turtles.* London: British Herpetological Society.

Mrosovsky, N. 1997. A general strategy for conservation through use of sea turtles. *Journal of Sustainable Use* 1:42–46.

Mrosovsky, N. 2001. The future of ridley arribadas in Orissa: from triple waste to triple win? *Kachhapa* 5:1–3.

Pritchard, P. C. H. 1979. *Encyclopedia of Turtles.* Neptune: TFH, Inc. Publications.

Pritchard, P. 1984. Guest editorial: Ostional management options. *Marine Turtle Newsletter* 31:2–4.

Pritchard, P. C. H. 1997a. Evolution, phylogeny and current status. In Musick, J. A., and Lutz, P. L. (Eds.). *The Biology of Sea Turtles.* Boca Raton, FL: CRC Press, pp. 1–28.

Pritchard, P. C. H. 1997b. A new interpretation of Mexican ridley population trends. *Marine Turtle Newsletter* 76:14–17.

Ross, J. P. 1996. Caution urged in the interpretation of trends at nesting beaches. *Marine Turtle Newsletter* 74:9–10.

Schulz, J. P. 1975. *Sea Turtles Nesting in Suriname.* Leiden: Rijksmuseum van Natuurlijke Historie.

Stewart, A. Y. 2001. Poached modernity: parks, people and politics in Nicaragua, 1975–2000. Ph.D. diss. Rutgers University, New Brunswick, NJ.

Trinidad, H., and Wilson, J. 2000. The bio-economics of sea turtle conservation and use in Mexico: history of exploitation and conservation policies for the olive ridley (*Lepidochelys olivacea*). In *Proceedings of the International Institute of Fisheries Economics and Trade Conference.* Corvallis, OR.

Valverde, R. A. 1999. Letter to the editors: On the Ostional affair. *Marine Turtle Newsletter* 86:6–8.

Von Mutius, A., and Van den Berghe, E. P. 2004. Hatching success of olive ridley sea turtles (*Lepidochelys olivacea*) in La Flor, Nicaragua. In Mosier, A., Foley, A., and Brost, B. (Compilers). *Proceedings of the Twenty-First Annual Symposium on Sea Turtle Biology and Conservation.* NOAA Technical Memorandum NMFS-SEFSC-477, p. 291.

Weinberg, B. 1991. *War on the Land: Ecology and Politics in Central America.* London: Zed Books.

Western, D., and Wright, M. A. 1994. The background to community-based conservation. In Western, D., and Wright, M. A. (Eds.). *Natural Connections: Perspectives in Community-Based Conservation.* Washington, DC: Island Press, pp. 1–12.

PETER C. H. PRITCHARD

3

Evolutionary Relationships, Osteology, Morphology, and Zoogeography of Ridley Sea Turtles

KEMP'S RIDLEY (*LEPIDOCHELYS KEMPII*) was described by Samuel Garman, of the Museum of Comparative Zoology at Harvard, in 1880 (Garman, 1880). He based the new species on two adult specimens from Key West, Florida, that had been sent to him by Richard M. Kemp, a man whose name has been immortalized in the name of the species. Garman actually named the new species *Thalassochelys kempii, Thalassochelys* being the generic name then in use for the loggerhead turtle (*Caretta caretta*), but his inconsistency of names even within a single paper has caused headaches among systematists for over a century. Despite his use of the binomial form, Garman thought that his new species might be distinct at a level above that of the species, which is why, on the next page, he proposed to place it in a separate subgenus (*Colpochelys*) and then proceeded to elevate it to full species status as *Colpochelys kempii*. Garman noted that it was a widespread belief that this new turtle was commonly considered a cross between a green turtle (*Chelonia mydas*) and a loggerhead, being known as "bastard" in the Gulf of Mexico, an epithet that persisted at least to the second half of the twentieth century in the German vernacular *Bastardschildkröte* (Wermuth and Mertens, 1961). This persistent belief in the hybrid origins of the ridley derived not just from the morphology of the animal, which in some respects might be superficially judged to be intermediate between the two species, but also because it was not known to nest anywhere, which seemed to support the "sterile hybrid" theory.

The hybrid theory would, of course, have evaporated rapidly if those who subscribed to it were aware that there was another species of *Lepidochelys,* namely, *L. olivacea* (olive ridley), abundant in the eastern Pacific and northern Indian Oceans, that indeed also had breeding colonies on both the African and South American sides of the Atlantic. Turtle fishermen operating in Florida and adjacent waters might be forgiven for being unaware that the world had other ridleys that reproduced normally, but for some reason even the scientific community took a long time to make the connection. Carr himself once admitted with some embarrassment that, when he saw olive ridleys nesting on the Pacific coast of Honduras in the 1940s, he assumed at first that they were green turtles. Carr (1942) was the first author since Baur (1890) to recognize that *kempii* and *olivacea* were closely related and belonged in the same genus. Carr (1956), in his book *The Windward Road,* had to report that the nesting habits of Kemp's ridley were still unknown, and the unusual breeding habits of the species were indeed not revealed until 1961, as reported by Hildebrand (1963) and Carr (1963). Garman (1880), however, while reporting the folk beliefs of the hybrid nature of the ridley, did not subscribe to them himself. In naming the species, he considered it distinct from the loggerhead in having a short, round shell, low "humps," marginal "plates," narrow head, and swollen jaws, and these are indeed good characters for recognizing the species.

Loggerheads have wider, more triangular heads and a hump (corresponding to thickened supracaudal bones) above the tail, and they lack the distinctive broadening of the shell, which in ridleys appears to assist in the nest-closing process. Garman also wrote that Kemp's ridley was distinguished from the olive ridley (then called *Thalassochelys olivacea*) by the shape of the head, swollen jaws, and the carapace scutes, although it is now known that the first two characters only separate the adults of the two species. However, the costal scute count, nearly always five on each side in *L. kempii* but generally six to nine, and only occasionally five, in *L. olivacea,* will distinguish even the hatchlings.

Garman (1880) correctly recognized all the species of sea turtle considered valid today and recognized none that today are considered invalid. Indeed, in addition to *L. kempii,* he described the noteworthy new Australian species *Chelonia (Natator) depressa,* although his professional interests were primarily ichthyological. However, his example was not followed by many subsequent authors. Not only did cheloniologists ignore *depressa* for nearly a century, but they confused olive ridley and the loggerhead for decades in the late nineteenth and early twentieth centuries, a misconception that perhaps reached its apogee in a paper by Gadow (1899), in which the author attempted to explain the apparent loss of scutes with ontogeny in loggerheads. Actually, his hatchlings were the multiscutate *L. olivacea,* and his adults the pentecostal *C. caretta.*

Today, the two ridleys are considered full species, *kempii* and *olivacea,* within the genus *Lepidochelys,* this generic name being considered senior to Garman's (1880) *Colpochelys.* The specific epithet cannot be displaced by the discovery of senior synonyms because it has been officially conserved by the International Commission for Zoological Nomenclature (Opinion 660, 1963). For example, one such senior synonym, *Testudo mydas minor* Suckow 1798, was thought by Wermuth (1956) to be based on a specimen of Kemp's ridley, although Pritchard and Trebbau (1984) summarize why it was probably based, in fact, on a specimen of the olive ridley.

Evolutionary Relationships

The genus *Lepidochelys* is placed within the order Chelonia or Testudines (or, increasingly, the Chelonii of Latreille, 1800, which has priority over Chelonia Macartney, 1802, and Testudinata Oppel, 1811). Testudines Linnaeus, 1758, is simply a Latin plural form of *Testudo* and thus lacks suprageneric nomenclatural rank. Testudines Batsh, 1788, was clearly presented as a familial, not an ordinal, name within the suborder Cryptodira. The latter allocation could be challenged because the cryptodires are defined on the basis of certain characteristics, such as the retractile neck, not shown by sea turtles. Zangerl (1969), for example, placed the Cheloniidae in a suborder that he called Metachelydia, whereas Gaffney (1984) proposed a complex new hierarchial classification in which the Cheloniidae were placed in the microorder Chelonioida, infraorder Procoelocryptodira, suborder Polycryptodira, parvorder Eucyptodira, hyperorder Daiocryptodira, megaorder Cryptodira, and gigaorder Casichelydia, of the order Testudines.

The family Cheloniidae is considered not to include natural subfamilies by many authors (Pritchard and Trebbau, 1984). Thirty-one cheloniid genera have been named, of which only five have living representatives. Some of the fossil genera are known from extremely fragmentary material, and in some cases names were published in the form of preliminary notices (Dollo, 1909) never followed up by a formal description.

There is some justification for allocating the living cheloniid genera (*Chelonia, Eretmochelys, Natator, Caretta,* and *Lepidochelys*) to two subfamilies, the Cheloniinae and the Carettinae (or two tribes, Chelonini and Carettini). The former would include, as a minimum, *Chelonia,* and the latter at least *Caretta* and *Lepidochelys.* The latter two genera show a number of distinctive common features including the triangular head, presence of two pairs of prefrontal scales, somewhat similar scalation of the crown of the head, and the insertion of an additional pair of costal scutes at the front of the series. The two are also linked by shared behavioral traits, such as the alternating terrestrial gait, similar egg size, and certain stereotyped features of the nesting process, such as the symmetrical, widely spaced posture of the hindlimbs during oviposition.

The placement of the hawksbill (*Eretmochelys imbricata*), however, is less clear. Some authorities place this genus close to *Chelonia* largely because of the presence of only four pairs of costal scutes and the clear-cut, nonfragmented head scales. The former character also correlates with certain other differences, such as the presence of additional anterior peripheral bones in *Caretta* and *Lepidochelys* and the nuchal scute having contact with the first costals. However, more than 60 years ago, Carr (1942) pointed out a series of characteristics that linked *Eretmochelys* with *Caretta* and *Lepidochelys* rather than with *Chelonia,* and his arguments remain persuasive (Pritchard and Trebbau, 1984).

The allocation of *Lepidochelys* to the family Cheloniidae appears to be unassailable. Other sea turtle families, such as the Dermochelyidae, Protostegidae, and Toxochelyidae, show different key characteristics, such as the layer of mosaic carapacial bones of the Dermochelyidae, the highly fontanelled shell, nasal bones, and often absent epiplastra of the Protostegidae, and the cruciform plastron of the Toxochelyidae. The rounded, thoroughly ossified carapace of *Lepidochelys* does show strong similarity to that of *Toxochelys* and some other toxochelyid genera, but this appears to be a case of parallelism rather than of evolutionary relationship.

The position of *Natator* is less clear. Long considered a species of *Chelonia,* the flatback turtle (*Natator depressus*) was recognized as a highly distinctive genus by Zangerl et al. (1988) and, in parallel, by Limpus et al. (1988), with a curious mosaic of features: a skull highly similar to that of *L. kempii* but a shell more similar to *Chelonia,* a terrestrial gait that shuffles between alternation and simultaneous heaving of the four limbs, and uniquely thin scutes like no other sea turtle in the world. Moreover, no other sea turtle species is confined to the continental shelf of a single nation, lays so few eggs in a clutch, or has fused sixth and seventh cervical vertebrae. It is also the only one in which the hatchlings have blue eyes. There is merit in the classification of *Natator* within a tribe or subfamily in its own right (Natatorini) alongside Chelonini and Carettini, although such an "unresolved trichotomy" is certain to be offensive to strict cladists. To add further confusion, Bowen and Karl (1996) found that, although mtDNA and nDNA analyses agreed in the placement of *Eretmochelys* with *Caretta* and *Lepidochelys,* the mtDNA analysis placed *Natator* with the *Carettini,* whereas nDNA placed it closer to *Chelonia,* perhaps one more argument in favor of the trichotomy.

As has been discussed by Zangerl (1980), tracing the evolutionary history of the Cheloniidae from the Cretaceous to the present leads to the odd conclusion that the most advanced forms occurred early in the fossil record, whereas the living cheloniids are remarkably generalized in their morphology and are seemingly primitive.

This suggests that the more specialized forms were ill adapted for changing environmental conditions and that the unspecialized central stem of the evolving cheloniids represented the formula for survival. Moreover, among the living genera, *Lepidochelys* seems to be the most primitive on morphological grounds, although not on behavioral grounds. In this genus, for example, the gigantic size of many of the specialized sea turtles is not expressed, and the rounded, ossified carapace and generalized food habits also mark this as an unspecialized genus. In connection with the generalized diet, *L. olivacea* appears to be quite opportunistic in its feeding habits (Márquez-M., 1990), often being almost exclusively herbivorous in the Indian Ocean (Biswas, 1982), whereas in the Eastern Pacific it is generally carnivorous (Pritchard and Trebbau, 1984). However, the fragmentation of the neural bones discussed below is an unusual development and almost certainly a specialized rather than a primitive one, even though a dominant theme in vertebrate evolution has been reduction rather than proliferation of discrete bony elements.

Few fossil turtles show fragmented neurals. The two fossil forms with this condition (the cheloniids *Glyptochelone suyckerbuyki* and *Procolpochelys grandaeva,* from the Cretaceous and the Eocene to Miocene, respectively, of Europe and North America) may or may not be monophyletic with *Lepidochelys.*

Osteology and Morphology

Carapace

The carapace of *L. kempii,* cordiform in the hatchlings, widens dramatically with growth, sometimes becoming wider than long in the half-grown, and even in adults the shell width is very close to the length. The mean carapace length and width of 50 hatchling *L. kempii* were 42.05 and 33.79 mm, respectively (Pritchard, 1980), giving a ratio of 1.24:1. The mean carapace length of nesting females was found to be 64.64 cm in 1966, 64.52 in 1979 and 64.85 in 1980 (Chávez et al., 1967; Pritchard, 1981). Pritchard (1969) found the shell width of adults to average 0.969 of shell length. The overall range of carapace length for nesting females was 59–71 cm in 1980 (Pritchard, 1981). The longest carapace lengths recorded for adult females measured 74–74.9 cm (Pritchard, 1969).

In general, the size of adult *olivacea* is similar to that of *kempii,* although the shape may be different, with a steep-sided, flat-topped carapace frequent (although not universal) in east Pacific populations, a feature possibly associated with the widespread habit of surface basking (with the top of the shell exposed to the sun) in these relatively cool waters. In my dissertation (Pritchard, 1969), I did not have access to adult *Lepidochelys* from Africa or anywhere in the Indian Ocean and thus did not recognize some of the size variation within *L. olivacea.* I found, in the western Atlantic (Suriname), the modal carapace length to be 69 cm, and the maximum 74; modal lengths were somewhat larger than in Pacific Honduras (65 cm) and Mexico (68 cm). The largest individual I found in this study was a 78-cm Mexican specimen. However, Fretey (2001) has now found unusually large *olivacea* in the tropical eastern Atlantic (São Thomé), with a range of 62–80 cm (mean 70.1 cm) (over curve) for 417 females, and 56 males measured 64–86 cm (mean 70.6 cm). Hatchlings in this population measured 38–44 mm (mean 41.8 mm), and two juveniles were both wider than long (51 × 54 cm and 33 × 34 cm), although possibly this disparity would not have been maintained if straight-line lengths had been used. My samples were entirely nesting females, whereas many of Fretey's were turtles taken at sea, often while copulating on the surface, by local fishermen.

Intercostal fontanelles are evident in the shells of subadult *L. kempii,* but these close as maturity is approached, and the carapace of the adult is thick and lacks any trace of fontanelles. There are 10 rib-bearing dorsal vertebrae. In the first pair the ribs are short and extend parallel to the second pair of ribs instead of converging and ankylosing with them as they do in many nonmarine turtles. Ribs of the first pair terminate distally in a broad but nonankylosing contact with the visceral surface of the first pleural bones.

The carapace includes 8 pairs of costal bones and 11–13 pairs of peripheral bones. Meristic variation in the peripheral series is concentrated in the area between the wide nuchal bone and the peripheral bones with gomphoses to receive the rib tips extending from each of the first pleural bones. This "free" section of the periphery of the carapace may be spanned by two to four peripheral bones on each side. In addition to these, the peripheral bones third from last in the series, situated between the rib tips of costals VI and VII, do not make contact with any of the rib tips, and this insertion corresponds to a wider than usual separation between the rib tips of costals VI and VII. Otherwise, each peripheral bone makes contact with a corresponding costal bone via the exposed rib tip.

The neural bone series is highly variable. In general, *Lepidochelys* shows a higher neural count than any other genus of living turtles. Typically, there is alternation between octagonal and squarish or rounded neural elements, but there is much variation in shape as well as in number. On occasion as many as 14 neurals are present. Sometimes asymmetric or fragmentary azygous bones may be intercalated at the sides of the neural series, or isolated neurals may be longitudinally divided. On rare occasions (Pritchard, 1969), there may be a slight posterior interruption of the neural series, with median contact between pleurals VIII, but, in sharp contrast to the situation in *Caretta,* the nuchal series is usually continuous from the nuchal bone to the suprapygals. In posthatchlings and juveniles, the neural series is strongly tuberculated, forming high, somewhat elongated prominences at positions corresponding to the rear of vertebral scutes I and II and a lower one at the rear of vertebral scute III. These eminences disappear entirely in adults, as do the serrations along the periphery of the posterior half of the carapace.

There are two suprapygals. The suture between them is anteriorly strongly convex. In juveniles, the suprapygals, especially the rear one, are narrow, but with growth the anterior suprapygal progressively embraces the posterior one between its extending posterolateral rami. Initially, the rear suprapygal makes sutural contact posteriorly only with the single pygal bone, but with growth it expands and makes progressively increasing sutural contact with the posteriormost pair of peripherals. The space posterior to the last pair of ribs is thus closed even when the intercostal fontanelles are still quite open.

Functional interpretation of the rounded, highly ossified carapace and the unique proliferation of the neural bones remains conjectural. Perhaps the former characteristics, as suggested above, are merely primitive. Certainly they are shared by such early sea turtles as the toxochelyids *Osteopygis, Thinochelys,* and *Toxochelys* as well as by the Eocene cheloniid *Puppigerus camperi* (Weems, 1974; Zangerl, 1980). However, these two features together created a very strong and unusually extended midperipheral section of the carapace, and those who have watched the vigorous nest-closing behavior of this species, in which the shell is rocked from side to side so that the lateral margins of the shell alternately thump the substrate, might conclude that the development indeed had considerable functional significance. The supernumerary neurals have no obvious function in adult *Lepidochelys* and may simply reflect the action of some harmless gene, perhaps associated with the tendency toward a parallel fragmentation of scutes and scales manifested in the carapace of the congener *L. olivacea* or in the head scales of all the Carettinae. However, it is conceivable that the additional neurals might play a role in stabilizing the vertebral tubercles that are such a striking characteristic of immature *Lepidochelys.* Certain fossil toxochelyids (e.g., *Prionochelys* and *Ctenochelys*) are known in which the vertebral tubercles were extremely well developed, and each bore an additional, seemingly neomorphic, "epithecal" bone near its tip (Zangerl, 1953, 1980).

Plastron

Like the carapace, the plastron of *Lepidochelys* is extensively ossified, so that in adults the only persistent cartilaginous areas are along the midlines of the hyo-, hypo-, and xiphiplastra, and each side of the posterior dagger-like process of the entoplastron. Retention of cartilage in these areas allows the plastron a modest degree of longitudinal flexibility along the midline and along the sides where the hyo- and hypoplastra meet the midperipheral bones. The entoplastron is distinctively shaped, the "blade" having straight, posteriorly convergent sides, forming a sharp posterior tip. In *Caretta* the flanks of the entoplastron are convex, in *Chelonia* they are concave, and in *Eretmochelys* the blade is narrow and parallel-sided (Ruckdeschel et al., 2000). The hyo-hypoplastral sutures in adult *Lepidochelys* develop distally to a degree that completely eliminates the lateral fontanelles present in the plastra of almost all other sea turtles. In mature *Lepidochelys,* these large plastral bones lose all of the proximal and distal interdigitations that characterize the hyo- and hypoplastra of most sea turtles, their flanks being separated from adjacent bones by only a narrow strip of gelatinous cartilage. In visceral view, the xiphiplastra can be seen to develop a process that penetrates deeply into the posterior margins of the hypoplastra. The medial junction of the epiplastra does not ankylose, presumably because of the need to retain some midline flexibility, but this junction is stabilized by considerable thickening, in a visceral direction, of the juxtaposed ends of these bones. The entoplastron remains thin and dagger-like throughout life, with a median visceral ridge in adults and extensive but nonankylosing anterior contact with the inner edge of the proximal ends of the epiplastra. A single pair of narrow, contiguous mesoplastra has been reported as an anomaly in *L. olivacea* (Pritchard, 1966), but this has not been reported for *L. kempii.*

The plastral formula for *L. kempii,* defined as 100 times the shortest distance across the ridge (between the axillary and inguinal notches) divided by half the width of the plastron, averaged 65.0 for 11 mature females (Pritchard, 1969). Zangerl (1958) found even lower indices for two immature specimens, 55.0 and 61.9. These are extremely low values for any cheloniid, and they provide a good index to the unusually short, broad shell of *L. kempii.* Other cheloniids, from the Cretaceous to the present, may show plastral formulas of more than 90 or even more than 100 (Zangerl, 1958). However, plastral formulas lower than those for *L. kempii* were shown by some of the toxochelyids, the cruciform plastron (with very narrow bridge) being an important diagnostic of this family.

The outer margins of the hyoplastra and hypoplastra of *Lepidochelys* are penetrated by a series of pores, each of which perforates one of the inframarginal scutes (which almost always number four pairs in the species; Pritchard, 1969) and opens into a small chamber within the bone of the plastron. These openings correspond to the Rathke's glands, present in the inguinal and/or axillary regions of many turtles but present as a complete series of glands spanning the width of the bridge only in *Lepidochelys.* The function of these glands in *Lepidochelys* is still uncertain, although their homologues in other turtles, including the chelydrids and kinosternids, chelids such as *Chelodina,* and even emydids such as *Melanochelys,* produce a very odiferous musky secretion, instantly detectable by humans. It has been speculated that the glands of *Lepidochelys,* even though their product has no obvious odor to humans, may play a role in the turtles' detection of conspecific animals in the vicinity, which in turn may assist in the coming together of turtles in the extraordinary simultaneous nesting assemblages or *arribadas* for which the genus is famous. The observation of a completely eyeless adult female *L. olivacea* participating in an arribada in Costa Rica (Mora and Robinson, 1982) certainly suggests that nonvisual senses play an important role in arribada formation.

Extremities

The forelimbs of *L. kempii* are relatively short compared to those of cheloniines or dermochelyids. Zangerl (1953) reported the femur of a near-adult *L. kempii* to be 92.1% as long as the humerus, a ratio much higher (i.e., reflecting relatively shorter forelimbs) than those (71.8–79.8%) found for representatives of three other families of sea turtles. However, it is important to remember that this ratio changes with ontogeny, juvenile sea turtles having relatively long forelimbs. In a 26.2-cm specimen (No. 1978 in the Chelonian Research Institute Collection) of *L. kempii* from Barnegat Bay, New Jersey, the proportion was much lower (79%).

Zangerl (1953) discussed other aspects of the relative limb proportions of *L. kempii* as well as details of the proportions of the shoulder girdles. He found the length of the scapular neck to be 46.7% of the length of the ventral prong of the scapular fork, and the length of the dorsal prong of the scapular fork and the maximum length of the coracoid to be 118.4% and 155.2%, respectively, of the length of the ventral prong of the scapular fork. The last ratio was much higher in the more streamlined and faster-swimming *Chelonia* and *Eretmochelys.*

The hindlimb skeleton of *L. kempii* is similar

to that of other living cheloniids and different from the fossil forms of the family in having these limbs modified to form steering rudders (and nest hole diggers) instead of paddles to assist in propulsion. Skeletally, such a change manifests itself in a near union of the two trochanters of the femur, which are separated by a deep valley in the fossil forms. The limbs of the living cheloniids also differ from those of the fossil ones in bony proportions and in organization of the carpus and tarsus.

A curious fact is that *L. kempii* has long been known from immature specimens of a wide range of sizes, with adults virtually unknown until recent decades, whereas the reverse is the case with *L. olivacea*. Despite the enormous arribadas that still occur in Mexico, Costa Rica, and India, juvenile *L. olivacea* remain extraordinarily cryptic and are very rarely encountered. Most published details of the osteology of subadult *Lepidochelys* thus refer to *L. kempii*. A complete explanation of this phenomenon has not been forthcoming, but part of the answer may be that *L. kempii* is usually a coastal, benthic-feeding species at nearly all life stages, whereas adult *L. olivacea* are much more pelagic (Byles and Plotkin, 1994), and the immatures perhaps even more so.

Skull and Vertebrae

The skull of *L. kempii* differs from that of the congener *L. olivacea* in many details (Pritchard, 1969). These include the generally more massive construction, with heavier rhamphothecae bearing a triturating ridge on each upper tomial surface that is also reflected in the underlying bone; the smaller orbit (perhaps correlated with the diurnal nesting habit) as well as the heavier maxillae; the slight bulging of the maxillary area, a feature noticed by Garman (1880) in his original description; the much narrower pterygoid "waist" (presumably corresponding to more bulky levator mandibulae muscles); the more extensive secondary palate; the heavier ventrally flanged supraoccipital process; and the greater relative width of the adult skull (width averages 0.988 of the basicranial length in adult *L. kempii* and 0.909–0.914 in different populations of *L. olivacea*, with all *L. kempii* having a ratio of more than 0.95 and all *L. olivacea* less than this).

The lower jaws of the two species of *Lepidochelys* show numerous differences. In *L. kempii*, characteristics of the lower jaw include

1. A bony point toward the rear of the mandibular symphysis
2. A concave bony alveolar surface (flat in *L. olivacea*)
3. Greater overall depth than in *L. olivacea*
4. A bluntly pointed tip of the coronoid bone (rounded off in *L. olivacea*)
5. A relatively small fossa meckelii
6. Articular surfaces somewhat backwardly directed (more vertical in *L. olivacea*)
7. Well-fused bones forming the articular surfaces (the articular, prearticular, and angular; loosely sutured and tending to be open at their anterodorsal end in *L. olivacea*)

The cervical vertebrae of *L. kempii* are illustrated by Williams (1950). The only ginglymoidous articulation is that between cervicals VII and VIII. However, the character of this joint is variable and sometimes ill defined. The anterior facet may be laterally cylindrical rather than truly doubled, and it is even possible for a double convex facet to articulate with a single (laterally elongated) concave facet, presumably because the ginglymoidous bony articular surface had a cartilaginous surface in life, whose contours masked the separation of the paired bony convexities. Other features of the morphology of *L. kempii* are presented elsewhere (Caillouet et al., 1986; Landry, 1989; Mast and Carr, 1989) and are not discussed further here.

Zoogeography

Most sea turtle species are nearly circumglobal in distribution, and *L. kempii* is thus a conspicuous exception. It apparently replaces *L. olivacea* in the Gulf of Mexico and northwestern Atlantic, much as *Chelonia agassizi* replaces *C. mydas* in the eastern Pacific. European *L. kempii* are not known to mingle with West African *L. olivacea* in the eastern Atlantic, although Fretey (2001) illustrated a possible zone of overlap between the two species in Macaronesia and West Africa, and recently there have been two records of *L. olivacea* within the foraging range of *L. kempii* in Florida (Foley et al., 2003).

As is well known, the nesting distribution of *L. kempii* is uniquely concentrated. Carr (1963) used the term "panspecific reproductive convergence" to refer to the extraordinary gathering of mature females of this species each spring along the coast near the village of Rancho Nuevo in the south of Tamaulipas, Mexico, to nest along a few tens of kilometers of the almost continuous sand beach that fronts the western Gulf of Mexico. Sporadic and individual nesting emergences have been reported from Padre Island, Texas, and the coast of Veracruz, Mexico, and some occur as far east as Isla Aguada, Campeche, Mexico (Ross et al., 1989), although the latter record appears to be based on a single but persistent individual that nested almost every year since 1990 (Márquez-M., 1990). Even taken collectively, these nestings represent a very low level of activity, although they may be increasing as the population of *L. kempii* as a whole increases. It remains possible that another arribada of *L. kempii* exists somewhere in the world. For example, the small group of nesting animals that emerged at Washington Beach just south of the Rio Grande (northern Tamaulipas) some years ago would have gone unnoticed had this not been a resort beach, as the tracks are so ephemeral and the nesting emergences themselves so brief that nesting easily can be overlooked. Moreover, the remote beaches of the enormous cape of Cabo Rojo, Veracruz, Mexico, are rarely surveyed. However, for the time being, we must proceed on the assumption that the Rancho Nuevo site is the only location where *L. kempii* nests regularly and in significant numbers.

Mature *L. kempii* are essentially confined to the Gulf of Mexico. The report by Chávez and Kaufmann (1974) of an individual nesting in Caribbean Colombia almost certainly represented a misidentified *L. olivacea* tagged in the Guianas (Pritchard and Trebbau, 1984). Moreover, such adults are not uniformly distributed in Gulf waters, as outside the nesting season, they are commonly encountered in biologically productive estuarine situations, such as the coast of Louisiana near the mouth of the Mississippi River or the Laguna del Carmen area of Campeche, as evidenced by tag returns of individuals tagged while nesting (Chávez, 1968; Pritchard and Márquez-M., 1973). The immature animals, however, frequently leave the Gulf of Mexico and are (or were) commonly found on the Atlantic coast of Florida and in appropriate embayments and protected coastal habitats as far north as New England (Pritchard, 1969). As an example, Ruckdeschel et al. (2000) documented 159 strandings of Kemp's ridleys between 1979 and 1998 on the shore of Cumberland Island, Georgia; they ranged in length from 22.8 to 60 cm and thus were all immatures. Bleakney (1955) even reported four specimens from Nova Scotia. Occasionally, individuals may reach Europe. Brongersma (1972) listed 25 records of *L. kempii* in European Atlantic waters, and it is of interest that two of the animals head-started in Galveston, Texas, were picked up in France and Morocco (Fontaine et al., 1989). Brongersma and Carr (1983) discussed the single known Mediterranean record, a 29.4-cm specimen from Malta.

Debate continues as to whether the animals in U.S. Atlantic waters, especially in the northern section, are lost to the population or are indulging in a normal (or at least optional) part of their developmental odyssey. Specimens as far north as Chesapeake Bay are known to travel out of the bay to Florida waters as part of a seasonal migration to escape winter temperatures (Musick et al., 1983). Musick et al. (1983) identified 47 *L. kempii* (as well as 664 *C. caretta* and 8 *Dermochelys coriacea*) in a study of sea turtles in Chesapeake Bay. Many of these animals were found dead. Fewer than 20% of the mortalities could be attributed positively to entanglement or incidental catch in fishing operations; doubtless this occurred in many other cases, but proof was lacking. Winter strandings were almost nonexistent, but there was a minor surge of strandings in June of each year, which Musick et al. (1983) considered possibly attributable to poor physical condition of many turtles resulting from suboptimal conditions during winter followed by an arduous migration to the bay. These authors found no evidence of turtles overwintering in Virginia waters, and although winter torpor has now been documented for both *C. caretta* (in Florida) and *C. agassizi* (in the Gulf of California), the evidence for this in *L. kempii* remains anecdotal and unproven. A series of 29 Virginia *L. kempii* showed an average carapace length of 40.0 cm, with 75% between 30 and 45 cm.

Ridleys occur regularly in waters north of Virginia. Carr (1957) reported an extraordinary observation made by W. Schevill of Woods Hole Oceanographic Institution of several dozen yearling ridleys, out of "a whole fleet of such turtles," stranded on Woods Hole beaches while traveling out of Buzzards Bay into Vineyard Sound. Lazell (1976) reported 16 specimens of *L. kempii* in Massachusetts waters and observed that their presence may be not only seasonal but also cyclic, adding that the last major Massachusetts ridley event was in 1961, a time, he postulated, that corresponded to the beginnings of the decline of the great arribadas at Rancho Nuevo.

Nevertheless, such invasions have occurred more recently and may still occur on a limited scale. A journalist (Anonymous, 1985) related that no fewer than 36 *L. kempii* as well as 3 *C. caretta* and 3 *C. mydas* were found in Long Island Sound, New York, in the fall of 1985, and doubtless more were found during the ensuing winter. The largest ridley was reported as less than 30 cm in "diameter." No mention was made of their being tagged, so they probably did not derive from the Galveston head-start project. The small size and very narrow size range of these juvenile Kemp's ridleys in Long Island Sound suggest that they represent a single year class, present for just the warmer months of the year and migrating southward (unless caught by a sudden cold front) in the fall.

Maigret (1983) gave a brief description of an enormous flotilla, numbering thousands of *L. kempii,* each about 30 cm in carapace length, of which 75 were captured for examination. This gathering was reported to have occurred in May 1982 at 33°N, 74°W, which corresponds to the northwestern Atlantic, between North Carolina and Bermuda. However, subsequent correspondence between Dr. Maigret and me first revealed to me that his concentration of turtles was on the other side of the Atlantic, off northwestern Africa (as the title of the paper implied). The given longitude of 74°W had been a misprint for 14°W, deriving perhaps from the European custom of writing the number "1" similar in form to the number "7" in U.S. or British orthography. Moreover, Dr. Maigret kindly agreed to seek confirmation of the identity of the turtles and was ultimately able to locate a specimen that had been kept and preserved. He sent me a series of photographs, and the specimen shown was unquestionably a loggerhead (*C. caretta*)!

In contrast to *L. kempii, L. olivacea* has a wide distribution in three oceans. Nevertheless, its occurrence is far from uniform even in tropical areas, and as mentioned above, we are still unable to specify where the vast hordes of immature *olivacea* that presumably exist actually reside. Nesting olive ridleys rarely utilize islands or shores with extensive coral, but it is still unclear which common features unite the beaches where they do nest preferentially or in large numbers. The species was first studied in the field by Deraniyagala (1939), who reported that olive ridleys nest regularly in Ceylon (Sri Lanka), and it since has been clarified that the species nests widely on the east coast of India, with tremendous arribadas forming at two or three sites in the state of Orissa. But elsewhere in the Indian Ocean, the species is decidedly scarce and in many areas is virtually unrecorded.

In the Pacific Ocean, there is nesting along a very extensive stretch of mainland shorelines of the Americas, from Sonora, Mexico south at least to Panama, with several arribada sites in Mexico (at least in recent times), two major ones in Costa Rica, and smaller arribadas recorded in southwestern Nicaragua and southern Panama. Turtles associated with these breeding assemblages may be found up to several hundred kilometers offshore (Plotkin et al., 1995) but are generally unknown in the mid-Pacific, and only sporadic or minor nesting has been reported for the western Pacific. In October 2003 I obtained a few olive ridley bones on southern Pinta Island, northwestern Galápagos, but normally the few ridleys that have been reported in the region have been far from land. Nevertheless, very large numbers of olive ridleys occur seasonally in waters near the Ecuadorian mainland.

In the western Atlantic, the distribution of nesting olive ridleys is dynamic, reflecting the shifting nature of the shoreline and beaches. Regular nesting occurred in northwestern Guyana in the 1960s, but this activity has now dwindled to almost nothing, and a small arribada at Eilanti, in extreme eastern Suriname, vigorous in the 1960s, has all but disappeared, although there has been a major outbreak of ridley nesting in extreme eastern French Guiana in recent years, especially on the beaches of the Ile de Cayenne, where leatherbacks nested in the 1960s. One assumes that the population has simply shifted its nesting as a response to shoreline and offshore erosional changes, but this suggestion has not been formally demonstrated. Nesting has been recorded at several Brazilian localities in recent decades, but again it is not clear whether these are traditional or ancient nesting sites or represent displacement of nesters from beaches in the Guianas.

Fretey (2001) documents quite extensive nesting by olive ridleys in West Africa, with a map showing many nesting localities between Guinea-Bissau and northern Angola. Nesting was relatively dense on southern Bioko, on Adonga Island (Bijagos Archipelago, Guinea-Bissau), and on the coasts of Gabon and Congo.

Whether *L. kempii* is evolutionarily closer to the ancestral generic stock than is *L. olivacea* is a matter for conjecture. However, although many of the differences between the species are a reflection of the more durophagous diet of *L. kempii,* the difference in scute configurations is more difficult to explain. Certainly the multiscutate condition of *L. olivacea* appears to be the derived condition: it is indeed unique, both in the high numbers of scutes (especially costals) borne by almost any representative of this species and in the extraordinary degree of intraspecific variability. The subject is discussed further by Pritchard (1969), Hill (1971), and Mast and Carr (1989). But this proliferation of scutes can occur in other turtle species, including freshwater and land forms, in cases where the animals were hatched artificially under stressful conditions of temperature or humidity.

Pritchard (1969) presented some conjecture as to how *L. kempii* may have arisen by isolation of an early *Lepidochelys* stock in the Gulf of Mexico following the closure of the Isthmus of Tehuantepec. Although the Caribbean route remained theoretically open to reinvasion (or escape), this path was not utilized because of the long reach of inappropriate habitat in the Caribbean, *Lepidochelys* flourishing in areas little frequented by *Caretta* and requiring productive estuarine conditions for optimal feeding. Between the mouth of the Orinoco River (an important feeding ground for *L. olivacea*) and the Laguna del Carmen in the Gulf of Mexico, a stretch of thousands of kilometers, there are no major estuaries, the rivers of the Caribbean Islands being too small to produce them, the mainland shores of western Venezuela, Colombia, and the Yucatan peninsula too dry, and the watersheds of Caribbean Central America too restricted by the narrowness of the coastal lowlands in these countries.

Hendrickson (1980) presented an intriguing hypothesis regarding the zoogeography of *L. kempii,* arguing that the species was isolated in Gulf waters following the relatively recent closure of the marine passage through Panama, 3.5–4 million years ago, rather than that of the Isthmus of Tehuantepec. Hendrickson (1980) also hypothesized that the closing of the Panamanian portal may have been accompanied by a major redirection of ocean currents, the Atlantic North Equatorial Current now being denied access to the Pacific and instead sweeping northward and clockwise with greatly increased force. He further theorized that the presence of this new current might have constituted a chronic stress on populations of *L. kempii* by sweeping relatively large numbers of immatures completely out of the Gulf of Mexico, to the eastern seaboard of the United States and ultimately, in some cases, to the British Isles and northern Europe. He expressed doubt that these individuals were ever able to make their way back to the Gulf of Mexico to join in the reproductive effort of the species.

The last question remains open. Certainly the adults of *L. kempii* seem to be able to resist drifting outside the Gulf of Mexico, and hardly a single adult has been reliably recorded from extra-Gulf waters. It is possible that juveniles that are not carried too far north on the Atlantic seaboard are ultimately able to return to the Gulf and reproduce, but perhaps specimens in Canadian or European waters are truly lost. If specimens in U.S. Atlantic waters were able to survive but were not able to return to the Gulf, the question arises as to what they would do when they reached maturity with regard to egg production and oviposition. In other words, why do they not nest alongside *Caretta* on southeastern Atlantic beaches?

Part of the answer, or perhaps part of a new question, may relate to the surprise observation of scattered Kemp's ridleys nesting on both coasts of Florida, and even as far north as the Carolinas, during 1989–1992. One of these nestings was on Florida's Gulf coast in St. Petersburg, Florida (Meylan et al., 1991), one at Daytona Beach on Florida's Atlantic coast, and one each in North and South Carolina (Anonymous, 1992). The last three are literally the first nestings ever recorded for *kempii* on the Atlantic coast. Possibly these turtles derived from head-started individuals released about a decade earlier in Florida Bay by the author under contract to NMFS. This hypothesis gains some support from the simultaneous discovery of some adult females, nesting at Rancho Nuevo, bearing "living tags" (i.e., visible grafts of epidermal and dermal tissue exchanged between carapacial and plastral locations), thus confirming at least the maturation time. Nevertheless, the origin of these Florida and Carolinas nesting records remains totally speculative at present, and the observations do not meet the criteria established by Caillouet (1998) and Caillouet et al. (1997) for positive identification of head-started animals. Still, it is noteworthy that many beaches in Florida and adjacent states had been intensively patrolled for loggerheads for decades before 1989, and these patrols brought to light growing evidence of nesting green turtles, significant numbers of nesting leatherbacks, and even very rare nesting hawksbills, but not a single observation of a nesting ridley. Thus, the hypothesis is untenable that Kemp's ridley once had a much wider nesting range in the Gulf of Mexico and that this range was restricted to the Rancho Nuevo and adjacent western Gulf beaches in the course of the decimation of the species in the 1950s and 1960s.

A fair proportion of the specimens of *L. kempii* found in Atlantic waters of Florida are healthy animals only a few centimeters short of mature carapace lengths, and it seems unlikely that such individuals would inevitably die before actually reaching maturity, having survived all dangers up to that point. However, because they do not occur in Atlantic waters as adults, I am inclined to believe that they are capable of returning to the Gulf to reproduce. Sea turtles have evolved with ocean currents as a constant factor in their lives ever since their ancestors first took to the sea, and it is improbable that *L. kempii* has not yet learned to cope with them. Intensified efforts to tag Kemp's ridleys on the Atlantic seaboard may help settle this longstanding issue.

Bowen and Karl (1996) examined mtDNA and nDNA sequence data for both *L. kempii* and *L. olivacea* to determine whether my hypotheses (Pritchard, 1969) relating to *Lepidochelys* evolution and zoogeography would be sustained or contradicted by the molecular data. In fact, the results were in full concordance. Both types of DNA analysis confirmed that the two taxa were distinct at the species level and were more distinct than any two populations within any other extant sea turtle genus. Moreover, the degree of divergence between olive ridleys (from the Atlantic and Pacific Oceans) and Kemp's ridleys was fully consistent with the 3–4 million years of isolation that would have resulted from the closing of the lacunae in the Central American isthmus.

ACKNOWLEDGMENTS

This contribution is a revision, updating, and broadening in scope of a paper by the author published in *Proceedings of the First International Symposium on Kemp's Ridley Sea Turtle Biology, Conservation and Management* (Caillouet and Landry, 1989). I gratefully acknowledge both the patience of the editor, Dr. Pamela Plotkin, as she awaited my tardy contribution and also the kind services of my volunteer assistant for many years, Marie Pomales Messner, in converting my original composition to electronic format by hand so that it could be reorganized, edited, and submitted for publication. Linh Uong kindly helped with proofreading.

LITERATURE CITED

Anonymous. 1985. Turtle. *The New Yorker.* December 30, pp. 16–18.

Anonymous. 1992. First Kemp's ridley nesting in South Carolina. *Marine Turtle Newsletter* 59:23.

Baur, G. 1890. The genera of the Cheloniidae. *American Naturalist* 24:486–487.

Biswas, S. 1982. A report on the olive ridley *Lepidochelys olivacea* (Testudines: Cheloniidae) of Bay of Bengal. *Records of the Zoological Survey of India* 79:275–302.

Bleakney, S. 1955. Four records of the Atlantic ridley turtle, *Lepidochelys kempi,* from Nova Scotia. *Copeia* 1955:137.

Bowen, B. W., and Karl, S. A. 1996. Population genetics, phylogeography, and molecular evolution. In Lutz, P., and Musick, J. (Eds.). *The Biology of Sea Turtles.* Boca Raton, FL: CRC Press, pp. 29–50.

Brongersma, L. D. 1972. *European Atlantic Turtles.* Leiden: Zoologische Verhandelingen. Rijksmuseum van Natuurlikje Historie.

Brongersma, L. D., and Carr, A. F. 1983. *Lepidochelys kempi* (Garman) from Malta. *Proceedings Koninklijke Nederlandse Akademie van Wetenschappen, Series C,* 86:445–454.

Byles, R. A., and Plotkin, P. T. 1994. Comparison of the migratory behavior of the congeneric sea turtles *Lepidochelys olivacea* and *L. kempii.* In Schroeder, B. A., and Witherington, B. E. (Compilers). *Proceedings of the Thirteenth Annual Sea Turtle Symposium.* NOAA Technical Memorandum NMFS-SEFSC-341, p. 39.

Caillouet, C. W., Jr. 1998. Testing hypotheses of the Kemp's ridley head-start experiment. *Marine Turtle Newsletter* 79:16–18.

Caillouet, C. W., Jr., and Landry, A. M. (Eds.). 1989. *Proceedings of the First International Symposium on Kemp's Ridley Sea Turtle Biology, Conservation and Management.* Texas A&M University Sea Grant, pp. 157–164.

Caillouet, C. W., Jr., Koi, D. B., Fontaine, C. T., Williams, T. D., Browning, W. J., and Harris, R. M. 1986. *Growth and survival of Kemp's ridley sea turtle,* Lepidochelys kempi, *in captivity.* NOAA Technical Memorandum NMFS-SEFC-186.

Caillouet, C. W., Jr., Robertson, B. A., Fontaine, C. T., Williams, T. D., Higgins, B. M., and Revera, B. 1997. Distinguishing captive-reared from wild Kemp's ridleys. *Marine Turtle Newsletter* 77:1–6.

Carr, A. F. 1942. Notes on sea turtles. *Proceedings of the New England Zoology Club* 21:1–16.

Carr, A. F. 1956. *The Windward Road.* New York: Alfred A. Knopf.

Carr, A. F. 1957. Notes on zoogeography of the Atlantic sea turtles of the genus *Lepidochelys. Revista de Biologia Tropical* 5:45–61.

Carr, A. F. 1963. Panspecific reproductive convergence in *Lepidochelys kempi. Ergebnisse der Biologie* 26:297–303.

Chávez, H. 1968. Marcado y recaptura de individuos de tortuga lora, *Lepidochelys kempi* (Garman). *Instituto Nacional de Investigaciones Biologico Pesqueras* 19:1–28.

Chávez, H., and Kaufmann, R. 1974. Informacion sobre la tortuga marina *Lepidochelys kempi* (Garman), con referencia a un exemplar marcado en Mexico y observado en Colombia. *Bulletin of Marine Science* 24:372–377.

Chávez, H., Contreras, M., and Hernandez, E. 1967. Aspectos biologicos y proteccion de la tortuga lora, *Lepidochelys kempi* (Garman), en las costa de Tamaulipas, Mexico. *Instituto Nacional de Investigaciones Biologico Pesqueras* 17:1–40.

Deraniyagala, P. E. P. 1939. *The Tetrapod Reptiles of Ceylon,* Volume 1: *Testudinates and Crocodilians.* Colombo: Colombo Museum Natural History Series.

Dollo, L. 1909. The fossil vertebrates of Belgium. *Annals of the New York Academy of Sciences* 19:99–119.

Foley, A. M., Dutton, P. H., Singel, K. E., Redlow, A. E., and Teas, W. G. 2003. The first records of olive ridleys in Florida, USA. *Marine Turtle Newsletter* 101:23–25.

Fontaine, C. T., Manzella, S. A., Williams, T. D., Harris, R. M., and Browning, W. J. 1989. Distribution, growth and survival of head started, tagged and released Kemp's ridley sea turtles (*Lepidochelys kempi*) from year-classes 1978–1983. In Caillouet, C. W., Jr., and Landry, A. M. (Eds.). *Proceedings of the First International Symposium on Kemp's Ridley Sea Turtle Biology, Conservation and Management.* Texas A&M University Sea Grant, pp. 124–144.

Fretey, J. 2001. *Biogeography and Conservation of Marine Turtles of the Atlantic Coast of Africa. CMS Technical Series, Publication No. 6.* Bonn: UNEP/CMS Secretariat.

Gadow, H. 1899. Orthogenetic variation in the shells of Chelonia. *Wiley Zoological Results, Cambridge University Press* 3:207–222.

Gaffney, E. S. 1984. Progress towards a natural hierarchy of turtles. *Studia Geologica Salmanticensia. Studia Palaeocheloniologica, Volume Especial* 1:125–132.

Garman, S. 1880. On certain species of Chelonioidae. *Bulletin of the Museum of Comparative Zoology* 6:123–126.

Hendrickson, J. R. 1980. The ecological strategies of sea turtles. *American Zoologist* 20:597–608.

Hildebrand, H. H. 1963. Hallazgo del area de anidacion de la tortuga marina "lora," *Lepidochelys kempi* (Garman), en la costa occidental del Golfo de Mexico. *Ciencia* 22:105–112.

Hill, R. L. 1971. Surinam turtle notes—1. Polymorphism of costal and vertebral laminae in the sea turtle (*Lepidochelys olivacea*). Stichting Natuurbehoud Suriname (STINASU). *Mededeling* 2:3–9.

Landry, A. M., Jr. 1989. Morphometry of captive-reared Kemp's ridley sea turtles. In Caillouet, C. W., Jr., and Landry, A. M. (Eds.). *Proceedings of the First International Symposium on Kemp's Ridley Sea Turtle Biology, Conservation and*

Management. Texas A&M University Sea Grant, pp. 220–231.

Lazell, J. D., Jr. 1976. *This Broken Archipelago. Cape Cod and the Islands, Amphibians and Reptiles*. New York: Demeter Press.

Limpus, C. J., Gyuris, E., and Miller, J. D. 1988. Reassessment of the taxonomic status of the sea turtle genus *Natator* McCulloch, 1908, with a redescription of the genus and species. *Transactions of the Royal Society of South Australia* 112:1–9.

Maigret, J. 1983. Repartition des tortues de mer sur les cotes ouest africaines. *Bulletin de la Société Herpetologique de France* 28:22–34.

Márquez-M., R. 1990. *FAO Species Catalogue. Sea Turtles of the World. FAO Fisheries synopsis No. 125, Vol. 11*. Rome: FAO.

Mast, R. B., and Carr, J. L. 1989. Carapacial scute variation in Kemp's ridley sea turtle (*Lepidochelys kempi*) hatchlings and juveniles. In Caillouet, C. W., Jr., and Landry, A. M. (Eds.). *Proceedings of the First International Symposium on Kemp's Ridley Sea Turtle Biology, Conservation and Management*. Texas A&M University Sea Grant, pp. 202–219.

Meylan, A. B., Castaneda, P., Coogan, C., Lozon, T., and Fletemeyer, J. 1991. First recorded nesting by Kemp's ridley in Florida, U.S.A. *Marine Turtle Newsletter* 48:8–9.

Mora, J. M., and Robinson, D. C. 1982. Discovery of a blind olive ridley turtle (*Lepidochelys olivacea*), nesting at Playa Ostional, Costa Rica. *Revista de Biologia Tropical* 30:178–179.

Musick, J. A., Byles, R., Klinger, R., and Bellmund, S. 1983. *Mortality and behavior of sea turtles in the Chesapeake Bay. Summary report, 1979 through 1983*. Gloucester Point, VA: Virginia Institute of Marine Science (unpublished).

Plotkin, P. T., Byles, R. A., Rostal, D. C., and Owens, D. W. 1995. Independent vs. socially facilitated migrations of the olive ridley, *Lepidochelys olivacea*. *Marine Biology* 122:137–143.

Pritchard, P. C. H. 1966. Occurrence of mesoplastra in a cryptodiran turtle *Lepidochelys olivacea*. *Nature* 210:652.

Pritchard, P. C. H. 1969. Studies of the systematics and reproductive cycles of the genus *Lepidochelys*. Ph.D. diss. University of Florida, Gainesville.

Pritchard, P. C. H. 1980. *Report on United States/Mexico conservation of Kemp's ridley sea turtle at Rancho Nuevo, Tamaulipas, Mexico in 1979*. Final Report on U.S. Fish and Wildlife Service Contract No. 14-16-0002-80-21 (unpublished).

Pritchard, P. C. H. 1981. *Report on United States/Mexico conservation of Kemp's ridley sea turtle at Rancho Nuevo, Tamaulipas, Mexico in 1980*. Preliminary Report on U.S. Fish and Wildlife Service Contract No. 14-16-0002-80-21 (unpublished).

Pritchard, P. C. H., and Márquez-M., R. 1973. Kemp's ridley or the Atlantic ridley, *Lepidochelys kempi*. *International Union of Conservation of Nature and Natural Resources Monograph No. 2 (Marine Turtle Series.)*, pp. 1–30.

Pritchard, P. C. H., and Trebbau, P. 1984. *The Turtles of Venezuela*. Oxford, OH: Society for the Study of Amphibians and Reptiles.

Ross, J. P., Beavers, S., Mundell, D., and Airth-Kindree, M. 1989. *The Status of Kemp's Ridley*. Washington, DC: Center for Marine Conservation.

Ruckdeschel, C., Shoop, C. R., and Zug, G. R. 2000. *Sea Turtles of the Georgia Coast*. Occasional Publication of the Cumberland Island Museum, No. 1.

Weems, R. E. 1974. Middle Miocene sea turtles (*Syllomus, Procolpochelys, Psephophorus*) from the Calvert Formation. *Journal of Paleontology* 48:278–302.

Wermuth, H. 1956. Versuch der Deutung einiger bisher übersehener Schildkröten Namen. *Zoologische Beitrage* 2:399–423.

Wermuth, H., and Mertens, R. 1961. *Schildkröten, Krokodile, Brückenechsen*. Jena: Gustav Fischer Verlag.

Williams, E. E. 1950. Variation and selection in the cervical central articulations of living turtles. *Bulletin of the American Museum of Natural History* 94:507–561.

Zangerl, R. 1953. The vertebrate fauna of the Selma Formation of Alabama, Part IV. The turtles of the family Toxochelyidae. Fieldiana. *Geology Memoirs* 3:137–274.

Zangerl, R. 1958. Die oligozänen Meerschildkröten von Glarus. *Schweizerischen Paläontologischen Abhandlungen* 73:5–55.

Zangerl, R. 1969. The turtle shell. In Gans, C., d'A. Bellairs, A., and Parsons, T. S. (Eds.). *Biology of the Reptilia, Morphology A*, Volume 1. London and New York: Academic Press, pp. 311–399.

Zangerl, R. 1980. Patterns of phylogenetic differentiation in the toxochelyid and cheloniid sea turtles. *American Zoologist* 20:585–596.

Zangerl, R., Hendrickson, L. P., and Hendrickson, J. R. 1988. A redescription of the Australian flatback turtle *Natator depressus*. *Bishop Museum Bulletins in Zoology* 1:1–69.

JOSEPH BERNARDO
PAMELA T. PLOTKIN

4

An Evolutionary Perspective on the *Arribada* Phenomenon and Reproductive Behavioral Polymorphism of Olive Ridley Sea Turtles (*Lepidochelys olivacea*)

RIDLEY TURTLES (Kemp's ridleys *Lepidochelys kempii* and olive ridleys *Lepidochelys olivacea*) are perhaps best known for their unusual reproductive behavior of synchronized mass nesting known as the *arribada* (Carr, 1967; Hughes and Richard, 1974). At a few specific beaches in the Indo-Pacific and western Atlantic, hundreds to tens of thousands of turtles emerge synchronously from the ocean in just a few days to nest in close proximity. This remarkable phenomenon has been filmed in many natural history documentaries and is well known among nonscientists, yet our understanding of the proximate and ultimate causes of this behavior still remains largely obscure.

Our goal in this chapter is to review what is known about arribada behavior and the hypotheses that have been offered to explain its evolution. We also examine this behavior in the context of evolutionary and behavioral ecology theory and propose other hypotheses to explain its evolution and maintenance. The chapter is organized in three parts. First, we define and describe the behavior and review what is known. Additionally, we review hypotheses that have been proposed to explain the arribada phenomenon and the work that has been conducted in this context, evaluating the arribada behavior in the context of the behavior of other turtles that do not nest synchronously, including other ridleys. This exercise reveals that ridleys exhibit a behavioral polymorphism because many individuals do not nest in arribadas but rather nest solitarily much like other sea turtles and freshwater turtles. This conclusion motivates the next part of the chapter in which we review examples of other synchronously reproducing organisms and the hypotheses that have been offered to explain the evolution and maintenance of synchronous reproduction. We apply these insights toward developing a better understanding of the arribada behavior, and we offer alternative explanations for the existence of the behavioral polymorphism in ridleys. In the next major section we examine the ecological and evolutionary significance of the behavioral polymorphism of *Lepidochelys*. In light of this critical consideration of both the evolution of synchronous reproduction and its rarity in turtles, coupled with the behavioral polymorphism of *Lepidochelys*, we offer a new mechanistic hypothesis that integrates these insights to explain the distinctiveness of the arribada.

Finally, we consider the types of research needed to evaluate the alternative hypotheses that we have proposed with a goal of motivating future research on ridleys in a far broader evolutionary and ecological context.

The Arribada Behavior: Overview

The seven species of sea turtles share a generalized nesting behavior (Carr and Ogren, 1960; Ehrenfeld, 1979; Miller, 1997) that is also widely shared by other aquatic turtles. Females crawl ashore on specific beaches during a distinct time of the year, select a location on the beach, clear sand away from the nest site, and form a body pit. They then excavate a flask-shaped hole deep in the sand using their hind flippers, oviposit into the hole, cover the eggs with sand, camouflage the nest site, and then return to sea.

Despite this conserved nesting and oviposition sequence, sea turtle species differ markedly in their spatiotemporal patterns of nesting. Some species (hawksbill, *Eretmochelys imbricata;* leatherback, *Dermochelys coriacea;* flatbacks, *Natator depressus*) are solitary nesters that emerge individually (Carr et al., 1966), a behavior that is common to many other aquatic turtles. Other sea turtle species (greens, *Chelonia mydas;* loggerheads, *Caretta caretta*) are colonial nesters that emerge individually but coincidentally with multiple conspecifics (Caldwell et al., 1959). This behavior, too, is shared with other aquatic turtles. The most unusual behavior both within sea turtles as a group and among all turtles in general is arribada nesting (Fig. 4.1), in which hundreds to thousands of conspecifics emerge synchronously, over a few days, en masse (Pritchard, 1969). Only sea turtles of the genus *Lepidochelys* exhibit arribada-nesting behavior.

The definition of an arribada has varied over time and has not always been applied consistently. Arribada nesting by *L. kempii* was first described by Hildebrand in 1963, whereas *L. olivacea* was still thought to be a solitarily nesting species late into the 1960s. For example, Carr (1967) stated: "The east Pacific ridley appears to follow a directly opposite plan. The east Pacific ridley spreads its nesting all the way

Fig. 4.1. A *Lepidochelys olivacea* arribada at Playa Nancite, Santa Rosa National Park, Area de Conservacion de Guanacaste, Costa Rica. In the foreground is a group of five female *L. olivacea* emerging together. Photograph by David Rostal.

from Ecuador to Baja California." Although long known to indigenous people, the arribada behavior of *L. olivacea* was first reported from Suriname (western Atlantic) and the Pacific coast of Mexico by Pritchard (1969), with the discovery of eastern Pacific arribadas in Costa Rica following in 1971, and reports of the northern Indian Ocean sites in Orissa, India, coming only in 1976 (Pritchard, 1979).

The arribada (a local name in Mexico) was initially characterized by Pritchard (1969) as "huge, simultaneous nesting aggregations" in reference to estimates of upward of 40,000 nesting turtles at Rancho Nuevo, Mexico, where the behavior was first recorded on film in 1947 (Pritchard, 1979). However, as additional sites of mass nesting came to light, the term arribada was loosely applied to groups as small as a few hundred nesting turtles or to nesting events that spanned multiple days to a month (Cornelius, 1986; Ballestero, 1996). The arribada is widely viewed as a mass synchronous nesting event, but there is little consensus on the spatial and temporal boundaries that define an arribada. Similarly, there is no consensus on how many turtles comprise an arribada. Cornelius (1986) referred to an arribada as a "group nesting phenomenon" with no restriction on the temporal span or group density.

Operationally, what distinguishes arribadas from other patterns of nesting in which large numbers of turtles nest over several weeks is the synchronicity of emergence and the relatively short interval (2–7 days) between the initiation and termination of nesting. Thus, the number of nesting turtles is less critical to recognizing that an arribada has occurred than is the issue of synchrony. This criterion is particularly important to keep in mind as populations have declined as a result of purposeful and incidental take, because nesting populations are much smaller now than historically. For example, when Kemp's ridley nesting numbers reached a low point in 1985, arribadas continued to occur and were recognized as such because, despite the hundred or fewer turtles participating, they exhibited synchronous emergence along an extensive area (see Social Facilitation below).

Thus, arribadas are unique because of their synchronicity. Ridleys migrate from distant areas of the ocean to nearshore waters of a few specific beaches (Plotkin, 2003), gathering there for several days or weeks. A day or two in advance of the arribada, turtles assemble within meters of the shore (Hughes and Richard, 1974; Cornelius, 1986; Mendonça and Pritchard, 1986; Plotkin et al., 1991) and have been observed swimming back and forth parallel to the beach and lying on the benthos until the arribada commences (Cornelius, 1986; Plotkin et al., 1991). Almost as if responding to a starter's pistol or an opening bell, turtles suddenly begin to emerge from the ocean to nest on the beach, and they continue to arrive for the next few days or weeks during a single arribada.

The timing of arribadas is highly variable and unpredictable. Their conduct varies between the two species, within a species, and within a nesting population. Changes occur on daily, monthly, and annual time scales. For example, arribadas may take place only during the day and cease at night, they may occur at night and cease during the day, or they might continue throughout an entire 24-hour cycle (Dash and Kar, 1990). In general, arribadas occur once a month during a discrete time of the year. However, there are instances when arribadas occur more frequently than once a month (Hirth, 1980; Ballestero, 1996) or less frequently when the turtles miss a month (Plotkin et al., 1997). Sometimes an individual arribada lasts for a single day (Plotkin, 1994) or as long as 30 days (Ballestero, 1996). In the eastern Pacific, arribadas coincide annually with the rainy season (Cornelius, 1986), but in the northern Indian Ocean, they coincide with the dry season (Dash and Kar, 1990).

Geographic Extent

Arribadas occur at only a few beaches worldwide in the eastern Pacific, western Atlantic, and northern Indian Oceans (Table 4.1; Figs. 4.2 and 4.3). However, the nesting range for the ridley species extends far beyond these select beaches because, as was discussed above, many ridleys nest solitarily (Table 4.2; Figs. 4.2–4.4). For example, in the eastern Pacific, arribadas occur annually at several beaches in Mexico and Nicaragua and at two beaches in Costa Rica from June through December. During the same time, solitary *L. olivacea* emerge individually to nest along nearly the entire coastline from Mexico to Ecuador

(Carr, 1967; Pritchard, 1969, 1979; R. Briseño and A. Abreu, personal communication).

The novelty of arribada nesting, whether considered in the context of just sea turtles or, more generally, in the context of all turtles, has prompted much attention and speculation, but this apparently derived behavior is surprisingly poorly understood. Two general areas of inquiry have examined the underlying proximate mechanisms of the behavior (causes) and its evolutionary implications (consequences).

Supposed Cues That Precipitate Arribadas

The proximate cues that initiate arribadas are unknown, but the literature is rife with speculation about the extrinsic and intrinsic factors responsible for regulating this behavior. These inferences are largely based on anecdotal or observed correlations between the phenology of arribadas and extrinsic factors.

METEOROLOGICAL CUES. One prevailing hypothesis asserts that arribadas begin when there is a strong onshore wind (Carr, 1967; Pritchard, 1969; Pritchard and Márquez-M., 1973; Schulz, 1975; Hendrickson, 1980). It is true that arribadas often begin when strong winds are present, but they may also begin when there is no wind at all (Hughes and Richard, 1974; P. T. Plotkin, personal observation). Using a time-series analysis, Jimenez-Quiroz et al. (2005) showed some relationship between the nesting

Table 4.1 Locations of olive ridley (*Lepidochelys olivacea*) *arribada* nesting beaches and estimates of the density of turtles or nests at each site

Country	Beach	Estimates of *arribada* size	Reference
West Atlantic			
Suriname	Galibi Nature Reserve	335 nests[a]	Hoeckert et al., 1996
East Pacific			
Mexico	Mismaloya	1,000–5,000 nests[a]	Briseño and Abreu, personal communication
Mexico	Tlacoyunque	500–1,000 nests[a]	Briseño and Abreu, personal communication
Mexico	Chacahua	10,000–100,000 nests[a]	Briseño and Abreu, personal communication.
Mexico	La Escobilla	1,000,000+ nests	Márquez-M. et al., 2005
Mexico	Morro Ayuta	10,000–100,000 nests[a]	Briseño and Abreu, personal communication.
Nicaragua	Chacocente	No estimate available	
Nicaragua	La Flor	1,300–9,000 turtles per arribada	Ruiz, 1994
Nicaragua	Masachapa	No estimate available	Cornelius, 1982
Nicaragua	Pochomil	No estimate available	Cornelius, 1982
Nicaragua	Boquita	No estimate available	Cornelius, 1982
Costa Rica	Nancite	2,000–12,000 turtles per arribada	Honarvar, personal communication
Costa Rica	Ostional	Average 50,000–200,000 turtles per arribada	Chaves et al., in press
Panama	Isla Cañas	5,000–12,000 turtles per arribada	Evans and Vargas, 1998
Northern Indian Ocean			
India	Gahirmatha	1,000–100,000+ turtles per arribada	Shanker et al., 2003
India	Devi River	No estimate available	Shanker et al., 2003
India	Rushikulya	10,000–200,000 turtles per arribada	Shanker et al., 2003

Note: Numbers provided are derived from the most recent estimate published or via personal communication.

[a] Large arribadas once occurred at these beaches but no longer do (Cliffton et al., 1982; Hoekert et al., 1996).

Table 4.2 Locations of olive ridley (*Lepidochelys olivacea*) solitary nesting beaches

Country	Beach	Reference
Western Atlantic		
Suriname		Pritchard, 1979
Guyana		Pritchard, 1979
French Guiana		Chevalier and Girondot, 1998
Brazil	Sergipe, Bahia, Ceará	da Silva et al., 2003
Eastern Atlantic		
The Gambia		Barnett et al., 2004
Guinea Bissau		Barbosa et al., 1998
Sierra Leone		Siaffa et al., 2003
Ivory Coast		Gomez et al., 2003
Ghana		Beyer, 2002
Togo		Hoinsoude et al., 2003
Benin		Doussou Bodjrenou et al., 2004
Boiko, São Thomé, Corisco, Mbanye, Hoco Islands		Fretey et al., 2004
Cameroon		Fretey et al., 2004
Equatorial Guinea		Fretey et al., 2004
Gabon		Fretey et al., 2004
Congo		Fretey et al., 2004
Angola		Fretey et al., 2004
Western Pacific		
Australia	Northern and northeast beaches	Limpus, 1975; Whiting, 1997
Brunei		Shanker and Pilcher, 2003
Malaysia	Sarawak	Tisen and Bali, 2002
Indonesia	Java	Suwelo, 1999
Vietnam		Shanker and Pilcher, 2003
East Pacific		
Mexico	Entire Pacific coast	Briseño and Abreu, personal communication
Guatemala	Hawaii Beach and others	Juarez and Muccio, 1997
Honduras	Punta Raton and others	Lagueux, 1991
El Salvador	Toluca, San Diego and others	Hásbun and Vasqúez, 1999
Nicaragua	Entire Pacific coast	Pritchard, 1979
Costa Rica	Entire Pacific coast	Pritchard, 1979
Panama		Pritchard, 1979
Colombia	La Cuevita	Martinez and Paez, 2000
Western Indian Ocean		
Mozambique		Pritchard, 1979
Madagascar		Pritchard, 1979
Kenya		Church, 2004
Tanzania		Frazier, 1976
Northern Indian Ocean		
India	Entire east coast; southwest coast	Tripathy et al., 2003; Krishna, 2004
Pakistan		Asrar, 1999
Sri Lanka	Northwest, west and southern coast	Amarasooriya and Jayathilaka, 2002
Andaman and Nicobar Islands		Andrews, 2000
Bangladesh		Sarker, 2004
Myanmar (Burma)		Shanker and Pilcher 2003
Thailand		Aureggi et al. 2004

Note: The most current available literature documenting turtles at each location was used in this table.

Fig. 4.2. Distribution of *Lepidochelys* nesting along coasts of the western Atlantic and eastern Pacific Oceans. Solid circles denote *arribada* beaches (Table 4.1); gray shading depicts the extent of documented solitary nesting (Table 4.2).

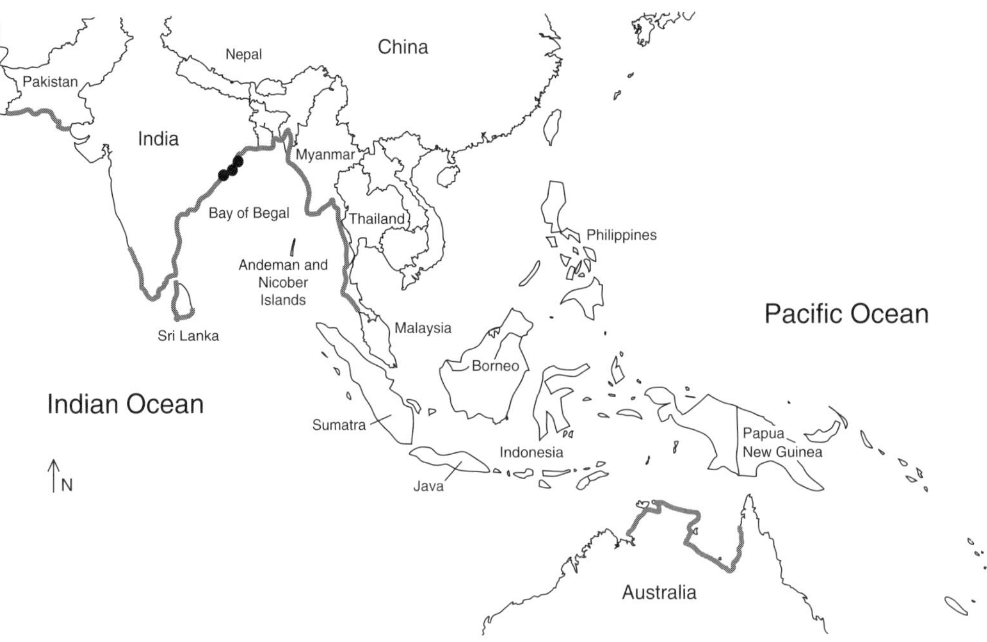

Fig. 4.3. Distribution of *Lepidochelys* nesting along coasts of the Indian Ocean basin and the northern coast of Australia. Solid circles denote *arribada* beaches (Table 4.1); gray shading depicts the extent of documented solitary nesting (Table 4.2).

Fig. 4.4. Distribution of *Lepidochelys* nesting in the eastern Atlantic Ocean. Gray shading depicts the extent of documented solitary nesting (Table 4.2).

cycle of Kemp's ridley and ambient temperature and wind but concluded that more data are necessary to strongly support this hypothesis.

LUNAR AND TIDAL CYCLES. Another popular explanation suggests that the lunar cycle is essential for initiating an arribada. Olive ridley arribadas in the eastern Pacific frequently coincide with the third-quarter moon, but they occur during other lunar phases as well (Hughes and Richard, 1974; Plotkin, 1994; Ballestero, 1996). Perhaps the most compelling evidence to contradict the lunar hypothesis is the occurrence of asynchronous arribadas on Pacific beaches in Costa Rica. Arribadas there occur at Nancite and Ostional, two beaches that are separated in distance by approximately 100 km and thus are under identical lunar cycles. The arribadas at Nancite and Ostional do not frequently occur simultaneously; in fact, they may occur days or even weeks apart (Hughes and Richard, 1974; P. T. Plotkin, personal observation). Mark–recapture studies have demonstrated that the vast majority of the ridleys nesting at these beaches are not the same turtles (Cornelius and Robinson, 1985), and thus, the asynchronous arribadas cannot be explained as turtles moving between two nesting beaches. These observations suggest that each arribada is in part regulated by cues present in the local environment.

On a finer time scale, we do know that arribadas are in part regulated by the day/night cycle and the daily tidal rhythm. The onset of Kemp's ridley arribadas coincides with daybreak; these turtles are exclusively diurnal nesters. In contrast, olive ridley arribadas occur primarily at night; their onset generally coincides with sunset. However, there are exceptions: olive ridley arribadas may also begin at sunrise and continue through the day, but this occurs infrequently (Hughes and Richard, 1974). There is also good evidence to show that arribadas begin around the time of the high tide: olive ridley arribadas in Costa Rica primarily occur just before or during the high tide (Plotkin, 1994; Ballestero, 1996).

SOCIAL FACILITATION. Another hypothesis is that arribada nesting is socially facilitated (Owens et al., 1982; Mora and Robinson, 1982). Social facilitation is defined as "an increase in the frequency or intensity of responses or the initiation of particular responses already in an animal's repertoire, when shown in the presence of others engaged in the same behavior and the same time" (Clayton, 1978). For many years it was thought that sea turtle migrations were socially facilitated because sea turtles often were observed together in groups in oceanic foraging areas and swimming to their nesting beaches (Oliver, 1946; Carr and Giovanolli, 1957; Leary, 1957; Caldwell et al., 1959; Carr, 1967; Meylan, 1982; Carr, 1986; Dash and Kar, 1990). For ridleys in particular, strong support for the existence of sociality comes from observations of olive ridleys aggregated in neritic and oceanic waters, the emergence of groups of females during an arribada (Fig. 4.1), and documented reemergence of groups of the same individual females nesting in close spatial proximity during subsequent arribadas (Pritchard, 1969; Hughes and Richard, 1974; Cornelius, 1986; Cornelius and Robinson, 1985; Dash and Kar, 1990).

Plotkin et al. (1995) tested the social facilitation hypothesis and found that olive ridleys from Nancite Beach, Costa Rica were not socially facilitated: turtles migrated independently of one another between arribadas and during postnesting migrations to oceanic feeding grounds. However, during subsequent arribadas, the groups of marked turtles returned to the beach at the same time and nested in close spatial proximity of one another (Plotkin et al., 1995). These results suggest that social facilitation plays no role in directing turtles to the vicinity of the nesting beach. However, it cannot be ruled out that, once the turtles are aggregated there, social stimulation (Clayton, 1978) might trigger the arribada (Plotkin, 1994).

A compelling observation that complicates acceptance of social stimulation as a cue that regulates the arribada is the conduct of the Kemp's ridley arribada in Mexico. Arribadas there were once dense aggregations, with tens of thousands of turtles nesting on the beach at the same time (see Shaver and Wibbels, Chapter 14). By the 1980s, the population declined (largely from human take), and only a few hundred adult female Kemp's ridleys remained (see Shaver and Wibbels, Chapter 14). Despite the current small population size, these turtles, spatially separated across approximately 100 km of beach, synchronously emerge to nest during arribadas (R. Byles, personal communication). This occurrence clearly demonstrates that these turtles receive the same cues even while they are not in close spatial proximity. Moreover, because the turtles are spatially dispersed, this also suggests that social stimulation does not play a role in Kemp's ridley arribadas.

In addition to cues that initiate an arribada, ridleys also apparently receive cues that delay an arribada. Plotkin et al. (1997) observed an arribada about to occur. Olive ridleys were aggregated in the nearshore waters of Nancite Beach, swimming back and forth parallel to the shore as had been observed in the preceding month on the day just before the arribada. A large storm (*temporale*) moved into the area on the day the turtles began to aggregate nearshore, and heavy rainfall began to fall. The turtles did not nest; instead, they left the immediate area and retreated further offshore. The rain fell continuously for several weeks, and the arribada was delayed until after the storm had dissipated. The turtles returned nearshore once conditions improved, and the arribada began the subsequent day. Clearly, ridleys have the ability to delay nesting when environmental conditions are unsuitable (Plotkin et al., 1997). Whatever the cues that initiate or retard an arribada, they apparently are received by almost all of the turtles simultaneously because there is little variance in the population responses.

SENSORY BIOLOGY OF CUE PROCESSING. The sensory modalities used by ridleys to detect these varied cues are also obscure. It is well established that sea turtles rely on auditory, visual, and olfactory stimuli from their environment (Moein Bartol and Musick, 2003), but their specific use in arribada behavior has not been investigated. Olfaction is the only sensory mechanism proffered as a potential arribada regulator (Owens et al., 1982; Robinson, 1987). Sea turtles do have a well-developed olfactory system (Owens et al., 1982) and the chemosensory ability to detect odors on land and in water (Owens et al., 1986). Robinson (1987) suggested that olive ridleys nesting in Costa Rica use olfactory signals obtained from the nearshore waters of nesting beaches to trigger their arribadas. These arribadas coincide with the rainy season, during which estuaries located behind the nesting beaches frequently open to the sea and release brackish water. The only problem with this hypothesis is that sometimes the estuaries do not open to the sea before the first arribadas of a nesting season, and sometimes they do not open to the sea at all (Plotkin, 1994).

Another suggestion that olfaction is involved in detecting cues for arribada behavior comes from specialized secretory glands known as Rathke's glands (Ehrenfeld and Ehrenfeld, 1973). Rathke's glands are typically embedded in fat (Wyneken, 2001), secrete a variety of compounds including lipids, carbohydrates, and proteins (Weldon and Tanner, 1990; Weldon et al., 1990), and are found in all families of turtles save the terrestrial Testudinidae (Ehrenfeld and Ehrenfeld, 1973). However, compared with other sea turtles, which generally have single Rathke's pores leading from the gland, *Lepidochelys* have enlarged Rathke's glands that lead to multiple pores along several inframarginal scutes (Rostal et al., 1991; Wyneken, 2001).

The function of these glands in ridleys is unknown (Wyneken, 2001), but their prominence has led to speculation that they may provide olfactory signals important in arribada behavior (Owens et al., 1982). During fieldwork at the arribada site at Nancite, Costa Rica, Plotkin (personal observation) made some observations relevant to this idea. In a series of about 10 *L. olivacea* caught in nearshore waters in the vicinity of the nesting beach during the nesting season, but several weeks before an arribada, turtles were found to have waxy plugs in their Rathke's pores. These plugs could be manually expelled by pressing on the sides of the pore, were 3–5 mm long, and had the consistency of lard or vegetable shortening. In contrast, about 10 turtles inspected for the presence of these waxy plugs after the arribada had commenced a few weeks later did not have them. Chin et al. (1996) chemically analyzed glycoproteins from the Rathke's gland secretions of *L. kempi* and found that they have 20% identity with esterase lipase protein families. This finding suggests that enzymatic breakdown of these waxy plugs may occur quickly, and the digested lipids would rapidly disperse on the ocean surface. Moreover, this temporal difference in the presence and absence of these waxy plugs coinciding with the timing of the arribada suggests that these secretions may relate to the synchrony of nesting behavior. Pritchard (1979) also speculated that Rathke's gland secretions might drive synchronization of the arribada.

Finally, one interesting observation used to support olfaction was made by Mora and Robinson (1982). They watched a blind olive ridley nest during an arribada at Ostional Beach. Clearly this observation indicates that visual stimuli are not critical to arribada behavior, but it does not indicate what other sensory mechanisms might have been operative at the time.

Much remains to be learned regarding the sensory mechanisms used by arribada-nesting ridleys and the stimuli that trigger arribadas. Well-planned experiments are needed to advance our understanding of the proximate cues regulating arribada behavior.

Synchronous Reproduction in Other Organisms

Whatever the proximate cues and regulators of this behavior, *Lepidochelys* arribadas are a remarkable and well-known example of synchronous reproduction. A wide variety of organisms reproduce in this way, and the evolutionary reasons for such reproductive behavior are discussed extensively. In the case of *Lepidochelys*, similarly, the evolutionary consequences of the arribada behavior are unresolved. In this section we use comparisons with other species to gain new insights into the adaptive significance, evolution, and maintenance of arribada behavior, beginning with comparisons with other turtles. We also critically evaluate ideas that have been suggested for why such a derived behavior has evolved in *Lepidochelys*, compared with other turtles.

Aggregative Nesting in Other Turtles and the Distinctiveness of *Lepidochelys* Arribadas

Several of the 305 other turtle species nest aggregatively, although none do so to the extent of the ridleys (Uetz and Etzold, 1996). One population of green sea turtles (Cheloniidae: *Chelonia mydas*) at Raine Island, Australia, nest in very large numbers, but these dense concentrations are thought to be a function of an isolated nesting beach with limited nesting habitat (Limpus et al., 2003). A similar argument was proposed to explain the formerly dense nesting aggregations of green turtles on Aves Island (Carr, 1967). All other green sea turtles nest individually and coincidentally, but at much lower densities.

Large Amazon River turtles (Podocnemidae: *Podocnemis expansa*) emerge rather synchronously after seasonal flooding and nest on large sandy riverbanks that were recently inundated (Vanzolini, 1967; Alho and Padua, 1982; Pritchard and Trebbau, 1984; Vanzolini, 2003). This behavior is unlike that of any of its five smaller congeners, which also nest on sandy riverbanks but not in any kind of social groups. One explanation for this behavior is that the large size of these turtles and their historically large populations limit the number of suitable oviposition beaches, resulting in group nesting (Pritchard and Trebbau, 1984). Another possibility is that because these are the largest turtles in the Amazon, their nests present a particularly rich food source, and so group nesting has the familiar predator-swamping effect. Whatever its cause(s), as far as is known, all *P. expansa* nest in the same manner; solitary nesting is unknown. Further, the nesting interval is unlike that seen in ridleys, in that turtles nest over as many as 60 days, so although there may be days in which large numbers of turtles nest, the temporal distribution of nesting is much broader (reviewed by Pritchard and Trebbau, 1984).

The extensive geographic distribution of arribadas and of solitarily nesting ridleys (Figs. 4.2–4.5) suggests that arribadas are not aggregations that result from lack of suitable nesting beaches as has been argued in these other species' large nesting aggregations.

Other turtle species such as diamondback terrapins (Emydidae: *Malaclemys terrapin*) and painted turtles (Emydidae: *Chrysemys picta*) may have locally high nest density, but none of this is especially synchronous or coordinated (e.g., Gibbons and Greene, 1991).

Thus, the rarity of synchronized nesting behavior among the other 305 species of testudines gives little insight into why it evolved in *Lepidochelys* and also illustrates how exclusive this behavior is to this genus. It is therefore necessary to expand our comparative exercise to other synchronously reproductive organisms.

Synchronous Reproduction in Nontestudines

The ecological and evolutionary dynamics of synchronous reproduction have long been studied (Allee, 1927) because synchronous reproduction occurs in a wide range of organisms including plants (Janzen, 1974; Augspurger, 1981), invertebrates (scleractinian corals: Babcock et al., 1986; palolo worms: Woodworth, 1903, 1907; Caspers, 1984; insects: Sweeney and Vannote, 1982; crabs: Morgan and Christy, 1995; Skov et al., 2005; myriapods: Mauck, 1901; horseshoe crabs, *Limulus polyphemus:* Botton et al., 1988; Chatterji, 1994), and vertebrates including fish such as salmon, grouper, grunion (Walker, 1949), and birds (Westneat, 1992; Smith, 2004), especially colonial species (Gochfeld, 1980; Findlay and Cook, 1982; Wittenberger and Hunt, 1985). These and many other examples give some insights into the diverse avenues of selection that lead to the evolution of reproductive synchrony.

One mechanism that explains reproductive synchrony is that it is simply an epiphenomenon of parallel responses among individuals to the

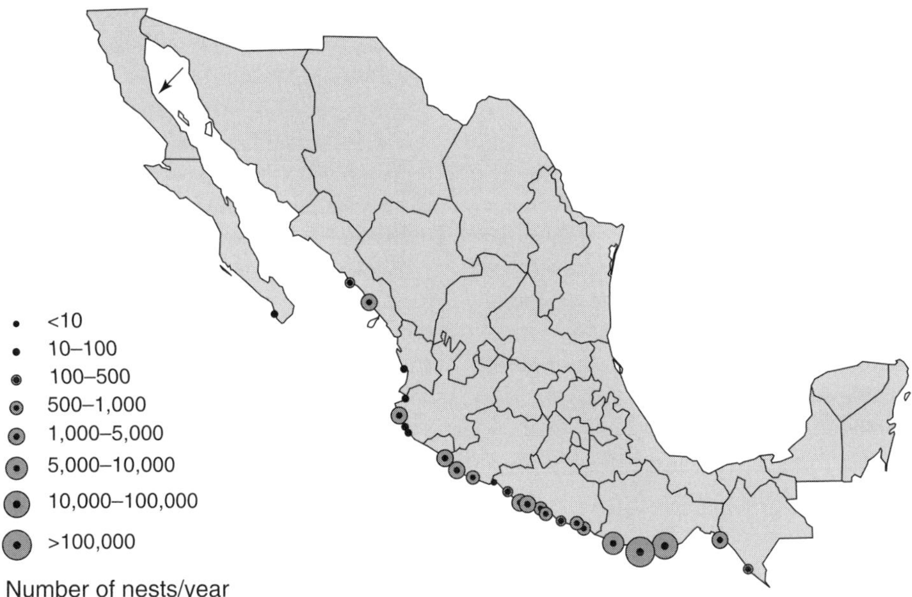

Fig. 4.5. Estimates of maximum number of nesting *L. olivacea* along the Pacific coast of Mexico, illustrating the continuous range of nesting density from <10 to >100,000 individuals per year. Map by R. Briseño and A. Abreu.

same migratory or breeding time cues (Findlay and Cooke, 1982; Rutberg, 1984; Godley et al., 2002). These cues may relate to changing temperatures, day lengths, and so on, which signal optimal breeding times, or they may relate to the availability of seasonally restricted food resources or breeding habitats. For instance, many species of marine birds use seasonally restricted food resources to support offspring production, so synchronous reproduction is being driven in part by spatially and temporally constrained resources (Findlay and Cooke, 1982). Another example occurs in explosively breeding amphibians that exploit ephemeral ponds, such as mole salamanders (e.g., *Ambystoma:* Tennessen and Zamudio, 2003) and many species of Anura.

The second hypothesis is that synchronized reproduction somehow directly influences fitness of participants compared with asynchronous reproducers. This reproductive strategy is believed to have evolved because of the individual fitness gained from being part of a group. For instance, nesting aggregations have been shown to facilitate reproduction (mate finding), increase food availability for offspring, and protect parents and their offspring from predators (Wilson, 1975). However, nesting aggregations can be disadvantageous because predators may in fact be more attracted to these aggregations than they would be to solitary nest sites (Vehrencamp et al., 1988).

It is important to note that whatever the proximate mechanisms of synchrony (whether individuals are responding to seasonal cues or seasonally available food resources, and so on), the synchronized responses mean that reproductive individuals are arriving in the same place at the same time, and this enhances mating opportunities. It may be difficult to separate the effects on fitness that such synchronous reproduction has as a result of its effects on the mating system (mating opportunities, potential for outcrossing, potential for multiple mating, potential for mate choice on the part of both sexes) from the effects on fitness arising from the use of common resources, and from the distinct effects on propagule survival, as well as hatchling survival (Jennions and Petrie, 2000). In mammals and many plants, in which mating and offspring production are widely separated in time because of gestation and seed maturation time, it may be possible to disentangle how synchronous behavior relates to selection on features of the mating system per se as distinct from effects upon offspring. However, in many organisms in which mating and gamete liberation occur close in time, understanding how selection on specific fitness components has driven the evolution of reproductive synchrony is challenging.

Whatever the cause of synchronous reproduction, a number of distinct fitness advantages accrue to participants. We first explore the fitness advantages that accrue directly to adults, bearing in mind that mating systems are necessarily affected by synchronous mating aggregations, and then examine those thought to accrue via offspring. We consider these issues from the joint perspective of other organisms and *Lepidochelys*.

Direct Effects of Mass Nesting on Adult Fitness

Compared with solitary nesting, arribadas afford several distinct reproductive advantages that accrue directly to adults, but this aspect of the arribada has been given almost no attention.

MATE FINDING. An obvious benefit of reproductive synchrony is simply mate finding. Mating aggregations permit species that are usually widely dispersed, and therefore have limited mate encounters, to locate mates reliably. Selection for mating aggregations in such species should be strong if mating opportunities are otherwise rather unlikely because aggregating individuals will have many more matings.

This logic applies directly to *Lepidochelys*. Olive ridleys are pelagic inhabitants (Plotkin et al., 1995) of the tropical waters of the largest and third largest ocean basins; the Pacific Ocean Basin occupies nearly one-third of the earth's surface, about 179.7 million km^2. Similarly, the Indian Ocean occupies about 20% of the earth's surface, about 73,556,000 km^2. There are also populations in the western and eastern Atlantic. Although *L. olivacea* principally utilizes the tropical and subtropical parts of these basins, even very large populations would be highly dispersed across this large area, meaning that mating opportunities at sea are likely to be rather rare occurrences. Such matings have been observed during oceanic transects in the eastern Pacific (Pitman, 1990) and generally involve individual pairs mating, suggesting that these are opportunistic unions (Plotkin et al., 1996). Plotkin et al. (1996) documented that male ridleys arrive in reproductive condition at nearshore waters just off Playa Nancite, Costa Rica (an arribada beach). These individuals have elevated testosterone levels and mature gonads, and many hold onto females even while not copulating, suggesting that breeding opportunities for males are critical in these nearshore waters.

Pandav and Choudhury (2000) conducted systematic searches in nearshore waters off the Orissa Coast of India near arribada beaches in an effort to document the magnitude of mating activity. They documented numerous matings over a 5,300-ha area, including documentation of multiple matings based on captures of animals in copulo. Thus, mating activity in waters in the vicinity of arribada beaches is nontrivial, and it should be studied in detail at other arribada sites.

MULTIPLE MATING. A second distinct reproductive advantage of synchronous arrival at breeding grounds is that it increases the potential for multiple mating. Multiple mating gives rise to numerous genetic benefits (reviews: Andersson, 1994; Keller and Reeve, 1995; Yasui, 1998; Jennions and Petrie, 2000) and has diverse fitness consequences for both males and females.

The reproductive strategies of males have been ignored in arguments concerning the evolution of the arribada, but male turtles also gain in distinct ways from the arribada behavior: male turtles have fitness too (Janzen, 1985). For males, especially in noncaregiving species, multiple mating directly increases fitness via higher offspring numbers. Importantly, males arrive in large numbers in waters adjacent to arribada beaches several weeks ahead of females and remain near shore until the peak nesting event begins (Plotkin et al., 1997). Because these animals are highly pelagic (Plotkin et al., 1995), the energetic expenditure by males to return to these nesting beaches cannot be ignored and strongly suggests that arribadas act to facilitate mate finding by both sexes.

For females, the effects of multiple mating are more varied and somewhat more subtle, but they are also somewhat controversial. A recent analysis of multiple mating in green turtles (Lee and Hays, 2004) concluded that females realize no benefits from multiple mating and that it is simply a consequence of female submission to male coercion. Although these authors noted that they did not exhaustively rule out all possible avenues of fitness gain arising from multiple mating, their conclusion is significant because, if accurate, it casts doubt on the varied avenues through which multiple mating has been thought to influence female fitness, both theoretically and from empirical analysis of a wide range of animals.

However, this conclusion seems overstated, given the extent to which multiple paternity has been found to typify turtle mating systems (Pearse and Avise, 2001; Pearse et al., 2002, 2006). Moreover, the conclusion of these authors suggests that the genetic benefits of multiple mating that have been detected in a wide diversity of other organisms simply do not apply to marine turtles, and they offer no explanation for this striking difference. For instance, multiple mating has been shown to increase offspring heterozygosity and fitness in insects (Bernasconi and Keller, 2001; Fox and Rauter, 2003; review: Arnqvist and Nilsson, 2000), a pseudoscorpion (Newcomber et al., 1999), fish (e.g., Garant et al., 2005), and birds (e.g., Lank et al., 2002; Foerster et al., 2003; reviews: Stutchbury and Morton, 1995; Petrie et al., 1998; Griffith et al., 2002). Given these and many other findings, the incompleteness of the Lee and Hays (2004) analysis, and the general inferential limitations of negative data, we give careful consideration to the range of genetic benefits that might accrue to female ridleys via the arribada behavior.

Females may multiply mate simply to avoid sperm limitation when mating opportunities are generally unpredictable (insurance hypothesis), as we have suggested is the case for pelagic *Lepidochelys.* Because of the unreliability of finding mates, as discussed earlier, it is possible that females might exhaust their sperm supply and thus have to forego oviposition opportunities if a mating opportunity did not present at the right time. Because females may store sperm for extended periods, multiple mating permits females to amass sperm in anticipation of numerous future oviposition events without the risk of missing oviposition opportunities because of chancy mating opportunities.

In philopatric species such as marine turtles including many *Lepidochelys,* the potential for inbreeding is high. Importantly, the costs of inbreeding are generally thought to be higher for females than for males (review: Lehmann and Perrin, 2003). Thus, another adaptive reason for female multiple mating is to reduce these costs. This may be especially important if kin recognition (and therefore, behavioral avoidance of inbreeding) is weak, which is unknown for *Lepidochelys.*

In addition to the insurance hypothesis and inbreeding avoidance, females may seek multiple mates because of various subtle genetic benefits to her offspring (reviews: Yasui, 1998; Jennions and Petrie, 2000). Sequential mating by females may allow for genetic "trading up," a scenario in which females hedge their bets by accepting a mating from the first available male but then mate with genetically superior males later if additional encounters occur. Related to this is the idea that multiple mating may also permit cryptic female choice (Eberhard, 1996), in which females distinguish among males' stored sperm at fertilization. They thus can exercise mate choice even if they are not able to do so before mating occurs. This important avenue of sexual selection by female animals remains poorly studied empirically, although the theoretical analysis of the issue is well developed (review: Pizzari and Birkhead, 2002).

For both sexes, multiple mating reduces the potential for inbreeding, which may be especially likely in species in which there is natal homing. The evolutionary basis for natal homing is unclear, but it too may relate to the predictability of mating aggregations that it affords to species that are otherwise solitary and widely dispersed. Multiple mating also increases effective population size (Sugg and Chesser, 1994; Sugg et al., 1996).

Sea turtles are well known to mate mutiply as evidenced by multiple paternity in every species studied (*L. kempii:* Kichler et al., 1999; *L. olivacea:* Hoekert et al., 2002; Jensen et al., in press; *C. mydas:* Parker et al., 1996; FitzSimmons, 1996, 1998; Peare et al., 1998; Ireland et al., 2003; Lee and Hays, 2004; *Caretta caretta:* Harry and Briscoe, 1988; Bollmer et al., 1992; Moore and Ball, 2002; *Dermochelys coriacea:* Crim et al., 2002), which accords with the prevalence of multiple paternity in turtles in general (review: Pearse and Avise, 2001). All the other sea turtle species save *D. coriacea* are continental shelf species; even green turtles, although they undertake cross-basin migrations between feeding and nesting sites, are mainly coastal forms. In contrast, *L. olivacea* are highly pelagic (Plotkin et al., 1995; Plotkin, 2003), as we discussed above, as are *D. coriacea* (Hays et al., 2004), which do not participate in mating aggregations as far as is known. The difference between these two pelagic turtle species, however, is their relative vagility. Leatherbacks appear to be far more vagile than ridleys (compare tracks in Plotkin et al., 1995, versus Hays et al., 2004) and thus probably have much higher pelagic mate encounter rates at sea than do ridleys. Thus, arribadas not only afford an opportunity to locate a mate but also provide opportunities for female choice and multiple mating.

Aggregative reproduction also may yield survivor benefits that accrue to offspring. It is under this category that predator swamping behavior belongs, which we now explore in detail.

The Predator Satiation Hypothesis

The evolution of arribada behavior in ridleys has been explained largely as a predator satiation strategy (Pritchard, 1969; Hendrickson, 1980; Eckrich and Owens, 1995). Despite hatchling frenzy behavior, which minimizes the time hatchlings are exposed to predators on the beach (Dial, 1987), intense predation occurs after hatchlings emerge from nests and before they reach the ocean (Ehrenfeld, 1979; Stancyk, 1982; Brown and Macdonald, 1995). The offspring of arribada nesters hatch synchronously and emerge en masse with tens of thousands to millions of hatchlings crawling seaward (Hughes and Richard, 1974; Cornelius, 1986; Cornelius et al., 1991). Terrestrial predators are quickly sated by this large pulse of prey, thereby allowing a large proportion of hatchlings to survive this vulnerable period in their early life history (Cornelius, 1986; Cornelius et al., 1991). Little is known of the importance of predation by fish. Hence, predator satiation appears to be a reasonable hypothesis for the evolution of arribada behavior.

Hughes and Richard (1974) agreed in principle with the predator satiation hypothesis (PSH) but suggested that olive ridleys are passively displaced by ocean currents to a general location near an arribada beach where they converge to nest and that this was advantageous to the turtles and their offspring. Determination of whether the use of arribada sites is a passive process versus an active decision by mothers is an important unresolved issue. The latter case would comprise a maternal effect if nesting beach selectivity by mothers relates to choice of optimal offspring rearing conditions or if these selected beaches advantageously influence offspring dispersal enhanced by site-specific currents (Bernardo, 1996).

In general, nests of all turtle species are subject to high levels of mortality. For example, in a study of three sympatric freshwater turtles, Burke et al. (1998) documented that more than 84% of 145 nests were destroyed by predators. Nests are large and conspicuous, contain many large highly nutritious eggs (eggs are protein and lipid dense; Congdon and Gibbons, 1990), and so are very profitable and attractive resources for predators. Turtle nesting strategies thus seem to reflect this mortality selection within the constraints of their ability to locate suitable oviposition sites. Most turtles nest at night when visual detection is less likely. Turtles also camouflage their nests after oviposition is complete. Finally, many turtle species disperse their nests in space and or time (Wilbur and Morin, 1988). Because turtles do not exhibit postovipositional parental care, these measures comprise the major avenues of maternal care (Pritchard, 1979; Wilbur and Morin, 1988). *Lepidochelys* eggs and nesting patterns are like those of other turtles in these respects; hence, there is no special reason to think that ridley nests are more attractive to predators or provide a better food resource than do nests of syntopically nesting heterospecifics.

In light of these general patterns of turtle oviposition behavior and the prevalence of predation on turtle nests, the distinct arribada behavior certainly appears to be a unique response to mortality selection due to nest predation. Indeed, the dominant explanation for the evolution of the arribada is that it evolved as a predator-swamping strategy (Pritchard, 1969; Cornelius, 1986; Ims, 1990a, 1990b). This explanation has been taken as a given, and a single empirical study seems to support this hypothesis. Working at the arribada beach at Nancite, Costa Rica, Eckrich and Owens (1995) compared predation rates on nests of solitary versus arribada nesters during the first 24 hours after oviposition. They found higher mortality of solitary nests and concluded that nesting en masse affords some reduction of per-nest mortality. An important caveat to this study, however, is that it was conducted at an arribada beach, and a fundamental premise of the predator-swamping strategy is that initial mortality at such sites is high because of predators cueing on the concentrated resources (Ims, 1990a, 1990b). At Nancite, local predator densities are high, and the predators are experienced hunters of turtle nests and hatchlings. Playa Nancite is only 1 km long, supports some of the densest turtle nesting in the world, but also supports an extremely dense predator assemblage (P. T. Plotkin, personal observation). Thus, selection against asynchronous nesters within a synchronous nesting beach is expected to be remarkably high (Ims, 1990a, 1990b), and these findings are equivocal (Ims, 1990a, 1990b). Thus, the Eckrich and Owens (1995) study is not an appropriate test of the PSH. Recall that most solitary nesters oviposit on solitary nesting beaches. Hence, a better approach to testing the PSH would be to compare relative fitness of arribada nesters using solitary nesters from a solitary beach instead of solitary nesters from an arribada beach.

This brief survey of synchronous reproductive behavior reveals a variety of ecological and behavioral reasons for the evolution of synchronous reproduction. Moreover, these diverse causes are not mutually exclusive. For example, predator swamping may be an additional benefit that arises from synchronous nesting that is mainly driven by mating system evolution or spatiotemporal constraints on availability of resources required for reproduction, or vice versa (Sweeney and Vannote, 1982).

Divergent Nesting Behavior in *L. olivacea*

Interestingly, olive ridleys *(Lepidochelys olivacea)* exhibit divergent oviposition behaviors: within a population some females nest solitarily; some nest in arribadas (Hirth, 1980; Ekrich and Owens, 1995; Kalb, 1999; Plotkin and Bernardo, 2003). As we have already discussed, there are also widespread solitary nesting beaches, interspersed with or even quite distant from arribada beaches (Figs. 4.2–4.5). Very little is known about this reproductive behavioral polymorphism and how it is maintained in populations.

The existence of two oviposition strategies within a single population is not easily explained by the PSH. Specifically, two questions arise. First, if offspring of arribada nesters do have enhanced fitness because of predator satiation, then why don't all olive ridley nest in arribadas? Conversely, why does solitary nesting persist even on the same beaches where arribada nesting and high predator densities occur? If there were a fitness advantage to arribada nesting, one would expect solitary nesting to be maintained in a population at low frequencies, if at all. But in fact, solitary nesters are very common and may outnumber arribada nesters worldwide (Cornelius and Robinson, 1985; Table 4.2; Figs. 4.2–4.5). The evolution and maintenance of these alternative reproductive tactics deserve careful scrutiny from both pure and applied perspectives, which is the focus of the balance of this chapter.

The arribada phenomenon is striking not only because of its rarity among turtles but because it is unusual in its magnitude even among large vertebrates. Only large herbivores such as wildebeest (Estes, 1976) rival these aggregations in terms of biomass of reproductive females. Most turtle species nest neither synchronously nor en masse. The rarity of this behavior among turtles also suggests that the conditions that permit the evolution of mass nesting behavior are rather restrictive. Our review of available data on the existence and magnitude of nesting on nonarribada beaches (Table 4.2; Figs. 4.2–4.5) forces us to conclude that solitary nesting is common and involves a nontrivial number of turtles. Thus, there is a reproductive behavioral polymorphism in ridley turtles, and we now explore how such a polymorphism might have evolved and is maintained by examining reproductive behavioral polymorphisms in other organisms.

The evolution and maintenance of polymorphisms in general (Moran, 1992; Brockmann, 2001) and geographic variation in behavior (Foster, 1999; Foster and Endler, 1999) and reproductive polymorphisms in particular are important areas of research in behavioral ecology and evolution. Reproductive polymorphisms exist in a wide array of species and relate to structural features, mate-garnering polymorphisms, and ovipositional polymorphisms (review: Gross, 1996). Most reproductive polymorphisms involve males. Examples include sneaker males in amphipods (Kurdziel and Knowles, 2002), insects (Emlen, 1997; Moczek and Emlen, 2000; Piennar and Greeff, 2003), horseshoe crabs (Brockmann, 1990, 2001), fish (Gross, 1991; Reichard et al., 2004), lizards (Zamudio and Sinervo, 2000), and birds (Lank et al., 2002). Behavioral polymorphisms in females are far less common; however, examples are known in lizards (Sinervo and Zamudio, 2001) and insects (Reguera and Gomendio, 2002; Richards et al., 2003).

The geographic locations of arribadas are of interest because they do not follow any obvious pattern, nor do they provide any clues to facilitate our understanding of the arribada phenomenon. As argued above, arribadas are not aggregations that result from lack of suitable nesting beaches as has been documented for other sea turtle species that aggregate on isolated beaches such as the Aves Island green turtles (Carr, 1967) or the Raine Island, Australia green turtles (Limpus et al., 2003). In fact, arribada beaches generally occur adjacent to other, seemingly suitable beaches. For example, arribadas occur at Playa Nancite, Costa Rica, a small stretch of sand approximately 1 km in length. Turtles aggregate to nest on this small beach but do not nest on the adjacent Playa Naranjo, which is much longer (Fig. 4.6). This is the case everywhere ridleys nest. Also, along the Pacific shores of the Americas, solitary nesters sometimes use the same beaches as arribada nesters (Pritchard, 1969;

Fig. 4.6. Aerial photograph of Playa Nancite and Playa Naranjo in Santa Rosa National Park, Area de Conservacion de Guanacaste, Costa Rica, illustrating the adjacency of an *arribada* beach (Nancite, located at bottom of photo) to an extensive, seemingly suitable beach (Naranjo, located at top of photo) at which *L. olivacea* only nests solitarily. Photograph courtesy of Daniel H. Janzen.

Cornelius, 1986; Plotkin and Bernardo, 2003) as well as many nonarribada beaches (Fig. 4.2).

The second question arising from the PSH explanation of arribada nesting is why don't all sea turtle species nest in arribadas? If there were an overwhelming fitness advantage to arribada nesting, one would expect the behavior to be more common among other sea turtles species as well as among freshwater and terrestrial turtles.

In the last three decades, the study of arribada-nesting behavior has dominated most aspects of research and conservation of this species; very little attention has been given to the solitary nesting strategy of the olive ridley. This is reflected in the literature wherein solitary nesting is not even mentioned in most review articles (Ehrhart, 1982; Reichart, 1993; VanBuskirk and Crowder, 1994; Miller, 1997) despite the fact that its extent was well known and described by Carr (1967). In contrast, substantial effort has been directed at studying arribada nesters and their offspring (Pritchard, 1969; Richard and Hughes, 1972; Hughes and Richard, 1974; Acuña-Mesen, 1983; Acuña-Mesen and Castillo, 1985; Cornelius and Robinson, 1985, 1986; Acuña-Mesen, 1988; Cornelius et al., 1991; Arauz-Almengor and Mo, 1994; Plotkin et al., 1995, 1997). The focus on arribada nesting has even extended to conservation efforts designed to promote recovery of olive ridley populations. Arribada nesting sites worldwide are protected, but little has been done to protect solitary nesting sites.

Relative Fitness of Arribada versus Solitary Nesters

As in any polymorphism, the maintenance of distinct reproductive strategies typically results from there being alternative fitness peaks on the adaptive landscape each of which confers relatively high fitness (Moran, 1992). In addition to the oft-hypothesized issue of predator satiation, we have described some of the other avenues by which fitness might be affected by arribada behavior (Table 4.3). So, the diversity of fitness effects of arribada nesting, we have suggested, has been underappreciated. But as we have also argued, the extent of solitary nesting, a contrasting strategy, has also been underappreciated if not essentially ignored, and there are a number of expectations for fitness differences between these reproductive strategies that remain mostly unstudied (Table 4.3 and below).

Clear behavioral differences exist between solitary and arribada nesters, yet many obvious questions have not been asked. For instance, are there morphological phenotypic differences between solitary and arribada nesters? Do solitary and arribada nesters differ significantly in maternal traits such as size or condition? Is variation in maternal traits correlated with reproductive traits such as fecundity, egg mass, or offspring size? Does variation in offspring traits affect offspring performance and fitness? Are arribada and solitary nesters even part of the

Table 4.3 Advantages and disadvantages of solitary versus *arribada* strategies

	Advantages	Disadvantages
Solitary	No interference competition No density-dependent mortality Low predation risk on solitary nesting beaches Relatively short nesting season Reproductive investment spatially distributed	High predation risk when nesting on *arribada* beaches Risk of total reproductive failure if predator cues in on nest or hatchlings High energetic cost during internesting period when females travel to other areas/nesting beaches
Arribada	Low predation risk Low energetic cost during internesting period when turtles remain near nesting beach Can delay nesting until environmental conditions are suitable (unknown for solitaries)	Interference competition Density-dependent mortality Attracts and sustains predators Comparatively long nesting season Reproductive investment at one site; risk of catastrophic loss

same genetic populations, or is there genetic substructuring of ridleys that is consonant with these reproductive differences.

Surprisingly few studies have compared solitary nesters with arribada nesters, but several differences have been identified (Table 4.4). Pritchard (1969), Kalb and Owens (1994), and Kalb (1999) found that solitary nesters oviposit on 14-day cycles, whereas arribada nesters oviposit every 28 days. Kalb (1999) also found that, within a nesting season, solitary nesters may use multiple beaches for oviposition, but arribada nesters display site fidelity.

Other analyses of nest survival and life history features of solitarily nesting versus arribada-nesting individuals at an arribada site (Playa Nancite, Costa Rica) have also been conducted. A first attempt at comparing relative fitness of offspring of solitary and arribada-nesting olive ridleys was made by Eckrich and Owens (1995), described above.

Phenotypic variation has been documented among and within olive ridley turtle populations. Maternal size, clutch size, egg size, and offspring size vary (Carr, 1952; Pritchard, 1969; Frazier, 1983; Reichart, 1993; VanBuskirk and Crowder, 1994); however, many studies provide no indication whether measurements were collected from solitary or arribada nesters and their offspring. Pritchard (1969) measured the straight-line carapace lengths of 199 solitary nesting olive ridleys in Honduras and found they ranged from 58 to 74 cm. He also found considerable variation in clutch size, from 48 to 151 eggs ($N =$ 106). An unpublished study (J. D. Maziarz and C. Mo, unpublished data, 1993) reported that 6 of 10 solitary nesters sampled at an arribada-nesting beach oviposited an average of 54 eggs, half the clutch size on average of arribada nesters (Cornelius et al., 1991). Variation in offspring phenotype has also been documented in the olive ridley. Carr (1952) measured the diameters of 50 eggs and reported a range of 3.21–4.54 cm. Pritchard (1969) measured the straight-line carapace lengths of 94 hatchlings and reported a range of 37.9–42.9 cm.

Table 4.4 Reproductive characteristics of solitary and *arribada L. olivacea* nesters

Attribute	Solitary	*Arribada*
Clutch frequency	2/year	2/year
Internesting interval	2 weeks	4+ weeks
Annual seasonality	All year	Rainy season
Site fidelity	Low	High
Testosterone titer at first nesting	Low	High

Plotkin and Bernardo (2003) examined life history features of females nesting at Nancite Beach, Costa Rica, during an arribada and straggler females nesting solitarily outside of the arribada to ascertain whether there were any significant differences between the groups. There were no differences between these groups in female size, egg size, or within-clutch variability in egg size, but arribada nesters produced significantly larger clutches.

Differential Selection in Solitary versus Arribada Beaches and Maintenance of Ovipositional Behavioral Polymorphism

There appear to be some life historical differences between solitary and arribada nesters. It is unknown whether there are also genetic differences that separate these groups, and the study of phenotypic differences between these nesting phenotypes, the relative fitness of their offspring, and their genetic relationships are all open questions.

It is possible, however, to develop an understanding of the relative fitness of these groups based on existing data and the differential selection that may relate to the maintenance of this polymorphism. We propose one scenario in Figure 4.7.

We have already reviewed the PSH as it relates to nest mortality, which is graphically represented in Figure 4.7A. But there is a distinct, contrasting source of nest mortality that has received little direct attention, which is also shown in Figure 4.7B: density-dependent nest mortality.

Nest destruction by later-nesting females has been observed in most sea turtle species, and the possibility that this process is density-dependent has been modeled for leatherbacks and greens (Bustard and Tognetti, 1969; Girondot et al., 2002; Caut et al., 2006; J. Bernardo and P. T. Plotkin, personal observation). For instance, Girondot et al. (2002) reported that leatherbacks destroyed

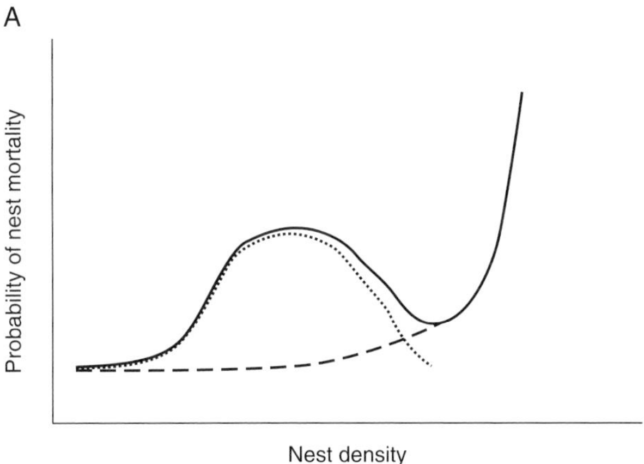

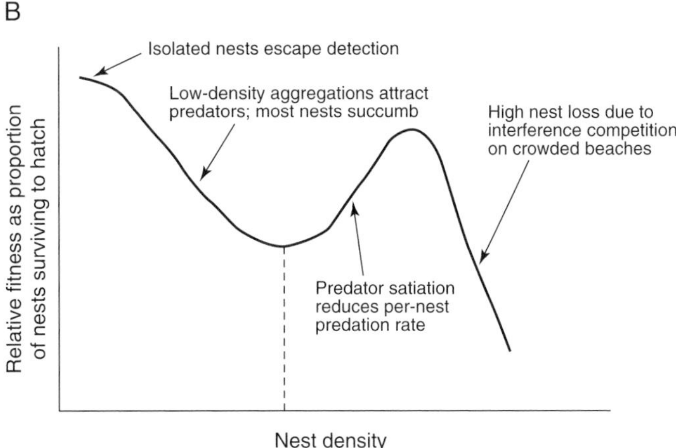

Fig. 4.7. Hypothetical density-dependent mortality functions relating nest loss to predation and interference competition that would result in two troughs in mortality (associated with two fitness peaks, part B), as an explanation of the maintenance of the nesting polymorphism in *L. olivacea.* (A) The dotted line illustrates mortality from predation, which is approximately a type III functional response. Predation is nil to low at low nest densities because of the difficulty that predators have cueing on nesting activity and nest placement. As nesting density increases, predation increases steeply because predators cue directly on heightened nesting activity as well as on the predatory behavior of other predators. As predators become sated, the probability of nest mortality begins to decline. Dashed line shows mortality from interference competition that occurs via nest destruction. This curve is more or less exponential, meaning that density-dependent interference is very low to low even through moderately high densities and then rapidly accelerates at high densities. Solid line shows the aggregate mortality curves and is simply the sum of these distinct sources of mortality. (B) Hypothetical fitness landscape illustrating how fitness arising from nest success depends nonlinearly on nesting density and the interaction of predation loss and interference competition. At extremely low densities (i.e., when turtles nest solitarily), the likelihood of nest detection by predators is low and random in part because predators are only stochastically cuing on nesting activity. Also, there is no mortality from interference competition because nest destruction caused by the nesting activities of other turtles rarely occurs in low-density situations. Vertical dashed line depicts the switching point due to maximal predator satiation, after which per-nest fitness begins to increase. The position of this switching point along the *x*-axis will be a function of predator density and hunger. Fitness increases until interspecific nest destruction becomes a major source of mortality.

almost 21% of existing nests at a major rookery in Guiana, at levels well below the hypothesized carrying capacity at that site. Importantly, these studies inferred that density-dependent nest destruction occurs at nest densities that are several orders of magnitude lower than those observed in arribada sites. In the case of ridleys, the sheer number of turtles nesting in spatially limited areas (1,000–500,000 turtles) must result in moderately high levels of density-dependent mortality of eggs during a single arribada. Moreover, turtles return approximately every month during a discrete nesting season (3–6 months), and nests that remained intact during the previous month are again at risk when new waves of turtles crawl ashore. In addition to nest disturbance, the existence of high nest densities over time apparently alters the nutrient composition of sand as well as the concentration of ammonia in the sand (McPherson and Kibler, in press). High ammonia concentrations and/or high concentrations of fungal and bacterial pathogens at beaches with high nest densities might also contribute to density-dependent nest loss, but these issues have barely been explored empirically.

In combination with the mortality due to predation, the combined mortality from these two distinct sources (Fig. 4.7A, solid line) results in a complex fitness function (Fig. 4.7B) with two distinct fitness peaks that relate to the two nesting strategies. Thus, we suggest that the evolution and maintenance of these alternative nesting phenotypes are consequences of this complex fitness function.

The impact of high density-dependent mortality at arribada sites, driven by multiple mechanisms just discussed (direct interference, site deterioration, pathogen accumulation), suggests that arribada beaches may actually deteriorate as optimal nesting sites over time, and it is possible that some, perhaps all, arribada beaches are ephemeral. Indeed, a mathematical model relating the degree of reproductive synchrony to the magnitude of density-dependent offspring mortalilty indicates that the reproduction should become more asynchronous as density-dependent offspring mortality increases (Richter, 1999). The brief period during which the existence of arribadas has actually been known (reviewed earlier) makes it impossible to gauge the dynamics of arribada nesting sites through time. Moreover, the existence of more or less continuous variation in the size of nesting populations across beaches indicates that the arribada phenomenon may be an extreme manifestation of a continuous process. For example, data from numerous Pacific beaches of Mexico indicate a continuum of nesting population sizes (Fig. 4.6).

Arribada Evolution and Its Implications: A New Hypothesis

Above we described a hypothetical but realistic scenario for the maintenance of the nesting behavioral polymorphism in *L. olivacea*. However, this scenario does not explain why arribada behavior evolved initially and why it is essentially unique to this genus. In other words, if other turtles were to evolve synchronous nesting, other aggregating nesters might experience a similar complex fitness function.

To solve this quandary, we return to the issue of multiple mating. It is now well established that multiple mating is a general feature of turtle mating systems, which suggests it has an important role in their reproductive success (review: Pearse and Avise, 2001). In the case of ridleys, although oceanic matings occur, the probability of multiple encounters at sea must be extremely small, so the possibility of multiple mating is restricted, as we previously argued. Because multiple mating has fitness implications for both males and females (reviewed above), and because multiple mating is a common feature of turtle mating systems, the ecological specialization by *Lepidochelys* to the pelagic realm may have selected for arribada aggregations to preserve these mating system features, which would otherwise be lost. Even when oceanic mating opportunities arise, the likelihood of mate choice is essentially nil, as minimally this would entail the simultaneous meeting of three highly pelagic individuals. We suggested before that arribada behavior does not occur in leatherbacks, the only other truly pelagic turtle, because they are so much more vagile than *Lepidochelys*. Genetic data show that leatherbacks do obtain many pelagic mating opportunities.

Specific Predictions of the Multiple Mating Hypothesis for Arribada Evolution

Several distinct predictions arise from our hypothesis that the arribada behavior evolved principally through selection for multiple mating. The first is that the average paternity of clutches of females from arribada beaches should exceed that of solitarily nesting females at nonarribada beaches. Importantly, solitarily nesting females at arribada beaches (e.g., stragglers that nest well ahead of or well after the main nesting peak) may not show this difference because they are still nesting at the arribada-nesting beaches at which male densities are very high, so they too will have had opportunities to multiply mate.

A second prediction arises from the large theoretical and empirical literature concerning the genetic benefits to both offspring and maternal fitness that accrue from polyandry and multiple paternity (reviewed above). Specifically, we predict that careful analysis of the relative fitness of hatchlings (e.g., as gauged by performance phenotypes such as crawling or swimming speed, stamina) from arribada beaches should be higher than those hatched on solitary nesting beaches.

Summary

In this chapter we have attempted to summarize as much as possible of what is known about the arribada nesting strategy, its determinants, and its evolutionary significance, both with respect to *Lepidochelys* as well as more generally among animals. We have also drawn attention to the largely ignored alternative, solitary nesting strategy, and we have offered ecological and evolutionary hypotheses to explain its existence and the maintenance of this reproductive polymorphism. Clearly there are more unresolved than resolved issues concerning the biology of *Lepidochelys* reproductive strategies, which are of great interest as a model system to study the evolution of polyandry and of reproductive behavioral polymorphisms. In addition to these general implications, these unresolved issues have largely unexplored implications for the conservation of these species as well. We have offered a number of hypotheses and predictions for future research to address these critical research gaps in the hope of stimulating well-thought-out research programs for future work.

The biology of *Lepidochelys* reproduction has been intriguing since its first description, yet there is a tremendous latent opportunity to use this system to address fundamental issues in behavioral ecology and evolution, evolution of sexual selection, and alternative ecological strategies. The relationship of these behavioral differences to the population genetic structure of ridleys is also an open question. Research programs in these areas will not only make important general contributions to these fields but will provide a more accurate and comprehensive perspective that is essential to the survival of *Lepidochelys* in the increasingly hostile human landscape.

LITERATURE CITED

Acuña-Mesen, R. A. 1983. El exito desarrollo de los huevos de la tortuga marina *Lepidochelys olivacea* Eschscholtz en Playa Ostional, Costa Rica. *Brenesia* 21:371–385.

Acuña-Mesen, R. A. 1988. Influencia del cautiverio, peso y tamaño en la migración de los neonatos de *Lepidochelys olivacea* Eschscholtz (Testudines: Cheloniidae). *Revista Biologia Tropical* 36:97–106.

Acuña-Mesen, R. A., and Castillo, D. 1985. Un obstaculo mas en la migracion hacia el mar de algunas neonatos de la tortuga marina *Lepidochelys olivacea* Eschscholtz. *Brenesia* 23: 323–334.

Alho, C. J. R., and Padua, L. F. M. 1982. Reproductive parameters and nesting behavior of the Amazon turtle *Podocnemis expansa* (Testudinata: Pelomedusidae) in Brazil. *Canadian Journal of Zoology* 60:97–103.

Allee, W. C. 1927. Animal aggregations. *Quarterly Review of Biology* 2:367–398.

Amarasooriya, K. D., and Jayathilaka, M. R. A. 2002. Marine turtle nesting on the beaches of the north-western, western, and southern provinces of Sri Lanka. In Mosier, A., Foley, A., and Brost, B. (Compilers). *Proceedings of the Twentieth Annual Symposium on Sea Turtle Biology and Conservation*. NOAA Technical Memorandum NMFS-SEFSC-477, pp. 93–95.

Andersson, M. 1994. *Sexual Selection*. Princeton, NJ: Princeton University Press.

Andrews, H. V. 2000. Current marine turtle situation in the Andaman and Nicobar Islands—an urgent need for conservation action. *Kachhapa* 3:19–23.

Arauz-Almengor, M., and Mo, C. L. 1994. Hatching rates of olive ridleys in a managed arribada beach, Ostional, Costa Rica. In Bjorndal, K. A., Bolten, A. B., Johnson, D. A., and Eliazar, P. J. (Compilers). *Proceedings of the Fourteenth Annual Symposium on Sea Turtle Biology and Conservation*. NOAA Technical Memorandum NMFS-SEFSC-351, pp. 7–8.

Arnqvist, G., and Nilsson, T. 2000. The evolution of polyandry:multiple mating and female fitness in insects. *Animal Behaviour* 60:145–164.

Asrar, F. F. 1999. Decline of marine turtle nesting populations in Pakistan. *Marine Turtle Newsletter* 83:13–14.

Augspurger, C. K. 1981. Reproductive synchrony of a tropical shrub: experimental studies on effects of pollinators and seed predators on *Hybanthus prunifolius* (Violaceae). *Ecology* 62:775–788.

Aureggi, M., Gerosa, G., and Chantrapornsyl, S. 2004. An update of sea turtle nesting along the

Andaman coast of Thailand, 1996–2000. In Coyne, M. S., and Clark, R. D. (Compilers). *Proceedings of the Twenty-First Annual Symposium on Sea Turtle Biology and Conservation.* NOAA Technical Memorandum NMFS-SEFSC-528, pp. 98–100.

Babcock, R. C., Bull, G. D., Harrison, P. L., Heyward, A. J., Oliver, J. K., Wallace, C. C., and Willis, B. L. 1986. Synchronous spawnings of 105 scleractinian coral species on the Great Barrier Reef. *Marine Biology* 90:379–394.

Ballestero, J. 1996. Weather changes and olive ridley nesting density in the Ostional Wildlife Refuge, Santa Cruz, Guanacaste, Costa Rica. In Keinath, J. A., Barnard, D. E., Musick, J. A., and Bell, B. A. (Compilers). *Proceedings of the Fifteenth Annual Symposium on Sea Turtle Biology and Conservation.* NOAA Technical Memorandum NMFS-SEFSC-387, pp. 26–30.

Barbosa, C., Broderick, A., and Catry, P. 1998. Marine Turtles in the Orango National Park (Bijagós Archipelago, Guinea-Bissau). *Marine Turtle Newsletter* 81:6–7.

Barnett, L. K., Emms, C., Jallow, A., Cham, A. M., and Mortimer, J. A. 2004. The distribution and conservation status of marine turtles in The Gambia, West Africa: a first assessment. *Oryx* 38:203–208.

Bernardo, J. 1996. Maternal effects in animal ecology. *American Zoologist* 36:83–105.

Bernasconi, G., and Keller, L. 2001. Female polyandry affects their sons' reproductive success in the red flour beetle *Tribolium castaneum*. *Journal of Evolutionary Biology* 14:186–193.

Beyer, K. 2002. Investigations into the Reproductive Biology and Ecology of Olive Ridley Turtles (*Lepidochelys olivacea*) Eschscholtz, 1829 at Old Ningo Beach, Ghana, West Africa. Thesis. University of Bremen.

Bollmer, J. L., Irwin, M. E., Rieder, J. P., and Parker, P. G. 1999. Multiple paternity in loggerhead turtle clutches. *Copeia* 1999:475–478.

Botton, M. L., Loveland, R. E., and Jacobsen, T. R. 1988. Beach erosion and geochemical factors: influence on spawning success of horseshoe crab (*Limulus polyphemus*) in Delaware Bay. *Marine Biology* 99:325–332.

Brockmann, H. J. 1990. Mating behavior of horseshoe crabs, *Limulus polyphemus*. *Behaviour* 114: 206–220.

Brockmann, H. J. 2001. The evolution of alternative strategies and tactics. *Advances in the Study of Behavior* 30:1–51.

Brown, L., and Macdonald, D. W. 1995. Predation on green turtle *Chelonia mydas* nests by wild canids at Akyatan Beach, Turkey. *Biological Conservation* 71:55–60.

Burke, V. J., Rathbun, S. L., Bodie, J. R., and Gibbons, J. W. 1998. Effect of density on predation rate for turtle nests in a complex landscape. *Oikos* 83:3–11.

Bustard, H. R., and Tognetti, K. P. 1969. Green sea turtles: a discrete simulation of density-dependent population regulation. *Science* 163: 939–941.

Caldwell, D. K., Berry, F. H., Carr, A., and Ragotzkie, R. A. 1959. Multiple and group nesting by the Atlantic loggerhead turtle. *Bulletin of the Florida State Museum* 4:308.

Carr, A. 1952. *Handbook of Turtles: The Turtles of the United States, Canada and Baja California.* Ithaca, NY: Cornell University Press.

Carr, A. 1967. *So Excellent a Fishe.* Garden City, NY: Natural History Press.

Carr, A. 1986. *New Perspectives on the Pelagic Stage of Sea Turtle Development.* NOAA Technical Memorandum NMFS-SEFSC-190.

Carr, A., and Giovannoli, L. 1957. The ecology and migrations of sea turtles. 2. Results of field work in Costa Rica, 1955. *American Museum Novitates* 1835:1–32.

Carr, A., and Ogren, L. 1960. The ecology and migrations of sea turtles. 4. The green turtle in the Caribbean Sea. *Bulletin of the American Museum of Natural History* 121:1–48.

Carr, A., Hirth, H., and Ogren, L. 1966. The ecology and migrations of sea turtles, 6. The hawksbill turtle in the Caribbean Sea. *American Museum Novitates* 2248:1–29.

Caspers, H. 1984. Spawning periodicity and habitat of the palolo worm *Eunice viridis* (Polychaeta: Eunicidae) in the Samoan Islands. *Marine Biology* 79:229–236.

Caut, S., Hulin, V., and Girondot, M. 2006. Impact of density-dependent nest destruction on emergence success of Guianan leatherback turtles (*Dermochelys coriacea*). *Animal Conservation* 9:189–198.

Chatterji, A. 1994. *The Horseshoe Crab—A Living Fossil.* Orissa, India: Jagannath Process.

Chaves, G., Morera, R., Aviles, J. R., Castro, J. C., and Alvarado, M. In press. Trends of the nesting activity of the "arribadas" of the olive ridley (*Lepidochelys olivacea,* Eschscholtz 1829), in the Ostional National Wildlife Refuge (1971–2005). *Chelonian Conservation and Biology.*

Chevalier, J., and Girondot, M. 1998. Nesting dynamics of marine turtles in French Guiana during the 1997 nesting season. Dynamique de pontes des Tortues luths en Guyane frangaise durant la saison 1997. *Bulletin de la Societe Herpetologique de France* 85–86:5–19.

Chin, C. C. Q., Krishna, R. G., Weldon, P. J., and Wold, F. 1996. Characterization of the disulfide bonds and the N-glycosylation sites in the glycoprotein from Rathke's gland secretions of Kemp's ridley sea surtle (*Lepidochelys kempi*). *Analytical Biochemistry* 233:181–187.

Church, J. 2004. Turtle conservation in Kenya. In Coyne, M. S., and Clark, R. D. (Compilers). *Proceedings of the Twenty-First Annual Symposium on Sea Turtle Biology and Conservation.* NOAA Technical Memorandum NMFS-SEFSC-528, pp. 139–140.

Clayton, D. A. 1978. Socially facilitated behavior. *Quarterly Review of Biology* 53:373–392.

Cliffton, K., Cornejo, D. O., and Felger, R. S. 1982. Sea turtles of the Pacific coast of Mexico. In Bjorndal, K. A. (Ed.). *Biology and Conservation of Sea Turtles.* Washington, DC: Smithsonian Institution Press, pp. 199–209.

Congdon, J. D., and Gibbons, J. W. 1990. Turtle eggs: their ecology and evolution. In Gibbons, J. W. (Ed.). *Life History and Ecology of the Slider Turtle.* Washington, DC: Smithsonian Institution Press, pp. 109–123.

Cornelius, S. E. 1982. Status of sea turtles along the Pacific coast of middle America. In Bjorndal, K. A. (Ed.). *Biology and Conservation of Sea Turtles.* Washington, DC: Smithsonian Institution Press, pp. 211–220.

Cornelius, S. E. 1986. *The Sea Turtles of Santa Rosa National Park.* San José, Costa Rica: Fundacion de Parques Nacionales.

Cornelius, S. E., and Robinson, D. C. 1985. *Abundance, distribution and movements of olive ridley sea turtles in Costa Rica, V.* Final report to U.S. Fish and Wildlife Service, Contract No. 14-16-0002-81-225.

Cornelius, S. E., and Robinson, D. C. 1986. Postnesting movements of female olive ridley turtles tagged in Costa Rica. *Vida Silvestre Neotropical* 1:12–23.

Cornelius, S. E., Ulloa, M. A., Castro, J. C., Del Valle, M. M., and Robinson, D. C. 1991. Management of olive ridley sea turtles nesting at Playas Nancite and Ostional, Costa Rica. In Robinson, J. G., and Redford, K. H. (Eds.). *Neotropical Wildlife Use and Conservation.* Chicago: The University of Chicago Press, pp. 111–135.

Crim, J. L., Spotila, L. D., Spotila, J. R., O'Connor, M., Reina, R., Williams, C. J., and Paladino, F. V. 2002. The leatherback turtle, *Dermochelys coriacea,* exhibits both polyandry and polygyny. *Molecular Ecology* 11: 2097–2106.

Dash, M. C., and Kar, C. S. 1990. *The Turtle Paradise: Gahirmatha (An Ecological Analysis and Conservation Strategy).* New Delhi, India: Interprint Press.

da Silva, A. C. C. D., Comin de Castilhos, J., Rocha, D. A. S., Oliveira, F. L. C., Weber, M. I., and Barata, P. C. R. 2003. Nesting biology and conservation of the olive ridley sea turtle (*Lepidochelys olivacea*) in the state of Sergipe, Brazil. In Seminoff, J. A. (Compiler). *Proceedings of the Twenty-Second Annual Symposium on Sea Turtle Biology and Conservation.* NOAA Technical Memorandum NMFS-SEFSC-503, p. 89.

Dial, B. E. 1987. Energetics and performance during nest emergence and the hatchling frenzy in loggerhead turtles (*Caretta caretta*). *Herpetologica* 43:307–315.

Doussou Bodjrenou, J., Montcho, J., and Sagbo, P. 2004. Challenges and prospects for sea turtle conservation in Benin, West Africa. In Coyne, M. S., and Clark, R. D. (Compilers). *Proceedings of the Twenty-First Annual Symposium on Sea Turtle Biology and Conservation.* NOAA Technical Memorandum NMFS-SEFSC-528, pp. 120–122.

Eberhard, W. G. 1996. *Female Control: Sexual Selection by Cryptic Female Choice.* Princeton, NJ: Princeton University Press.

Eckrich, C. E., and Owens, D. W. 1995. Solitary versus arribada nesting in the olive ridley sea turtle (*Lepidochelys olivacea*): A test of the predator-satiation hypothesis. *Herpetologica* 51:349–354.

Ehrenfeld, D. W. 1979. Behavior associated with nesting. In Harless, M., and Morlock, H. (Eds.). *Turtles: Perspectives and Research.* Malabar, FL: Robert E. Krieger Publishing Co., pp. 417–434.

Ehrenfeld, J. G., and Ehrenfeld, D. W. 1973. Externally secreting glands of fresh water and sea turtles. *Copeia* 1973:305–314.

Ehrhart, L. M. 1982. A review of sea turtle reproduction. In Bjorndal, K. A. (Ed.). *Biology and Conservation of Sea Turtles.* Washington, DC: Smithsonian Institution Press. pp. 29–38.

Emlen, D. J. 1997. Alternative reproductive tactics and male dimorphism in the horned beetle *Onthophagus acuminatus. Behavioral Ecology and Sociobiology* 41:335–341.

Estes, R. D. 1976. The significance of breeding synchrony in the wildebeest. *East African Wildlife Journal* 14:135–152.

Evans, K. E., and Vargas, A. R. 1998. Sea turtle egg commercialization in Isla de Cañas, Panama. In Byles, R., and Fernandez, Y. (Compilers). *Proceedings of the Sixteenth Annual Symposium on Sea Turtle Biology and Conservation.* NOAA Technical Memorandum NMFS-SEFSC-412, p. 45.

Findlay, C. S., and Cooke, F. 1982. Synchrony in the lesser snow goose (*Anser caerulescens caerulescens*): II. The adaptive value of reproductive synchrony. *Evolution* 36:786–799.

FitzSimmons, N. N. 1996. Use of microsatellite loci to investigate multiple paternity in marine turtles. In Bowen, B. W., and Witzell, W. N. (Eds.). *Proceedings of the International Symposium on Sea Turtle Conservation Genetics.* NOAA Technical Memorandum NMFS-SEFSC-396, pp. 69–77.

FitzSimmons, N. N. 1998. Single paternity of clutches and sperm storage in the promiscuous green turtle (*Chelonia mydas*). *Molecular Ecology* 7:575–584.

Foerster, K., Delhey, K., Johnson, A., Lifjeld, J. T., and Kempenaers, B. 2003. Females increase offspring heterozygosity and fitness through extrapair matings. *Nature* 425:714–717.

Foster, S. A. 1999. The geography of behaviour: an evolutionary perspective. *Trends in Ecology and Evolution* 14:190–195.

Foster, S. A., and Endler, J. A. (Eds.). 1999. *Geographic Variation in Behavior: Perspectives and Evolutionary Mechanisms.* New York: Oxford University Press.

Fox, C., and Rauter, C. 2003. Bet-hedging and the evolution of multiple mating. *Evolutionary Ecology Research* 2003:273–286.

Frazier, J. 1976. Sea turtles in Tanzania. *Tanzania Notes and Records* 77–78:11–14.

Frazier, J. G. 1983. Analisis estadistico de la tortuga golfina *Lepidochelys olivacea* (Eschscholtz) de Oaxaca, Mexico. *Ciencia Pesquera* 4:49–75.

Fretey, J., Formia, A., Tomas, J., Dontaine, J.-F., Billes, A., and Angoni, H. 2004. Presence, nesting and conservation of *Lepidochelys olivacea* in the Gulf of Guinea. In Coyne, M. S., and Clark, R. D. (Compilers). *Proceedings of the Twenty-First Annual Symposium on Sea Turtle Biology and Conservation.* NOAA Technical Memorandum NMFS-SEFSC-528, p. 172.

Garant, D., Dodson, J. J., and Bernatchez, L. 2005. Offspring genetic diversity increases fitness of female Atlantic salon (*Salmo salar*). *Behavioral Ecology and Sociobiology* 57:240–244.

Gibbons, J. W., and Greene, J. L. 1991. Reproduction in the slider and other species of turtles. In Gibbons, J. W. (Ed.). *Life History and Ecology of the Slider Turtle.* Washington, DC: Smithsonian Institution Press, pp. 124–134.

Girondot, M., Tucker, A. D., Rivalan, P., Godfrey, M. H., and Chevalier, J. 2002. Density-dependent nest destruction and population fluctuations of Guianan leatherback turtles. *Animal Conservation* 2002(5):75–84.

Gochfeld, M. 1980. Mechanisms and adaptive value of reproductive synchrony in colonial seabirds. In Burger, J., Olla, B. L., and Winn, H. E. (Eds.). *Behavior of Marine Animals: Current Perspectives in Research. Vol. 4. Marine Birds.* New York: Plenum Press, pp. 207–270.

Godley, B. J., Broderick, A. C., Frauenstein, R., Glen, F., and Hays, G. C. 2002. Reproductive seasonality and sexual dimorphism in green turtles. *Marine Ecology Progress Series* 226:125–133.

Gomez, J., Sory, B., and Mamadou, K. 2003. A preliminary survey of sea turtles in the Ivory Coast. In Seminoff, J. A. (Compiler). *Proceedings of the Twenty-Second Annual Symposium on Sea Turtle Biology and Conservation.* NOAA Technical Memorandum NMFS-SEFSC-503, p. 146.

Griffith, S. C., Owens, I. P. F., and Thuman, K. A. 2002. Extra pair paternity in birds: a review of interspecific variation and adaptive function. *Molecular Ecology* 11:2195–2212.

Gross, M. R. 1991. Evolution of alternative reproductive strategies: frequency-dependent sexual selection in male bluegill sunfish. *Philosophical Transactions of the Royal Society of London B* 332:59–66.

Gross, M. R. 1996. Alternative reproductive strategies and tactics: diversity within sexes. *Trends in Ecology and Evolution* 11:92–98.

Harry, J. L., and Briscoe, D. A. 1988. Multiple paternity in the loggerhead turtle (*Caretta caretta*). *Journal of Heredity* 79:96–99.

Hasbún, C. R., and Vásquez, M. 1999. Sea turtles of El Salvador. *Marine Turtle Newsletter* 85:7–9.

Hays, G. C., Houghton, J. D. R., and Myers, A. E. 2004. Pan-Atlantic leatherback turtle movements. *Nature* 429:522.

Hendrickson, J. R. 1980. The ecological strategies of sea turtles. *American Zoologist* 20:597–608.

Hildebrand, H. H. 1963. Hallazgo del area de anidacion de la tortuga marina "lora" *Lepidochelys kempi* (Garman), en la costa occidental del Golfo de Mexico. *Ciencia, Mexico* 22(4):105–112.

Hirth, H. 1980. Nesting biology of sea turtles. *American Zoologist* 20:507–523.

Hoekert, W. E. J., Schouten, A. D., Van Tienen, L. H. G., and Weijerman, M. 1996. Is the Suriname olive ridley on the eve of extinction? First census data for olive ridleys, green turtles and leatherbacks since 1989. *Marine Turtle Newsletter* 75:1–4.

Hoekert, W. E. J., Neufeglise, H., Schouten, A. D., and Menken, S. B. J. 2002. Multiple paternity and female-biased mutation at a microsatellite locus in the olive ridley sea turtle (*Lepidochelys olivacea*). *Heredity* 89:107–113.

Hoinsoude, G. S., Bowessidjaou, J. E., Kokouvi, G. A., Iroko, F., and Fretey, J. 2003. Plan for sea turtle conservation in Togo. In Seminoff, J. A. (Compiler). *Proceedings of the Twenty-Second Annual Symposium on Sea Turtle Biology and Conservation.* NOAA Technical Memorandum NMFS-SEFSC-503, p. 117.

Hughes, D. A., and Richard, J. D. 1974. The nesting of the Pacific ridley turtles *Lepidochelys olivacea* on Playa Nancite, Costa Rica. *Marine Biology* 24:97–107.

Ims, R. A. 1990a. On the adaptive value of reproductive synchrony as a predator-swamping strategy. *American Naturalist* 136:485–498.

Ims, R. A. 1990b. The ecology and evolution of reproductive synchrony. *Trends in Ecology and Evolution* 5:135–140.

Ireland, J. S., Broderick, A. C., Glen, F., Godley, B. J., Hays, G. C., Lee, P. L. M., and Skibinski, D. O. F. 2003. Multiple paternity assessed using microsatellite markers, in green turtles *Chelonia mydas* (Linnaeus, 1758) of Ascension Island, South Atlantic. *Journal of Experimental Marine Biology and Ecology* 291:149–160.

Janzen, D. H. 1974. Why bamboos wait so long to flower. *Annual Review of Ecology and Systematics* 7:347–391.

Janzen, D. H. 1985. Male dogs have fitness. *Biotropica* 17:205.

Jennions, M. D., and Petrie, M. 2000. Why do females mate multiply? A review of the genetic benefits. *Biological Review* 75:21–64.

Jensen, M. P., Loeschcke, V., and Abreu-Grobois, A. In press. The mating system of the Ostional (Costa Rica) olive ridley rookery studied through microsatellites. In *Proceedings of the Twenty-Fifth Annual Symposium on Sea Turtle Conservation and Biology.* Savannah, Georgia, 2005.

Jiménez-Quiroz, M. C., Filonov, F., Tereshchenko, I., and Márquez-M., R. 2005. Time-series analyses of the relationship between nesting frequency of the Kemp's ridley sea turtle and meteorological conditions. *Chelonian Conservation and Biology* 4:774–780.

Juarez, R., and Muccio, C. 1997. Sea turtle conservation in Guatemala. *Marine Turtle Newsletter* 77: 15–17.

Kalb, H. J. 1999. Behavior and physiology of solitary and arribada nesting olive ridley sea turtles (*Lepidochelys olivacea*) during the internesting period. Ph.D. diss. Texas A&M University, College Station, TX.

Kalb, H. J., and Owens, D. W. 1994. Differences between solitary and arribada nesting olive ridley females during the internesting period. In Bjorndal, K. A., Bolten, A. B., Johnson, D. A., and Eliazar, P. J. (Compilers). *Proceedings of the Fourteenth Annual Workshop on Sea Turtle Biology and Conservation.* NOAA Technical Memorandum NMFS-SEFSC-351, p. 68.

Keller, L., and Reeve, H. K. 1995. Why do females mate with multiple males—the sexually selected sperm hypothesis. *Advances for the Study of Behaviour* 24:291–315.

Kichler, K., Holder, M. T., Davis, S. K., Márquez-M., R., and Owens, D. W. 1999. Detection of multiple paternity in the Kemp's ridley sea turtle with limited sampling. *Molecular Ecology* 8:819–830.

Krishna, S. 2004. Sporadic nesting of olive ridley turtles (*Lepidochelys olivacea*) and efforts to conserve them along the coast of Karnataka (South West India). In Coyne, M. S., and Clark, R. D. (Compilers). *Proceedings of the Twenty-First Annual Symposium on Sea Turtle Biology and Conservation.* NOAA Technical Memorandum NMFS-SEFSC-528, pp. 218–219.

Kurdziel, J. P., and Knowles, L. L. 2002. The mechanisms of morph determination in the amphipod *Jassa:* implications for the evolution of alternative male phenotypes. *Proceedings of the Royal Society of London B* 269:1749–1754.

Lagueux, C. J. 1991. Economic analysis of sea turtle eggs in a coastal community on the Pacific coast of Honduras. In Robinson, J. G., and Redford, K. H. (Eds.). *Neotropical Wildlife Use and Conservation.* Chicago: The University of Chicago Press, pp. 136–144.

Lank, D. B., Smith, C. M., Hanotte, O., Ohtonen, A., Bailey, S., and Burke, T. 2002. High frequency of polyandry in a lek mating system. *Behavioral Ecology* 13:209–215.

Leary, T. R. 1957. A schooling of leatherback turtles, *Dermochelys coriacea coriacea,* on the Texas coast. *Copeia* 1957(3):232.

Lee, P. L., and Hays, G. C. 2004. Polyandry in a marine turtle: females make the best of a bad job. *Proceedings of the National Academy of Science USA* 101:6530–6535.

Lehmann, L., and Perrin, N. 2003. Inbreeding avoidance through kin recognition: choosy females boost male dispersal. *American Naturalist* 162: 638–652.

Limpus, C. J. 1975. The Pacific Ridley, *Lepidochelys olivacea* (Eschscholtz) and other sea turtles in northeastern Australia. *Herpetologica* 31:444–445.

Limpus, C. J., Miller, J. D., Parmenter, C. J., and Limpus, D. J. 2003. The green turtle, *Chelonia mydas,* population of Raine Island and the northern Great Barrier Reef: 1843-2001. *Memoirs of the Queensland Museum* 49:349–440.

Márquez-M., R., Carrasco, M. A., Jiménez, M. C., Peñaflores-S., C., and Bravo-G., R. 2005. Kemp's and olive ridley sea turtles population status. In Coyne, M. S., and Clark, R. D. (Compilers). *Proceedings of the Twenty-First Annual Symposium on Sea Turtle Biology and Conservation.* NOAA Technical Memorandum NMFS-SEFSC-528, pp. 237–239.

Martinez, L. M., and Paez, V. P. 2000. Nesting ecology of the olive ridley turtle (*Lepidochelys olivacea*) at La Cuevita, Chocoan Pacific coast, Colombia, in 1998. *Actualidades Biologicas Medellin* 22(73):131–143.

Mauck, A. V. 1901. On the swarming and variation in a myriapod (*Fontaria virginiensis*). *American Naturalist* 35:477–478.

McPherson, D., and D. Kibler. In press. Ecological impact of olive ridley nesting at Ostional, Costa Rica. In *Proceedings of the Twenty-Fifth Annual Symposium on Sea Turtle Conservation and Biology.* Savannah, Georgia, 2005.

Mendonça, M. T., and Pritchard, P. C. H. 1986. Offshore movements of post-nesting Kemp's Ridley sea turtles (*Lepidochelys kempi*). *Herpetologica* 42:373–381.

Meylan, A. 1982. Estimation of population size in sea turtles. In Bjorndal, K. A. (Ed.). *Biology and Conservation of Sea Turtles.* Washington, DC: Smithsonian Institution Press, pp. 135–138.

Miller, J. D. 1997. Reproduction in sea turtles. In Lutz, P. L., and Musick, J. A. (Eds.). *The Biology of Sea Turtles.* Boca Raton, FL: CRC Press, pp. 51–58.

Moczek, A. P., and Emlen, D. J. 2000. Male horn dimorphism in the scarab beetle *Onthophagus taurus:* do alternative tactics favor alternative phenotypes? *Animal Behaviour* 59:459–466.

Moein Bartol, S., and Musick, J. A. 2003. Sensory biology of sea turtles. In Lutz, P. L., Musick, J. A., and Wyneken, J. (Eds.). *The Biology of Sea Turtles,* Volume 2. Boca Raton, FL: CRC Press, pp. 79–102.

Moore, M. K., and Ball, R. M., Jr. 2002. Multiple paternity in loggerhead turtle (*Caretta caretta*) nests on Melbourne Beach, Florida: a microsatellite analysis. *Molecular Ecology* 11:281–288.

Mora, J. M., and Robinson, D. C. 1982. Discovery of a blind olive ridley turtle (*Lepidochelys olivacea*) nesting at Playa Ostional, Costa Rica. *Revista Biologia Tropical* 30:178–179.

Moran, N. A. 1992. The evolutionary maintenance of alternative phenotypes. *American Naturalist* 139:971–989.

Morgan, S. G., and Christy, J. H. 1995. Adaptive significance of the timing of larval release by crabs. *American Naturalist* 145:457–479.

Newcomer, S. D., Zeh, J. A., and Zeh, D. W. 1999. Genetic benefits enhance the reproductive success of polyandrous females. *Proceedings of the National Academy of Sciences USA* 96:10236–10241.

Oliver, J. A. 1946. An aggregation of Pacific turtles (*Lepidochelys olivacea*). *Copeia* 1946(2):103.

Owens, D. W., Grassman, M. A., and Hendrickson, J. H. 1982. The imprinting hypothesis and sea turtle reproduction. *Herpetologica* 38:124–135.

Owens, D. W., Commuzie, D. C., and Grassman, M. 1986. Chemoreception in the homing and orientation behavior of amphibians and reptiles with special reference to sea turtles. In Duvall, D., Muller-Schwarze, D., and Silverstein, R. M. (Eds.). *Chemical Signals in Vertebrates 4.* New York: Plenum Publishing Co., pp. 341–355.

Pandav, B., and Choudhury, B. C. 2000. Conservation and management of olive ridley sea turtle (*Lepidochelys olivacea*) in Orissa. Final Report, Wildlife Institute of India.

Parker, P. G., Waite, T. A., and Peare, T. 1996. Paternity studies in animal populations. In Smith, T. B., and Wayne, R. K. (Eds.). *Molecular Genetic Approaches in Animal Conservation.* New York: Oxford University Press, pp. 413–423.

Peare, T., Parker, P. G., and Irwin, M. E. 1998. Paternity analysis in the green sea turtle. In Byles, R., and Fernandez, Y. (Compilers). *Proceedings of the Sixteenth Annual Symposium on Sea Turtle Biology and Conservation.* NOAA Technical Memorandum NMFS-SEFSC-412, p. 116.

Pearse, D. E., and Avise, J. C. 2001. Turtle mating systems: behavior, sperm storage and genetic paternity. *Journal of Heredity* 92:206–211.

Pearse, D. E., Janzen, F. J., and Avise, J. C. 2002. Multiple paternity, sperm storage, and reproductive success of female and male painted turtles (*Chrysemys picta*) in nature. *Behavioral Ecology and Sociobiology* 51:164–171.

Pearse, D. E., Brigham Dastrup, R., Hernandez, O., and Sites, J. W., Jr. 2006. Paternity in an Orinoco population of endangered Arrau river turtles, *Podocnemis expansa* (Pleurodira; Podocnemididae), from Venezuela. *Chelonian Conservation and Biology* 5(2).

Petrie, M., Doums, C., and Moller, A. P. 1998. The degree of extra-pair paternity increases with genetic variability. *Proceedings of the National Academy of Sciences USA* 95:9390–9395.

Pienaar, J., and Greeff, J. M. 2003. Different male morphs of *Otitesella pseudoserrata* fig wasps have equal fitness but are not determined by different alleles. *Ecology Letters* 6:286–289.

Pitman, R. L. 1990. Pelagic distribution and biology of sea turtles in the eastern tropical Pacific. In Richardson, T. H., Richardson, J. I., and Donnelly, M. (Compilers). *Proceedings of the Tenth Annual Workshop on Sea Turtle Biology and Conservation.* NOAA Technical Memorandum NMFS-SEFC-278, pp. 143–148.

Pizzari, T., and Birkhead, T. R. 2002. The sexually-selected sperm hypothesis: sex-biased inheritance and sexual antagonism. *Biology Reviews* 77:183–209.

Plotkin, P. T. 1994. Migratory and reproductive behavior of the olive ridley turtle *Lepidochelys olivacea* (Eschscholtz, 1829), in the eastern Pacific Ocean. Ph.D. diss. Texas A&M University, College Station, TX.

Plotkin, P. T. 2003. Adult migrations and habitat use. In Lutz, P. L., Musick, J. A., and Wyneken, J. (Eds.). *Biology of Sea Turtles,* Volume 2. Boca Raton, FL: CRC Press, pp. 225–242.

Plotkin, P., and Bernardo, J. 2003. Investigations into the basis of the reproductive behavioral polymorphism in olive ridley sea turtles. In Seminoff, J. A. (Compiler). *Proceedings of the Twenty-Second Annual Symposium on Sea Turtle Biology and Conservation,* NOAA Technical Memorandum NMFS-SEFSC-503, p. 29.

Plotkin, P., Polak, M., and Owens, D. W. 1991. Observations on olive ridley sea turtle behavior prior to an arribada at Playa Nancite, Costa Rica. *Marine Turtle Newsletter* 53:1.

Plotkin, P. T., Byles, R. A., Rostal, D. C., and Owens, D. W. 1995. Independent vs. socially facilitated migrations of the olive ridley, *Lepidochelys olivacea. Marine Biology* 122:137–143.

Plotkin, P. T., Owens, D. W., Byles, R. A., and Patterson, R. 1996. Departure of male olive ridley turtles (*Lepidochelys olivacea*) from a nearshore breeding ground. *Herpetologica* 52:1–7.

Plotkin, P. T., Rostal, D. C., Byles, R. A., and Owens, D. W. 1997. Reproductive and developmental synchrony in female *Lepidochelys olivacea. Journal of Herpetology* 31:17–22.

Pritchard, P. C. H. 1969. Studies of the systematics and reproduction of the genus *Lepidochelys.* Ph.D. diss. University of Florida, Gainesville.

Pritchard, P. C. H. 1979. *Encyclopedia of Turtles.* Neptune, NJ: T.F.H. Publications, Inc.

Pritchard, P. C. H., and Márquez-M., R. 1973. *Kemp's Ridley Turtle or Atlantic Turtle.* IUCN Monograph No. 2, Morges, Switzerland.

Pritchard, P. C. H., and Trebbau, P. 1984. *The Turtles of Venezuela.* Oxford, OH: Society for the Study of Amphibians and Reptiles.

Reguera, P., and Gomendio, M. 2002. Flexible oviposition behavior in the golden egg bug (*Phyllomorpha laciniata*) and its implications for offspring survival. *Behavioral Ecology* 13:70–74.

Reichard, M., Smith, C., and Jordan, W. C. 2004. Genetic evidence reveals density-dependent mediated success of alternative mating behaviours in the European bitterling (*Rhodeus sericeus*). *Molecular Ecology* 13:1569–1578.

Reichart, H. A. 1993. *Synopsis of Biological Data on the Olive Ridley Sea Turtle Lepidochelys olivacea (Eschscholtz, 1829) in the Western Atlantic.* NOAA Technical Memorandum NMFS-SEFSC-336.

Richard, J. D., and Hughes, D. A. 1972. Some observations of sea turtle nesting activity in Costa Rica. *Marine Biology* 16:297–309.

Richards, M. H., von Wettberg, E. J., and Rutgers, A. C. 2003. A novel social polymorphism in a primitively eusocial bee. *Proceedings of the National Academy of Sciences USA* 100:7175–7180.

Richter, T. A. 1999. The effects of density-dependent offspring mortality on the synchrony of reproduction. *Evolutionary Ecology* 13:167–172.

Robinson, D. 1987. Two hypotheses on arribada behavior. In Serino, J. L. (Compiler). *Seventh Annual Workshop on Sea Turtle Biology and Conservation.* Unpublished proceedings, p. 19.

Rostal, D. C., Williams, J. A., and Weldon, P. J. 1991. Rathke's gland secretion by loggerhead (*Caretta caretta*) and Kemp's ridley (*Lepidochelys kempi*) sea turtles. *Copeia* 1991:1129–1132.

Ruiz, G. A. 1994. Sea turtle nesting population at Playa La Flor, Nicaragua: An olive ridley "Arribada" beach. In Bjorndal, K. A., Bolten, A. B., Johnson, D. A., and Eliazar, P. J. (Compilers). *Proceedings of the Fourteenth Annual Symposium on Sea Turtle Biology and Conservation.* NOAA Technical Memorandum NMFS-SEFSC-351, pp. 129–130.

Rutberg, A. T. 1984. Birth synchrony in American bison (*Bison bison*): response to predation or season? *Journal of Mammalogy* 65:418–423.

Sarker, S. 2004. Sea turtle nesting, threat to survival and conservation efforts in Bangladesh. In Coyne, M. S., and Clark, R. D. (Compilers). *Proceedings of the Twenty-First Annual Symposium on Sea Turtle Biology and Conservation.* NOAA Technical Memorandum NMFS-SEFSC-528, pp. 307–308.

Schulz, J. P. 1975. Sea turtles nesting in Surinam. *Zoologische Verhandelingen (Leiden)* 143:3–172.

Shanker, K., and Pilcher, N. J. 2003. Marine turtle conservation in South and Southeast Asia: hopeless cause or cause for hope? *Marine Turtle Newsletter* 100:43–51.

Shanker, K., Pandav, B., and Choudhury, B. C. 2003. An assessment of the olive ridley turtle (*Lepidochelys olivacea*) nesting population in Orissa, India. *Biological Conservation* 115:149–160.

Siaffa, D. D., Aruna, E., and Fretey, J. 2003. Presence of sea turtles in Sierra Leone (West Africa). In Seminoff, J. A. (Compiler). *Proceedings of the Twenty-Second Annual Symposium on Sea Turtle Biology and Conservation.* NOAA Technical Memorandum NMFS-SEFSC-503, p. 285.

Sinervo, B., and Zamudio, K. R. 2001. The evolution of alternative reproductive strategies: fitness differential, heritability, and genetic correlation between the sexes. *Journal of Heredity* 92:198–205.

Skov, M. W., Hartnoll, R. G., Ruwa, R. K., Jude, P., Shunula, J. P., Vannini, M., and Canniccie, S. 2005. Marching to a different drummer: crabs synchronize reproduction to a 14-month lunar-tidal cycle. *Ecology* 86:1164–1171.

Smith, H. G. 2004. Selection for synchronous breeding in the European starling. *Oikos* 105:301–311.

Stancyk, S. E. 1982. Non-human predators of sea turtles and their control. In Bjorndal, K. A. (Ed.). *Biology and Conservation of Sea Turtles.* Washington, DC: Smithsonian Institution Press, pp. 139–152.

Stutchbury, B. J., and Morton, E. S. 1995. The effect of breeding synchrony on extra-pair mating systems in songbirds. *Behaviour* 132:675–690.

Sugg, D. W., and Chesser, R. K. 1994. Effective population sizes with multiple paternity. *Genetics* 137:1147–1155.

Sugg, D. W., Chesser, R. K., Dobson, F. S., and Hoogland, J. L. 1996. Population genetics meets behavioral ecology. *Trends in Ecology and Evolution* 11:338–342.

Suwelo, I. S. 1999. Olive ridley records from South Banyuwangi, East Java. *Marine Turtle Newsletter* 85:9.

Sweeney, B. W., and Vannote, R. L. 1982. Population synchrony in mayflies: A predator satiation hypothesis. *Evolution* 36:810–821.

Tennessen, J. A., and Zamudio, K. R. 2003. Early male reproductive advantage, multiple paternity and sperm storage in an amphibian aggregate breeder. *Molecular Ecology* 12:1567–1576.

Tisen, O. B., and Bali, J. 2002. Current status of marine turtle conservation programmes in Sarawak, Malaysia. In Mosier, A., Foley, A., and Brost, B. (Compilers). *Proceedings of the Twentieth Annual Symposium on Sea Turtle Biology and Conservation.* NOAA Technical Memorandum NMFS-SEFSC-477, pp. 12–14.

Tripathy, B., Shanker, K., and Choudhury, B. C. 2003. Important nesting habitats of olive ridley turtles *Lepidochelys olivacea* along the Andhra Pradesh coast of eastern India. *Oryx* 37:454-463.

Uetz, P., and Etzold, T. 1996. The EMBL/EBI Reptile Database. *Herpetological Review* 27:174—175. <www.reptile-database.org.>

Van Buskirk, J., and Crowder, L. B. 1994. Life-history variation in marine turtles. *Copeia* 1994:66–81.

Vanzolini, P. E. 1967. Notes on the nesting behaviour of *Podocnemis expansa* in the Amazon Valley (Testudines, Pelomedusidae). *Papeis Avulosos de Zoologia* 20(17):191–215.

Vanzolini, P. E. 2003. On clutch size and hatching success of the South American turtles *Podocnemis expansa* (Schweigger, 1812) and *P. unifilis* Troschel, 1848 (Testudines, Podocnemididae). *Anais da Academia Basileira de Ciencias* 75: 415–430.

Vehrencamp, S. L., Koford, R. R., and Bowen, B. S. 1988. The effect of breeding-unit size on fitness components in groove-billed Anis. In Clutton-Brock, T. H. (Ed.). *Reproductive Success: Studies of Individual Variation in Contrasting Breeding Systems.* Chicago: University of Chicago Press, pp. 291–304.

Walker, B. W. 1949. Periodicity of Spawning by the Grunion, *Louresthes tenuis,* an Atherine Fish. Ph.D. diss. University of California, Los Angeles.

Weldon, P. J., and Tanner, M. J. 1990. Lipids in the Rathke's gland secretions of hatchling loggerhead sea turtles (*Caretta caretta*). *Copeia* 1990: 575–578.

Weldon, P. J., Mason, R. T., Tanner, M. J., and Eisner, T. 1990. Lipids in the Rathke's gland secretions of hatchling Kemp's ridley sea turtles (*Lepidochelys kempi*). *Comparative Biochemistry and Physiology* 96B:705–708.

Westneat, D. F. 1992. Nesting synchrony by female red-winged blackbirds: effects on predation and breeding success. *Ecology* 73:2284–2294.

Whiting, S. D. 1997. Observations of a nesting olive ridley turtle in the Northern Territory. *Herpetofauna* 27(2):39-42.

Wilbur, H. M., and Morin, P. J. 1988. Life history evolution in turtles. In Gans, C., and Huey, R. B. (Eds.). *Biology of the Reptilia,* Volume 16. New York: Alan R. Liss, pp. 387–439.

Wilson, E. O. 1975. *Sociobiology.* Boston: Harvard University Press.

Wittenberger, J. F., and Hunt, G. L., Jr. 1985. The adaptive significance of coloniality in birds. In Farner, D. S., King, J. R., and Parkes, K. C. (Eds.). *Avian Biology VIII.* New York: Academic Press, pp. 1–78.

Woodworth, W. McM. 1903. Preliminary report on the "palolo" worm of Samoa, *Eunice viridis* (Gray). *American Naturalist* 37:875–881.

Woodworth, W. McM. 1907. The palolo worm, *Eunice viridis* (Gray). *Bulletin of the Museum of Comparative Zoology* 51:3–21.

Wyneken, J. 2001. *The Anatomy of Sea Turtles.* NOAA Technical Memorandum NMFS-SEFSC-470.

Yasui, Y. 1998. The 'genetic benefits' of female multiple mating reconsidered. *Trends in Ecology and Evolution* 13:246–250.

Zamudio, K. R., and Sinervo, B. 2000. Polygyny, mate-guarding, and posthumous fertilization as alternative male mating strategies. *Proceedings of the National Academy of Sciences USA* 97:14427–14432.

MELISSA L. SNOVER
ALETA A. HOHN
LARRY B. CROWDER
SELINA S. HEPPELL

5

Age and Growth in Kemp's Ridley Sea Turtles

Evidence from Mark–Recapture and Skeletochronology

RIDLEYS ARE THE SMALLEST SEA TURTLES, reaching just 60–70 cm straight carapace length (SCL) as adults (Márquez-M., 1994; Van Buskirk and Crowder, 1994). Captive (head-started) Kemp's ridleys (*Lepidochelys kempii*) show rapid growth in their first year (Caillouet et al., 1995, 1997a), and limited data from wild Kemp's ridleys tagged and released as hatchlings and later recaptured corroborate this (B. Higgins, personal communication). In general, ridleys are on the "fast" end of the growth rate continuum for sea turtles, with the earliest estimated age at sexual maturation (ASM) for any of the hardshell species (Heppell et al., 2003).

Although growth rates are suspected to be fast in Kemp's ridleys as compared to other hardshell sea turtles, they are poikilothermic, and growth rates are inexorably linked to environmental conditions including temperature and resource availability (Bjorndal, 2003). Such a link can create large differences in individual growth rates and size-at-age among individuals in a population, especially for animals that take a decade or more to reach reproductive maturity. Most of what is known of sea turtle growth rates comes from mark–recapture studies (Chaloupka and Musick, 1997). However, with such high variability in sea turtle growth rates, large sample sizes are needed for mark–recapture studies to describe how regional and age-specific variability contribute to the mean growth rates for the species (Heppell et al., 2003).

Mark–recapture is the most common method of collecting growth information on sea turtles, yet there are other methods used to gain information on age and growth, including skeletochronology and studies of captive or recaptured known-age individuals. All of these methods have been applied to Kemp's ridley sea turtles (Caillouet et al., 1995; Schmid and Witzell, 1997; Zug et al., 1997; Snover and Hohn, 2004). In particular, advances in applying skeletochronology to sea turtles (Snover and Hohn, 2004) open the possibility of rapidly accumulating size-at-age and growth rate data as well as individual growth trajectories.

Chaloupka and Musick (1997) present a thorough review of methodologies used to estimate age and growth in sea turtles. In this chapter, we review how mark–recapture and skeletochronology studies of both known-age and unknown-age individuals have been applied to Kemp's ridley sea turtles. We also present the findings of a new skeletochronology study of Kemp's ridleys, focusing on both interindividual and intraindividual variation in growth.

Estimating Age and Growth with Mark–Recapture Data

Background and Methodology

Tagged Kemp's ridleys have come from several coastal in-water projects that captured and tagged wild Kemp's ridleys as well as other sea turtle species (Epperly et al., 1995; Schmid, 1995, 1998). These studies generally have used internal passive integrated transponder (PIT) tags injected into the fore flippers and external Inconel or Jumbo Roto plastic tags attached to the foreflippers (Epperly et al., 1995; Schmid, 1995, 1998). The turtles were measured and tagged at initial capture and measured again when recaptured.

Variability in the recapture interval is a primary problem in generalizing growth rates from mark–recapture data. Growth in Kemp's ridleys is not constant over seasons and is reduced in the winter months (Snover and Hohn, 2004). Hence, if annual growth rates are estimated based on growth over a portion of the summer, growth rates will be overestimated. Similarly, growth will be underestimated if the recapture interval occurred over the winter or early spring. Ideally, only recapture intervals of a year or multiples of a year should be used, but such data are rare. In an extensive mark–recapture study of wild juvenile Kemp's ridleys in Florida, Schmid (1998) noted that both the mean and variance in growth rates decreased with increasing time at large when only recapture intervals greater than 180 days were considered.

A second problem in mark–recapture growth rate studies is the tendency for researchers to exclude negative growth rates from analyses. It seems unlikely that turtles actually shrink in carapace length between captures, so such measurements are obviously caused by measurement errors, and hence the temptation arises to exclude them. However, measurement errors occur in both directions with equal probability, and removing negative growth rates will truncate the error in a dataset, biasing it toward errors that overestimate growth. Hence, all measurement data, including those of negative growth, should be kept in any analyses of mark–recapture data.

Another source of data from tagged Kemp's ridleys results from intensive conservation efforts, including the head-start experiment and coded wire tag (CWT) experiment, from which large numbers of tagged, known-age turtles were released. Both experiments were part of a binational Kemp's Ridley Recovery Program operated by state and federal U.S. and Mexican agencies, including the National Marine Fisheries Service (NMFS), Galveston Laboratory, the Instituto Nacional de la Pesca (INP) of Mexico, the U.S. Fish and Wildlife Service, the National Park Service, Texas Parks and Wildlife Department, Florida Audubon Society, and the Gladys Porter Zoo (Klima and McVey, 1995; Caillouet et al., 1997b). The head-start experiment used a variety of tagging methods for individual identification through adulthood (10–15 years old; Shaver and Wibbels, Chapter 14). Between 1978 and 1992, over 22,000 Kemp's ridleys were released as part of this experiment (Shaver and Wibbels, Chapter 14). In the CWT experiment, CWTs were injected into the fore- and rear flippers of hatchlings, and the turtles were released at the nesting beach in Mexico. The placement of the tags indicates the turtle's year class (Caillouet et al., 1997b; Higgins et al., 1997). The CWT experiment resulted in the release of over

43,000 hatchlings between 1996 and 2000 (B. Higgins, personal communication). These studies developed a unique dataset of actual size-at-age data and the potential for continued collection of these valuable data. With the head-start experiment, caution must be made in applying the resulting size-at-age information from these turtles to the wild population because head-start turtles were captive-raised for a year or more (Shaver and Wibbels, Chapter 14). Consequently, their subsequent growth rates in the wild may not represent normal growth for their ages or sizes.

A common method in analyzing mark–recapture growth data is to fit the capture and recapture lengths and time intervals to a growth interval model based on the von Bertalanffy or logistic growth models (Fabens, 1965; Frazer, 1987; Chaloupka and Musick, 1997). Unfortunately, most mark–recapture growth data sets do not span the entire life cycle, and hence, curves created from those data can only imply average growth trajectories over the size range of data used (Chaloupka and Musick, 1997). Such curves should not be used to estimate ASM unless data from mature animals were in the dataset, although the expected number of years spent in the stage represented by the dataset can be estimated.

Recently, sea turtle researchers have been trying to determine how much variability in growth is attributable to external factors. A common approach taken is to apply generalized additive models (GAM) to growth rates from sea turtle mark–recapture data (Chaloupka and Limpus, 1997; Limpus and Chaloupka, 1997; Seminoff et al., 2002; Balazs and Chaloupka, 2004). The datasets typically include one or multiple growth records from numerous individual turtles. These studies have highlighted the importance of considering variables such as size, year, sex, and recapture interval when analyzing growth rate information because these factors can all contribute to growth rates, even though such GAM analyses have not yet been applied to Kemp's ridley sea turtles.

Estimates of Growth Rates

One of the big questions regarding growth rates in Kemp's ridleys is the difference in growth between juveniles in the Gulf of Mexico and those that feed along the Atlantic coast of the United States. Schmid and Witzell (1997) compared growth data from mark–recapture studies on the east (Atlantic Ocean) and west (Gulf of Mexico) coasts of Florida. Although somewhat inconclusive because of small sample size, their data suggest slower growth rates in the Gulf of Mexico. This observation is based on both estimated an-

Table 5.1 Von Bertalanffy parameter and age at sexual maturity estimates for Kemp's ridleys

Region[a]	Size at maturation	Time to maturation	Asymptotic length	Growth coefficient k	Sample size	Method[b]	Reference
GOM[c]	60.0 cm SCL	10 years	62.3 cm SCL	0.317	117	M-R	Caillouet et al., 1995
Atlantic	—	N.E.[d]	61.1 cm SCL	0.577	12	M-R	Schmid, 1995[e]
Atlantic/GOM[c]	65.0 cm SCL	15.7 years	79.4 cm SCL	0.130	70	S	Zug et al., 1997
GOM[c]	—	N.E.[d]	91.4 cm SCL	0.085	24	M-R	Schmid, 1998[e]
Atlantic	64.0 cm SCL	12–13 years	73.2 cm SCL	0.167	38	M-R, S	Turtle Expert Working Group, 2000
GOM[c]	64.0 cm SCL	10–11 years	71.1 cm SCL	0.210	58	M-R, S	Turtle Expert Working Group, 2000
Atlantic	60.0 cm SCL	10–17 years	74.9 cm SCL	0.115	109	S	Present study

[a]Region where tagging studies occurred or where skeletochronology samples were taken.

[b]M-R indicates mark recapture; S indicates skeletochronology.

[c]Gulf of Mexico.

[d]Not estimated.

[e]Numbers indicate curves where all recapture data were used.

nual growth rates (5.9–8.8 cm/yr Atlantic and 3.6–5.4 cm/yr Gulf of Mexico) and growth modeled using von Bertalanffy equations (Table 5.1).

In a more thorough analysis of Gulf of Mexico Kemp's ridleys, Schmid (1998) reports size-based annual growth rates for Kemp's ridleys at a mean of 4.6 cm/yr for turtles in the 30–40 cm SCL size, 6.2 cm/yr for turtles 40–50 cm SCL, and 4.6 cm/yr for turtles 50–60 cm SCL. For Kemp's ridleys at large for more than 180 days, Schmid (1998) found growth rates ranging from 1.2 to 5.4 cm/yr over all size categories.

In an analysis of head-start recaptures, Fontaine et al. (1989) found extreme differences in growth rate (weight) between Kemp's ridleys recaptured in the Gulf of Mexico and those recaptured in the Atlantic. Growth rates of Kemp's ridleys between the two habitats diverged after 1 year at large (approximately age 2).

Age Estimates

The primary reason to fit a von Bertalanffy growth curve to growth increment data, either from mark–recapture or from skeletochronology, is to approximate mean time to sexual maturity in sea turtles. Such information is critical to studies of population dynamics and recovery rates. For studies in which an appropriate range of values has been included in the model, estimates of Kemp's ridley ASM from mark–recapture studies range from 10 years to 13 years (Table 5.1; Caillouet et al., 1995; Schmid and Witzell, 1997; Turtle Expert Working Group, 2000).

Estimating Age and Growth with Skeletochronology

Background and Methodology

Skeletochronology is the technique of estimating age from growth marks found in cross sections of long bones. This technique has been widely applied to reptiles and amphibians (Castanet, 1994) and was first applied to Kemp's ridley sea turtles by Zug et al. (1997). In current skeletochronology studies of sea turtles, primarily the humerus bone has been used. Zug et al. (1986) determined that this was the optimal bone for observing growth marks in cross section. A site on the humerus just distal to the deltopectoral crest has the largest proportion of compact bone (where growth marks are retained) and allows for a consistent sampling site from bone to bone (Snover and Hohn, 2004). The humerus also has the advantage of being relatively easy to sample from a carcass, which is an important consideration for large-scale sample collection.

The technique of skeletochronology is based on the fact that bone growth is often annually cyclic and that there is a predictable periodicity in which bone formation ceases or slows before new, relatively rapid bone formation resumes (Simmons, 1992; Castanet et al., 1993; Klevezal, 1996). This interruption of bone formation is evidenced within cross sections of the humerus by histological features, which take two forms in decalcified and stained thin sections. The more common feature is a thin line that appears darker than the surrounding tissue, termed the "line of arrested growth" (LAG) (Castanet et al., 1977). The second, less common, feature is a broader and less distinct line that also stains darker, referred to as an annulus (Castanet et al., 1977). Both of these features indicate a slowing or cessation of growth. Alternating with LAG or annuli are broad zones that stain homogeneously light, indicating areas of active bone formation. Together, a broad zone followed by either a LAG or an annulus comprises a skeletal growth mark (Fig. 5.1; Castanet et al., 1993).

Chaloupka and Musick (1997) identify three obstacles to skeletochronology that must be overcome before it can be considered a valid method for age estimation. The first of these is to establish that the observed growth marks are deposited on an annual, or otherwise predictable, basis. In sea turtle bones, there is considerable resorption and remodeling of periosteal tissue resulting in loss of the earliest growth marks. Hence, the other two obstacles identified by Chaloupka and Musick (1997) both deal with methods used to estimate the number of layers lost to resorption, the development of a robust method for modeling the number of growth marks lost, and the validation of a predictable proportionality (either linear or nonlinear) between bone dimensions and body size, in this case, carapace length. An added benefit to establishing a predictable proportionality between

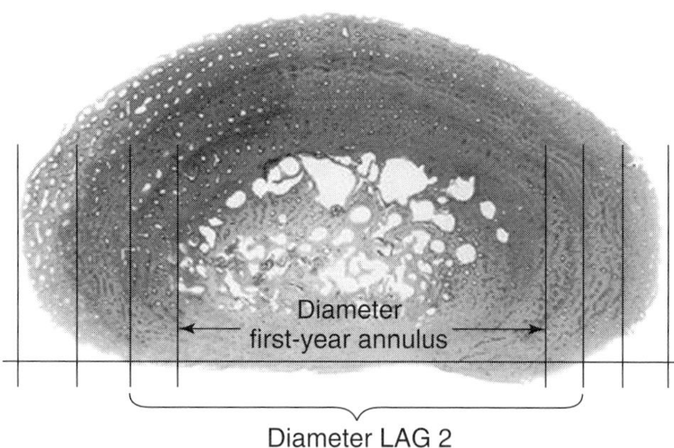

Fig. 5.1. Image of a thin-sectioned and stained Kemp's ridley humerus cross section. This animal was 3.25 years old; age was estimated from counts of the lines of arrested growth (LAGs) and stranding date. Diagram demonstrates how LAG and annulus diameters were measured and ages estimated; a total of four diameters were measured on this humerus, including the full outer diameter. The 1-year growth rate used in the estimation of von Bertalanffy parameters from this animal was the difference between the carapace lengths estimated from the diameters of LAGs 3 and 2. The scale bar represents 1 mm.

bone dimension and body size is the ability to estimate body size from early growth marks (Bjorndal et al., 2003; Snover and Hohn, 2004). With this methodology, individual growth trajectories over periods of years can be assessed. In addition, the methodology allows for the rapid accumulation of size-at-age data with constant (1-year) time intervals.

Validation

Snover and Hohn (2004) analyzed 4 head-start and 13 CWT Kemp's ridleys that had been recovered stranded on beaches. From these turtles, they were able to validate the annual nature of the growth marks. In addition, an analysis of wild Kemp's ridleys demonstrated that growth marks are deposited in the spring (Snover and Hohn, 2004).

For interpretation of the growth marks, the known-age CWT Kemp's ridleys revealed that the first-year growth mark is different from subsequent marks (Snover and Hohn, 2004). Characterization of the first-year mark allowed us to distinguish it from supplemental marks that would otherwise interfere with age assignments. Closely spaced LAGs have been interpreted differently in skeletochronology studies (see review in Snover and Hohn, 2004). Snover and Hohn (2004) confirmed that in Kemp's ridleys, LAGs that appear very close to one another in humerus cross sections represent distinct years and indicate years of little or no growth.

Snover and Hohn (2004) also established that there is a constant proportionality between dimensions of the humerus and carapace length. Regressions of seven measurements of the humerus with carapace length for Kemp's ridleys reveal strong correlations. The best metric is between carapace length (L) and diameter of the humerus (D) at the sectioning site for skeletochronology. This relationship is described by ($r^2 = 0.96$, $P < 0.005$)

$$L = 2.48 \cdot D + 2.74 \tag{5.1}$$

Application to Kemp's Ridleys

Although Zug et al. (1997) were the first to apply skeletochronology to Kemp's ridleys, they were not able to offer any validation to justify that the growth marks observed in cross sections of the humerus were annual and correctly interpreted. Contrary to the findings of Schmid and Witzell (1997), the authors found some evidence that Kemp's ridleys in the Atlantic grow more slowly than similar-sized Kemp's ridleys in the Gulf of Mexico but cautioned that the size range of samples from the two regions may confound this interpretation. They estimated ASM between 11 years and 16 years.

In a further analysis of the Kemp's ridley size-at-age data from Zug et al. (1997), Chaloupka and Zug (1997) identified a potential polyphasic relationship between size and age that could be indicative of an ontogenetic shift in growth rates, where individual movement to new feeding areas, a shift in diet, or physiological changes may increase growth rates for particular age classes. The authors used nonparametric smoothing techniques to visualize the functional form of the size-at-age data and subsequently fit a polyphasic parametric function to the data. The resulting curve suggests different phases of growth, with a slowing of growth rates followed by a surge in growth rates. Parametric growth curves such as the von Bertalanffy, logistic, and Gompertz cannot detect these different growth compartments because these functions model growth rate with increasing length as either linearly declining or hump-shaped. Such curves will not detect multiple ontogenetic shifts in growth rates; however, the resulting estimate of ASM may be comparable for both parametric and nonparametric curve-fitting techniques.

Using the findings from the validation study of Snover and Hohn (2004), we were able to apply skeletochronology to Kemp's ridleys with unknown histories. For this study, our samples were obtained primarily from the mid-Atlantic states of the United States. To expand our size range and sample size of large subadults and adults, we included samples from the Gulf of Mexico that were in these size categories.

Methods

We received front flippers from dead Kemp's ridleys stranded along the coasts of North Carolina, Virginia, and Maryland. Because Kemp's ridleys are listed as endangered under the U.S. Endangered Species Act, sacrificing individuals for aging studies is inconsistent with conservation efforts; hence, all samples were collected from stranded turtles found dead or that died after attempts at rehabilitation. For most of the turtles, SCL was recorded, measured as standard straight-line length from the nuchal notch to the posterior end of the posterior marginal. If only curved carapace length (CCL) was recorded, it was converted to SCL using the regression equation ($r^2 = 0.99$; $P < 0.001$)

$$\mathrm{SCL} = 0.957 \cdot \mathrm{CCL} - 0.696 \tag{5.2}$$

This equation was generated from 309 paired SCL and CCL measurements of Kemp's ridleys between 18.4 cm and 66.2 cm SCL.

We used a subsample of 144 Kemp's ridleys between 21.7 cm and 50.5 cm SCL and for which age could be estimated with a high degree of confidence for this study. The sample contained 21 males and 54 females with sex confirmed by visual examination of the gonads during necropsy. The sex of the remaining 69 specimens was not determined. An additional 13 samples between 50.6 cm and 62.0 cm SCL were used to estimate growth rates in large individuals for completion of a growth curve. These turtles were recovered from mid-Atlantic states and Texas, and because of resorption, accurate ages could not be assigned to them.

A complete description of the histological methods for bone preparation is presented by Snover and Hohn (2004). We took a digital image of a stained cross section from each bone and compared the digital image to the actual cross section viewed under a stereoscopic microscope. Each LAG or annulus was identified on the digital image, and the total number of LAGs and annuli were quantified to assign age (Fig. 5.1). Kemp's ridley nests hatch between late May and mid-July (Márquez-M., 1994). Such a narrow range of hatch dates allowed for partial years to be incorporated in age assignments as follows: turtles that died from September to November were assigned ages equal to the LAG count plus a quarter year; from December to February, LAG count plus a half year; March to May, LAG count plus three-quarters of a year; and June to August were assigned integer age numbers equal to the number of LAGs.

On the digital images of each of the 144 specimens, we measured the diameter of every LAG for which it was possible to measure the full lateral LAG diameter. Measurements were made of LAG diameters on the lateral axis of the bone, parallel to the dorsal edge (Fig. 5.1). The first growth mark in a Kemp's ridley appeared as a diffuse annulus that was not always clearly visible (Fig. 5.1; Snover and Hohn, 2004). When the boundary of the mark could be clearly distinguished, this annulus was measured at the outermost points of the darkly stained area (Fig. 5.1). Because this mark is not always clearly visible, we were not able to measure it on every bone. Using the digital images from the 13 large turtles for which age could not be assigned, we measured the diameters of the two outermost LAGs.

To estimate carapace lengths from LAG diameters, we used a back-calculation technique from the fisheries literature, the body proportional hypothesis (BPH) (Francis, 1990), incorporating the relationship between humerus diameter and carapace length (equation 5.1):

$$L_i = (2.48) \cdot D_i + 2.74)(L_{\text{final}})/ (2.48 \cdot D_{\text{final}} + 2.74) \quad (5.3)$$

Here L_i is the estimated carapace length at the time the *i*th LAG was deposited, D_i is the diameter of the *i*th LAG, and L_{final} and D_{final} are the carapace length and humerus diameter at death. Final sample sizes for each age and for each sex varied (Table 5.2). In total, we had 319 back-calculated lengths at age for 144 aged turtles.

All of the resulting back-calculated lengths and one-year growth increments from the turtles were used to determine the mean and variance in size-at-age and age-specific growth rates. To fit a growth curve to the data, we used only one yearly growth increment from each turtle (including the 13 large turtles), estimated from the outermost two LAGs or LAG and annulus, so as not to bias the curve with multiple measurements from individual turtles (Fig. 5.1).

Because not every bone had two measurable LAGs or annuli, the resulting sample size for the 1-year growth interval per turtle was 109. The von Bertalanffy growth interval equation was fit to the data from 109 pairs of carapace lengths using the following equation (Fabens, 1965):

$$L_1 = L_\infty (L_\infty L_2)e^{-k} \quad (5.4)$$

In this equation, L_1 and L_2 represent the carapace length estimated from the two outermost LAG diameters, with L_1 being the last LAG and L_2 the second-to-last LAG. Asymptotic length and intrinsic rate of growth are represented by L_∞ and *k*, respectively. Once the L_∞ and *k* parameters were solved for by the curve fit, the curve was plotted with size as a function of age using the von Bertalanffy growth function (VBGF; von Bertalanffy, 1938):

$$L_x = L_\infty - (L_\infty - L_0)e^{-kx} \quad (5.5)$$

In this equation, *x* is age in years and L_0 is initial size at hatching. Chaloupka and Musick (1997) caution against substituting parameters estimated from a growth-at-size function to a growth-at-age function. Here we are able to compare how well the size-at-age VBGF (equation 5.5), using parameters estimated from the growth-at-size VBGF (equation 5.4), describe actual size-at-age data. Assuming a mean length at first reproduction of 60 cm SCL (Turtle Expert Working Group, 2000), we use the results of the VBGF to approximate ASM. In addition, we fit cubic smoothing splines through the data. These smoothers do not assume an a priori parametric function such as the VBGF but rather fit a curve through data locally; at any one point, the curve depends only on the data at that point and specified neighboring points. As a result, smoothers can highlight underlying trends in data, such as polyphasic growth, that are lost when parametric functions that specify a certain shape are fit (Chaloupka and Zug, 1997). We visually compare the fitted VBGF with the cubic smoothing splines.

Table 5.2 Resulting sample sizes for back-calculated lengths in the skeletochronology study

Age (years)	Size-at-age (cm) All	Size-at-age (cm) Male	Size-at-age (cm) Female	Growth increment (cm) All	Growth increment (cm) Male	Growth increment (cm) Female
1	90	14	28	56	13	26
2	105	19	50	66	13	30
3	68	14	29	30	6	14
4	30	6	14	16	4	6
5	16	4	6	7	2	3
6	7	2	3	2	1	0
7	2	1	0	1	1	0
8	1	1	0	0	0	0

Note: Size-at-age values indicate the number of turtles for which we could back-calculate a length at that age. Growth increments indicate the number of turtles for which we could estimate the amount of growth for the year following the assigned age from back-calculating length the following year.

Results and Discussion

Size-at-Age

The size-at-age data for the 144 Kemp's ridleys for which age could be assigned with a high degree of confidence corresponded well to the size-at-age data where carapace length was estimated from LAG diameters (Fig. 5.2). On the basis of the average size of Kemp's ridley at hatching (4.2 cm SCL; van Buskirk and Crowder, 1994), growth from age 0 to age 1 appeared to be a separate phase from growth after age 1 (mean = 16.9 cm ± 2.3 SD). A function such as the VBGF that models growth rates as declining monotonically from birth to adult will not describe such a shift in growth rates. Therefore, we set L_0 in equation 5.5 at 21 cm SCL, which was calculated as the average size at age 1 from the back-calculation results (Table 5.3). This value compares favorably with the mean length at age for the 1-year-old turtles with age assigned (Table 5.3; $n = 11$; mean = 23.5 cm SCL). The mean of the age-assigned turtles is larger than the back-calculated mean because these age-assigned turtles were between 1.0 and 1.5 years old.

The VBGF was fit to growth rate data only, similar to the type of data collected from mark–recapture studies with the exception that all time intervals are 1 year. When the parameters from the growth-at-size VBGF were applied to the size-at-age VBGF, the resulting curve coincided well with the estimated size-at-age data (Fig. 5.2). The cubic smoothing spline indicated a relatively constant increase in length-at-age and coincided closely with the fitted VBGF (Fig. 5.2), indicating that VBGFs generated from appropriate growth rate data can be used to estimate average size-at-age. However, neither curve is capable of capturing the high degree of variability in growth rates that can occur. This type of variability can be determined only through the examination of actual size-at-age data.

When mean length-at-age was considered, mean lengths increased in a relatively linear, or constant, manner up to age 6 (Fig. 5.3). After age 6, sample sizes are too small to make any inferences (Table 5.2). Though not a perfect fit, the VBGF described the trend in increasing mean length-at-age relatively well, although it appears to overestimate length in older turtles (Fig. 5.3). It is possible that the older turtles for which age could be assigned were growing more slowly than the average in the population as the interior of the humerus was not yet remodeled to an extent where all of the LAG could not be identified.

One benefit of fitting a VBGF to growth data that span all lengths is to use the curve to estimate ASM. Age at sexual maturity is very poorly understood in sea turtles and yet is critical to studies of population dynamics. Hence, if reasonable approximations of this parameter can be generated from VBGF, it is important to do so. In this case we had growth data ranging from 14 cm

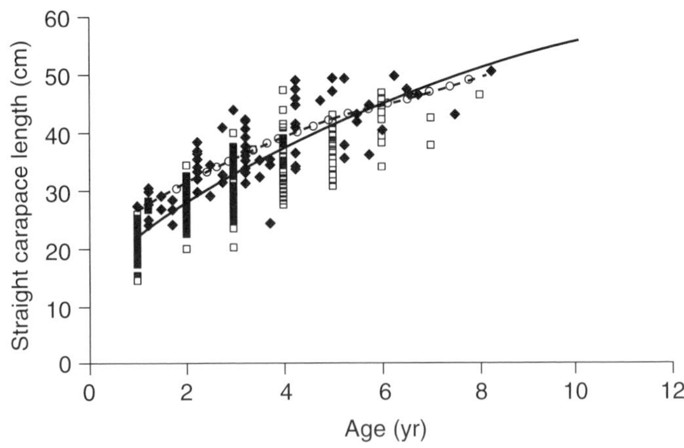

Fig. 5.2. Relationship between carapace length and age for Kemp's ridley sea turtles (*Lepidochelys kempii*). Filled diamonds are the length and skeletochronology age estimates for 144 stranded Kemp's ridleys recovered along the mid-Atlantic coast of the United States. Open squares are the back-calculated carapace lengths from all LAGs that could be measured on 144 humerus cross sections. See Table 5.2 for sample sizes. The solid line is the von Bertalanffy growth curve prepared by using the last full-year growth increment from 109 humerus cross sections, including increments from large turtles consistent with length at sexual maturity. The solid line with open circles is a cubic smoothing spline fit through the 144 size-at-age data.

Table 5.3 Mean length-at-age of Kemp's ridleys using three different methods in the skeletochronology study

	Length of age–estimated turtles (SCL)			Length estimated from back-calculation of LAG diameters			von Bertalanffy model fit		
Age (years)	*n*	Mean length (cm)	SD	*n*	Mean length (cm)	SD	Mean length (cm)	95% CI	
0–1.0	11	23.5	2.08	90	20.9	2.1	—	—	—
1.25–2.0	31	27.6	2.08	105	26.7	2.6	26.9	26.4	27.2
2.25–3.0	37	35.4	3.05	68	30.1	3.7	32.1	31.2	32.9
3.25–4.0	36	36.5	3.71	30	34.1	5.2	36.7	35.4	38.0
4.25–5.0	15	42.0	5.42	16	36.1	3.8	40.9	39.1	42.6
5.25–6.0	9	41.5	4.53	7	41.2	4.4	44.6	42.4	46.8
6.25–7.0	3	48.0	1.62	2	40.1	3.3	47.9	45.2	50.6
7.25–8.0	1	43.1	—	1	46.5	—	50.8	47.7	54.0
8.25–9.0	1	50.5	—	0			53.4	49.9	57.1
10.0							55.8	51.8	59.9
11.0							57.8	53.5	62.5
12.0							59.7	55.0	64.8
13.0							61.3	56.3	66.8

Note: Age ranges in the first column represent assigned ages from lines at growth (LAG) counts estimated to the nearest 0.25 year. Ages associated with the back-calculated lengths and the von Bertalanffy model fit are integer values. Age-estimated turtles are turtles for which age could be assigned by direct count of LAGs. Back-calculated lengths are back-calculated from LAG diameters. The last three columns are the results of the von Bertalanffy growth function (VBGF) with the upper and lower 95% confidence intervals around the fitted parameters.

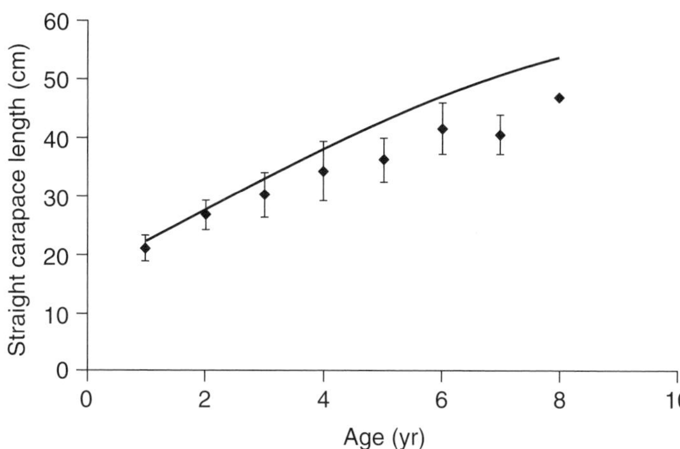

Fig. 5.3. Relationship between mean carapace length and age for Kemp's ridley sea turtles. Data were generated from back-calculated lengths at age based on LAG diameters (see Table 5.2 for sample sizes). Error bars indicate standard error. The solid line is the von Bertalanffy growth curve prepared by using the last full-year growth increment from 109 humerus cross sections, including increments from large turtles consistent with length at sexual maturity.

SCL to 61 cm SCL with which to fit the curve. Size of adult Kemp's ridleys is defined as greater than 60 cm SCL based on data collected from nesting females (Turtle Expert Working Group, 2000). Based on the 95% confidence intervals of the VBGF curve, this size was reached between age 9.9 and 16.7 years, with a mean of 12.0 years.

We used the size-at-age data from the 144 aged Kemp's ridleys to estimate age based on length for Kemp's ridley (Table 5.4). The probabilities were calculated directly from the number of turtles of a given age in each size class by using all of the back-calculated lengths. Because growth rates are linked to environmental conditions, these results are applicable only to Kemp's ridleys from the mid-Atlantic coast of the United States. We also estimated mean lengths at age from each of the three methods used in this study: direct count of LAGs, back-calculated length at age, and the size-at-age VBGF (Table 5.3).

Table 5.4 Probability of age given straight carapace length (SCL) for Kemp's ridleys from the Atlantic coast of the United States

	Age (years)						
Size range (cm)	1	2	3	4	5	6	7
15–19.99	0.90	0.07	0.03	0	0	0	0
20–24.99	0.67	0.30	0.03	0	0	0	0
25–29.99	0.01	0.63	0.30	0.06	0	0	0
30–34.99	0	0.18	0.48	0.21	0.11	0.02	0
35–39.99	0	0	0.23	0.32	0.32	0.09	0.04
40–44.99	0	0	0.1	0.3	0.3	0.2	0.1

Note: These data are based on 144 stranded Kemp's ridleys for which carapace length was recorded, and age could be assigned from direct counts of lines of arrested growth (LAGs).

Growth Rates

We were able to assign ages to turtles from 1 to 8 years. Within that range, the variability in size at age made it difficult to detect a trend in growth rates at age (Fig. 5.4A). However, when mean growth rates at age were considered, growth after age 1 was high, followed by reduced growth at age 2. Mean growth rates appeared relatively constant from age 3 to age 5 (Fig. 5.4B).

As with the size-at-age data, there was a great deal of variability in annual growth rate at size (Fig. 5.5), with a declining trend as length at sexual maturity is approached (Fig. 5.5). Hence, the derivative of the size-at-age VBGF described the data well. The cubic smoothing spline through the data indicated more dynamics in the growth rates of small turtles than the VBGF and was likely a result of the decrease in growth rates noted between ages 2 and 3 (Fig. 5.4B).

In their first year of life, Kemp's ridleys from the mid-Atlantic grew an average of 16.7 cm SCL based on a mean hatchling length of 4.2 cm (Van Buskirk and Crowder, 1994). Subsequent annual growth rates are much lower, and we found a mean of 4.4 cm/yr ± 0.3 SE for 20–30 cm SCL, 5.1 cm/yr ± 0.5 SE for 30–40 cm SCL, and 5.0 cm/yr ± 1.0 SE for 40–50 cm SCL.

Individual Growth Patterns

One striking result from the plotted size-at-age data is the high level of variability that can occur in size at a given age (Fig. 5.2). Bjorndal et al. (2003) suggest that compensatory growth in small juvenile loggerheads (*Caretta caretta*) minimizes variability in size-at-age. Compensatory growth is a mechanism by which individuals that are small for their age as a result of poor resources or environmental conditions exhibit rapid growth when exposed to more favorable conditions, "catching up" in size to their conspecifics that have experienced more favorable conditions (Bjorndal et al., 2003). One way Bjorndal et al. (2003) demonstrated compensatory growth had occurred was by comparing length at age with the amount of subsequent annual growth. They compared the slopes of regression lines through each age class to a regression line through all age classes. When the regression line through an age class declined more steeply than the regression through all of the data (exclusive of the data from the age class being compared), compensatory growth was inferred. As in this study, Bjorndal et al. (2003) back-calculated lengths from LAG dimensions to assess length at age. The researchers found that the slopes through the first two ages of loggerheads were significantly different from the slope through all of the data, which indicates that turtles that were small for their age grew more than turtles that were large for their age. From this they inferred that during the early posthatchling and juvenile years, loggerheads exhibit compensatory growth.

We applied the same method to our Kemp's ridley data, looking first at all of the data and then at a single cohort for which there was a

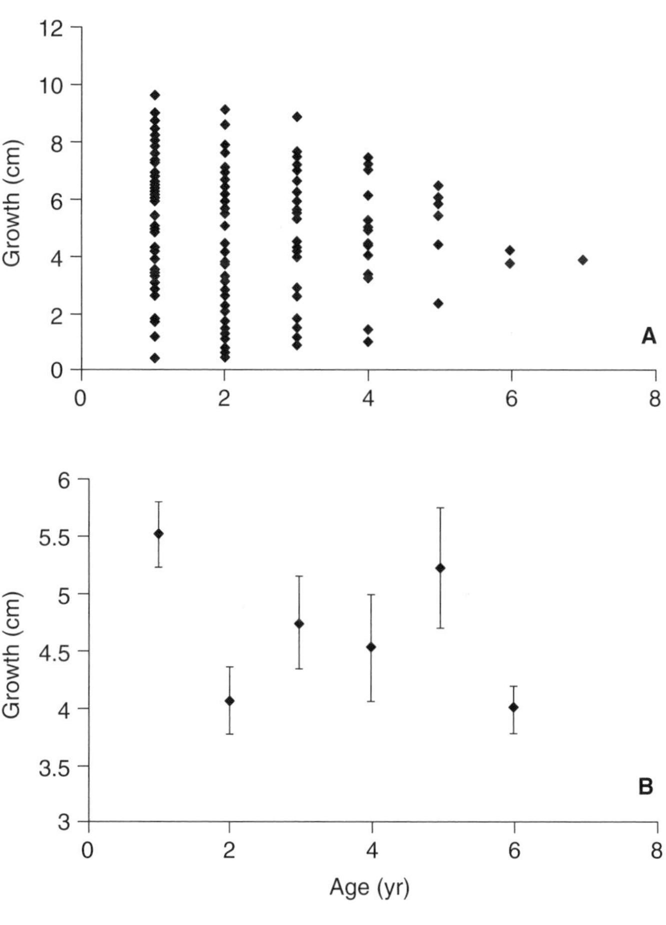

Fig. 5.4. Relationship between age and the growth increment that occurred the following year. All data are from the back-calculated lengths at age based on LAG diameters (n = 178): (A) all of the data; (B) mean and standard error of the data.

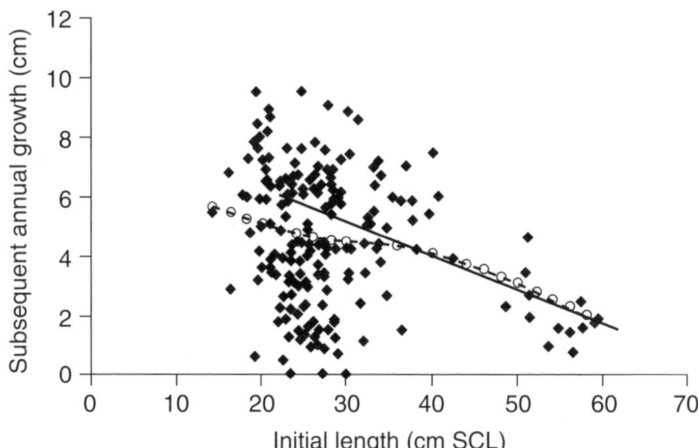

Fig. 5.5. Relationship between initial length and the subsequent amount of annual growth (n = 191). Data are from all of the back-calculated lengths at age based on LAG diameters. The solid line is the derivative of the von Bertalanffy growth curve prepared by using the last full-year growth increment from 109 humerus cross sections, including increments from large turtles consistent with length at sexual maturity. The curved line with open circles is a cubic smoother through the data.

large sample size (Fig. 5.6). None of the regressions was significantly different from zero, and only one showed a declining trend (Fig. 5.6). In addition, the slope of the single declining trend was not significantly different from the slope of the regression through ages 2, 3, and 4 combined ($P > 0.20$) (Zar, 1999). To ensure that these results were not confounded by interannual variability in growth rates, we looked at just the cohort from 1997, for it had a larger sample size, even though it was comprised only of age classes 1 and 2. Slopes of regression lines through both

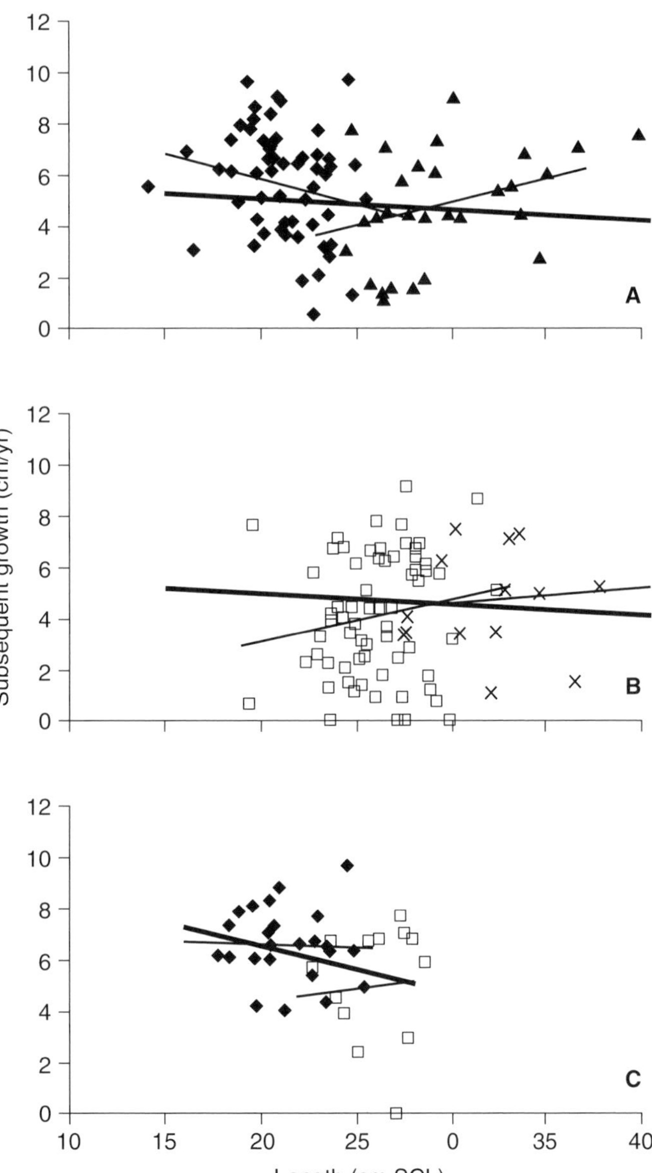

Fig. 5.6. Relationships between initial length at age and the subsequent annual growth. (A) Ages 1 (n = 55) and 3 (n = 29) for all of the data from the back-calculated lengths at age. Thick solid line is the regression through all of the data (ages 1–4; r^2 = 0.006; P = 0.30); thin lines are the regressions through age 1 (r^2 = 0.05; P = 0.11) and age 3 (r^2 = 0.11; P = 0.20). (B) Ages 2 (n = 65) and 4 (n = 16) for all of the data from the back-calculated lengths at age. Thick solid line is the regression through all of the data (same as shown in a); thin lines are the regressions through age 2 (r^2 = 0.03; P = 0.20) and age 4 (r^2 = 0.01; P = 0.65). (C) Ages 1 (n = 25) and 2 (n = 13) for the 1997 cohort. Thick solid line is the regression through all of the data (ages 1 and 2; r^2 = 0.08; P = 0.07); thin lines are the regressions through age 1 (r^2 = 0.003; P = 0.78) and age 2 (r^2 = 0.006; P = 0.79).

ages 1 and 2 were positive, although not significantly different from 0 (Fig. 5.6).

Hence, we found no indication of compensatory growth in Kemp's ridleys that would act as a mechanism to reduce variability in size at age. This apparent difference in early growth between loggerheads and Kemp's ridleys is evidenced by the greater overlap in size-at-age that we observed in Kemp's ridleys as compared to that observed by Bjorndal et al. (2003) for similar-aged loggerheads.

Inflection points in individual growth curves are usually related to ontogenetic shifts, either behavioral or physiological (von Bertalanffy, 1960; Eisen, 1975; Atchley, 1984). Surges in growth rates often coincide with ontogenetic shifts to more profitable habitats (Werner and Gilliam, 1984; Gilliam and Fraser, 1988), and a decline in the profitability of a habitat for a given size class results in increased variance in size-at-age and growth rates (Lomnicki, 1988). In a reanalysis of the age data from Zug et al. (1997), Chaloupka and Zug (1997) fit a polynomial regression to the size-at-age data and detected a polyphasic nature in growth rates through time. With our larger data set, we could not identify a similar polyphasic growth pattern that was consistent among individuals (Figs. 5.3 and 5.7A–C). It was relatively common for Kemp's ridleys in our study to experience decreased growth between ages 2 and 3, as evidenced by the lower mean growth rate for that age (Fig. 5.4B) as compared to the spread of the data (Fig. 5.4A). The only time zero growth was detected was between 2 years and 3 years. However, decreased growth between those ages was not observed in all turtles (Fig. 5.4A). Growth trajectories for turtles that had at least 4 years of data demonstrate that decreased growth between ages 2 and 3 is common but not universal (Fig. 5.7). Growth after age 3 most often appeared to be linearly increasing; however, some turtles displayed decreased growth over a year, giving the appearance of polyphasic growth (Fig. 5.7). These shifts in growth rates did not

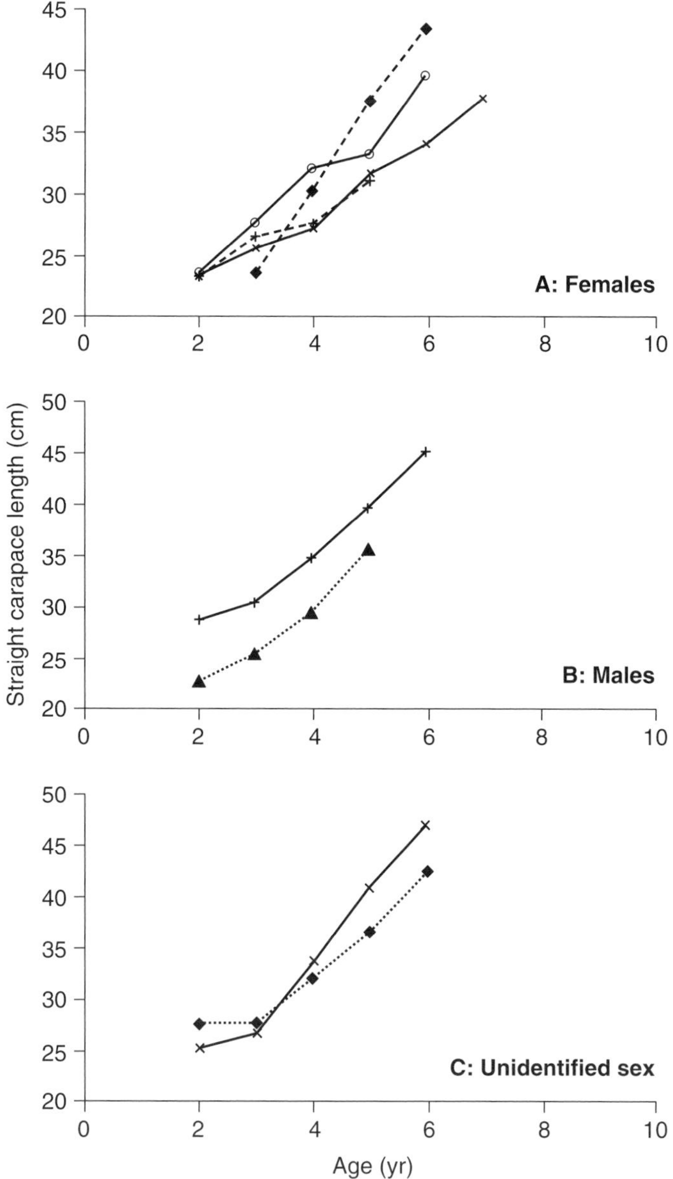

Fig. 5.7. Individual growth trajectories of Kemp's ridley sea turtles based on skeletochronology data. Profiles of growth trajectories are shown for turtles that had at least four LAGs that could be measured. Carapace lengths were back-calculated based on LAG diameters.

Fig. 5.8. Relationship between mean length and age for male (dashed line) and female (solid line) Kemp's ridley sea turtles based on skeletochronology data. Bars on data points indicate standard error.

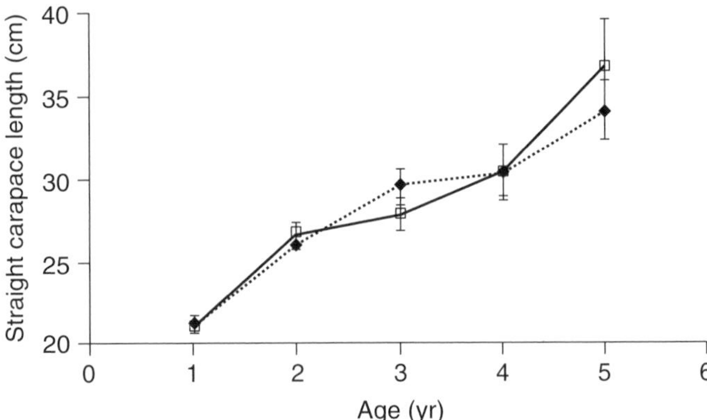

appear to occur at a specific size or age in these turtles. When the data were considered by sex, mean lengths at age suggest polyphasic growth trajectories with different ages for the growth rate shift (Fig. 5.8). These generalities are not corroborated by the individual growth trajectories of males and females (Fig. 5.7); the apparent polyphasic growth profile may be a function of small sample size.

Conclusions and Future Research Needs

Although size-at-age is poorly understood in sea turtles, more is known about the Kemp's ridley than any of the other sea turtle species, especially with the addition of these new skeletochronology data. Our advanced understanding of Kemp's ridley age and growth resulted from extensive conservation efforts, including the head-start experiment and coded wire tag experiment (Klima and McVey, 1995; Caillouet et al., 1997a, 1997b; Shaver and Wibbels, Chapter 14), which released over 65,000 tagged, known-age Kemp's ridleys into the wild. Recaptured turtles from these experiments allowed direct analyses of valuable size-at-age data (Caillouet et al., 1995). Turtles recovered as strandings from both of these experiments have proven vital in validation studies for skeletochronology techniques that can then be applied to untagged turtles (Snover and Hohn, 2004).

It is interesting to note that the mean estimated length of 1-year-old Kemp's ridleys from this study is very comparable to, and slightly higher than, the mean length of 1-year-old captive Kemp's from the head-start experiment (21 cm SCL versus 19.5 cm SCL) (Caillouet et al., 1997a). It is usually thought that sea turtle growth rates in captivity are much greater than what would be expected in the wild (Caillouet et al., 1995; Chaloupka and Musick, 1997). However, in the case of the first year of life for Kemp's ridleys, apparently this is not the case.

That growth rates and size-at-age are highly variable in sea turtles has long been assumed. The skeletochronology work from this study highlights how extreme the variability can be. Posthatchling Kemp's ridleys follow the currents of the Gulf of Mexico and either remain within the Gulf of Mexico or enter the Florida Current and, eventually, the Gulf Stream (Collard, 1990; Collard and Ogren, 1990). Subsequently, juvenile Kemp's ridleys use the coastal benthic habitats of either the Gulf of Mexico or the Atlantic Ocean. This study focused on stranded Kemp's ridleys from the mid-Atlantic coast region of the United States, and the growth rates and potential ASM were representative of small to large juvenile turtles that entered the Gulf Stream. We observed substantial variability in annual growth rates in turtles within the same habitat. Habitats of the Gulf of Mexico versus the mid- and northern Atlantic coast will present different environments for growth, including temperature, and some difference in size-at-age between the two habitats is likely.

For a more complete understanding of Kemp's ridleys' life history, similar analyses are needed for juvenile Kemp's ridleys that strand in the Gulf of Mexico. There was no substantial difference between our skeletochronology growth rates for juveniles from the mid-Atlantic and those reported by Schmid (1998) for the Gulf of Mexico: 5.1 cm versus 4.6 cm for 30–40 cm SCL (Atlantic versus Gulf of Mexico) and 5.0 cm versus 6.2 cm for 40–50 cm SCL. Because growth rates of head-start recaptures were reported in weights, it is difficult to compare the results of this study with those of head-start recaptures from the Atlantic coast (Fontaine et al., 1989). So the question remains: Are growth rates in the Gulf of Mexico higher, lower, or similar to growth rates along the Atlantic coast for Kemp's ridleys of similar sizes and ages?

Our estimate of ASM between 9.9 and 16.7 years coincides with previous estimates of 10–15.7 years, which derived from both mark–recapture and skeletochronology studies (Table 5.1) as well as the ages of head-start Kemp's ridleys found nesting (10–15 years; Shaver and Wibbels, Chapter 14). It is not yet known what proportion of Kemp's ridley hatchlings remain in the Gulf of Mexico and what proportion emigrate to the Gulf Stream. Therefore, to gain a complete understanding of growth and ASM for the population as a whole, information on what fraction of a cohort is utilizing each habitat is necessary. It is possible that recaptures and recoveries of turtles from the individual cohort releases of the CWT experiment will shed light on this question. Results of this study emphasize the value of integrating skeletochronology with other techniques for a better understanding of the life history of the Kemp's ridley and of sea turtles in general.

Some of the individual growth trajectories demonstrated polyphasic growth, but others did not. Whether these shifts in growth rates are caused by extrinsic (environmental) factors or intrinsic (genetic or developmental) factors or both remains unclear. When separated by sex, mean growth rates appeared to decline between ages 2 and 3 in males and between ages 3 and 4 in females (Fig. 5.8); however, observations of sex-specific individual growth trajectories (Fig. 5.7) suggest that this pattern is not universal. The accumulation of additional individual growth trajectories from turtles in both the Gulf of Mexico and the Atlantic will help to clarify any general patterns in growth rates that are related to sex and environment.

Skeletochronology has proven to be a valuable tool in understanding growth and size-at-age in Kemp's ridley sea turtles, thanks in large part to the availability of known-age turtles. These encouraging findings suggest that we should investigate similar techniques for other species, including the olive ridley (*Lepidochelys olivacea*), in order to increase our understanding of growth rates and ASM in sea turtles.

ACKNOWLEDGMENTS

We thank L. Avens, A. Read, D. Rittschof, and two anonymous reviewers for their valuable comments on earlier versions of this manuscript. A. Gorgone, B. Brown, and J. Weaver provided assistance with the preparation of turtle humeri. Most of the humeri were received through the Sea Turtle Stranding and Salvage Network, a cooperative endeavor among the National Marine Fisheries Service, other federal and state agencies, many academic and private entities, and innumerable volunteers. Special thanks go to R. Boettcher and W. Teas for their assistance with the sample collections. In addition, humeri were received from B. Higgins at the National Marine Fisheries Service, Galveston Lab, the Virginia Marine Science Museum Stranding Program, the Maryland Department of Natural Resources, and the Massachusetts Audubon Society in Wellfleet. Funding was provided by the National Marine Fisheries Service. All work was done under and complied with the provisions of Sea Turtle Research Permit TE-676379-2 issued by the U.S. Fish and Wildlife Service. S.S.H. was supported in part through the Oregon Agricultural Experiment Station under project ORE00102 and a contract from the National Marine Fisheries Service, Southeast Fisheries Science Center.

LITERATURE CITED

Atchley, W. R. 1984. Ontogeny, timing of development, and genetic variance-covariance structure. *American Naturalist* 123:519–540.

Balazs, G. H., and Chaloupka, M. Y. 2004. Spatial and temporal variability in somatic growth of green sea turtles (*Chelonia mydas*) resident in the

Hawaiian Archipelago. *Marine Biology* 145: 1043–1059.

Bjorndal, K. A. 2003. Roles of loggerhead sea turtles in marine ecosystems. In Bolten, A. B., and Witherington, B. E. (Eds.). *Loggerhead Sea Turtles.* Washington, DC: Smithsonian Institution Press, pp. 235–254.

Bjorndal, K. A., Bolten, A. B., Dellinger, T., Delgado, C., and Martins, H. R. 2003. Compensatory growth in oceanic loggerhead sea turtles: response to a stochastic environment. *Ecology* 84:1237–1249.

Caillouet, C. W., Jr., Fontaine, C. T., Manzella-Tirpak, S. A., and Williams, T. D. 1995. Growth of head-started Kemp's ridley sea turtles (*Lepidochelys kempi*) following release. *Chelonian Conservation Biology* 1:231–234.

Caillouet, C. W., Jr., Fontaine, C. T., Williams, T. D., and Manzella-Tirpak, S. A. 1997a. Early growth in weight of Kemp's ridley sea turtles (*Lepidochelys kempii*) in captivity. *Gulf Research Reports* 9:239–246.

Caillouet, C. W., Jr., Robertson, B. A., Fontaine, C. T., Williams, T. D., Higgins, B. M., and Revera, D. B. 1997b. Distinguishing captive-reared from wild Kemp's ridleys. *Marine Turtle Newsletter* 77:1–6.

Castanet, J. 1994. Age estimation and longevity in reptiles. *Gerontology* 40:174–192.

Castanet, J., Meumier, F. J., and de Ricqles, A. 1977. L'enregistrement de la croissance cyclique par le tissu osseux chez les vertébrés poïkilothermes: données comparatives et essai de synthése. *Bulletin Biologique de la France et de la Belgique* 111:183–202.

Castanet, J., Francillon-Viellot, H., Meunier, F. J., and De Ricqles, A. 1993. Bone and individual aging. In Hall, B. K. (Ed.). *Bone,* Volume 7: *Bone Growth—B.* Boca Raton, FL: CRC Press, pp. 245–283.

Chaloupka, M. Y., and Limpus, C. J. 1997. Robust statistical modeling of hawksbill sea turtle growth rates (southern Great Barrier Reef). *Marine Ecology Progress Series* 146:1–8.

Chaloupka, M. Y., and Musick, J. A. 1997. Age, growth, and population dynamics. In Lutz, P. L., and Musick, J. A. (Eds.). *The Biology of Sea Turtles.* Boca Raton, FL: CRC Press, pp. 233–276.

Chaloupka, M. Y., and Zug, G. R. 1997. A polyphasic growth function for the endangered Kemp's ridley sea turtle, *Lepidochelys kempii. Fishery Bulletin* 95:849–856.

Collard, S. B. 1990. The influence of oceanographic features in post-hatching sea turtle distribution and dispersion in the pelagic environment. In Richardson, T. H., Richardson, J. I., and Donelly, M. (Compilers). *Proceedings of the 10th Annual Workshop on Sea Turtle Biology and Conservation.* NOAA Technical Memorandum NMFS-SEFC-278, pp. 111–114.

Collard, S. B., and Ogren, L. H. 1990. Dispersal scenarios for pelagic post-hatchling sea turtles. *Bulletin of Marine Science* 47:233–243.

Eisen, E. J. 1975. Results of growth curve analysis in mice and rats. *Journal of Animal Science* 42: 1008–1023.

Epperly, S. P., Braun, J., and Veishlow, A. 1995. Sea turtles in North Carolina waters. *Conservation Biology* 9:384–394.

Fabens, A. J. 1965. Properties and fitting of the von Bertalanffy growth curve. *Growth* 29:265–289.

Fontaine, C. T., Manzella, S. A., Williams, T. D., Harris, R. M., and Browning, W. J. 1989. Distribution, growth and survival of head started, tagged and released Kemp's ridley sea turtles (*Lepidochelys kempi*) from year-classes 1978–1983. In Caillouet, C. W., Jr., and Landry, A. M., Jr. (Eds.). *Proceedings of the First International Symposium on Kemp's Ridley Sea Turtle Biology, Conservation and Management.* Texas A & M University Sea Grant College Program, Galveston, TAMU-SG-89-105, pp. 124–144.

Francis, R. I. C. C. 1990. Back-calculation of fish length: a critical review. *Journal of Fish Biology* 36:883–902.

Frazer, N. B. 1987. Preliminary estimates of survivorship for wild juvenile loggerhead sea turtles (*Caretta caretta*). *Journal of Herpetology* 21:232–235.

Gilliam, J. F., and Fraser, D. F. 1988. Resource depletion and habitat segregation by competitors under predation hazard. In Ebenman, B., and Persson, L. (Eds.). *Size-Structured Populations.* New York: Springer-Verlag, pp. 173–184.

Heppell, S. S., Snover, M. L., and Crowder, L. B. 2003. Sea turtle population ecology. In Lutz, P. L., Musick, J. A., and Wyneken, J. (Eds.). *The Biology of Sea Turtles,* Volume 2. Boca Raton, FL: CRC Press, pp. 275–306.

Higgins, B. M., Robertson, B. A., and Williams, T. D. 1997. *Manual for Mass Wire Tagging of Hatchling Sea Turtles and the Detection of Internal Wire Tags.* NOAA Technical Memorandum NMFS-SEFSC-402.

Klevezal, G. A. 1996. *Recording Structures of Mammals: Determination of Age and Reconstruction of Life History.* Rotterdam: A. A. Balkema.

Klima, E. F., and McVey, J. P. 1995. Headstarting the Kemp's ridley turtle, *Lepidochelys kempi.* In Bjorndal, K. A. (Ed.). *The Biology and Conservation of Sea Turtles.* Washington, DC: Smithsonian Institution Press, pp. 481–487.

Limpus, C., and Chaloupka, M. Y. 1997. Nonparametric regression modeling of green sea turtle growth rates (southern Great Barrier Reef). *Marine Ecology Progress Series* 149:23–34.

Lomnicki, A. 1988. *Population Ecology of Individuals.* Princeton, NJ: Princeton University Press.

Márquez-M., R. 1994. *Synopsis of Biological Data on the Kemp's Ridley Turtle,* Lepidochelys kempi *(Garman, 1880).* NOAA Technical Memorandum NMFS-SEFSC-343.

Schmid, J. R. 1995. Marine turtle populations on the east-central coast of Florida: results of tagging studies at Cape Canaveral, Florida, 1986–1991. *Fishery Bulletin* 93:139–151.

Schmid, J. R. 1998. Marine turtle populations on the west-central coast of Florida: results of tagging studies at the Cedar Keys, Florida, 1986–1995. *Fishery Bulletin* 96:589–602.

Schmid, J. R., and Witzell, W. N. 1997. Age and growth of wild Kemp's ridley turtles (*Lepidochelys kempi*): cumulative results from tagging studies in Florida. *Chelonian Conservation Biology* 2:532–537.

Seminoff, J. A., Resindiz, A., Nichols, W. J., and Jones, T. T. 2002. Growth rates of wild green turtles (*Chelonia mydas*) at a temperate foraging area in the Gulf of California, México. *Copeia* 2002:610–617.

Simmons, D. J. 1992. Circadian aspects of bone biology. In Hall, B. K. (Ed.). *Bone,* Volume 6: *Bone Growth—A.* Boca Raton, FL: CRC Press, pp. 91–128.

Snover, M. L., and Hohn, A. A. 2004. Validation and interpretation of annual skeletal marks in loggerhead (*Caretta caretta*) and Kemp's ridley (*Lepidochelys kempii*) sea turtles. *Fishery Bulletin* 102:682–692.

Turtle Expert Working Group. 2000. *Assessment Update for the Kemp's Ridley and Loggerhead Sea Turtle Populations in the Western North Atlantic.* NOAA Technical Memorandum NMFS-SEFSC-444.

Van Buskirk, J., and Crowder, L. B. 1994. Life-history variation in marine turtles. *Copeia* 1994: 66–81.

von Bertalanffy, L. 1938. A quantitative theory of organic growth (inquiries on growth laws II). *Human Biology* 10:181–213.

von Bertalanffy, L. 1960. Principles and theory of growth. In Nowinski, W. (Ed.). *Fundamental Aspects of Normal and Malignant Growth.* Amsterdam: Elsevier, pp. 137–259.

Werner, E. E., and Gilliam, J. F. 1984. The ontogenetic niche and species interactions in size-structured populations. *Annual Review of Ecology and Systematics* 15:393–425.

Zar, J. H. 1999. *Biostatistical Analysis, Fourth Edition.* Upper Saddle River, NJ: Prentice Hall, pp. 360–364.

Zug, G. R., Wynn, A. H., and Ruckdeschel, C. 1986. Age determination of loggerhead sea turtles, *Caretta caretta,* by incremental growth marks in the skeleton. *Smithsonian Contributions to Zoology* 427:1–34.

Zug, G. R., Kalb, H. J., and Luzar, S. J. 1997. Age and growth in wild Kemp's ridley sea turtles *Lepidochelys kempii* from skeletochronolog-ical data. *Biological Conservation* 80:261–268.

KRISTINA KICHLER HOLDER
MARK T. HOLDER

6

Phylogeography and Population Genetics

SEA TURTLES CAN BE HARD TO STUDY. They are morphologically conservative; it is hard to tell what population an individual belongs to simply by looking at it. The animals spend the overwhelming majority of their lives at sea, where they are inaccessible for direct observation. Hatchlings are quite small and grow considerably before reaching adulthood, when, if they are females, they reappear on the nesting beach. This makes it difficult to effectively tag hatchlings in order to monitor the early years of their lives and difficult to learn much at all about males. For all these reasons, genetic analyses have become an important tool of sea turtle biology. DNA provides us with information about the identity of individuals and the evolutionary history of species. In the study of the genus *Lepidochelys,* genetic analyses have been used in projects with broadly divergent goals, such as determining the taxonomic status of a species or ascertaining the mating system of a population. Genetic analyses have had a great impact on the conservation of the species by strengthening the position of Kemp's ridleys (*L. kempii*) on the endangered species list, improving our understanding of the genetic diversity of nesting populations and how best to protect this diversity, increasing our knowledge of critical behaviors, and enabling forensic scientists to assist efforts to enforce restrictions on the trade in endangered species.

Phylogenetics

Morphological (Zangerl et al., 1988) and molecular analyses (Bowen et al., 1993; Dutton et al., 1996) have recognized the ridleys as the sister group to loggerheads (*Caretta caretta*), but the species status of the Kemp's ridley has received a great deal of attention. Are Kemp's ridleys and olive ridleys separate species? The fact that they can be discriminated based on morphology is insufficient, by itself, to warrant species status. It has been proposed, based on morphology, that black turtles (*Chelonia agassizi*) are a sister species to green turtles (*Chelonia mydas*). Molecular evidence, however, has clearly indicated that Pacific green turtle populations and black turtles are more closely related to each other than to Atlantic green turtle populations (Karl and Bowen, 1999). The same situation could exist with ridleys. Kemp's ridleys may be phylogenetically nested within olive ridleys instead of being a sister taxon to all olive ridleys. If that is the case, are Kemp's ridleys more closely related to east Pacific or Atlantic olive ridley populations?

Bowen et al. (1991) used restriction site analyses of mitochondrial DNA (mtDNA) to address the species status of *L. kempii*. That study included samples from two olive ridley populations, an eastern Pacific (Costa Rican) population and an Atlantic (Suriname) population. Atlantic and Pacific loggerhead populations were used as outgroups. They detected a low level of polymorphism within the Kemp's ridleys, whereas the olives were monomorphic. Although there is no level of sequence divergence at which groups are automatically elevated to species status, the level of divergence between Kemp's and olive ridleys ($p = 1.2 \pm 0.3\%$) exceeded the within-species divergences that had been observed in global surveys of loggerheads (Bowen et al., 1991) and green turtles (Bowen et al., 1989). Subsequent DNA sequencing (Bowen et al., 1998) of the highly variable mitochondrial control region has resulted in estimates of the divergence between the ridley species ($p = 5.2$–6.9%) that fall within the range of intraspecific divergences for other sea turtle species. The maximum divergences between control region haplotypes within single species are 4.4% (Encalada et al., 1996) in *C. mydas* (Dutton et al., 1996, report 7% for a longer sequence from the control region) and 5% in loggerheads (Encalada et al., 1998). Dutton et al. (1999) report a much lower maximum divergence between haplotypes (1.4%) for *Dermochelys coriacea*. On the other end of the spectrum, the Brazilian populations of *Eretmochelys imbricata* display haplotypes that are 10–12% divergent from the haplotypes found in the Caribbean (Bass et al., 1996). Although the recent comparisons of the levels of sequence divergence do not strengthen the case for full species status for the Kemp's ridley, analyses of both the cytochrome *b* gene (Bowen et al., 1993) and the control region (Bowen et al., 1998) indicate reciprocal monophyly of Kemp's and olive ridley mitochondrial genomes (i.e., all of the Kemp's haplotypes group together on the phylogeny and are sister to all of the olive ridley haplotypes). In addition, the sequence divergence between Kemp's and olives is more than twice as large as the deepest divergence seen within the olives. Both of these results are consistent with the species designation for the Kemp's ridley.

Phylogeography and Population Genetics

In addition to helping to delineate species boundaries, molecular genetic analyses can be used to answer many questions about the biogeographic history of a species. What are the evolutionary relationships between olive ridley nesting populations? How have they become globally distributed, and how long ago did it happen? Is gene flow, via migration, sufficiently restricted that these populations are evolving independently? If not, which populations are connected into evolutionary units?

The answers to these questions have far-reaching conservation implications. Because of the limited means available for funding conservation programs, it is necessary to prioritize which species and which populations will receive the most help. It is easier to argue the importance of protecting a genetically distinct population than one that appears to be a random sampling of individuals from a widely distributed species.

Phylogeography

In one of the earliest discussions of the origin of the two ridley species, Pritchard (1969) proposed that the Isthmus of Panama separated olive ridleys from Kemp's ridleys and that olive ridleys later expanded their range westward, only recently colonizing the Atlantic by way of the Cape of Good Hope. The presence of a single genotype in both the East Pacific and Atlantic populations of olive ridleys analyzed by Bowen et al. (1991) was consistent with Pritchard's hypothesis of recent colonization. However, the same result could have been obtained by migration between long-established populations.

To address ridley biogeography more thoroughly, Bowen et al. (1998) constructed a phylogeny of haplotypes of the mtDNA control region. Samples were taken from Kemp's ridleys and seven olive ridley nesting beaches: three from the Atlantic Ocean (Suriname, Brazil, Guinea-Bissau), one eastern Pacific population (Costa Rica), and three from the Indian Ocean (northern Australia, Malaysia, and Sri Lanka). Subsequent work by Shanker et al. (2004) provided data from the major nesting sites along the east coast of India (Orissa and Madras). No samples from Indian Ocean beaches west of Sri Lanka have been sequenced. These studies revealed a total of 4 *L. kempii* and 17 *L. olivacea* haplotypes. Figure 6.1 shows an estimate of the maximum likelihood phylogeny for all ridley haplotypes with branch lengths estimated independently and under a molecular clock (the molecular clock cannot be rejected for these data, using the Akaike (1973) Information Criteria.

As in the previous studies (Bowen et al., 1991, 1993), the olive ridleys are monophyletic with respect to the Kemp's ridleys. On the basis of a molecular clock of 2% sequence divergence per million years (taken from a study of green turtles; Encalada et al., 1996), the split between Kemp's and olive ridleys dates to approximately 2.5–3.5 million years BP (Bowen et al., 1998). This is consistent with Pritchard's hypothesis that the separation of the ancestral ridley populations was caused by the closure of the Isthmus of Panama (Pritchard, 1969).

Within olive ridleys, there appear to be two primary lineages (Fig. 6.1), which diverged approximately 1–1.5 million years ago (based on the aforementioned molecular clock). All of the mitochondrial sequences seen in olive ridleys appear to be recently derived from either haplotype J or haplotype K4. From the molecular clock of 2% divergence per million years, the haplotypes within each of the two lineages diverged approximately 200,000–400,000 years BP (Bowen et al., 1998; Shanker et al., 2004). Haplotype J was seen only in the Indian–West Pacific populations (Bowen et al., 1998), but descendants of this haplotype are found worldwide. In fact, all of the olives from the Atlantic and the eastern Pacific have mitochondrial haplotypes that are derived from the J allele. Haplotypes from the K4 clade have been observed only in India and Sri Lanka (Bowen et al., 1998; Shanker et al., 2004). Thus, the center of haplotype diversity for olive ridleys appears to be the Indian–West Pacific populations.

Bowen et al. (1998) pointed out that their data supported another aspect of Pritchard's (1969) explanation of ridley biogeography: The Atlantic populations of olive ridleys appear to be the result of recent immigration from the Indian Ocean. All of the individuals in the three Atlantic populations displayed haplotype F or a haplotype (E) that is derived from F by a single substitution. The F haplotype has not been found in any other population, but it appears to be a descendant of haplotype J, which is found exclusively in the Indian Ocean.

The presence of Kemp's ridleys in the Gulf of Mexico implies that *Lepidochelys* was in the eastern Pacific when the Isthmus of Panama closed, but the olives that currently nest in Costa Rica display haplotype N or one of its descendants (Bowen et al., 1998). Haplotype N appears to be a form of haplotype J with two additional mutations. As mentioned above, the Atlantic populations of olives are almost fixed for haplotype F, which is also a descendant of J. Thus, in about 400,000 years, one maternal lineage appears to have spread from the Indian or western Pacific Ocean and become fixed in the eastern Pacific and the Atlantic. The relative climatic stability of the region (Kotilainen and Shackleton, 1995) may be a reason that it has served as a source for all mitochondrial haplotypes found in the world (Bowen et al., 1998; Shanker et al., 2004). However, with

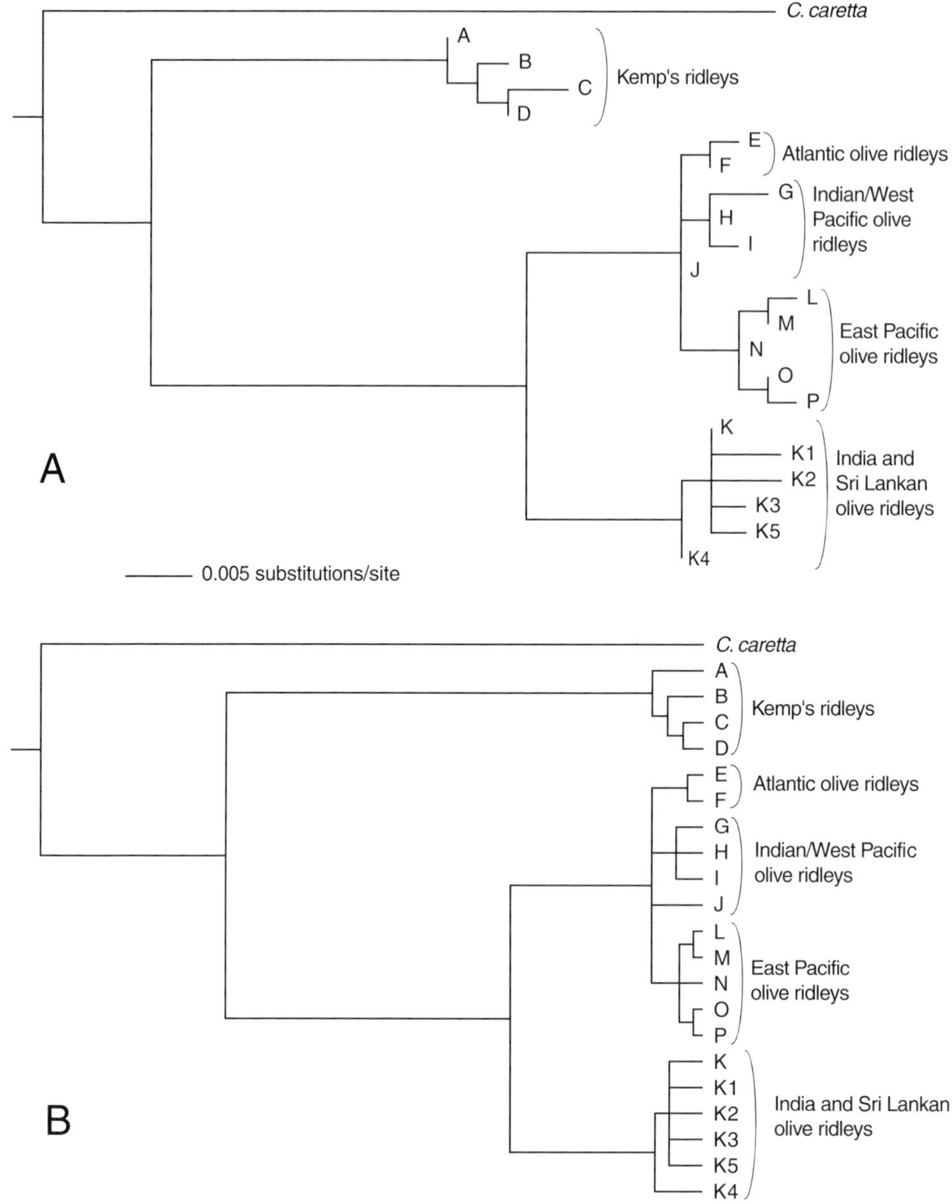

Fig. 6.1. A phylogenetic tree, inferred using maximum likelihood, of the 4 Kemp's ridley haplotypes (A–D) and 17 olive ridley haplotypes (E–P, K1–K5) that have been reported by Bowen et al. (1998) and Shanker et al. (2004). Geographic regions associated with the haplotypes are indicated (note that Shanker et al. [2004] have also observed one individual of haplotype N in India). Branch lengths are drawn in proportion to their maximum-likelihood estimates under the Hasegawa-Kishino-Yano (HKY) + G (Hasegawa et al., 1985; Yang, 1994) model of sequence evolution as implemented in PAUP*4.0b10 (Swofford, 2002). (A) Branch length estimates when the molecular clock is not enforced (these are the conditions under which the tree search was performed). (B) Branch length estimates of the same tree under a molecular clock. The Kemp's ridley haplotypes form a monophyletic group with respect to the olive ridleys, supporting the species designation of *L. kempii*. The olive ridley haplotypes are divided into two clusters of closely related haplotypes that appear to be descendants of haplotype J and haplotype K4.

data from only one locus, it is not possible to determine whether migrants from the Indian Ocean established the nesting populations in the Atlantic and recolonized the eastern Pacific or if they have simply interbred with existing populations.

Figure 6.2 depicts an evolutionary explanation for the observed distribution of haplotypes. Note that the relative timing of some events in the figure is actually unknown. For example, the figure depicts migration of the J haplotype into the eastern Pacific followed by the acquisition of two mutations to convert it to the N haplotype. Under this scenario the individual of haplotype N observed in India by Shanker et al. (2004) would represent another migration event (not shown on the figure) back to the Indian Ocean. Another plausible scenario would involve the N haplotype existing first in the Indian Ocean and then migrating to the eastern Pacific (followed by the haplotype's extinction in most Indian Ocean populations). Again, data at additional loci are needed to distinguish between these possibilities.

An effect of the maternal inheritance of mitochondrial genes is a shorter coalescence time when compared to nuclear genes. If nuclear alleles prove to have deeper divergences than what has been observed at the mitochondrial genes, they would provide additional molecular information on that period of evolutionary history following the initial separation of the ridley species that can only be speculated on at present. As Figure 6.2 indicates, the history of olive ridley populations before the coalescence point for the mitochondrial genome (including the distribution of ridleys when the Kemp's and olives diverged) is currently a mystery.

In any case, the populations in Sri Lanka and eastern India are of particular conservation interest. The Sri Lankan population is the only sampled population in the world with appreciable proportions of both the J and K4 clades (Bowen et al., 1998). The Indian populations are dominated by alleles from the K4 clade, but three individuals from the J clade were also found (two individuals with the J haplotype and one with the N haplotype that had been known only from Costa Rica). Shanker et al. (2004) argue that the Indian population merits recognition as an evolutionarily significant unit (sensu Moritz, 1994). Certainly the preservation of the Indian and Sri Lankan populations is of particular importance in maintaining the full range of mitochondrial genetic diversity in olive ridleys.

Population Genetics

Olive ridleys show strong evidence of population subdivision, with low estimates of migration rates ($Nm < 1$) between most pairs of populations in the Bowen et al. (1998) study (note that the populations from Guinea-Bissau and Malaysia were excluded from the calculation of migration rates because of the small number of individuals sampled). The exception to this pattern was the pair of populations in Suriname and Brazil; these populations were almost monomorphic for the F haplotype. The high value for Nm (4.5) calculated for this pair of populations could reflect either the recent establishment of the nesting populations in the Atlantic or a high level of migration (Bowen et al., 1998). The intensive sampling by Shanker et al. (2004) found no subdivision along 2,000 km of the East Indian coast. The K haplotype dominated all of these populations, although the low-frequency K3 haplotype was found in three of the four nesting beaches studied. Their conclusion that the entire eastern coast of India represents a single population is also supported by microsatellite (Aggarwal et al., 2006) and tagging data (Pandav, 2000). Interestingly, the Shanker et al. (2004) study also revealed a low level of migration ($Nm = 0.36$) between the nesting populations in Madras and nearby Sri Lanka (a distance of only 500 km).

It must be kept in mind that the studies discussed here (Bowen et al., 1998; Shanker et al., 2004) are restricted to analyses of the mitochondrial genome. Nuclear genes may generate a very different picture by reflecting male-mediated gene flow. Natal homing in females, which has been verified in other sea turtle species (Allard et al., 1994; Schierwater and Schroth, 1996; Dutton et al., 1999), will appear as genetic isolation of nesting populations in analyses of mitochondrial genes. Actual migration rates could be much higher if males are mating with females from more than one nesting population. Shanker et al. (2004) report the haplotypes for eight mating pairs that were sampled off the nesting beaches

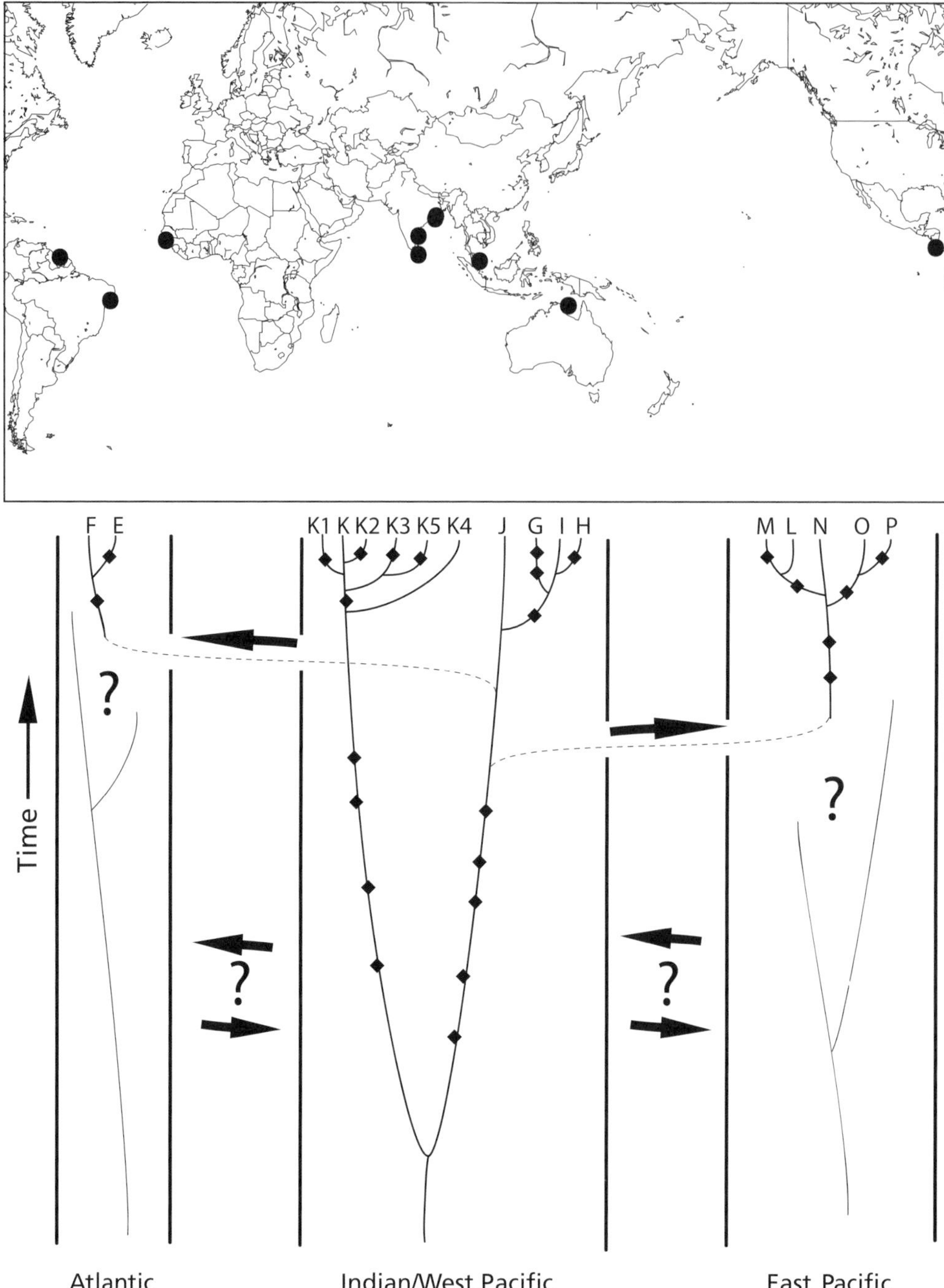

Fig. 6.2. A map of the world showing the location of olive ridley nesting beaches that have been sampled by Bowen et al. (1998) or Shanker et al. (2004). Below the map the haplotypes observed in each of the oceans are shown with the depiction of an evolutionary process that could have produced the pattern. Dark vertical lines delimit the Atlantic, Indian–West Pacific, and East Pacific populations. Solid lines connecting haplotypes correspond to the inferred tree for the alleles (see Fig. 6.1, but note that polytomies have been arbitrarily resolved in Fig. 6.2). The migrations inferred from the data are shown by arrows and broken lines connecting the haplotypes. Diamonds indicate mutations that characterize the haplotypes. Question marks highlight the fact that we do not know if populations existed in the Atlantic and East Pacific when the recent migrations occurred, nor do we know the early history of migration between oceans.

in East India; the haplotype frequencies of these mating pairs are very similar to the frequencies of females nesting (i.e., most of the males also have the K haplotype). Without good estimates of the worldwide haplotype frequencies (small nesting beaches may be overrepresented if the current data are simply concatenated), it is difficult to determine whether this finding indicates that the male olive ridleys display a similar level of philopatry.

Mating Systems

Studying the reproductive biology of endangered species is of particular importance. The value of reproductive animals in maintaining or increasing population size is greater per individual than at all other life stages, making their protection critically important. Further, knowledge of mating systems can improve predictions of the effect of conservation strategies on the genetic variation that is essential to the long-term survival of endangered species. Two studies have made use of microsatellite markers to investigate the mating behavior of *L. kempii* (Kichler et al., 1999) and *L. olivacea* (Hoekert et al., 2002). In both studies the approach was to generate multilocus genotypes of nesting females and their offspring in order to infer paternity.

Hoekert et al. (2002) studied the mating system of the olive ridley rookery in Suriname, once considered the largest Atlantic population of olive ridleys. The number of clutches laid each year dropped by nearly 90% between 1968 and 1995. Such a population in decline becomes vulnerable to loss of genetic variability that may be critical to its long-term survival. Simply put, smaller populations maintain less genetic variation. However, factors other than just the total population size have an effect on the rate at which genetic diversity is lost. For instance, the effects of a small or decreasing population size can be mitigated by a mating system that increases the effective population size (N_e). Polygyny, males mating with multiple females, and polyandry, females mating with multiple males, are examples of mating systems that result in an increase in N_e. Sea turtle populations may be both polygynous and polyandrous, but it is difficult to observe mating directly, as the ocean provides a particularly large expanse in which to search for mating pairs. The most effective way of studying the mating system of a sea turtle population is to make indirect observations by genotyping the very accessible nesting females and their offspring. Comparing mother and offspring genotypes at highly polymorphic loci may allow unique paternal genotypes to be inferred. These genotypes can then be used to determine the number of males that fertilized each clutch and the proportion of offspring sired by each male.

Hoekert et al. (2002) analyzed nuclear DNA from blood samples collected from 10 nesting female olive ridleys and a total of 703 offspring (15–103 offspring sampled per clutch) at two microsatellite loci, Ei8 and Cm84. Population allele frequencies at these loci were determined from the genotypes of 25 unrelated adult turtles. Multiple paternity can be detected based on the number of alleles and/or deviations from the Mendelian expectations under a null model of a single father. Because of the high level of polymorphism in these markers, the probability of detecting multiple paternity in the Suriname olive ridley population, if it existed, was calculated as 0.903. Multiple paternity was detected in 2 of the 10 clutches examined. In these two clutches, it was possible to assign unique genotypes to the fathers and determine the proportion of offspring sired by each male. The authors defined the primary male for each clutch as the male that sired the most offspring. The primary male of one clutch sired 92% of the clutch; the primary male of the other clutch sired 73% of the clutch. Although the probability of multiple paternity in a clutch going undetected was 10%, the observed level of multiple paternity in this *L. olivacea* population was considerably lower than the level observed in *L. kempii* (Kichler et al., 1999).

The most extensive study of the reproductive behavior of Kemp's ridleys (Rostal, 1991) made use of the captive population at the Cayman Island Turtle Farm. In fact, in the literature, there is only one report of a Kemp's ridley mating pair being observed in the wild (Shaver, 1992). Kichler et al. (1999) were able to conduct a paternity study of the nesting population at Rancho Nuevo, Tamaulipas, Mexico, based primarily on samples salvaged from inviable embryos and dead hatchlings. DNA from 26 nesting females and their offspring were analyzed at three microsatellite loci.

Population allele frequencies were calculated from the genotypes of an additional 196 individuals. In stark contrast to the findings of Hoekert et al. (2002) in Suriname, multiple paternity was confirmed by the presence of three or more paternal alleles at one or more of the loci for the offspring of 58% of the nesting females included in the study. This is a much higher rate of multiple paternity than has been observed in at least four other sea turtle species (Table 6.1). In each of those clutches where multiple paternity was confirmed, a majority of the sampled hatchlings were sired by a single male. Small sample sizes in combination with unequal paternal contributions could have caused some multiple paternity to go undetected. The number of offspring sampled per clutch ranged from 1 to 10; 4 or fewer offspring had been sampled for 7 of the 11 females for which multiple paternity of offspring could not be verified. Maximum-likelihood modeling by Kichler et al. (1999) preferred a hypothesis in which all females mated multiple times and the primary father (as defined by Hoekert et al., 2002) sired 75% of the clutch. The hypothesis of equal paternal contribution was a significantly worse fit to the data. The results from ridleys (Kichler et al., 1999; Hoekert et al., 2002) along with work done on green turtles (FitzSimmons, 1998), where clutches were sampled extensively, raise questions of sperm competition and sperm storage in sea turtles. Differential mating success may be an important part of understanding the evolution of sea turtle reproductive behavior.

The high level of multiple paternity observed in Kemp's ridleys begs the question: What is different about this population? It has been suggested that high levels of multiple paternity are most consistent with mass courtship behavior (Harry and Briscoe, 1988). A paternity study of a captive population of green turtles, where the high density of animals generates mass courtship conditions, found multiple paternity in all clutches (Craven, 2001). The *arribada* nesting of ridleys often coincides with a high density of males and females mating offshore and may provide the opportunity for multiple matings. Such an offshore assembly of males and females has been observed for the olive ridley population nesting at Nancite in Guanacaste, Costa Rica (Kalb et al., 1992). Many observations of mating among these animals have been made. This line of reasoning suggests that high levels of multiple paternity may be expected in wild populations of *Lepidochelys* that nest in arribadas. Unlike the Kemp's ridleys, the olive ridley population nesting in Suriname does not currently nest in arribadas. The level of multiple paternity observed in the Suriname population may then be unusually low because the reduction in population size may have disrupted normal reproductive behavior.

However, there are other possible explanations for different species and different populations exhibiting different mating systems. Certainly no offshore assembly of Kemp's ridleys has been observed, and a very low level of multiple paternity has been found in a population of green turtles where females have been observed mating multiple times. A full understanding of reproductive behavior will require more studies integrating behavioral observations and genetic analyses.

Table 6.1 Variation of observed levels of multiple paternity among sea turtle species

Species	Location	Observed frequency of multiple paternity	Reference
Lepidochelys kempii	Mexico	0.58[a]	Kichler et al., 1999
L. olivacea	Suriname	0.2	Hoekert et al., 2002
Caretta caretta	Australia	0.33[a]	Harry and Briscoe, 1988
Chelonia mydas	Australia	0.09	FitzSimmons, 1998
Dermochelys coriacea	Caribbean	0.09	Curtis et al., 2000

[a]These values may significantly underestimate actual levels of multiple paternity because of limited sampling (*L. kempii*) or low levels of polymorphism at the markers used (*C. caretta*).

The work that has been done to date does not allow us to draw general conclusions. One population in one species shows low levels of multiple paternity; one population in another species shows high levels of multiple paternity. What these studies represent, however, is a demonstration of the applicability of genetic analysis to the study of the reproductive behavior of sea turtle populations. Analyses such as the ones described in the studies of Hoekert et al. (2002) and Kichler et al. (1999) may be the only realistic way of obtaining information about the mating systems of wild populations so that we can improve our understanding of how to protect the genetic diversity at both population and species levels.

Prospects for Future Work

As the previous sections have attempted to convey, molecular genetic techniques have already improved our understanding of *Lepidochelys*, but genetic studies will continue to be an important tool for sea turtle researchers. For example, our understanding of the history, population structure, migration rates, and effective population sizes could be improved dramatically by using a multilocus approach.

Currently almost all of our information about the genetic history of the species is from mitochondrial sequences (microsatellite allele frequencies have been examined in only a few isolated populations). Nuclear DNA sequences might provide additional information about evolutionary history following the separation of the *L. kempii* and *L. olivacea* species. As discussed, the mitochondrial haplotypes in the Atlantic and in the eastern Pacific appear to be the result of migration out of the Indian–West Pacific region. Data from several additional markers might reveal whether these migration events were episodes followed by interbreeding with populations that were already in existence or whether the current populations in the Atlantic and eastern Pacific are fixed for alleles that are recently derived from the Indian Ocean at every locus. If the latter scenario were supported, it might imply bouts of extinction and recolonization. Such a result would clearly affect conservation strategies by reinforcing the need to protect as many populations as feasible (or to identify those populations that might be ephemeral in nature).

Collecting data from multiple loci would allow for the application of maximum-likelihood analyses based on coalescence theory (e.g., Beerli and Felsenstein, 2001). Such approaches represent powerful analysis tools that can simultaneously infer the history of changes in population sizes and migration. Recent application of these tools (Roman and Palumbi, 2003) to humpback and fin whale data has demonstrated that previous estimates of historical population sizes for these species were dramatically underestimated. Additionally, DNA analysis has become a high-profile aspect of forensic science, and its usefulness is not restricted to cases involving human victims. Forensic scientists with the National Marine Fisheries Service use DNA analyses to investigate violations of the restrictions on trade in endangered species (Woodley and Ball, 1996). Restriction enzymes that generate species-specific fragment patterns have been identified from cytochrome *b* sequence data. These enzymes are used in tests to determine the species composition of confiscated turtle products quickly and cheaply. Moore et al. (2003) demonstrated that these RFLPs can be used to identify turtle tissue even when it has been cooked. Work done in their laboratory has resulted in at least eight convictions for violations of the Endangered Species Act. As hypervariable markers, such as microsatellites, are scored on more populations of olive ridleys around the world, it may be possible to use forensic techniques to determine which population is being exploited. Such markers have been developed by FitzSimmons (1998), Kichler et al. (1999), Hoekert et al. (2002), and Aggarwal et al. (2004).

In addition to being useful for law enforcement purposes, the ability to identify the source of an individual from a blood sample could provide information on migratory routes. Of course, such applications only have the power to discriminate between and among populations if there is appreciable population subdivision. If male-mediated gene flow is sufficient to disrupt the structure resulting from natal homing in females, then it may not be possible to precisely identify source populations for olive ridleys. The data collected thus far suggest the possibilities of what additional data may reveal.

ACKNOWLEDGMENTS

We thank Steve Morreale and an anonymous reviewer for helpful comments on the draft of this chapter. We also thank Kartik Shanker for his comments and for generously sharing his unpublished results.

LITERATURE CITED

Aggarwal, R. K., Velvan, T. P., Udaykumar, D., Hendre, P. S., Shanker, K., and Singh, L. 2004. Development and characterization of novel microsatellite markers from the olive ridley sea turtle (*Lepidochelys olivacea*). *Molecular Ecology Notes* 4:77–79.

Aggarwal, R. K., Shanker, K., Ramadevi, J., Velvan, T. P., Choudhury, B. C., and Singh, L. 2006. Genetic analysis of olive ridley (*Lepidochelys olivacea*) populations from the East coast of India using microsatellite markers and mitochondrial d-loop haplotypes. In Pilcher, N. J. (Ed.). *Proceedings of the Twenty-third Annual Symposium on Sea Turtle Biology and Conservation*. NOAA Technical Memorandum NMFS-SEFSC-536, p. 109.

Akaike, H. 1973. Information theory and an extension of the maximum likelihood principle. In Petrov, B. N., and Csaki, F. (Eds.). *Proceedings of the Second International Symposium on Information Theory*. Budapest: Akademiai Kiado, pp. 267–281.

Allard, M. W., Miyamoto, M. M., Bjorndal, K. A., Bolten, A. B., and Bowen, B. W. 1994. Support for natal homing in green turtles from mitochondrial DNA sequences. *Copeia* 1994:34–41.

Bass, A. L., Good, D. A., Bjorndal, K. A., Richardson, J. I., Hillis, Z.-M., Horrocks, J. A., and Bowen, B. W. 1996. Testing models of female reproductive migratory behaviour and population structure in the Caribbean hawksbill turtle, *Eretmochelys imbricata*, with mtDNA sequences. *Molecular Ecology* 5:321–328.

Beerli, P., and Felsenstein, J. 2001. Maximum likelihood estimation of a migration matrix and effective population size in *n* subpopulations by using a coalescent approach. *Proceedings of the National Academy of Sciences USA* 98:4563–4568.

Bowen, B. W., Meylan, A. B., and Avise, J. C. 1989. An odyssey of the green sea turtle: Ascension Island revisited. *Proceedings of the National Academy of Sciences USA* 86:573–576.

Bowen, B. W., Meylan, A. B., and Avise, J. C. 1991. Evolutionary distinctiveness of the endangered Kemp's ridley sea turtle. *Nature* 352:709–711.

Bowen, B. W., Nelson, W. S., and Avise, J. C. 1993. A molecular phylogeny for marine turtles: trait mapping, rate assessment, and conservation relevance. *Proceedings of the National Academy of Sciences USA* 90:5574–5577.

Bowen, B. W., Clark, A. M., Abreu-Grobois, A. F., Chaves, A., Reichart, H. A., and Ferl, R. J. 1998. Global phylogeography of the ridley sea turtles (*Lepidochelys* spp.) as inferred from mitochondrial DNA sequences. *Genetica* 101: 179–189.

Craven, K. S. 2001. The roles of fertility, paternity and egg yolk lipids in egg failure of the green sea turtle, *Chelonia mydas*. Ph.D. diss. Texas A&M University, College Station.

Curtis, C., Williams, C. J., and Spotila, J. R. 2000. Mating system of Caribbean leatherback turtles as indicated by analysis of microsatellite DNA from hatchlings and adult females. In Abreu-Grobois, F. A., Briseño-Dueñas, R., Márquez-M., R., and Sarti, L. (Eds.). *Proceedings of the Eighteenth Annual Workshop on Sea Turtle Biology and Conservation*. NOAA Technical Memorandum NMFS-SEFC-436, p. 155.

Dutton, P. H., Davis, S. K., Guerra, T. M., and Owens, D. W. 1996. Molecular phylogeny for marine turtles based on sequences of the ND4-Leucine tRNA and control regions of mitochondrial DNA. *Molecular Phylogenetics and Evolution* 5:511–521.

Dutton, P. H., Bowen, B. W., Owens, D. W., Barragan, A., and Davis, S. K. 1999. Global phylogeography of the leatherback turtle (*Dermochelys coriacea*). *Journal of Zoology* 248:397–409.

Encalada, S. E., Lahanas, P. N., Bjorndal, K. A., Bolten, A. B., Miyamoto, M. M., and Bowen, B. W. 1996. Phylogeography and population structure of the green turtle (*Chelonia mydas*) in the Atlantic Ocean and Mediterranean Sea: a mitochondrial DNA control region sequence assessment. *Molecular Ecology* 5:473–484.

Encalada, S. E., Bjorndal, K. A., Bolten, A. B., Zurita, J. C., Schroeder, B., Possardt, E., Sears, C. J., and Bowen, B. W. 1998. Population structure of loggerhead turtle (*Caretta caretta*) nesting colonies in the Atlantic and Mediterranean as inferred from mitochondrial DNA control region sequences. *Marine Biology* 130:567–575.

FitzSimmons, N. N. 1998. Single paternity of clutches and sperm storage in the promiscuous green turtle (*Chelonia mydas*). *Molecular Ecology* 7:575–584.

Harry, J. L., and Briscoe, D. A. 1988. Multiple paternity in the loggerhead turtle (*Caretta caretta*). *Journal of Heredity* 79:96–99.

Hasegawa, M., Kishino, H., and Yano, T. 1985. Dating the human–ape splitting by a molecular clock of mitochondrial DNA. *Journal of Molecular Evolution* 22:160–174.

Hoekert, W. E. J., Neuféglise, H., Schouten, A. D., and Menken, S. B. J. 2002. Multiple paternity and female-biased mutation at a microsatellite locus in the olive ridley sea turtle (*Lepidochelys olivacea*). *Heredity* 89:107–113.

Kalb, H., Valverde, R., and Owens, D. 1992. What is the reproductive patch of the olive ridley sea turtle? In Richardson, J. A., and Richardson, T. H. (Eds.). *Proceedings of the Twelfth Annual Workshop on Sea Turtle Biology and Conservation.* NOAA Technical Memorandum NMFS-SEFC-361, pp. 57–60.

Karl, S. A., and Bowen, B. W. 1999. Evolutionary significant units versus geopolitical taxonomy: molecular systematics of an endangered sea turtle (genus *Chelonia*). *Conservation Biology* 13:990–999.

Kichler, K., Holder, M. T., Davis, S. K., Márquez-M., R., and Owens, D. W. 1999. Detection of multiple paternity in Kemp's ridley sea turtle with limited sampling. *Molecular Ecology* 8:819–830.

Kotilainen, A. T., and Shackleton, N. J. 1995. Rapid climate variability in the North Pacific Ocean during the past 95,000 years. *Nature* 377: 323–326.

Moore, M. K., Bemiss, J. A., Rice, S. M., Quattro, J. M., and Woodley, C. M. 2003. Use of restriction fragment length polymorphisms to identify sea turtle eggs and cooked meats to species. *Conservation Genetics* 4:95–103.

Moritz, C. 1994. Defining "evolutionary significant units" for conservation. *Trends in Ecology and Evolution* 9:373–375.

Pandav, B. 2000. Conservation and management of olive ridley sea turtles on the Orissa coast. Ph.D diss. Utkal University, Bhubaneshwar.

Pritchard, P. C. H. 1969. Studies of the systematics and reproductive cycles of the genus *Lepidochelys.* Ph.D. diss. University of Florida, Gainesville.

Roman, J., and Palumbi, S. R. 2003. Whales before whaling in the North Atlantic. *Science* 301: 508–510.

Rostal, D. C. 1991. The reproductive behavior and physiology of the Kemp's ridley sea turtle, *Lepidochelys kempi* (Garman, 1880). Ph.D. diss. Texas A&M University, College Station.

Schierwater, B., and Schroth, W. 1996. Molecular evidence for precise natal homing in loggerhead sea turtles. *Bulletin of the Ecological Society of America* 77:393.

Shanker, K., Aggarwal, R. K., Ramadevi, J., Choudhury, B. C., and Singh, L. 2004. Phylogeography of olive ridley turtles (*Lepidochelys olivacea*) on the east coast of India: implications for conservation theory. *Molecular Ecology* 13:1899–1909.

Shaver, D. J. 1992. *Lepidochelys kempii* (Kemp's ridley sea turtle) reproduction. *Herpetological Review* 23:59.

Swofford, D. L. 2002. *PAUP* Phylogenetic Analysis Using Parsimony (*and Other Methods) Version 4.* Sunderland, MA: Sinauer Associates.

Woodley, C. M., and Ball, R. M. 1996. Identification of marine turtle species: when your science becomes forensic. In Bowen, B. W., and Witzell, W. N. (Eds.). *Proceedings of the International Symposium on Sea Turtle Conservation Genetics.* NOAA Technical Memorandum NMFS-SEFSC-396, pp.163–173.

Yang, Z. 1994. Maximum likelihood phylogenetic estimation from DNA sequences with variable rates over sites: approximate methods. *Journal of Molecular Evolution* 39:306–314.

Zangerl, R., Hendrickson, L. P., and Hendrickson, J. R. 1988. A redescription of the Australian flatback sea turtle, *Natator depressus. Bishop Museum Bulletin in Zoology* 1:1–69.

ROLDÁN A. VALVERDE
ERICH K. STABENAU
DUNCAN S. MACKENZIE

7

Respiratory and Endocrine Physiology

RIDLEY SEA TURTLES ARE REPRESENTED by two species, the olive ridley (*Lepidochelys olivacea*) and the Kemp's ridley (*L. kempii*). Each species exhibits distinct distribution and abundance, with the olive ridley being significantly more abundant and much more widely distributed. However, these species share the remarkable phenomenon of mass nesting, or *arribada*. An arribada is the synchronized nesting of large numbers of females over a discrete time interval, typically lasting a few days. This mass nesting behavior is limited to the genus *Lepidochelys*. Some olive ridley arribadas may include over 100,000 turtles in a single event (Valverde et al., 1998), in contrast with Kemp's ridley arribadas, which are well under the historical numbers recorded in the 1940s (Márquez-M. et al., 1996). Most of our knowledge of ridley sea turtle biology comes from studies conducted on this terrestrial, ephemeral phase of the turtles' life cycle because of intense interest in this unusual reproductive behavior and the fact that it is easier in the wild to study these migratory animals while on land than during their less accessible, pelagic stages. Despite the technical difficulties of working with large, pelagic turtles in captivity, intense interest in the conservation of ridley turtles has provided unique opportunities for an examination of their metabolic and endocrine physiology. In this chapter, we review our current knowledge of these areas and suggest further directions that might profitably be pursued in nonreproductive ridley physiology.

Respiratory and Acid–Base Balance

Much of the available information on the blood respiratory and acid–base physiology of ridley sea turtles resulted from research with captive-reared Kemp's ridleys. Animals for these studies were raised at the NMFS Galveston Laboratory as part of the Head Start Experiment or were reared in captivity for periods ranging from 1 to 4 years for use as test subjects in turtle excluder device (TED) certification trials. Most of the initial research focused on quantifying the physiological stress associated with involuntary forced submersion. However, the availability of the NMFS Galveston turtles led to additional research on Kemp's ridley sea turtle physiology, on such diverse topics as the anesthetic management of Kemp's ridleys subjected to surgical procedures (Moon and Stabenau, 1996) and identification and characterization of erythrocyte anion exchange in Kemp's ridley (Stabenau et al., 1991b). The following section summarizes the information compiled during the course of these studies.

Submergence Studies

It was suggested in the late 1980s and early 1990s that the commercial shrimp fishery was responsible for significant numbers of sea turtle deaths as a result of incidental capture of turtles during trawling (Henwood and Stuntz, 1987; National Research Council, 1990). In fact, annual mortality estimates ranged from 5,500 to 50,000 Kemp's ridley and loggerhead turtles (*Caretta caretta*) killed in commercial shrimping-related activities. It was proposed that at-sea mortality would be negligible if tow times were reduced to 60 minutes or less (Henwood and Stuntz, 1987; National Research Council, 1990). No information was available in the literature, however, on the physiological consequences of prolonged submergence of sea turtles. Thus, initial studies were designed to quantify the physiological effects of prolonged, forced submergence in Kemp's ridley turtles.

Extended submergence can cause severe acid–base imbalances that could reduce turtle survival. Therefore, lengthy forced submergence experiments were conducted on small numbers of 2- to 4-year-old Kemp's ridleys with congenital flipper or shell deformities. Deformed turtles and numbers used were limited in the experiments because of the possibility that extended submergence might cause severe acid–base imbalances that could lead to death. Although some of the submergence experiments were conducted with noncannulated turtles, for six turtles, the right carotid artery was occlusively cannulated to permit blood sampling without repeated handling. Because a 90-minute tow time would cause significant mortality (National Research Council, 1990), submergence durations of 20, 40, 60, and 80 minutes were planned. All experiments were conducted under appropriate state and federal threatened/endangered species permits to the NMFS.

To perform the first series of experiments, presubmergence blood samples were collected from arterial cannulas, and the turtles were individually submerged in a weighted canvas bag for 20 minutes. Blood samples were then collected immediately postsubmergence and at 50 and 250 minutes postsubmergence. In this experiment and in all of the others below, no more than 4–6% of total blood volume was collected during the serial sampling. Blood pH and blood gases (P_{CO_2} and P_{O_2}) were analyzed with a blood gas analyzer with electrodes thermostatted to turtle body temperature. Intracellular pH was determined by centrifuging an aliquot of blood, removing the supernatant, and freezing the resulting pellet. The sample was then repeatedly thawed and frozen to lyse the erythrocytes and permit measurement of intracellular pH. Blood lactate was determined enzymatically after deproteinating blood with perchloric acid. Plasma Na^+ and K^+ were analyzed by flame photometry, and plasma Cl^- was measured with a chloridometer. Erythrocyte cell water was determined by adding an aliquot of blood to preweighed aluminum foil and drying the sample to a constant weight in an oven at 80°C. The dried blood samples were then reconstituted with nitric acid to permit measurement of the intracellular Cl^- concentrations.

After only 20 minutes, involuntary submergence of 4-year-old 20-kg Kemp's ridley turtles produced a significant and severe respiratory and metabolic acidosis (Table 7.1). Extracellular and intracellular pH and plasma bicarbonate significantly decreased, and significant increases in P_{CO_2} (average increase of 24.8 mm Hg) and lactate (average increase of 16.6 mM) were measured (Table 7.1). For comparison, Lutz and Bentley (1985) reported similar lactate increases following a 30-minute submergence of loggerhead turtles. Hochachka et al. (1975) reported lactate increases greater than 35 mM following 2 hours of forced submergence of green sea turtles (*Chelonia mydas*). As shown in Table 7.1, increasing the length of the forced submersion to 40 or 60 minutes in two turtles produced profound and severe respiratory and acid–base imbalances, with extracellular and intracellular pH dropping below 6.5 and P_{CO_2} increasing to almost 150 mm Hg. The consequence of extended submergence was that the turtles were incapable of staying at the surface to ventilate. Thus, they had to be held out of water during the postsubmergence recovery period. Although the turtles eventually recovered from the severe blood acid–base and respiratory imbalance, the blood acid–base data and the turtle postsubmergence behavior suggest that a comparably submerged, incidentally captured turtle at sea that was released "alive" would not survive. Moreover, the physiological consequences of the long-term exposure to the various organs and organ systems were not examined. Subsequent submergence of one cannulated 4-kg Kemp's ridley revealed that turtle size also influenced the magnitude of the acid–base disturbance. Blood pH decreased from 7.36 to 6.5, P_{CO_2} increased from 37.1 to 159.1 mm Hg, and lactate increased from 4.6 to 20.2 mM. The 0.86 drop in blood pH measured in the 4-kg juvenile turtle is over two times the decrease in blood pH measured in 20-kg turtles (average 0.41 units). After 50 minutes, the small Kemp's ridley still had a blood pH of 6.84, suggesting incomplete recovery from the acid–base disturbance. The severity of the acid–base changes measured in this juvenile turtle clearly suggests that a turtle of this size cannot tolerate a 40- or 60-minute submergence without its survival being affected.

It should be noted that the hematological values shown in Table 7.1 are comparable to those found in subsequent studies with non–congenitally deformed Kemp's ridley turtles (E. K. Stabenau, unpublished data). However, definitive conclusions about healthy wild animals cannot be made from these forced submergence studies because these experiments used only a limited number of congenitally deformed turtles. Nevertheless, these results suggest that reduction of shrimp trawl tow times, by itself, would be an ineffective management strategy for a number of reasons. The National Research Council (1990) suggested that restricting tow times to 60 minutes or less in winter and 40 minutes or less in summer may be sufficient to reduce incidental capture mortality. However, the data discussed above reveal that involuntary submergence and turtle size clearly influence the magnitude of the blood acid–base imbalance. Second, the physiological consequences to turtles that may be exposed to multiple periods of extended forced submergence have not been examined. Extended periods of forced submergence of Kemp's ridley turtles would clearly predispose the turtles to dying if the turtles were returned to the water immediately following submergence. Third, enforcing trawl times of any magnitude during any season would be difficult.

Continued at-sea mortality caused by incidental capture of sea turtles in commercial shrimp fishing trawls led the U.S. government to pass regulations in 1987 requiring commercial shrimping vessels to use nets equipped with TEDs (see National Research Council, 1990, for a discussion of TEDs and TED modifications). This also prompted further physiological research projects that were designed to quantify the blood respiratory and acid–base status of Kemp's ridley turtles subjected to involuntary submersion in TED-equipped nets (Stabenau et al., 1991a). TED testing or certification involves exposing turtles to control and candidate TEDs. Through the mid-1990s, 2- and 3-year-old captive-reared Kemp's ridley turtles were used as test subjects in TED trials. The test involved placing a turtle inside a weighted canvas or mesh bag that is transferred from the water's surface to the trawl by means of a messenger line attached to the trawl headrope. Divers then released the turtle into the mouth of the trawl. Each turtle was given 5 minutes to escape the trawl voluntarily; turtles remaining in the trawl after 5 minutes were removed by divers. To quantify the physiological stress associated with TED exposure, presubmergence blood samples were collected from the dorsal cervical sinus of

Table 7.1 Effects of prolonged forced submergence on the blood respiratory, acid–base, and ionic status of 4-year-old, 20.3 ± 2.7–kg Kemp's ridley turtles at 25.8 ± 1.2°C

Characteristic		Blood-sampling intervals			
		Presubmergence	Postsubmergence	50-minute recovery	250-minute recovery
pH_o	20 min	7.50 ± 0.07	7.09 ± 0.20	7.31 ± 0.20	7.57 ± 0.07
	40 min	7.57	6.76	7.21	7.65
	60 min	7.33	6.49	6.90	7.47
pH_i	20 min	7.25 ± 0.09	6.92 ± 0.13	7.05 ± 0.07	7.26 ± 0.03
	40 min	7.21	6.59	6.93	7.29
	60 min	7.10	6.46	6.75	7.21
RBC water (%)	20 min	64.9 ± 0.67	64.7 ± 1.35	65.0 ± 1.41	64.4 ± 0.83
	40 min	64.5	66.1	66.2	64.2
	60 min	66.0	33.0	31.0	35.0
HCO_3^- (mM)	20 min	30.0 ± 6.37	19.5 ± 4.27	17.2 ± 4.59	29.3 ± 5.39
	40 min	30.8	16.6	12.2	34.4
	60 min	40.6	17.6	7.6	22.7
Cl_o^- (mM)	20 min	111.5 ± 3.14	116.6 ± 4.16	114.1 ± 5.61	115.2 ± 4.25
	40 min	115.6	115.3	109.7	109.9
	60 min	116.9	119.0	121.7	121.0
Cl_i^- (mM)	20 min	72.4 ± 1.94	75.4 ± 1.71	74.1 ± 3.60	74.2 ± 3.64
	40 min	74.5	76.2	72.6	70.5
	60 min	87.2	105.9	105.7	84.4
Hematocrit (%)	20 min	29.2 ± 0.87	30.7 ± 3.05	30.4 ± 3.25	29.5 ± 4.04
	40 min	29.1	28.5	30.5	28.0
	60 min	30.5	35.0	35.5	28.5
P_{CO_2} (mm Hg)	20 min	32.8 ± 4.74	57.6 ± 18.78	29.6 ± 6.21	27.0 ± 4.20
	40 min	28.7	105.3	27.3	26.7
	60 min	51.5	149.4	37.3	32.7
Lactate (mM)	20 min	1.35 ± 0.87	17.9 ± 5.65	18.4 ± 6.30	5.2 ± 4.05
	40 min	0.9	29.7	27.3	6.9
	60 min	4.1	20.8	20.7	16.8
Na_o^+ (mM)	20 min	139.8 ± 2.45	148.2 ± 3.22	142.9 ± 5.29	141.5 ± 5.70
	40 min	146.4	155.6	147.2	142.7
	60 min	139.9	161.8	146.9	144.3
K_o^+ (mM)	20 min	3.5 ± 0.62	5.9 ± 1.04	4.2 ± 1.12	3.8 ± 0.45
	40 min	3.4	6.0	4.0	3.7
	60 min	5.7	14.2	9.3	8.1
P_{O_2} (mm Hg)	20 min	82.6 ± 11.78	57.2 ± 27.70	75.1 ± 22.20	74.4 ± 24.48
	40 min	85.0	35.2	97.7	67.4
	60 min	42.9	6.5	38.4	30.1

Note: Turtles were confined in weighted canvas bags and submerged for 20 ($n = 4$), 40 ($n = 1$), or 60 ($n = 1$) minutes. The subscripts i and o represent intracellular and extracellular, respectively. Where appropriate, data are expressed as mean ± SD.

Kemp's ridley turtles as described by Owens and Ruiz (1980). The shortest submersion duration in the original study was under 2.7 minutes, and the maximum time any turtle spent underwater was 7.3 minutes. This included the time to get the turtle to the headrope, the maximum exposure time to escape the TED, and the time to reach the surface after exiting the TED. Turtles were immediately returned to the trawling vessel for collection of postsubmergence blood samples. Blood pH and P_{CO_2} were analyzed with a commercial blood gas analyzer with electrodes thermostatted at 37°C. The data were adjusted to turtle body temperature with requisite correction factors for sea turtle blood (Stabenau and Heming, 1993). Blood and plasma lactate and Na^+, K^+, and Cl^- were analyzed as described above.

As a result of these experiments, it soon became apparent that even TEDs did not fully alleviate the physiological impacts of trawling on sea turtles. Short-term involuntary submergence of Kemp's ridley turtles in TED-equipped nets produced significant blood respiratory and metabolic derangements. Specifically, submergence stress produced significant decreases in blood pH and significant increases in blood lactate, P_{CO_2}, and plasma K^+. Although recovery from the acid–base disturbance was not measured in these studies, subsequent experiments in 1993 and 1994 revealed that a period of at least 3 hours was required for recovery of blood variables to prestress levels. The physiological data collected during the TED certification trials indicated that short-term (i.e., 7.5 minutes or less) forced submergence of Kemp's ridley turtles exceeded the animals' aerobic capacity. It is unclear, however, if there were long-term consequences to forced submergence. For example, no information is available on whether turtles become more susceptible to repeated submergence in TED-equipped nets or whether turtles resume normal diving and feeding behaviors following forced submergence (for a description of wild Kemp's ridley diving submergence duration and swimming speed, see Renaud [1995]).

In 1994, additional TED certification trials were conducted with yearling Kemp's ridley and loggerhead turtles. The objective was to assess whether loggerhead turtles could be used as surrogates for Kemp's ridley turtles in annual TED tests and certification trials and to determine the physiological stress of smaller turtles during TED-test forced submersion. Forced submergence experiments (submergence duration = 7.5 minutes) were conducted as described above, with the exception that the turtles were held in large in-water pens 30 days before the TED trials to simulate semiwild conditioning. In addition, blood samples were collected 3 and 6 hours after submergence. Yearling Kemp's ridley and loggerhead turtles exhibited nearly identical pre- to postsubmergence changes in blood pH and lactate. Presubmergence pH in Kemp's ridley and loggerhead turtles was 7.53 ± 0.06 (mean ± SD) and 7.57 ± 0.05, respectively. The postsubmergence blood pH measured immediately after the turtle surfaced decreased to 7.06 ± 0.03 in ridleys and 7.16 ± 0.11 in loggerheads. Three hours after submergence, the blood pH was 7.55 ± 0.04 and 7.52 ± 0.04 in ridley and loggerhead turtles, respectively. No differences in blood pH were detected 6 hours postsubmergence when compared to that measured in presubmergence samples, suggesting that turtles exhibited a full recovery from the forced submergence. These data, in combination with comparable blood lactate loads following submergence, indicated that yearling loggerhead turtles could serve as surrogates for yearling Kemp's ridley turtles during TED testing and certification. However, extending this species comparison beyond yearling turtles has not been investigated. More importantly, no study has examined the physiological effects of forced submergence in TED-equipped nets as a function of turtle size.

Recently, Stabenau and Vietti (2003) examined the physiological effects of multiple submergences on the blood respiratory, acid–base, and ionic status of 6- to 7-kg loggerhead turtles. The purpose of these experiments was to determine whether repeated forced submergence induces progressive, significant blood acid–base disturbances. Experiments were conducted initially under laboratory conditions by confining turtles in weighted canvas bags for three 7.5-minute submergences with a 10-, 42-, or 180-minute "rest" interval between successive submergences. Field experiments were also conducted by exposing turtles under TED-test conditions, with the exception that divers held the exit door closed for 5 minutes. Thus, the total time underwater for each turtle was 7.5 minutes. Blood samples were collected before and immediately after each submergence and 3 hours after the last submergence. No turtles died during the course of these studies. The data revealed that (1) the initial submergence produced a severe metabolic and respiratory acidosis in all turtles, (2) successive submergences produced significant changes in blood variables (e.g., lactate, pH, P_{CO_2}), although the magnitudes of the imbalances were reduced as the number of submergences increased, (3) significant water movement into and out of red blood cells occurred during and after multiple forced submergence, (4) increasing the interval between successive submergences permitted greater recovery of blood homeostasis, (5) similar changes were not observed in nonsubmerged control turtles that had the same serial blood sampling regimen, and (6) repetitive submergence in TED-equipped nets (assuming proper installation and use) would not cause death provided that the animals had an adequate rest interval at the surface between submergences (Stabenau and Vietti, 2003). Comparable information is not available for any other size of turtle or any other sea turtle species and warrants investigation.

Additional Physiological Studies

Much of the information described above was concerned with the physiological effects of forced submersion, but additional experiments have been conducted with blood and other tissues from Kemp's ridley turtles. The following section summarizes the results of these studies.

Red Cell Ion Transport

The presence of erythrocyte anion-exchange protein (i.e., Band 3) permits membrane HCO_3^- transport in a number of species primarily through Na-independent HCO_3^-/Cl^- exchange. Erythrocyte HCO_3^- in many species plays a critical role in CO_2 transport and regulation of plasma and intracellular pH. Although the mechanism of HCO_3^-/Cl^- exchange has been described in a wide variety of vertebrates, less information is available on HCO_3^-/Cl^- exchange in reptilian red blood cells. In turtles, the presence of Band 3 protein for erythrocyte HCO_3^-/Cl^- exchange has been demonstrated in the slider turtle *Pseudemys scripta* (Drenckhahn et al., 1987) and the Kemp's ridley (Stabenau et al., 1991b). In the latter study, it was determined that Kemp's ridley turtle erythrocytes contain 4×10^6 copies of Band 3 protein per cell, or 8,000 copies of Band 3 per square micrometer (Stabenau et al., 1991b). For comparison, human and trout erythrocytes possess approximately 7,000 and 30,000 copies of Band 3 per square micrometer, respectively (Knauf, 1979; Romano and Passow, 1984). Kinetic analysis of the HCO_3^-/Cl^- transporter revealed that erythrocyte anion exchange in Kemp's ridley turtles may be a potentially rate-limiting step for capillary CO_2 exchange (Stabenau et al., 1991b). This limitation may be directly related to the results of the previous studies that showed that short- or long-term forced submersion of Kemp's ridley turtles leads to substantial increases in blood P_{CO_2} from normal values of 35–40 mm Hg to values well above 100 mm Hg (Table 7.1). If capillary HCO_3^-/Cl^- exchange is rate-limiting in the elimination of elevated blood CO_2 following submersion stress, then postcapillary erythrocyte anion exchange would continue and produce significant changes to arterial blood P_{CO_2} and pH (Stabenau et al., 1991b). More recently, Stabenau and Vietti (2003) proposed that loggerhead turtle erythrocyte HCO_3^-/Cl^- and Na^+/H^+ exchangers function as regulatory volume-increase mechanisms. Na^+ and Cl^- enter osmotically shrunken erythrocytes through the respective transporters, while H^+ and HCO_3^- leave the cells as the countertransported ions. Extracellular H^+ and HCO_3^- then combine to form CO_2 and H_2O. The net result is osmotically obliged water entry into the shrunken erythrocytes, which causes the cells to swell. It is possible that similar ion transport mechanisms are present in Kemp's ridley turtle erythrocytes.

HENDERSON–HASSELBALCH CONSTANTS. A common theme in many of the studies mentioned thus far is the measurement of blood P_{CO_2} or plasma HCO_3^-. However, it is not always possible to measure plasma HCO_3^- concentration or to utilize blood gas analyzers with pH, P_{CO_2}, and P_{O_2} electrodes thermostatted to turtle body temperature. Thus, temperature correction factors of known physiological constants are required for dealing with cold-blooded animals. Mammalian constants are commonly used in analyses of the acid–base status of nonmammalian species despite evidence that these practices produce misleading results (Stabenau and Heming, 1993). Stabenau and Heming (1993) determined the constants of the Henderson–Hasselbalch equation, α_{CO_2} and pK_a, over a 20–30°C temperature range for Kemp's ridley turtle blood and plasma. These constants are typically used in the Henderson–Hasselbalch equation, $pH = pK_a + \log(HCO_3^-/\alpha_{CO_2} \cdot P_{CO_2})$ to calculate blood pH, P_{CO_2}, or HCO_3^- or to back-correct values based on temperature differences. It was found that use of classical mammalian-derived values of α_{CO_2} and pK_a was not appropriate for Kemp's ridley turtle plasma. Specifically, mammalian-derived constants would confound analyses of the effects of temperature or pH on sea turtle blood and plasma.

BLOOD AND MUSCLE O_2-CARRYING PROPERTIES. Although substantial information is available in the literature on the blood O_2 storage properties of green, loggerhead, and leatherback (*Dermochelys coriacea*) turtles (Lapennas and Lutz, 1982; Lutcavage et al., 1990), less is known about these variables in Kemp's ridley turtles. Stabenau and Heming (1994) determined that the blood O_2 dissociation curves from Kemp's ridley turtles exhibited a classic sigmoidal shape with a P_{50} of 31.2 ± 0.3 at a hematocrit of 29%. The Kemp's ridley turtle P_{50} was similar to that reported for green turtles (Lapennas and Lutz, 1982) but was considerably less than that measured for loggerhead turtles (Lapennas and Lutz, 1982). Nevertheless, the blood hemoglobin concentration is similar among Kemp's ridley, loggerhead, and green turtles (Stabenau and Heming, 1994; Lapennas and Lutz, 1982) and is significantly less than that of leatherback turtles (Lutcavage et al., 1990). The muscle myoglobin concentration is 3.1 ± 0.84 $mg \cdot g^{-1}$ of tissue in the Kemp's ridley turtle (Stabenau and Heming, 1994) and 2.9 and 4.9 $mg \cdot g^{-1}$ of tissue in the loggerhead (Lutz and Bentley, 1985) and leatherback (Lutcavage et al., 1990) turtles, respectively. These studies revealed that O_2 stores in Kemp's ridley turtle muscle, blood, and lung were 3.6–4.7%, 24.8–35.0%, and 60.3–75.2%, respectively. For comparison, Lutz and Bentley (1985) reported that O_2 stores in the loggerhead turtle were 3.6% in the muscle, 24.8% in the blood, and 71.6% in the lung. Taken together, these results indicate that the diving capacity of shallow-water coastal sea turtle species, such as the Kemp's ridley and loggerhead turtles, is limited by low blood and tissue O_2 stores as compared to deep-diving leatherback turtles and that these animals have a reliance on lung O_2 stores (Stabenau and Heming, 1994). The authors also found that Kemp's ridley turtles possess a nonbicarbonate buffer capacity of 19.7 slykes. Many reptile species have significantly higher blood-buffering capacities (Butler and Jones, 1983). The relatively low buffer capacity in Kemp's ridleys may reveal why this species exhibits such substantial and significant changes in blood pH following forced submergence, even for relatively short periods. It is unknown, however, if significant changes in blood pH are observed following voluntary diving by Kemp's ridleys.

ANESTHESIA. Although Butler et al. (1984) and Shaw et al. (1992) utilized the inhalational anesthetics halothane and isoflurane in green and loggerhead turtles, respectively, limited information was available on the anesthetic management of Kemp's ridley turtles. Thus, Moon and Stabenau (1996) examined the anesthetic and postanesthetic management of Kemp's ridley turtles undergoing a fairly invasive surgical procedure. Anesthetic induction with isoflurane (3.4 ± 0.3% in 2 L of carrier gas/min) followed orotracheal intubation. Induction occurred rapidly in 7 ± 1 minutes. The carrier gases during induction were either 100% O_2, 5% CO_2:95% O_2, or 21–40% O_2 (79% or 60% N_2, respectively). No differences in induction time were detected with the various carrier gases. The right carotid artery was occlusively cannulated with polyethylene tubing during anesthesia, and a second surgery was performed 7–10 days later to remove the cannula. Pulse rate was monitored throughout the study with an ultrasonic Doppler flow probe (Parks Medical Electronics, Cherry Hill, NJ) placed over the femoral triangle, thoracic inlet, caudodorsal aspect of the front flippers, or dorsal cervical sinus (Moon and Stabenau, 1996). The preoperative pulse rate was 34 ± 3 beats/min, 15 ± 1 beats/min intraoperatively and during initial recovery, and 50 ± 2 beats/min when awake. Sudden tachycardia was an excellent predictor of the "awake" state in Kemp's ridley turtles. Blood pressures (systolic/diastolic in mm Hg) were 31 ± 6/20 ± 4, 40 ± 4/25 ± 3, and 46 ± 4/39 ± 2, respectively, during the intraoperative, recovery, and awake phases. The turtles became significantly acidotic as a result of an elevated blood lactate concentration during the early and late recovery phases. The duration of the recovery phase (e.g., time of unresponsiveness before being classified as awake) was 241 ± 31 minutes. This study revealed that inhalational anesthetics could be used for safe and rapid anesthetic induction of Kemp's ridley turtles, and the authors also provided previously unavailable intraoperative vital signs.

ENTANGLEMENT NET CAPTURE STRESS. It must be mentioned that all of the physiological respiratory and acid–base studies described in this review thus far utilized captive-reared Kemp's ridley turtles. More recently, Hoopes et al. (2000) examined the physiological effects of capturing wild Kemp's ridley turtles in entanglement nets. In this study, turtles captured in entanglement nets were placed into in-water cages or on-shore holding tanks. Recovery from capture was then monitored via blood samples collected from turtles immediately postcapture, and at 1, 3, 6, 10, 24, and 48 hours postcapture, depending on turtle size. Larger turtles had more blood samples collected as a result of increased availability. Blood lactate and catecholamine concentrations were measured from each turtle in addition to plasma Na^+, K^+, and Cl^- concentrations. The data revealed that capture in entanglement nets produced a substantial blood metabolic disturbance as indicated by elevated plasma lactate concentrations. Increased plasma norepinephrine, epinephrine, and K^+ concentrations were also detected in capture samples. Although it may have been anticipated that entanglement netting would produce significant changes in blood parameters, an unanticipated result was that the recovery protocol had a significant influence on blood parameters. Specifically, placement of captured turtles into on-shore holding tanks resulted in additional significant increases in the plasma lactate concentrations, whereas turtles in the in-water cages exhibited no changes to the plasma lactate after capture. In addition, the lactate concentrations declined to less than 1 mM by 6 and 10 hours postcapture for turtles in the in-water cages and holding tank treatments, respectively. The decline in lactate over time suggests that repetitive serial blood sampling of wild Kemp's ridleys did not adversely affect results. Thus, Hoopes et al. (2000) clearly showed that sea turtle biologists must be cognizant of physiological disturbances caused by capture protocols and that recovery of blood homeostasis is influenced by postcapture holding protocols.

Summary

The availability of captive-reared Kemp's ridley sea turtles has provided respiratory and reproductive physiologists with ample opportunities to perform many valuable experiments that otherwise would not have been possible. In fact, other than the study by Hoopes et al. (2000), there are no studies on the respiratory and acid–base status of blood and other tissues from wild Kemp's ridley or olive ridley turtles. The data from the present review indicate that ridleys have a remarkable ability to tolerate severe acid–base disturbances. In the forced submergence studies, a drop in blood pH of almost 1 full unit and an increase in blood P_{CO_2} to values over 150 mm Hg were measured. However, the long-term physiological consequences of extended forced submergence are unknown. Surviving an extended forced submergence is only part of the battle. The turtle must also have the capacity to resume normal activities such as feeding, diving, and reproduction, in a timely fashion. Importantly, turtles must not be subjected to resubmergence. Finally, comparable physiological information is unavailable for wild Kemp's ridley and olive ridley turtles and warrants investigation.

Endocrinology

Endocrinology, the study of the synthesis, secretion, and physiological actions of hormones, has historically focused on model species that are readily available for experimental study. Hormones are stored in endocrine glands and circulate in the blood in minuscule amounts. Therefore, large numbers of animals generally are needed to supply adequate tissue for extraction and purification of hormones as well as for biochemical studies of hormone actions at target cells. Because blood hormone concentrations change dynamically over minutes to hours, frequent blood sampling from captive animals is necessary to most effectively describe hormone response to experimental manipulations. Sea turtles, comprising rare and endangered species living in remote locations with limited availability in captivity, would thus seem to represent a poor choice for studies of endocrine physiology.

It is therefore surprising that sea turtles are among the best-studied reptiles in terms of their endocrinology. This has been motivated in part by the importance to conservation of the understanding of the regulation of reproduction, metabolism, and stress responses (Owens, 1997). Additionally, the large body size of sea turtles facilitates collection of blood samples (Owens and Ruiz, 1980), which are readily obtained from nesting females and captive-reared animals. Captive rearing programs have served as a source of blood samples for hormone measurement and pituitary tissue for hormone purification. Animals from the Cayman Turtle Farm, for example, have provided tissue as a source for purification of pituitary hormones (Licht and Papkoff, 1985; Yasuda et al., 1989, 1990), leading to major advances in our understanding of the regulation of growth and reproduction in both marine and freshwater turtles. They have also served as subjects for studies of behavioral endocrinology (Rostal et al., 1998). Development and application of endocrine techniques have thus progressed more rapidly for sea turtles of the family Cheloniidae than for other reptilian families. However, their status as threatened or endangered species and the remote location of many natural populations have limited application of the invasive techniques of endocrinology primarily to measurement of circulating hormones, particularly in ridley turtles. Because animals are generally unavailable for surgical ablation, hormone administration, or tissue collection for in vitro studies, no information is available on hormone biosynthesis, metabolism, or receptor physiology.

Techniques, primarily radioimmunoassay (RIA), have been developed for the purification or measurement of six sea turtle pituitary hormones: the neurohypophyseal peptide hormone arginine vasotocin (AVT) and its associated carrier protein neurophysin (NP; Licht et al., 1984; Figler et al., 1989), the three glycoprotein hormones, follicle-stimulating hormone (FSH), luteinizing hormone (LH), and thyrotropin (TSH, MacKenzie et al., 1981; Licht and Papkoff, 1985), and the two peptide hormones, growth hormone (GH; Yasuda et al., 1989) and prolactin (PRL; Chang and Papkoff, 1985; Yasuda et al., 1990). Of these six, only the RIA for AVT was developed specifically for ridley turtles. The other RIAs (developed for *Chelonia* of *Chelydra* hormones) can detect hormones in blood or pituitary tissue of several turtle genera (both freshwater and marine species), presumably because of the structural similarity of the pituitary hormones among turtles (Licht, 1978). The RIAs for FSH and LH have been used to study endocrine regulation of sea turtle reproduction (Rostal, Chapter 8). The dynamic changes in circulating levels of these hormones associated with ovulation appear conserved in all sea turtle species examined, including ridleys (Owens, 1997; Rostal, Chapter 8). Although an RIA for turtle GH has been successfully employed in the study of the regulation of GH secretion from pituitaries of freshwater turtles (Denver and Licht, 1989, 1990), comparable studies have not been performed in sea turtles. Likewise, the PRL assay can be used to detect in vitro PRL secretion from adult (but not juvenile) freshwater turtle pituitaries (Preece and Licht, 1987; Denver and Licht, 1988) but has not been applied for sea turtle blood or tissues. These studies have demonstrated the practicality of development of techniques for pituitary hormone measurement in sea turtles. With the advent of modern molecular approaches, it should be possible to develop methods for the measurement of pituitary hormones using minute amounts of tissue from any sea turtle species.

Fundamental studies of pituitary hormone function may benefit from an examination of unique aspects of sea turtle physiology and behavior. As an example, the mass nesting behavior of ridley turtles (arribada) has provided an ideal opportunity to examine the endocrine regulation of egg laying in reptiles. Because a large number of olive ridleys nested within 24 hours on a single beach, Figler et al. (1989) were able to obtain replicated blood samples from nine defined stages of oviposition. These samples were used to demonstrate that AVT undergoes a transient elevation in the blood coincident with its proposed physiological action, oviductal contraction (Figler et al., 1989). This study produced such a clear picture that it is included in a leading textbook of endocrinology (Hadley, 1999) as an illustration of the function of AVT in oviposition. The longevity of sea turtles makes them particularly intriguing for the examination of the role of GH in the regulation of growth in reptiles, whereas unique aspects of their reproductive biology (migration, high fecundity, nest site fidelity, mass nesting behavior) make them of interest in the study of the evolution of the role of PRL in reproduction. PRL has been implicated in the regulation of the behavior associated with nesting in precocial bird species (Goldsmith, 1983), suggesting it as an intriguing candidate in the regulation of mass nesting behavior in ridleys. PRL has also been implicated in the regulation of ion transport and water permeability of membranes. The endocrine regulation of salt and water balance, including the regulation of calcium metabolism, is unknown in sea turtles (Lutz, 1997). Studies of the osmoregulatory endocrinology of sea turtles, when compared to marine birds that share similar salt excretory organs, would be of interest in identifying common endocrine adaptations to the marine environment.

Techniques for measurement of protein hormones have been limited in their application to sea turtles because of their requirement for collections of fresh tissues and laborious biochemical purification, but other hormones that are conserved in structure across vertebrate classes are more easily measured using commercially available reagents. Specifically, techniques developed for the measurement of steroid hormones, produced by reproductive organs and the adrenal gland, and thyroid hormones, from the thyroid gland, have been employed for sea turtle blood following relatively simple validation. Reproductive steroid hormones in ridley turtles are discussed in detail in Chapter 8. Here, we focus on two endocrine glands, the adrenal and thyroid, that represent the only nonreproductive peripheral endocrine glands examined to date in ridley turtles. Studies in diverse vertebrate species, including freshwater turtles, indicate that these two endocrine glands are controlled by similar endocrine pathways, initiating with neurohormones produced in the hypothalamus, that regulate production and secretion of pituitary hormones (Fig. 7.1).

Adrenal Physiology

STRESS SYSTEM AND STRESS RESPONSE. Survival of a species is dependent on the ability of the individuals to monitor changes in the environment and respond adaptively. In vertebrates, monitoring of environmental changes is under strict control of a well-developed and complex central nervous system (CNS). The CNS is ca-

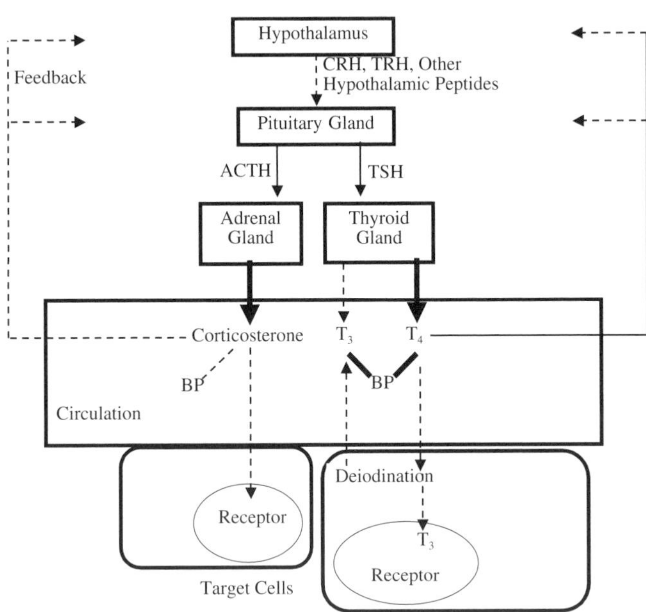

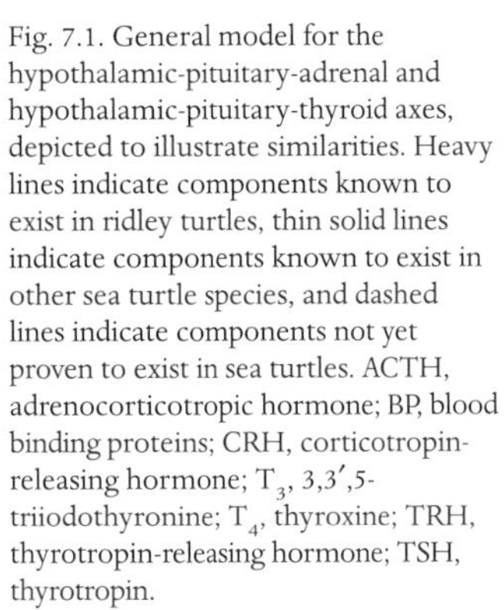
Fig. 7.1. General model for the hypothalamic-pituitary-adrenal and hypothalamic-pituitary-thyroid axes, depicted to illustrate similarities. Heavy lines indicate components known to exist in ridley turtles, thin solid lines indicate components known to exist in other sea turtle species, and dashed lines indicate components not yet proven to exist in sea turtles. ACTH, adrenocorticotropic hormone; BP, blood binding proteins; CRH, corticotropin-releasing hormone; T_3, 3,3′,5-triiodothyronine; T_4, thyroxine; TRH, thyrotropin-releasing hormone; TSH, thyrotropin.

pable of sensing changes in the external as well as in the internal environment, integrating this information, and producing an appropriate response. Frequently, environmental changes disrupt the homeostasis of the organism, requiring modifications in cellular activity to ensure appropriate continued function. Thus, CNS output, manifested through the activation of peripheral effectors such as the adrenal medulla and cortex, has as its primary objective the regulation of homeostatic mechanisms that allow the organism to respond adaptively to environmental changes. The disruption of homeostasis is termed "stress," and the specific stimuli that elicit such disruption are termed "stressors."

Before we engage in a discussion of ridley stress endocrinology, it is important to generate an appropriate perspective by describing the broader stress response of vertebrates, much of which is known from mammalian studies. In the past two decades the field of stress endocrinology has seen substantial conceptual advances that allow us to better understand the physiological and adaptive significance of the stress response. One such advance is the concept of the "stress system." The stress system comprises specific brain areas and peripheral effectors that elicit adaptive behavioral, neuroendocrine, and physiological responses aimed at promoting adaptation and survival (Chrousos and Gold, 1992). When a stressor is perceived, a specific cascade of neural networks is activated. Neural activation in response to a stressor has been demonstrated by the increased transcription of immediate early genes, such as the proto-oncogenes c-*fos* and c-*jun*, among others (Senba and Ueyama, 1997). Activated brain areas include diencephalic and brainstem nuclei, of which the hypothalamic paraventricular nuclei play a central role in regulating the activity of the hypothalamo-pituitary-adrenal (HPA) axis (Fig. 7.1). The HPA axis, along with the brain–adrenal medulla (catecholamine) branch, is one of the most important peripheral components of the stress system.

In classical stress endocrinology, the HPA axis is activated by a broad array of internal and external environmental variables. The unifying factor that stimulates the activation of the axis is the disruption of systemic homeostasis (Chrousos and Gold, 1992). Homeostatic disruptors constitute a stressor and induce glucocorticoid release. Glucocorticoids exert multiple effects in the organism given the ubiquitous presence of their receptors. A simplified list of these effects includes antiinflammatory, antiimmune, antireproductive, and hyperglycemic functions (Widmaier, 1990; Sapolsky et al., 2000). On the basis of these actions, it has been suggested that glucocorticoids suppress the general stress response to prevent the organism from overreacting in a pathological fashion to the stressor (Munck et al., 1984; Sapolsky et al., 2000). Aside from how glucocorticoids achieve their effects, it is widely accepted that one of their main functions is the promotion of survival (Darlington et al., 1990; Sapolsky et al., 2000; Wingfield and Kitaysky, 2002). Thus, increases in blood concentration of glucocorticoids represent a signal that the organism has triggered its adaptive mechanisms in response to a stressor.

As one of the first neuroendocrine signals during the stress response, hypothalamic corticotropin-releasing hormone (CRH) is secreted into the hypothalamo-pituitary portal circulation (Fig. 7.1). Here CRH stimulates pituitary corticotropes to synthesize and release adrenocorticotropic hormone (ACTH) into the systemic circulation. ACTH then stimulates the release of glucocorticoids from the adrenal cortex (Fig. 7.1). In most mammals and fish, cortisol is the main glucocorticoid, whereas in rodents, birds, amphibians, and reptiles, corticosterone appears to be the main glucocorticoid (reviewed by Sandor et al., 1976). First attempts to identify glucocorticoids in ridley turtles were inconclusive (Chester Jones et al., 1959), and more recent attempts to measure cortisol in sea turtle blood have been unsuccessful (A. Aguirre, personal communication). However, corticosterone is well established as the primary corticosteroid in reptiles (Sandor et al., 1976) and has now been measured in response to a variety of putative stressful stimuli in ridley turtles.

STRESS AND RIDLEY TURTLES. Because the genus *Lepidochelys* includes the most, as well as the least, abundant populations of sea turtles, it is striking that so little is known of their stress endocrinology. It is important that we understand how these turtles adapt to stress not only to advance our knowledge of their biology but also to develop more complete and effective approaches to protecting these reptiles. One reasonable approach would be to conduct stress studies on the more abundant olive ridley for comparison or extrapolation to the more endangered Kemp's ridley. Fortunately, the availability of juvenile Kemp's ridleys at the NMFS Galveston Laboratory used in the head-start experiment has provided rare opportunities for studies of adrenal physiology in this species as well. The following pages review our current knowledge of the ridley endocrine stress physiology based on published and unpublished reports and data. Reviewed accounts on ridley stress physiology, combining field and laboratory-based approaches as well as captive-raised and wild sea turtles, have produced an interesting but as of yet incomplete picture of this field.

An important first step in piecing together the endocrine stress response in ridleys was the characterization of basal circulating glucocorticoid concentrations in the wild. Because corticosterone plays an important role in the control of metabolic functions and undergoes daily fluctuations in response to endogenous rhythms (Mizock, 1995), adrenal output may change significantly in a dynamic fashion. To prevent misinterpretation of experimental outcomes, it is also essential to determine basal fluctuations in blood corticosteroid concentration in each particular species. To address this concern, daily basal corticosteroid concentration was measured in adult, free-ranging olive ridleys off Nancite, an arribada nesting beach in Costa Rica (Valverde et al., 1999). In this initial study, 10 females were captured in the water at each of four times during the day: 0600, 1200, 1800, and 2400. Mean corticosterone concentration remained below 0.2 ng/ml throughout this 24-hour period, with no indication of daily variability. These results were further supported by the sampling of over 100 female olive ridleys randomly captured off Nancite Beach at different times during the peak nesting season. Nearly all turtles exhibited blood corticosterone concentration under 0.6 ng/ml (Fig. 7.2). These data are in agreement with basal blood corticosterone concentration measured in green turtles during the nesting process (Jessop et al., 1999). Interestingly, these basal values for ridley and green sea turtles are well below those described for the gopher tortoise, *Gopherus polyphemus* (Ott et al., 2000), and suggest a hypoactivity of the ridley and green HPA axes. It is important to note that only few published studies provide evidence for, or of lack of, a daily cycle in basal circulating corticosteroid in sea turtles.

The first study of adrenal corticosteroids in the context of the stress response in wild olive ridleys is that by Schwantes (1986). In this study conducted at La Escobilla Beach, in Oaxaca, Mexico, three to eight females were sampled at each of nine different stages of nesting: stranding, beginning nest, body pit, egg chamber, first egg, midclutch, last egg, covering, and returning. Circulating corticosterone concentration remained below assay detectability (<1 ng/ml) throughout the nesting process. These results are similar to basal concentration of corticosterone in the green during the same nesting phases (Jessop et al., 1999). In addition, Schwantes (1986) provided the first demonstration of the functionality of the ridley HPA axis by sampling wild adult olive ridleys held at the former slaughterhouse in San Agustinillo in Oaxaca. Males and females held at the slaughterhouse exhibited mean corticosterone concentrations ranging from 4 to 10 ng/ml, 3- to 50-fold greater than those concentrations observed in animals nesting, mating, or basking offshore. Circulating corticosterone was significantly higher in mating males than in mating females, and in stressed males relative to stressed females held at the slaughterhouse. However, sexual differences were not apparent in basking animals. It is important to point out that it was not possible to determine whether males and females had been exposed to the same treatment at the slaughterhouse. This makes it difficult to ascertain whether differences in corticosterone concentration found in these turtles represent a form of sexual dimorphism. Nevertheless, these results suggest that stress, but not nesting, is an effective activator of the ridley HPA axis.

A contrasting study (Valverde et al., 1999) indicated that nesting may also induce a stress response in female olive ridleys. Circulating basal corticosterone was significantly elevated in a

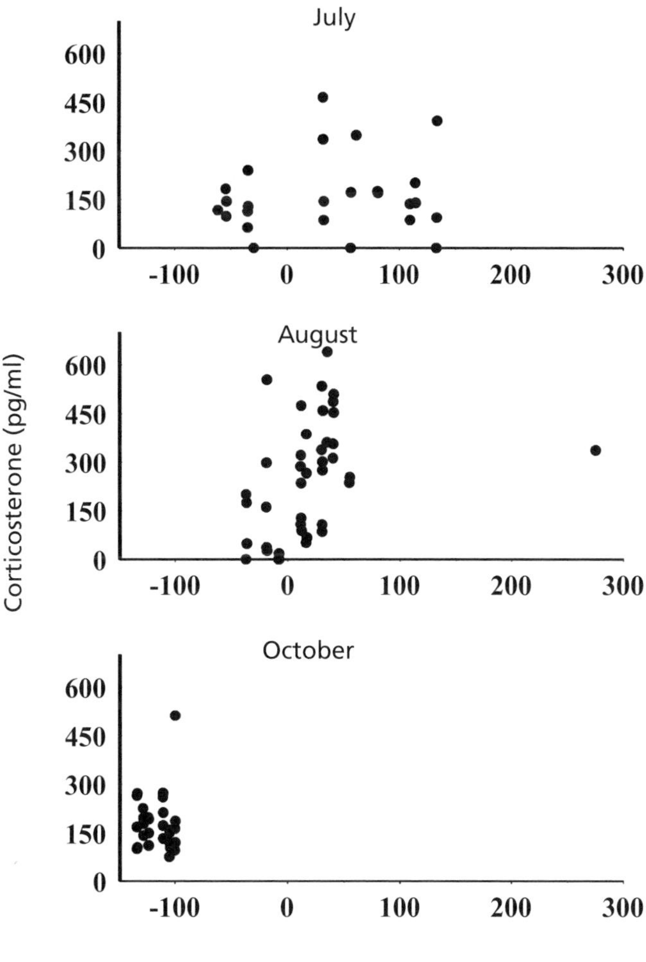

Fig. 7.2. Serum corticosterone levels in basking female olive ridley sea turtles captured throughout the day. Time 0 represents the onset of the *arribada* each month.

group of nesting arribada females randomly sampled at Nancite after completion of nesting, compared to basking females at the same location (Valverde et al., 1999). This may be attributed to AVT, a neurohormone secreted by the neurohypophysis, which has been shown to increase sharply in blood concentration in olive ridleys at the beginning of the nesting process, reaching a peak at the time of oviposition and returning to basal levels on the return of the turtles to the ocean (Figler et al., 1989). The role of AVT is presumably to regulate oviductal contractions by which eggs are forced out of the oviduct. It is possible that secretion of neurohypophyseal hormones such as AVT, which is also known to potentiate CRH-induced ACTH pituitary release in many vertebrate species (Fryer and Leung, 1982; Lilly et al., 1989; Harvey and Hall, 1990), may activate the pituitary-adrenal axis during nesting, resulting in the observed rise in corticosterone in nesting female turtles. If true, this evidence would support the idea that the ridley pituitary gland is sensitive to hypothalamic CRH during arribada nesting.

More recently juvenile Kemp's ridleys were shown to respond to the stress of capture and handling with increased corticosteroid output and increased blood glucose (Gregory and Schmid, 2001). In this study, wild juvenile Kemp's ridleys were captured by entanglement nets. Blood samples were taken at 0, 30, and 60 minutes after release from the nets. Both circulating corticosterone and glucose concentrations increased over time following capture. The concentration of corticosterone at time zero in this study was 10-fold higher than basal levels measured in other studies of juvenile Kemp's ridleys and adult olive ridleys (Valverde et al., 1999; Ortiz et al., 2000). This suggests that concentrations observed at time zero after net capture were already well above basal values. The actual time that turtles spent in the net was not known, with a reported maximum possible time of 15 minutes before samples were taken. Because the HPA axis of other reptiles has been shown to be significantly activated within 10 minutes after capture (Moore et al., 1991; Dunlap and Wingfield, 1995), and the adrenal gland can respond more robustly to stimulation after ACTH priming by prior stress (Ehrhart-Bornstein et al., 1998), it is possible that the HPA axis of juvenile ridleys also was rapidly activated by capture stress.

A comprehensive study on the stress response of arribada olive ridleys at Nancite Beach, Costa Rica, was undertaken by Valverde (1996). During arribadas, females tend to concentrate in a relatively small area, often encountering many physical obstacles including their conspecifics and beach debris. In spite of such high interaction frequency, turtles complete their nesting successfully. This has led to the hypothesis that arribada ridleys might have evolved a central mechanism that increases their sensory threshold, resulting in a lower response to environmental stimulation (Valverde, 1996; Valverde et al., 1999). This hypothesis predicted that nesting arribada turtles would exhibit a hyporesponsive HPA axis. To test the hypothesis, nesting arribada turtles were captured and turned on their backs for blood sampling over the subsequent 6 hours, after the completion of egg laying. Solitary nesting and basking females, as well as males, were also captured as controls and subjected to the same restraint and sampling protocol. The "turning stress" was very instrumental in the stimulation of the ridley HPA axis; turtle groups exhibited a corticosterone response to turning. However, males responded more rapidly (significant increase over basal levels by 20 minutes) than any female group, with the arribada turtles responding at the slowest rate (significant increase by 120 minutes). By the end of the 6 hours, all groups had reached similar circulating mean corticosterone concentration of approximately 4 ng/ml (Valverde et al., 1999). These data support the hypothesis that the arribada females undergo an inhibition of their HPA axis.

Two aspects of interest in this study are that solitary turtles tended to exhibit a slightly higher adrenal response than arribada and basking turtles, indicating that solitary turtles were more sensitive to the physical stressor. The other interesting aspect is that a subgroup of the female turtles in every group studied exhibited no detectable responsiveness to turning stress (refractory females) (Valverde et al., 1999). The former data suggest the interesting possibility that solitary turtles may be too sensitive to physical stimulation to nest at the high densities that characterize arribada events. The latter data suggest that a subpopulation of nesting females at Nancite is experiencing active inhibition of their HPA axis, an inhibition that was not observed in any male olive ridley. A caveat in the latter observation is that the males included in the study were mating at the time of capture. It has been shown that mating olive ridley males exhibit higher circulating corticosterone concentration than mating females (Schwantes, 1986). It is therefore possible that the sensory systems of mating males must remain alert to environmental stimulation because other males may try to dislodge them during the mating process (Booth and Peters, 1972; Alvarado and Figueroa, 1989). Nevertheless, in general these results support the hypothesis that inhibition of the arribada female HPA axis allows them to complete the nesting process in spite of extensive physical contact with and disturbance from conspecifics during arribadas.

An important question related to this hypothesis is whether the sea turtle HPA axis exhibits periods of higher and lower activity during key life history events, such as reproductive and migratory activities, as has been amply demonstrated in birds (the so-called "adrenal modulation," e.g., Wingfield and Kitaysky, 2002). To this effect it has been shown that female green and hawksbill sea turtles exhibit a depressed adrenal response to a physical stressor during reproductive periods with respect to animals that are not reproductively active (Jessop, 2001). It has been suggested that this reduced adrenal response may play an important role in ensuring that reproduction will indeed carry on in spite of environmental stressors (Wingfield et al., 1998).

In an attempt to identify possible mechanisms responsible for the inhibition of the HPA axis in arribada females, several experiments were performed. To test the hypothesis that a steroid hormone from the developed ovary suppresses the HPA axis, circulating testosterone concentrations were measured in the same turtles (Valverde et al., 1999). Testosterone production in female ridleys has been shown to be maximal in fully developed ovaries and minimal in the quiescent ovary (Rostal et al., 1997, 1998). Testosterone concentration did not differ between responsive and refractory females, indicating that refractoriness was not related to ovarian testosterone production or to clutch number.

To test the hypothesis that ridley pituitary sensitivity was suppressed, basking and nesting turtles were subjected to turning for 3 hours (to identify potentially refractory females) (Valverde, 1996) and then injected into the cervical sinus with an approximate average dose of 1.22 μg ovine CRH/kg body mass; control turtles received saline. Blood samples were obtained 0, 2, and 6 hours postinjection. Adrenal response showed that all animals responded to turning stress with increased blood corticosterone (Fig. 7.3). Although this confirms the functionality of the ridley HPA axis, the lack of refractory females precluded a full testing of the hypothesis. However, these data do suggest a lack of sensitivity of the HPA axis to ovine CRH, but it is possible also that the CRH did not reach the pituitary or that the system was already maximally stimulated by the turning stress alone. The lack of sensitivity of ridleys to CRH would be in contrast to the human pituitary, which is capable of responding to intravenous CRH injections in a dose as low as 0.03 μg/kg body mass (Coiro et al., 1995). Further study, ideally utilizing a dose-response approach, is required to address the issue of the role of CRH in regulating ridley ACTH secretion.

Finally, to test the hypothesis that the adrenal gland was refractory to ACTH stimulation in nesting olive ridleys, arribada females were first subjected to turning stress for 4 hours, then injected with either porcine ACTH (0.6 IU pACTH/kg of body mass) or saline solution. Turtles were held in a special corral for 24 hours postinjection in which they were able to move freely. Results showed that all turtles responded to turning stress with equivalent increased blood corticosterone, again precluding testing the refractory HPA axis hypothesis. However, ACTH-injected females did show a significant increase in corticosterone with respect to controls at all times after injection. Thus, although hypothalamic control of pituitary ACTH has yet to be established, the ridley adrenal gland is capable of rapid activation by exogenous ACTH of greater magnitude than elicited by turning stress. This supports the suggestions that the ridley pituitary is

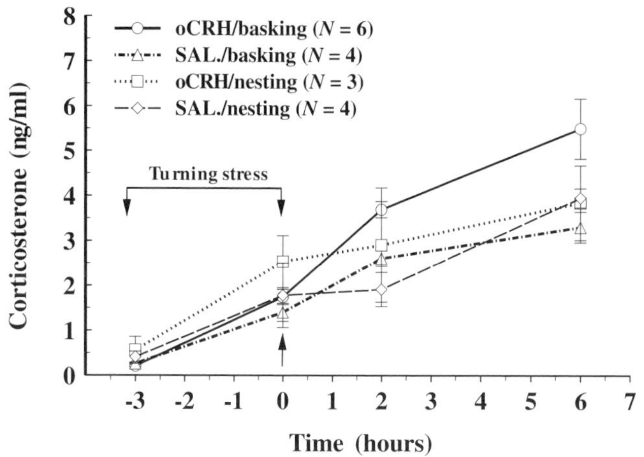

Fig. 7.3. Adrenal response to turning and ovine corticotropin-releasing hormone (CRH) or saline injections in nesting and basking olive ridley sea turtles. At time 3 hours, animals were turned on their backs; at time 0, animals were turned on their plastrons, injected in the cervical sinus, and then allowed to wander in an enclosure for the remainder of the experiment. Figure shows mean corticosterone values ± SEM. Single arrow indicates time of injection.

not producing enough ACTH to effectively control the adrenal gland and that the HPA axis is centrally inhibited during nesting.

From the previous discussion, at least two experiments are suggested. First, it would be necessary to test the hypothesis that only breeding olive ridleys undergo inhibition of their HPA axis, as has been shown for green turtles (Jessop, 2001). This hypothesis was addressed by searching for olive ridleys in international waters of the eastern tropical Pacific Ocean, away from nesting grounds (Owens, 1993), to capture only reproductively inactive adult olive ridleys (based on blood testosterone concentrations <20 pg/ml) (Plotkin et al., 1995). Two mature males and one female (testosterone = 35 pg/ml with small developing follicles) were captured and subjected to turning stress, with blood sampling for a period of 6 hours (Fig. 7.4). Males were further subjected to laparoscopy for gonadal examination and subsequently blood-sampled. Data showed that all animals responded to capture and turning stress with increased blood corticosterone, in a manner similar to animals near or at the nesting beach (Valverde, 1996). Males responded with higher production of corticosterone than the female, but the small sample size precluded statistical analysis. Thus, it was not possible to test adequately the hypothesis that reproductively quiescent turtles would exhibit a more robust stress response. Further work in this area is needed.

The second suggested experiment would be important to determine whether only arribada species, which are subjected to much greater disturbance on nesting beaches, exhibit diminished adrenal responsiveness. To address this hypothesis, nesting loggerhead turtles were also subjected to turning stress and compared with solitary nesting olive ridleys (Valverde, 1996; Valverde et al., 1996). If HPA axis inhibition was exclusive to ridleys, loggerheads (a solitary nesting species) should exhibit a robust response to turning. However, loggerheads not only responded sluggishly to the turning stress, as do olive ridleys, but also exhibited a lower magnitude of response than the solitary olive ridleys by the end of the experiment (Fig. 7.5). Interestingly, it has been shown that green sea turtles do not exhibit increased adrenal output in response to increased nesting density (Jessop et al., 1999). This suggests that green turtles may also be hyporesponsive to physical stimulation during nesting. Thus, the hypothesis of the exclusiveness of the hypoactive ridley HPA axis is not supported.

The data discussed up to now suggest that the HPA axis of the olive ridley, and perhaps that of sea turtles in general, responds sluggishly to physical stress. This is particularly clear in com-

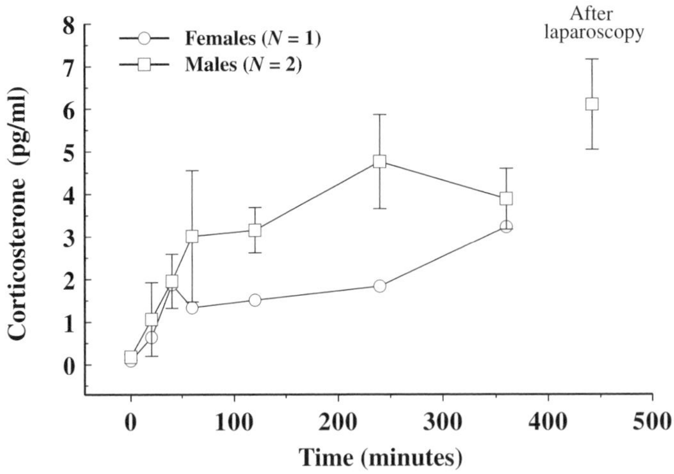

Fig. 7.4. Mean circulating corticosterone levels (± SEM) in two male and one female olive ridleys captured in the open ocean and subjected to turning stress. Last mean value for males represents corticosterone levels measured in samples obtained after laparoscopic surgery.

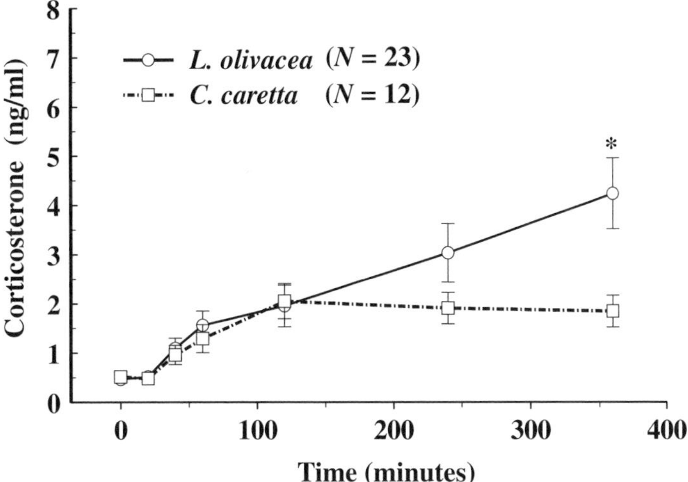

Fig. 7.5. Mean circulating corticosterone levels (± SEM) in nesting female olive ridley and loggerhead sea turtles subjected to turning stress. Asterisk indicates significant differences between mean corticosterone levels of loggerhead and olive ridley females at 360 minutes from initiation of the experiment ($P = 0.008$).

parison to other vertebrates, such as birds, fish, and lizards, that exhibit significant elevations within 10 minutes in response to capture and handling (Dunlap and Wingfield, 1995; Kakizawa et al., 1995; Wada and Shimizu, 2004). The ridley adrenal gland, however, does not appear to be insensitive to stimulation, suggesting that the sluggishness of the system is likely the result of a central mechanism that operates to increase sensory thresholds of the turtles during nesting. Recently, an evolutionarily ancient CRH binding protein was detected in the hypothalamus of the turtle (Seasholtz et al., 2002). This protein has been characterized in vertebrates as a high-affinity binding protein, capable of modulating the bioactivity of CRH and, presumably, the activity of the HPA axis. It is possible that the production of this protein is elevated during reproduction in the female turtle, which would in turn prevent fast activation of the HPA axis. If supported, this mechanism may be linked to the high disturbance threshold (HDT), a trance-like behavior that all nesting sea turtle species exhibit, particularly during egg laying (Valverde, 1996). This HDT behavior has been described in the literature as a period when turtles appear to be quiescent, oblivious to physical (visual, auditory, tactile) stimulation (Hughes and Richard, 1974; Ehrenfeld, 1979). Such a common link could explain why the HPA axis of the loggerhead sea turtle may respond similarly to that of the olive ridley.

GLUCOSE AND STRESS. Blood glucose concentration has also been used to assess the impact of a stressor on the systemic homeostasis of the organism (Widmaier, 1990; Chrousos and Gold, 1992). Thus, circulating glucose measurements may be used as a diagnostic tool for stress in ridleys. Indeed, wild juvenile Kemp's ridleys have been reported to respond to capture and handling with elevated blood glucose within 1 hour of capture (Gregory and Schmid, 2001), indicating that ridleys can exhibit a hyperglycemic response to a stressor. However, it is not known whether glucose increases are caused by adrenocorticosteroid activity or another factor, such as catecholamine hormones. Interestingly, blood glucose has been shown to remain unchanged in the face of variations in the external osmotic environment in captive juvenile Kemp's ridley sea turtles (Ortiz et al., 2000). This, and the lack of adrenal responsiveness to osmotic changes, suggest that the external osmotic environment does not represent a stressor to the ridley's physiology. However, it is important to keep in mind that glucose concentration can vary independently as a result of carbohydrate metabolism, which is not well understood in sea turtles.

In contrast to juvenile ridleys, free-ranging adult male and female olive ridleys did not respond to capture and turning stress with increased concentration of blood glucose within 6 hours (Valverde et al., 1999). In a subsequent study, arribada olive ridleys subjected to turning stress for 4 hours exhibited basal glucose levels similar to those observed in the 6-hour study (Valverde, 1996). However, when these animals were injected with saline or ACTH and subsequently held in captivity on land for a period of 24 hours, blood glucose dropped significantly to about half initial levels. Thus, the results of these two studies do not support a hyperglycemic role for glucocorticoids in reproductively active male and female olive ridleys. An interesting aspect of this study is that basal blood glucose concentrations of these turtles were significantly below those of turtles captured in feeding areas for the population in the eastern tropical Pacific Ocean. It has been suggested that sea turtles do not feed actively during the reproductive months (Owens, 1976, 1980). Evidence for this hypothesis has recently been reported (Tucker and Read, 2001). It is possible that this hypophagia might limit the energy stores of the turtle. This energy limitation may in turn impact not only the gluconeogenic capacity of glucocorticoids by decreasing the availability of noncarbohydrate substrates but also the hyperglycemic effect of other stress hormones such as catecholamines. From the above discussion, it seems that the use of blood glucose is not a reliable tool to describe the stress response of the actively reproductive ridley turtle.

Salinity and Adrenocortical Response

Ridleys, like all other sea turtles, live in a hyperosmotic environment. In order to help maintain ion homeostasis, these marine reptiles have evolved a highly efficient salt gland (Reina et al., 2002). Secretory activity of the sea turtle salt gland is thought to be partially under the control of adrenal steroids (Holmes and McBean, 1964). Using captive-raised Kemp's ridleys, Morris (1982) studied adrenal responsiveness to a change in ambient salinity in the context of salt gland function. In a series of experiments, 3-year-old Kemp's ridleys were subjected to changes in salinity. Turtles were transferred from salt water at 34 parts per thousand (ppt) to either 34 ppt (control), 17 ppt, or 0 ppt salinity. An hour after transfer, all treatment groups showed significant, equivalent elevations in circulating corticosterone (up to a mean of approximately 9 ng/ml), which then declined to basal levels (less than 2 ng/ml) by 6 hours. Two weeks after transfer, all turtles were transferred back to 34 ppt. Following this second transfer, no increases in corticosterone were observed at 5 or 24 hours, although, based on results of the first experiment, this sampling interval would have missed acute corticosterone changes in the first hour. In a second experiment, Kemp's ridleys were intraperitoneally injected with 0.25 ml/kg of either 30% or 0.9% (physiological) saline. Both groups of turtles exhibited elevated corticosterone by 30 and 60 minutes, but no differences were found between the two treatments. Although these two sets of experiments do suggest activation of the adrenal axis with handling or sampling, as has been shown more recently for captive Kemp's ridleys (Stephenson et al., 2000), they do not support a direct role of adrenal glucocorticoids in the adaptation to osmotic challenges. To our knowledge, the studies by Morris (1982) were the first to demonstrate that the ridley HPA axis can respond to physical stressors with robust production of corticosterone.

In a similar study, Ortiz et al. (2000) examined water flux and osmotic and adrenal responses to acute salinity challenge in four juvenile, captive-raised Kemp's ridleys. Full-strength seawater-acclimated animals were switched to fresh water for 4 days and then back to full strength for another 7 days. Rapid water changes (<6 minutes) in the holding tank were achieved without handling animals, and blood samples were taken at 2-day intervals for measurement of aldosterone and corticosterone. No changes were found in either hormone following transfers in spite of significant changes in water flux, plasma ion composition, and plasma osmolality. Although these data suggest that these steroids do not play an important role in the osmoregulatory response of ridley turtles, it is again possible that more frequent sampling might have revealed transient hormone changes. Both Morris (1982) and Ortiz et al. (2000) suggest that acute changes in ambient salinity do not constitute a stressor to Kemp's ridleys.

Catecholamine Hormones and Stress

Chromaffin cells of the adrenal medulla, a peripheral component of the sympathetic nervous system, release the catecholamine hormones epinephrine and norepinephrine (also known as adrenaline and noradrenaline). In mammals, catecholamine hormones induce rapid energy mobilization as well as increased cardiac output and blood pressure, which are thought to mediate adaptation to acute stress (Chrousos and Gold, 1992). Studies in freshwater turtles have found a significant (less than 1 hour) elevation of blood epinephrine, norepinephrine, and glucose during diving hypoxia (e.g., Wasser and Jackson, 1991), consistent with a role for these hormones in glucose mobilization. The only report on ridley catecholamine hormones is that of Hoopes et al. (2000). This study has been described in more detail in the previous section on Entanglement Net Capture Stress. Wild, juvenile Kemp's ridleys were retrieved after spending less than 10 minutes in entanglement nets, and blood was sampled for up to 48 hours while they were held in in-water cages and on-shore tanks. Blood concentrations of both epinephrine and norepinephrine were elevated at the initial sample and then declined over 6–10 hours, reaching minimal levels at 48 hours. Hoopes et al. (2000) suggest that postcapture holding conditions influenced the rate of recovery from stress because catecholamine levels declined more rapidly in turtles held in in-water cages than in turtles held in on-land tanks. The only other study of catecholamine hormones in sea turtles found no change in circulating epinephrine and norepinephrine in postnesting female green turtles restrained for up to 10 minutes (Hamann et al., 2003). Hamann et al. (2003) suggest that this lack of elevation may be a result of desensitization of catecholamine production during nesting, similar to the situation discussed above for glucocorticoids. Although these two studies are difficult to compare because of differences in capture protocols, age, and reproductive condition of the turtles, both suggest that further study of the dynamics of blood catecholamine hormones will provide information on the sensitivity of sea turtles to natural and human-imposed stressors. It will be challenging to conduct such studies in wild ridley turtles. To characterize the activation of catecholamine release, initial blood samples should be obtained from undisturbed animals. Furthermore, repetitive sampling to characterize recovery will require restraint of animals, for possibly as long as 24–48 hours.

Summary

Current available data indicate that ridley turtles possess a functional HPA axis. Moreover, the axis has been shown to function much like that of other vertebrate species. Specifically, the axis is activated by physical stressors and by pharmacological manipulation. Interestingly, the ridley HPA axis of the adult turtle exhibits a higher activation threshold than that of many other reptiles, and that of vertebrates in general, in response to physical stimulation. In addition, some ridley turtles can exhibit nearly complete endocrine (corticosterone) refractoriness to physical stressors. These aspects of the ridley HPA axis may assist the animals during nesting, particularly at high densities. The nature of this HPA axis modulation is not understood and warrants further investigation. In order to establish whether this phenomenon is related to reproductive condition or energy stores, it is important to conduct experiments with smaller ridleys located in the feeding grounds. The modulatory phenomenon of the ridley HPA axis suggests the possibility that these animals possess alternative homeostatic mechanisms that allow them to withstand the disruptive effects of stressors. To understand the role of glucocorticoids in the physiology of the ridley, the molecular and biochemical characterization of the ridley glucocorticoid receptor would be most informative. In addition, the evidence available on blood glucose supports the idea that the adult ridley effectively undergoes a period of hypophagia. This phenomenon seems to be restricted to the reproductive months and may impinge on the ability of glucocorticoids and other hyperglycemic hormones to promote an elevation in blood glucose. More studies are required to elucidate the relationship between energy balance and the dynamic nature of the physiology of the HPA axis.

Thyroid Physiology

The regulation of thyroid hormone secretion and the actions of thyroid hormones at target tissues are not well understood for most commonly studied reptilian species. Because thyroid hormones consistently have been implicated as playing a role in supporting energetically demanding processes such as nutrient assimilation, growth, development, reproduction, and migration (Eales, 1979; McNabb, 1992), thyroid function in reptiles is of basic interest because of their evolutionary position at the transition to endothermy. More recently, reptilian thyroid glands have been studied because of the possibility of environmentally induced endocrine disruption (Crain et al., 1998). Studies of thyroid function in ridley turtles may thus contribute both to our basic understanding of the evolution of thyroid function and to a more practical understanding of the potential impact of anthropogenic chemicals on threatened species.

Studies in diverse vertebrate species have led to a consistent picture of the basic organization of the hypothalamo-pituitary-thyroid axis (Fig. 7.1; McNabb, 1992; Hulbert 2000): hypothalamic peptide hormones regulate secretion of pituitary TSH; TSH in turn activates the synthesis and secretion of the thyroid hormones thyroxine (tetraiodothyronine or T_4) and 3,3′,5-triiodothyronine (T_3) from the thyroid gland. TSH stimulates unique processes in thyroid cells, including iodide uptake and organification, thyroglobulin synthesis, thyroid hormone synthesis and cleavage from thyroglobulin, and thyroid hormone secretion. TSH also promotes thyroid cell growth and differentiation. In most vertebrates examined, T_4 is the predominant hormone released to the circulation from the thyroid gland (McNabb, 1992).

Data available for ridley turtles address relatively few aspects of thyroid function. Indeed, the first comprehensive review of reptilian thyroid function (Lynn, 1970) makes only a single reference to the genus *Lepidochelys*. This is a description of the histological appearance of the thyroid gland of the olive ridley (Yamamoto, 1960), which demonstrates the characteristic follicular appearance consistent with the general vertebrate structure supporting thyroid hormone synthesis. More recent studies on circulating thyroid hormones, described below, lead us to believe that thyroid function in *Lepidochelys* resembles the generalized vertebrate model. Additionally, comparative studies have confirmed the existence of the basic elements of the hypothalamic-pituitary-thyroid axis in turtles. A TSH homologous to that found in other vertebrates has been purified from the green turtle (MacKenzie et al., 1981), and we expect that a similar protein exists in ridley pituitaries. This TSH has been used for the development of an RIA capable of measuring TSH in blood and pituitary tissue of freshwater and marine turtles. Use of the RIA has confirmed that thyroid hormones exert a negative feedback on pituitary TSH production in green turtles (MacKenzie et al., 1981) and in slider turtles (Denver and Licht, 1988). It has also identified several hypothalamic peptides, including thyrotropin-releasing hormone (TRH) and others conventionally associated with the regulation of pituitary hormones such as GH or ACTH, capable of stimulating pituitary TSH secretion in slider turtles (Denver and Licht, 1989, 1990). The ability of CRH to stimulate TSH release in slider turtles suggests a linkage between the thyroid and adrenal systems (Denver and Licht, 1989). Further study of hypothalamic regulation of TSH in sea turtles would thus provide a broader comparative perspective on the regulation of TSH secretion and a better understanding of their integrated endocrine response to stress. Presently, no information is available on the nature of TSH in ridley turtles or its hypothalamic control.

The slider turtle thyroid axis is now the most intensively studied of any reptilian species. In contrast, the limited availability of sea turtle species for experimental or descriptive examination of thyroid function has hindered our understanding of how applicable slider turtle data are to nonemydid turtles. In the case of thyroid hormone blood transport, for example, emydid turtles appear quite unusual among ectothermic vertebrates. Thyroid hormones circulate in the blood bound to several binding proteins that serve to facilitate their transport to distant tissues (Fig. 7.1). In slider turtles, a unique vitamin D binding protein binds T_4 with high affinity in the circulation (Licht, 1994). Circulating T_4 stimulates the production of this binding protein, resulting in periods of the year when blood binding capacity, and therefore circulating T_4, is among the highest observed in any vertebrate, including mammals (Licht et al., 1990). Phylogenetic studies of the distribution of this binding protein in reptiles show that it is not present in nonemydid species, including olive ridley (Licht et al., 1991). This was confirmed by Haynes (1990) in a detailed analysis of thyroid hormone binding to sea turtle serum proteins. In comparison to humans and *Trachemys,* serum from sea turtles, including Kemp's ridley, showed a diminished ability to bind both T_3 and T_4. Scatchard analysis of binding to Kemp's ridley serum demonstrated the presence of a single, high-affinity, moderate-capacity T_4 binding site with an affinity similar to the human thyroid hormone transporter transthyretin (Table 7.2). Binding affinity of T_3 in Kemp's ridley blood was about 10-fold lower than that of T_4. This study concluded that Kemp's ridley turtles possess a single, moderate-affinity thyroid hormone transport protein in serum. Phylogenetic studies of transthyretin (Power et al., 2000) have indicated that it is not present in reptile blood, suggesting that a different, unknown protein functions as the primary thyroid hormone transport protein in ridley turtles. Further studies are needed to determine whether modulation of the blood content of this protein serves to alter thyroid hormone delivery to peripheral tissues.

At peripheral tissues, T_4 enters target cells and is converted to deiodinated metabolites by intracellular deiodinase enzymes (Fig. 7.1). Outer ring deiodinases (ORD) convert T_4 to T_3, which, because it generally has a higher affinity for the nuclear thyroid hormone receptor, is considered the active intracellular form of thyroid hormone (McNabb, 1992; Hulbert, 2000). Inner ring deiodinases (IRD) can also convert T_4 to a variety of deiodinated metabolites, many of which are considered to be inactive breakdown products of thyroid hormone (Hulbert, 2000). The T_3 generated by ORD can be made available to nuclear thyroid hormone receptors to regulate the transcription of specific genes, can be broken down

Table 7.2 Characteristics of thyroid hormone binding to serum proteins in sea turtles

	T_3			T_4		
Species	Site	Affinity[a]	Capacity[b]	Site	Affinity[a]	Capacity[b]
Lepidochelys kempii	1	0.855	2.43	1	6.29	2.48
Caretta caretta	1	0.554	1.96	1	7.81	0.604
Chelonia mydas	1	1.20	2.17	1	4.89	4.73
Trachemys scripta	1	1.614	0.144	1	83.9	0.188
Homo sapiens	1	40.91	0.096	1	255.7	0.0484
	2	1.79	0.117	2	8.43	0.665

Note: Data from Haynes (1990). Serum samples were pooled, and endogenous thyroid hormone was removed using ion-exchange resin. Serum was then subjected to saturation analysis using radioiodinated thyroid hormones on Sephadex minicolumns. All assays were performed at 20°C. Data were analyzed using the LIGAND computer program to determine the number of thyroid hormone binding sites in each species, the affinity of each binding site for thyroid hormones, and the total binding capacity of the serum for each thyroid hormone. Data for *T. scripta* and *H. sapiens* are included for comparison to sea turtle serum.

[a]Affinity units: 10^7 M^{-1}.

[b]Capacity units: 10^{-6} M.

further intracellularly to inactive products, or can be released back to the blood, in considerable amounts in some species, where it presumably is capable of activating additional thyroid hormone receptors at distant targets (Fig. 7.1; Eales and Brown, 1993). Eventually, thyroid hormone metabolites are conjugated and excreted by the kidney and the liver, although evidence exists in some species for an active enterohepatic cycling system that may return significant amounts of thyroid hormone to the circulation through deconjugation by bacterial flora in the gut (Eales and Brown, 1993). We have no information for any sea turtle species on any of these peripheral processes critical to the activation and reception of thyroid hormones at their target tissues. The threatened/endangered status of ridley turtles has precluded directed or opportunistic collection of fresh tissue needed for in vitro enzyme and receptor assays. However, molecular biological techniques now permit characterization of the expression of the genes for these proteins in minute tissue samples and could be applied to ridley turtle tissue to confirm the existence of deiodination enzymes and thyroid hormone receptors.

Currently, the most commonly used method for evaluation of thyroid function in reptiles is measurement of circulating thyroid hormone concentrations. This can be achieved relatively easily through the collection of blood samples, followed by measurement of thyroid hormones using a validated RIA. Although it gives a static picture of the total thyroid hormone concentration in blood at a particular point in time, this technique provides little information on the nature of the binding of thyroid hormones to blood proteins or to the actual rate of delivery of thyroid hormones to receptors in target tissues. Nonetheless, this approach is useful in identifying periods of hypothalamic-pituitary-thyroid axis activation, potentially representing those times at which thyroid hormone secretion or hormone stimulation of target tissues is greatest. This is normally the first step in initiating a more detailed examination of thyroid function. For example, a dramatic midsummer peak in T_4 in slider turtles (Licht et al., 1985) suggested that environmental temperature may activate the thyroid axis and that thyroid hormones may be involved in somatic growth, both later confirmed by laboratory experimentation (Licht et al., 1989; Denver and Licht, 1991). The power of hormone measurement is enhanced with an increased frequency of sampling, giving a more precise picture of the dynamics of activation of the pituitary-thyroid axis. A common criticism of blood measurement studies is that sampling at one time of day may fail to detect dynamic circadian changes in blood thyroid hormones, as has been noted for a number of fish species (Leatherland, 1994). In this regard, it is important to note that Moon et al. (1999) measured circulating T_4 in seven blood samples taken over 48 hours from captive Kemp's ridleys, held at constant temperature and light cycle, and detected no evidence of a daily fluctuation. However, thyroid hormones might not be as stable in the wild, under conditions of variable temperature, photoperiod, and feeding. Channel catfish, which exhibit no daily changes in thyroid hormones when held in the laboratory (Gaylord et al., 2001), display dramatic increases in both T_3 and T_4 in the afternoon when held in outdoor ponds (Loter, 1998).

Several studies have described circulating thyroid hormones in ridley turtles (Owens, 1997). Moon et al. (1998) have provided the only data on T_4 in wild ridley turtles, reporting a mean of 6.7 ng/ml for nesting and 3.3 ng/ml for swimming olive ridleys in Mexico. Although samples can be collected easily during the brief period of nesting in females, this limited perspective on thyroid hormones in wild turtles is of minimal value without comparative data from other times or physiological conditions. Our most detailed information on seasonal thyroid hormones in ridley turtles comes from captive animals. Moon et al. (1998) found seasonal changes in circulating T_4 in adult female, but not adult male, Kemp's ridleys held in indoor tanks under artificial simulated natural photoperiod and natural temperature in Galveston, Texas. In females, the distinct annual cycle showed elevated T_4 in March to May, as water temperatures were increasing, followed by a decline in July to October, when temperatures were maximal. A second peak in T_4 occurred in late November to December. Mean T_4 in both sexes ranged from 5 to 13 ng/ml, substantially lower than that in slider turtles but at the higher end of the range reported for most wild reptiles (e.g., Bona-Gallo et al., 1980; John-Alder, 1984; Naulleau et al., 1987; Kohel et al., 2001). Rostal et al. (1998) also found seasonal changes in a captive population of Kemp's ridleys held in large outdoor ponds under natural photoperiod and temperature at the Cayman Turtle Farm. Although the history and environmental conditions were quite different for this group of animals, the range of T_4 was similar to that in the Galveston ridleys. Once again, males exhibited a low magnitude of T_4 changes with the exception of a significant elevation of T_4 (over 5 ng/ml) found only in March. Females showed more dynamic annual T_4 changes, with a major T_4 peak (near 11 ng/ml) in December, followed by a smaller peak in March and a gradual decline in summer.

The sexual dimorphism in T_4 cycles at both locations suggests an interaction between reproductive condition and thyroid function. A reciprocal relationship between T_4 and testosterone, noted in prior reptilian studies (Bona-Gallo et al., 1980; Licht et al., 1985; Naulleau et al., 1987), was not apparent in Kemp's ridleys. It is possible that the relatively low sampling frequency (once every 1–3 months) may have missed some transient T_4 elevations, yet similarities between the two locations are intriguing. In both groups of females, maximal T_4 was observed at the time of initiation of ovarian recrudescence. This coincided with elevated blood vitellogenin (as indicated by blood calcium) (Rostal et al., 1998). A second T_4 elevation coincided with mating activity. The consistent results between these two captive populations suggest that T_4 participates in the energetically demanding process of vitellogenesis, possibly promoting the mobilization of lipid and protein in the liver for vitellogenin synthesis. Alternatively, changes in blood composition associated with vitellogenesis may alter T_4 transport by increasing thyroid hormone binding to blood proteins. However, Heck et al. (1997) did not find support for binding of T_3 or T_4 to vitellogenin in Kemp's ridleys; pharmacologically elevated blood vitellogenin had no effect on plasma thyroid hormone binding up to 75 days following estrogen treatment. Elevated T_4 during vitellogenesis, whether associated with vitellogenin or bound to other blood proteins, raises the possibility that it can move into the yolk of ridley eggs, as has been noted for avian and teleost species (Wilson and McNabb, 1997; Tagawa and Brown, 2001). Additional studies of thyroid hormone transport and incorporation into yolk are needed to establish a role for maternal hormones in the regulation of sea turtle embryogenesis. If thyroid hormones are incorporated into ridley eggs and play a significant role in the regulation of embryonic development, as has been proposed for slider turtle steroid hormones (Bowden et al., 2002), disruption of maternal thyroid function in wild populations may influence embryonic development. Such disruption might include alteration of maternal thyroid hormone secretion, deiodination, blood transport, or receptor binding (Brucker-Davis, 1998; Eales et al., 1999). Exposure of marine birds to aromatic hydrocarbons has been found to alter thyroid mass, circulating thyroid hormone levels, and egg yolk composition (Rolland, 2000), suggesting that animals living in contaminated marine habitats may be vulnerable to endocrine disruption.

Blood T_3 in captive or wild ridleys is normally at or below the sensitivity of the RIAs used (<0.1 ng/ml) (Rostal et al., 1998; Moon et al., 1998, 1999). Low or nondetectable T_3 has been noted in a number of reptilian species (John-Alder, 1984; Licht et al., 1990; Kohel et al., 2001). Circulating T_3 reflects a combination of factors, including rates of formation, blood protein binding, and clearance (McNabb, 1992). Active deiodinases have been characterized in a variety of tissues from slider turtles (Hugenberger and Licht, 1999), a species with similar nondetectable T_3, demonstrating that low T_3 does not necessarily reflect a lack of peripheral deiodination. Low blood T_3 in Kemp's ridleys is likely caused in large part by a low affinity of T_3 binding to serum proteins (Table 7.2). In this regard, ridley turtles resemble mammals, which exhibit similar low circulating T_3 in spite of active outer ring deiodination in many peripheral tissues. Barely detectable T_3 (up to 0.9 ng/ml) has been noted in some captive Kemp's ridleys maintained on constant warm temperatures and daily feeding of a high-protein diet (Moon et al., 1998, 1999). However, additional T_3 data from wild animals under a variety of physiological conditions are clearly needed for a better understanding of the significance of circulating T_3.

Seasonal T_4 cycles in ridley turtles are likely a result of multiple influences. One variable that may seasonally modulate thyroid hormone secretion and transport is body temperature (Licht et al., 1989). To determine if temperature influences blood thyroid hormones in Kemp's ridleys, Moon et al. (1997) subjected a group of 10 immature Kemp's ridleys to a controlled decrease in temperature. Although animals remained active as temperature dropped from 25°C to 10°C, time of submergence increased and food consumption decreased below 15°C. A dramatic decline in circulating T_4 also was observed at all temperatures below 25°C (Moon, 1992). Because studies of slider turtles have shown that alterations in thyroid status or environmental conditions that diminish blood binding protein capacity result in substantial reductions in circulating thyroid hormone concentrations (Licht et al., 1990), we evaluated these Kemp's ridley blood samples for thyroid hormone binding to serum proteins at the acclimation temperatures of the turtles, using a Sephadex column index of total thyroid hormone binding (Haynes, 1990). In Kemp's ridley serum pooled from animals acclimated to 30°C, both T_3 and T_4 binding were relatively stable over the range of temperatures from 10°C to 40°C. In contrast, the blood samples taken from turtles acclimated to low temperatures showed significantly lower binding of T_4, but not T_3, at 15°C (Haynes, 1990). These studies support a role for thyroid hormones in promoting energy utilization during periods of elevated temperature or increased activity in Kemp's ridleys. They also indicate that decreases in body temperature may reduce thyroid hormone delivery to target tissues (Haynes and MacKenzie, 1990). Alterations in thyroid hormone delivery may thus be one facet of the deleterious changes that contribute to physiological impairment associated with cold temperature (e.g., cold stunning) (Milton and Lutz, 2003).

Because turtles that are cold also cease feeding, Moon et al. (1999) evaluated the possibility that the decline of circulating thyroid hormones might be caused in part by reduced food intake. However, no changes in T_4 or T_3 were observed in captive Kemp's ridleys during 2 weeks of food deprivation. In a reciprocal experiment, in which feeding was increased 250%, satiated turtles also showed no significant changes in T_4 or T_3 in comparison to control (fed) turtles. The lack of response to alterations in ration may be a species-specific phenomenon in ridley turtles because captive green sea turtles in the same experiment did exhibit significant changes in both T_3 and T_4. The thyroid response to food intake is complex, and previous nutritional history may influence the magnitude of response to food deprivation (MacKenzie et al., 1998). Kemp's ridley studies do not provide support for the activation of thyroid hormone production by nutrient intake, as has been suggested for other vertebrates (MacKenzie et al., 1998). However, well-fed captive animals maintain significant nutritional stores, particularly fat (Owens, 1997), which may minimize the effects of this relatively brief food deprivation. Again, studies of thyroid function in wild animals undergoing natural cycles of food intake may help elucidate the significance of these findings.

Summary

The limited data available suggest that ridley turtles have a thyroid gland capable of increased T_4 production during times of anabolic activation. Reproductive activity and increased temperature may both enhance thyroid hormone secretion and transport. Distinct annual T_4 cycles have only been observed in captive female Kemp's ridleys; further clarification of the role of thyroid hormones in the physiology of ridley turtles must await a more detailed evaluation of wild animals. In addition, more information is needed on the dynamics of thyroid hormone secretion to determine whether relatively infrequent sampling protocols, often an unavoidable limitation in sea turtle studies, provide representative data. Because of the stability of thyroid hormones in well-preserved serum, blood samples already stored in freezers could be analyzed relatively easily for T_3 and T_4. Such studies would provide us with needed baseline data on the normal range of thyroid hormones under diverse physiological and environmental conditions. In particular, it would be of interest to determine whether the elevation in circulating T_4 observed in captivity during vitellogenesis and mating occurs in wild animals. The thyroid hormone content of ridley eggs should also be determined to establish whether maternal hormones appear in yolk. Identification of the location of deiodinases and receptors, useful for establishing thyroid hormone targets and their sensitivity to stimulation, could come from classical studies with tissue salvaged from animals killed for other purposes or from molecular studies of tissue biopsies. Further studies of hypothalamic control of thyroid function, although useful in establishing a linkage between the pituitary-adrenal axis and thyroid, can be achieved only with more invasive in vivo or in vitro techniques. All of this information would provide more effective tools for the evaluation of the impact of human activities on an endocrine gland that is likely involved in the regulation of development and metabolism. Ridley turtles, as fully aquatic species often found near shore, are potentially exposed to a number of anthropogenic chemicals that may impact thyroid function. To evaluate the effects of endocrine-disrupting chemicals on adult, juvenile, or embryonic animals, it is essential that thyroid activity be characterized in populations from relatively pristine habitats as well as in potentially impacted wild populations.

Final Remarks

Despite the challenges of working with ridley turtles, significant progress has been made in the examination of their metabolic and endocrine physiology. Mass nesting behavior and captive rearing programs have provided unique opportunities for investigations of stress, respiratory, and endocrine physiology, which should serve as the foundation for future investigations. Much is left to discover that will help to fill the substantial gaps in our knowledge of the biology of these fascinating reptiles. Of special interest is the arribada phenomenon, unique to the genus *Lepidochelys*. These unexplained large aggregations may have played an evolutionary role in ensuring the survival of the species by conferring individuals with the protection against predators inherent to a large group of conspecifics. However, in modern times these same aggregations could lead to the decimation of the species because animals are particularly vulnerable to anthropogenic activities when on the beach or in coastal waters. The proximity of such large numbers of animals to nesting beaches provides a rare opportunity for an examination of the role of endocrine systems in the regulation of behavior of wild reptiles. Because modern analytical techniques require minuscule amounts of blood for hormone and metabolite measurement, investigators collecting blood samples from wild animals should seek out collaborations that promote the broadest possible analysis of sizable sea turtle blood samples.

It is our hope that the study of the physiology of both species of *Lepidochelys* helps to enhance our understanding of their life histories and to facilitate the design and implementation of improved conservation practices. For example, an improved understanding of the respiratory physiology of the Kemp's ridley may lead to effective protocols to help these animals recover from capture in trawling nets of shrimp boats. Similarly, understanding the role of thyroid hormones in metabolism or thermoregulation may contribute to our ability to assess the impact of exposure to cold waters or to endocrine-disrupting chemicals. In addition, characterization of the endocrine stress response in ridley turtles may help us understand the potential impact of tagging programs, turtle-based tourism programs, incidental fisheries, and egg-harvesting practices on the reproductive biology and behavior of these animals. Importantly, we have no direct data on the relationship between reproduction and stress in ridley turtles. Throughout these studies, it is important to bear in mind the population status of these animals. We must ensure that the need to protect the species is never secondary to the need to understand their physiology.

LITERATURE CITED

Alvarado, J., and Figueroa, A. 1989. Breeding dynamics in the black turtle (*Chelonia agassizi*) in

Michoacan, Mexico. In Eckert, S., Eckert, K., and Richardson, T. (Compilers). *Proceedings of the Ninth Annual Workshop on Sea Turtle Biology and Conservation.* NOAA Technical Memorandum NMFS-SEFC-0232, pp. 5–7.

Bona-Gallo, A., Licht, P., MacKenzie, D. S., and Lofts, B. 1980. Annual cycles in levels of pituitary and plasma gonadotropin, gonadal steroids, and thyroid activity in the Chinese cobra, *Naja naja. General and Comparative Endocrinology* 42:477–493.

Booth, J., and Peters, J. A. 1972. Behavioral studies on the green turtle (*Chelonia mydas*) in the sea. *Animal Behaviour* 20:808–812.

Bowden, R. M., Ewert, M. A., and Nelson, C. E. 2002. Hormone levels in yolk decline throughout development in the red-eared slider turtle (*Trachemys scripta elegans*). *General and Comparative Endocrinology* 129:171–177.

Brucker-Davis, F. 1998. Effects of environmental synthetic chemicals on thyroid function. *Thyroid* 8:827–856.

Butler, P. J., and Jones, D. R. 1983. The comparative physiology of diving in vertebrates. In Lowenstein, O. (Ed.). *Advances in Comparative Physiology and Biochemistry.* New York: Academic Press, pp. 180–364.

Butler, P. J., Milsom, W. K., and Woakes, A. J. 1984. Respiratory, cardiovascular, and metabolic adjustments during steady state swimming in the green turtle, *Chelonia mydas. Journal of Comparative Physiology* 154B:167–174.

Chang, Y.-S., and Papkoff, H. 1985. Isolation and properties of sea turtle (*Chelonia mydas*) pituitary prolactin. *General and Comparative Endocrinology* 60:372–379.

Chester Jones, I., Phillips, J. G., and Holmes, W. N. 1959. Comparative physiology of the adrenal cortex. In Gorbman, A. (Ed.). *Comparative Endocrinology.* New York: Wiley, pp. 582–612.

Chrousos, G. P., and Gold, P. W. 1992. The concepts of stress system disorders: overview of behavioral and physical homeostasis. *JAMA* 267: 1244–1252.

Coiro, V., Volpi, R., Capretti, L., Speroni, G., Caffarra, P., Scaglioni, A., Malvezzi, L., Castelli, A., Caffarri, G., and Rossi, G. 1995. Low-dose ovine corticotropin-releasing hormone stimulation test in diabetes mellitus with or without neuropathy. *Metabolism: Clinical and Experimental* 44:538–542.

Crain, D. A., Guillette, L. J., Pickford, D. B., Percival, H. F., and Woodward, A. R. 1998. Sex-steroid and thyroid hormone concentrations in juvenile alligators (*Alligator mississippiensis*) from contaminated and reference lakes in Florida, USA. *Environmental Toxicology and Chemistry* 17:446–452.

Darlington, D. N., Chew, G., Ha, T., Keil, L. C., and Dallman, M. F. 1990. Corticosterone, but not glucose, treatment enables fasted adrenalectomized rats to survive moderate hemorrhage. *Endocrinology* 127:766–772.

Denver, R. J., and Licht, P. 1988. Thyroid status influences in vitro thyrotropin and growth hormone responses to thyrotropin-releasing hormone by pituitary glands of hatchling slider turtles (*Pseudemys scripta elegans*). *Journal of Experimental Zoology* 246:293–304.

Denver, R. J., and Licht, P. 1989. Neuropeptides influencing in vitro pituitary hormone secretion in hatchling turtles. *Journal of Experimental Zoology* 251:306–315.

Denver, R. J., and Licht, P. 1990. Modulation of neuropeptide-stimulated pituitary hormone secretion in hatchling turtles. *General and Comparative Endocrinology* 77:107–115.

Denver, R. J., and Licht, P. 1991. Dependence of body growth on thyroid activity in turtles. *Journal of Experimental Zoology* 258:48–59.

Drenkhahn, D., Oelmann, M., Scaaf, P., Wagner, M., and Wagner, S. 1987. Band 3 is the basolateral anion exchanger of dark epithelial cells of turtle urinary bladder. *American Journal of Physiology* 252:C570–C574.

Dunlap, K. D., and Wingfield, J. C. 1995. External and internal influences on indices of physiological stress. I. Seasonal and population variation in adrenocortical secretion of free-living lizards, *Sceloporus occidentalis. Journal of Experimental Zoology* 271:36–46.

Eales, J. G. 1979. Thyroid functions in cyclostomes and fishes. In Barrington, E. J. W. (Ed.). *Hormones and Evolution,* Volume 1. New York: Academic Press, pp. 341–436.

Eales, J. G., and Brown, S. B. 1993. Measurement and regulation of thyroidal status in teleost fish. *Reviews in Fish Biology and Fisheries* 3:299–347.

Eales, J. G., Brown, S. B., Cyr, D. G., Adams, B. A., and Finnson, K. R. 1999. Deiodination as an index of chemical disruption of thyroid hormone homeostasis and thyroidal status in fish. In Henshel, D. S., Black, M. C., and Harrass, M. C. (Eds.). *Environmental Toxicology and Risk Assessment: Standardization of Biomarkers for Endocrine Disruption and Environmental Assessment:* Volume 8. ASTM STP 1364. West Conshohocken, PA: American Society for Testing and Materials, pp. 136–164.

Ehrenfeld, D. W. 1979. Behavior associated with nesting. In Harless, M., and Morlock, H. (Eds.). *Turtles: Perspectives and Research.* New York: Wiley-Liss, pp. 417–434.

Ehrhart-Bornstein, M., Hinson, J. P., Bornstein, S. R., Scherbaum, W. A., and Vinson, G. P. 1998. Intraadrenal interactions in the regulation of adrenocortical steroidogenesis. *Endocrinology Review* 19:101–143.

Figler, R. A., MacKenzie, D. S., Owens, D. W., Licht, P., and Amoss, M. S. 1989. Increased levels or arginine vasotocin and neurophysin during nesting in sea turtles. *General and Comparative Endocrinology* 73:223–232.

Fryer, J. N., and Leung, E. 1982. Neurohypophyseal control of cortisol secretion in the teleost *Carassius auratus. General and Comparative Endocrinology* 48:425–431.

Gaylord, T. G., MacKenzie, D. S., and Gatlin, D. M. 2001. Growth performance, body composition, and plasma thyroid hormone status of channel catfish (*Ictalurus punctatus*) in response to short-term feed deprivation and refeeding. *Fish Physiology and Biochemistry* 24:73–79.

Goldsmith, A. R. (1983). Prolactin in avian reproductive cycles. In Balthazart, J., Pröve, E., and Gilles, R. (Eds.). *Hormones and Behavior in Higher Vertebrates.* Berlin: Springer-Verlag, pp. 375–387.

Gregory, L. S., and Schmid, J. R. 2001. Stress responses and sexing of wild Kemp's ridley sea turtles (*Lepidochelys kempii*) in the northeastern Gulf of Mexico. *General and Comparative Endocrinology* 124:66–74.

Hadley, M. E. 1999. *Endocrinology.* Princeton, NJ: Prentice Hall.

Hamann, M., Limpus, C. J., and Whittier, J. M. 2003. Seasonal variation in plasma catecholamines and adipose tissue lipolysis in adult female green sea turtles (*Chelonia mydas*). *General and Comparative Endocrinology* 130:308–316.

Harvey, S., and Hall, T. R. 1990. Hormones and stress in birds: Activation of the hypothalamo-pituitary-adrenal axis. In Epple, A., Scanes, C. G., and Stetson, M. H. (Eds.). *Progress in Comparative Endocrinology.* New York: Wiley-Liss, pp. 453–460.

Haynes, S. P. 1990. The effects of temperature on thyroid hormone binding to serum proteins in sea turtles. M.S. thesis, Texas A&M University, College Station.

Haynes, S. P., and MacKenzie, D. S. 1990. Thyroid hormone binding to serum proteins in sea turtles. *American Zoologist* 30:124A.

Heck, J., MacKenzie, D. S., Rostal, D., Medlar, K., and Owens, D. W. 1997. Estrogen induction of plasma vitellogenin in the Kemp's ridley sea turtle (*Lepidochelys kempii*). *General and Comparative Endocrinology* 107:280–288.

Henwood, T. A., and Stuntz, W. E. 1987. Analysis of sea turtle captures and mortalities during commercial shrimp trawling. *Fishery Bulletin* 85:813–817.

Hochachka, P. W., Owen, T. G., Allen, J. F., and Whittow, G. C. 1975. Multiple end products of anerobiosis in diving vertebrates. *Comparative Biochemistry and Physiology* 50B:17–22.

Holmes, W. N., and McBean, R. L. 1964. Some aspects of electrolyte excretion in the green turtle, *Chelonia mydas mydas. Journal of Experimental Biology* 41:81–90.

Hoopes, L. A., Landry, A. M., and Stabenau, E. K. 2000. Physiological effects of capturing Kemp's ridley sea turtles, *Lepidochelys kempii,* in entanglement nets. *Canadian Journal of Zoology* 78: 1941–1947.

Hugenberger, J. L., and Licht, P. 1999. Characterization of thyroid hormone 5′-monodeiodinase activity in the turtle (*Trachemys scripta*). *General and Comparative Endocrinology* 113:343–359

Hughes, D. A., and Richard, J. D. 1974. The nesting of the Pacific Ridley turtle *Lepidochelys olivacea* on Playa Nancite, Costa Rica. *Marine Biology* 24:97–107.

Hulbert, A. J. 2000. Thyroid hormones and their effects. A new perspective. *Biological Reviews of the Cambridge Philosophical Society* 75:519–631.

Jessop, T. S. 2001. Modulation of the adrenocortical stress response in marine turtles (Cheloniidae): evidence for a hormonal tactic maximizing maternal reproductive investment. *Journal of Zoology London* 254:57–65.

Jessop, T. S., Limpus, C. J., and Whittier, J. M. 1999. Plasma steroid interactions during high-density green turtle nesting and associated disturbance. *General and Comparative Endocrinology* 115: 90–100.

John-Alder, H. B. 1984. Seasonal variations in activity, aerobic energetic capacities, and plasma thyroid hormones (T_3 and T_4) in an iguanid lizard. *Journal of Comparative Physiology B* 154: 409–419.

Kakizawa, S., Kaneko, T., Hasegawa, S., and Hirano, T. 1995. Effects of feeding, fasting, background adaptation, acute stress and exhaustive exercise on the plasma somatolactin concentrations in rainbow trout. *General and Comparative Endocrinology* 98:137–146.

Knauf, P. A. 1979. Erythrocyte anion exchange and the Band 3 protein: transport kinetics and molecular structure. In Bronner, F., and Kleinzeller,

A. (Eds.). *Current Topics in Membranes and Transport,* Volume 12. New York: Academic Press, pp. 249–363.

Kohel, K. A., MacKenzie, D. S., Rostal, D. C., Grumbles, J. C., and Lance, V. A. 2001. Seasonality in plasma thyroxine in the desert tortoise, *Gopherus agassizii. General and Comparative Endocrinology* 121:214–222.

Lapennas, G. N., and Lutz, P. L. 1982. Oxygen affinity of sea turtle blood. *Respiratory Physiology* 48:59–74.

Leatherland, J. F. 1994. Reflections on the thyroidology of fishes: from molecules to humankind. *Guelph Ichthyology Review* 2:1–67.

Licht, P. 1978. Studies on the immunological relatedness among tetrapod gonadotropins and their subunits with antisera to sea turtle hormones. *General and Comparative Endocrinology* 36:68–78.

Licht, P. 1994. Thyroxine-binding protein represents the major vitamin D-binding protein in the plasma of the turtle, *Trachemys scripta. General and Comparative Endocrinology* 93:82–92.

Licht, P., and Papkoff, H. 1985. Reevaluation of the relative activities of the pituitary glycoprotein hormones (follicle-stimulating hormone, luteinizing hormone, and thyrotropin) from the green sea turtle, *Chelonia mydas. General and Comparative Endocrinology* 58:443–451.

Licht, P., Pickering, B. T., Papkoff, H., Pearson, A., and Bona-Gallo, A. 1984. Presence of a neurophysin-like precursor in the green turtle (*Chelonia mydas*). *Journal of Endocrinology* 103:97–106.

Licht, P., Breitenbach, G. L., and Congdon, J. D. 1985. Seasonal cycles in testicular activity, gonadotropin, and thyroxine in the painted turtle, *Chrysemys picta,* under natural conditions. *General and Comparative Endocrinology* 59:130–139.

Licht, P., Denver, R. J., and Pavgi, S. 1989. Temperature dependence of in vitro pituitary, testis, and thyroid secretion in a turtle, *Pseudemys scripta. General and Comparative Endocrinology* 76:274–285.

Licht, P., Denver, R. J., and Stamper, D. L. 1990. Relation of plasma thyroxine binding to thyroidal activity and determination of thyroxine binding proteins in a turtle, *Pseudemys scripta. General and Comparative Endocrinology* 80:238–256.

Licht, P., Denver, R. J., and Herrera, B. E. 1991. Comparative survey of blood thyroxine binding proteins in turtles. *Journal of Experimental Zoology* 259:43–52.

Lilly, M. P., DeMaria, E. J., Bruhn, T. O., and Gann, D. S. 1989. Potentiated cortisol response to paired hemorrhage: Role of angiotensin and vasopressin. *American Journal of Physiology* 257:R118–R126.

Loter, T. C. 1998. Regulation of seasonal changes in circulating thyroid hormones in channel catfish. M.S. thesis, Texas A&M University, College Station.

Lutcavage, M. E., Bushnell, P. G., and Jones, D. R. 1990. Oxygen transport in the leatherback sea turtle, *Dermochelys coriacea. Physiological Zoology* 63:1012–1024.

Lutz, P. 1997. Salt, water, and pH balance. In Lutz, P. L., and Musick, J. A. (Eds.). *The Biology of Sea Turtles.* Boca Raton, FL: CRC Press, pp. 343–361.

Lutz, P. L., and Bentley, T. B. 1985. Respiratory physiology of diving in the sea turtle. *Copeia* 1985: 671–679.

Lynn, W. G. 1970. The thyroid. In Gans, C., and Parsons, T. S. (Eds.). *Biology of the Reptilia, Volume 3, Morphology C.* London: Academic Press, pp. 201–234.

MacKenzie, D. S., Licht, P., and Papkoff, H. 1981. Purification of thyrotropin from the pituitaries of two turtles: the green sea turtle and the snapping turtle. *General and Comparative Endocrinology* 45:39–48.

MacKenzie, D. S., VanPutte, C. M., and Leiner, K. A. 1998. Nutrient regulation of endocrine function in fish. *Aquaculture* 161:3–25.

Márquez-M., R., Byles, R. A., Burchfield, P., Sánchez, M., Diaz P., J. F., Carrasco A., M. A., Leo P., A. S., and Jiménez O., C. 1996. Good news! Rising numbers of Kemp's ridleys nest at Rancho Nuevo, Tamaulipas, Mexico. *Marine Turtle Newsletter* 73:2–5.

McNabb, F. M. A. 1992. *Thyroid Hormones.* Englewood Cliffs, NJ: Prentice Hall.

Milton, S. L., and Lutz, P. L. 2003. Physiological and genetic responses to environmental stress. In Lutz, P. L., Musick, J. A., and Wyneken, J. (Eds.). *The Biology of Sea Turtles,* Volume 2. Boca Raton, FL: CRC Press, pp. 163–197.

Mizock, B. A. 1995. Alterations in carbohydrate metabolism during stress: A review of the literature. *American Journal of Medicine* 98:75–84.

Moon, D.-Y. 1992. The response of sea turtles to temperature changes: behavior, metabolism, and thyroid hormones. Ph.D. diss. Texas A&M University, College Station.

Moon, D.-Y., MacKenzie, D. S., and Owens, D. W. 1997. Simulated hibernation of sea turtles in the laboratory: I. Feeding, breathing frequency, blood pH, and blood gases. *Journal of Experimental Zoology* 278:372–380.

Moon, D.-Y., MacKenzie, D. S., and Owens, D. W. 1998. Serum thyroid hormone levels in wild and captive sea turtles. *Korean Journal of Biological Science* 2:177–181.

Moon, D.-Y., Owens, D. W., and MacKenzie, D. S. 1999. The effects of fasting and increased feeding on plasma thyroid hormones, glucose, and total protein in sea turtles. *Zoological Science* 16:579–586.

Moon, P. F., and Stabenau, E. K. 1996. Anesthetic and postanesthetic management of sea turtles. *Journal of the American Veterinary Medical Association* 208:720–726.

Moore, M. C., Thompson, C. W., and Marler, C. A. 1991. Reciprocal changes in corticosterone and testosterone levels following acute and chronic handling stress in the tree lizard, *Urosaurus ornatus. General and Comparative Endocrinology* 81:217–226.

Morris, Y. A. 1982. Steroid dynamics in immature sea turtles. M.S. thesis, Texas A&M University, College Station.

Munck, A., Guyre, P. M., and Holbrook, N. J. 1984. Physiological functions of glucocorticoids in stress and their relation to pharmacological actions. *Endocrine Review* 5:25–44

National Research Council. 1990. *Decline of Sea Turtles: Causes and Prevention.* Washington, DC: National Academy Press.

Naulleau, G., Fleury, F., and Boissin, J. 1987. Annual cycles in plasma testosterone and thyroxine in the male aspic viper *Vipera aspis* L. (Reptilia, Viperidae), in relation to the sexual cycle and hibernation. *General and Comparative Endocrinology* 65:254–263.

Ortiz, R. M., Patterson, R. M., Wade, C. E., and Byers, F. M. 2000. Effects of acute fresh water exposure on water flux rates and osmotic responses in Kemp's ridley sea turtles (*Lepidochelys kempii*). *Comparative Biochemistry and Physiology* 127A:81–87.

Ott, J. A., Mendonça, M. T., Guyer, C., and Michener, W. C. 2000. Seasonal changes in sex and adrenal steroid hormones of gopher tortoises *Gopherus polyphemus. General and Comparative Endocrinology* 117:299–312.

Owens, D. W. 1976. Endocrine control of reproduction and growth in the green sea turtle *Chelonia mydas.* Ph.D. diss. University of Arizona, Tucson.

Owens, D. W. 1980. The comparative reproductive physiology of sea turtles. *American Zoologist* 20: 547–564.

Owens, D. W. 1993. The research vessel *Gyre* cruise in the eastern tropical Pacific. *Marine Turtle Newsletter* 62:8–9.

Owens, D. W. 1997. Hormones in the life history of sea turtles. In Lutz, P. L., and Musick, J. A. (Eds.). *The Biology of Sea Turtles.* Boca Raton, FL: CRC Press, pp. 315–341.

Owens, D. W., and Ruiz, G. J. 1980. New methods of obtaining blood and cerebrospinal fluid from marine turtles. *Herpetologica* 36:17–20.

Plotkin, P. T., Byles, R. A., Rostal, D. C., and Owens, D. W. 1995. Independent versus socially facilitated oceanic migrations of the olive ridley, *Lepidochelys olivacea. Marine Biology* 122:137–143.

Power, D. M., Elias, N. P., Richardson, S. J., Mendes, J., Soares, C. M., and Santos, C. R. A. 2000. Evolution of the thyroid hormone-binding protein transthyretin. *General and Comparative Endocrinology* 119:241–255.

Preece, H., and Licht, P. 1987. Effects of thyrotropin-releasing hormone in vitro on thyrotropin and prolactin release from the turtle pituitary. *General and Comparative Endocrinology* 67:247–255.

Reina, R. D., Jones, T. T., and Spotila, J. R. 2002. Salt and water regulation by the leatherback sea turtle *Dermochelys coriacea. Journal of Experimental Biology* 205:1853–1860.

Renaud, M. L. 1995. Movements and submergence patterns of Kemp's ridley turtles (*Lepidochelys kempii*). *Journal of Herpetology* 29:370–374.

Rolland, R. M. 2000. A review of chemically-induced alterations in thyroid and vitamin A status from field studies of wildlife and fish. *Journal of Wildlife Diseases* 36:615–635.

Romano, L., and Passow, H. 1984. Characterization of anion transport system in trout red blood cells. *American Journal of Physiology* 246:C330–C338.

Rostal, D. C., Grumbles, J. S., Byles, R. A., Márquez-M., R., and Owens, D. W. 1997. Nesting physiology of the Kemp's ridley sea turtle, *Lepidochelys kempii,* at Rancho Nuevo, Tamaulipas, Mexico, with observation on population estimates. *Chelonian Conservation and Biology* 2:538–547.

Rostal, D. C., Owens, D. W., Grumbles, J. S., MacKenzie, D. S., and Amoss, M. S. 1998. Seasonal reproductive cycle of the Kemp's ridley sea turtle (*Lepidochelys kempii*). *General and Comparative Endocrinology* 109:232–243.

Sandor, T., Fazekas, A. G., and Robinson, B. H. 1976. The biosynthesis of corticosteroids throughout the vertebrates. In Chester Jones, I., and Henderson, I. W. (Eds.). *General, Comparative, and Clinical Endocrinology of the Adrenal Cortex.* New York: Academic Press, pp. 125–142.

Sapolsky, R. M., Romero, L. M., and Munck, A. U. 2000. How do glucocorticoids influence the stress response? Integrating permissive, suppressive, stimulatory, and preparative actions. *Endocrine Review* 21:55–89.

Schwantes, N. 1986. Aspects of circulating corticosteroids in sea turtles. M.S. thesis, Texas A&M University, College Station.

Seasholtz, A. F., Valverde, R. A., and Denver, R. J. 2002. Corticotropin-releasing hormone-binding protein: biochemistry and function from fishes to mammals. *Journal of Endocrinology* 175:89–97

Senba, E., and Ueyama, T. 1997. Stress-induced expression of immediate early genes in the brain and peripheral organs of the rat. *Neuroscience Research* 29:183–207.

Shaw, S. L., Leone-Kabler, S., Lutz, P. L., and Schulman, A. 1992. Isoflurane: a safe and effective anesthetic for marine and freshwater turtles. In *Proceedings of the 1992 International Wildlife Rehabilitation Council Conference,Omnipress, Madison,* pp. 112–119.

Stabenau, E. K., and Heming, T. A. 1993. Determination of the constants of the Henderson-Hasselbach equation, α_{CO_2} and pK_a, in sea turtle plasma. *Journal of Experimental Biology* 180:311–314.

Stabenau, E. K., and Heming, T. A. 1994. The in vitro respiratory and acid–base properties of blood and tissue from the Kemp's ridley sea turtle, *Lepidochelys kempii. Canadian Journal of Zoology* 72:1403–1408.

Stabenau, E. K., and Vietti, K. R. N. 2003. The physiological effects of multiple forced submergences in loggerhead sea turtles (*Caretta caretta*). *Fishery Bulletin* 101:889–899.

Stabenau, E. K., Heming, T. A., and Mitchell, J. F. 1991a. Respiratory, acid–base and ionic status of Kemp's ridley sea turtles (*Lepidochelys kempii*) subjected to trawling. *Comparative Biochemistry and Physiology A* 99:107–111.

Stabenau, E. K., Vanoye, C. G., and Heming, T. A. 1991b. Characteristics of the anion transport system in sea turtle erythrocytes. *American Journal of Physiology* 261:R1218–R1225.

Stephenson, K., Vargas, P., and Owens, D. W. 2000. Diurnal cycling of corticosterone in a captive population of Kemp's ridley sea turtles (*Lepidochelys kempii*). In Abreu-Grobois, F. A., Briseño-Dueñas, R., Márquez-M., R., and Sartí-Martínez, L. (Compilers). *Proceedings of the Eighteenth International Sea Turtle Symposium.* NOAA Technical Memorandum NMFS-SEFSC-436, pp. 232–233.

Tagawa, M., and Brown, C. L. 2001. Entry of thyroid hormones into tilapia oocytes. *Comparative Biochemistry and Physiology B* 129:605–611.

Tucker, A. D., and Read, M. 2001. Frequency of foraging by gravid green turtles (*Chelonia mydas*) at Raine Island, Great Barrier Reef. *Journal of Herpetology* 35:500–503.

Valverde, R. A. 1996. Corticosteroid dynamics in a free-ranging population of olive ridley sea turtles (*Lepidochelys olivacea* Eschcholtz, 1829) at Playa Nancite, Costa Rica as a function of their reproductive behavior. Ph.D. diss. Texas A&M University, College Station.

Valverde, R. A., Provancha, J. A., Coyne, M. S., Meylan, A., Owens, D. W., and MacKenzie, D. S. 1996. Stress in sea turtles. In Keinath, J. A., Barnard, D. E., Musick, J. A., and Bell, B. A. (Compilers). *Proceedings of the Fifteenth Annual Symposium on Sea Turtle Biology and Conservation.* NOAA Technical Memorandum NMFS-SEFSC-387, pp. 326–329.

Valverde, R. A., Cornelius, S. E., and Mo, C. L. 1998. Decline of the olive ridley sea turtle (*Lepidochelys olivacea*) nesting assemblage at Nancite Beach, Santa Rosa National Park, Costa Rica. *Chelonian Conservation and Biology* 3:58–63.

Valverde, R. A., Owens, D. W., MacKenzie, D. S., and Amoss, M. S. 1999. Basal and stress-induced corticosterone levels in olive ridley sea turtles (*Lepidochelys olivacea*) in relation to their mass nesting behavior. *Journal of Experimental Zoology* 284:652–662.

Wada, M., and Shimizu, T. 2004. Seasonal changes in adrenocortical responses to acute stress in polygynous male bush warblers (*Cettia diphone*). *General and Comparative Endocrinology* 135: 193–200.

Wasser, J. S., and Jackson, D. C. 1991. Effects of anoxia and graded acidosis on the levels of circulating catecholamines in turtles. *Respiratory Physiology* 84:363–377.

Widmaier, E. P. 1990. Glucose homeostasis and hypothalamic-pituitary-adrenocortical axis during development in rats. *American Journal of Physiology* 259:E601–E613.

Wilson, C. M., and McNabb, F. M. A. 1997. Maternal thyroid hormones in Japanese quail eggs and their influence on embryonic development. *General and Comparative Endocrinology* 107: 153–165.

Wingfield, J. C., and Kitaysky, A. S. 2002. Endocrine responses to unpredictable environmental events: Stress or anti-stress hormones? *Integrative and Comparative Biology* 42:600–609.

Wingfield, J. C., Maney, D. L., Breuner, C. W., Jacobs, J. D., Lynn, S., Ramenofsky, M., and Richardson, R. D. 1998. Ecological bases of hormone–behavior interactions: the emergency life history stage. *American Zoologist* 38:191–206.

Yamamoto, Y. 1960. Comparative histological studies of the thyroid gland of lower vertebrates. *Folia Anatomica Japonica* 34:353–387.

Yasuda, A., Yamaguchi, K., Papkoff, H., Yokoo, Y., and Kawauchi, H. 1989. The complete amino acid sequence of growth hormone from the sea turtle (*Chelonia mydas*). *General and Comparative Endocrinology* 73:242–251.

Yasuda, A., Kawauchi, H., and Papkoff, H. 1990. The complete amino acid sequence of prolactin from the sea turtle (*Chelonia mydas*). *General and Comparative Endocrinology* 80:363–371.

DAVID C. ROSTAL

8

Reproductive Physiology of the Ridley Sea Turtle

RIDLEY SEA TURTLES SHARE many characteristics with the other sea turtles, but their reproductive biology is unique. They display the unique mass nesting behavior called *arribada* in which tens of thousands of females nest in a single night or day at one beach (Bernardo and Plotkin, Chapter 4). *Lepidochelys kempii* is also unique as it is the only sea turtle to nest during the day (Pritchard and Márquez-M., 1973). *L. olivacea* also may nest during daylight hours, but the majority of its nesting occurs at night. The arribada behavior is associated with a longer-than-average internesting interval of approximately 25 days (Pritchard, 1990; Rostal et al., 1997; Plotkin et al., 1997; Kalb, 1999; Rostal, 2005). Although this mass nesting behavior may have had evolutionary advantages for ridleys, it also made them vulnerable to excessive harvest of the adults and eggs by humans. This harvest, in addition to the incidental take of ridleys in fisheries, has led to drastic declines in some populations and the complete loss of two of three major nesting populations of *L. olivacea* on the Pacific coast of Mexico (National Research Council, 1990). *Lepidochelys kempii* was nearly lost; it was classified as endangered following its decline from 40,000 nesting females in 1947 to fewer than 1,000 nesting females in the early 1990s (National Research Council, 1990; Rostal et al., 1997). As a result of its classification as an endangered species, *L. kempii* is probably one of the best-studied sea turtles, whereas *L. olivacea* remains one of the least studied.

Many aspects of the reproductive biology of ridley sea turtles remain a mystery, such as the primary stimulus or trigger for mass nesting (arribadas). What cues are used by turtles to congregate offshore and wait for a specific night or day to emerge en masse to nest? What mechanisms allow fertile eggs with developing embryos to be retained in the oviduct for up to double the normal internesting period (i.e., 14–28 days) and still remain viable? What hormonal triggers are involved in stimulating or delaying the normal nesting event so that this highly synchronous nesting can occur?

Our current knowledge of seasonal reproductive chronology, nesting biology, oviductal egg development, fecundity, and arribada behavior of ridley turtles are reviewed in this chapter, along with discussion of the similarities to and differences from the other sea turtle species. The information reviewed here is primarily based on the results from studies of captive and wild populations (Plotkin et al., 1996; Plotkin et al., 1997; Rostal et al., 1997; Kalb, 1999; Rostal, 2005) from three nesting beaches that have been instrumental in our learning about the biology of ridley turtles: Rancho Nuevo, Mexico (*L. kempii*) and Oaxaca, Mexico and Nancite Beach, Costa Rica (*L. olivacea*).

Reproductive Biology of Sea Turtles

Sea turtles represent the most fecund amniotic vertebrates. They are iteroparous, produce large clutches and multiple clutches within a season, lay large eggs, and have a short internesting period (9–14 days). Green turtles (*Chelonia mydas*) are capable of producing 150+ eggs per clutch and laying in excess of six clutches in a single nesting season (Hendrickson, 1958, cited in Moll, 1979). The leatherback (*Dermochelys coriacea*) is capable of producing 70+ eggs per clutch and laying in excess of 10 clutches in a single nesting season (Tucker, 1989; Rostal et al., 2001). Other species such as the loggerhead (*Caretta caretta*) and the hawksbill (*Eretmochelys coriacea*) are similarly fecund (Moll, 1979).

The reproductive biology of five sea turtle species, in particular their reproductive physiology, has been studied to varying degrees. There have been three major reviews of sea turtle reproduction in the past decade (Miller, 1997; Owens, 1997; Hamman et al., 2003). A model for understanding sea turtle reproduction was developed from studies of captive *Chelonia mydas* (Owens, 1997). This model provided a template to study and compare other sea turtles. Studies with captive and wild populations, both in the water and at the nesting beach, have expanded our understanding of sea turtle reproductive physiology (*Chelonia mydas,* Licht et al., 1979; Owens, 1980; *L. kempii,* Rostal et al. 1997, 1998; *L. olivacea,* Licht et al., 1982; *C. caretta,* Wibbels et al., 1990; Whittier et al., 1997; *D. coriacea,* Rostal et al., 1996, 2001). The circannual endocrine patterns are now known for three species, *C. mydas* (Licht et al., 1979), *C. caretta* (Wibbels et al., 1990), and *L. kempii* (Rostal et al., 1998). Other studies have focused on nesting physiology (Licht et al., 1982; Figler et al., 1989; Wibbels et al., 1992; Rostal et al., 1997; Whittier et al., 1997; Hamman et al., 2002; Rostal, 2005).

Male Ridley Turtles

Male Reproductive System

Hamann et al. (2003) reviewed the male sea turtle reproductive system and spermatogenesis. Similar to other sea turtles, the male ridley has a paired reproductive tract with the testes located intraperitoneal to the kidney. The epididymis is coiled and is adjacent to the testis. Fully mature sperm are stored in the epididymis until mating. Male ridleys display secondary sex characteristics such as an elongated tail with enclosed penis and enlarged claws on the front flippers similar to other sea turtles (Owens, 1997). In addition, males display a decornified region on the midline of the plastron, which is hypothesized to be a secondary sex characteristic resulting from increased androgen levels (Wibbels et al., 1991; Owens, 1997). Mating is seasonal and observed shortly before the beginning of the nesting season and may continue into the early nesting season. Our present knowledge of the male reproductive cycle is largely from studies of captive male *L. kempii* at the Cayman Turtle Farm, Grand Cayman, Cayman Islands (Rostal et al., 1998; Rostal, 2005) with data from wild male

L. olivacea largely coming from in-water studies from Nancite Beach, Costa Rica (Plotkin et al., 1996; Kalb, 1999).

Male Testicular Cycle

Male *L. kempii* exhibit seasonal patterns in testicular development, spermatogenesis, and testosterone production (Rostal et al., 1998; Fig. 8.1). Captive male *L. kempii* at the Cayman Turtle Farm displayed a rise in testosterone 4–5 months before mating, during which time testicular recrudescence and spermatogenesis occur. Circulating testosterone rises during the fall and winter when water temperatures begin to decline (August to January). During this period, males do not display reproductive behavior (courtship or mounts) or increased swimming activity. As water temperatures increase from a low of 26°C in January to a high of 31°C in August, testosterone declines during the mating (March) and nesting (April to July) periods. Spermatogenesis was confirmed to be seasonal in male *L. kempii* using laparoscopy (Rostal, 1991, 2005). Spermatogenesis is testosterone dependent, and the response has been correlated to decreasing temperature in a variety of turtles as well as other vertebrates. Elevated levels of testosterone are observed before mating (November) when spermatogenesis has progressed to stages 4 and 5 of the classification of McPherson et al. (1982; Fig. 8.2). Seminiferous tubules are enlarged with abundant spermatocytes and spermatids. Mating is observed during March, at which time circulating testosterone levels have begun to decline. During the nesting season (April to July), circulating testosterone levels are near their nadir, seminiferous tubules have regressed and the lumina are filled with debris from the previous cycle (Fig. 8.2). Similar seasonal reproductive patterns have been observed in male captive *C. mydas* (Licht et al., 1985) and wild *C. caretta* (Wibbels et al., 1990). Histological data from *C. caretta* support the conclusion that spermatogenesis lasts for 9 months (Hamann et al., 2003).

Observations of wild male *L. kempii* and *L. olivacea* support observations from captive animals. Male *L. kempii* have been observed mating in nearshore waters off Rancho Nuevo, Mexico during the early nesting season (April and May), when females are potentially still receptive to mating. In the eastern Pacific Ocean, male *L. olivacea* mate in summer and early fall, before the nesting season that occurs from June to December (Plotkin et al., 1996; Kalb, 1999). Plotkin et al. (1996) collected blood samples and testis biopsies from male *L. olivacea* during summer months (July to September) off Nancite Beach. These males had soft plastra as reported by Wibbels et al. (1991) and Owens (1997). Testosterone levels of male *L. olivacea* were elevated to levels similar to those observed in captive male *L. kempii* during the mating period. Testis biopsies of male *L. olivacea* demonstrated active spermatogenesis with mature sperm in the lumina of the seminiferous tubules. These same male *L. olivacea* were fitted with satellite transmitters to track their migrations; they subsequently departed the nearshore waters off Nancite Beach following mating and before the onset of major arribadas in the fall (September to November) (Plotkin et al., 1996).

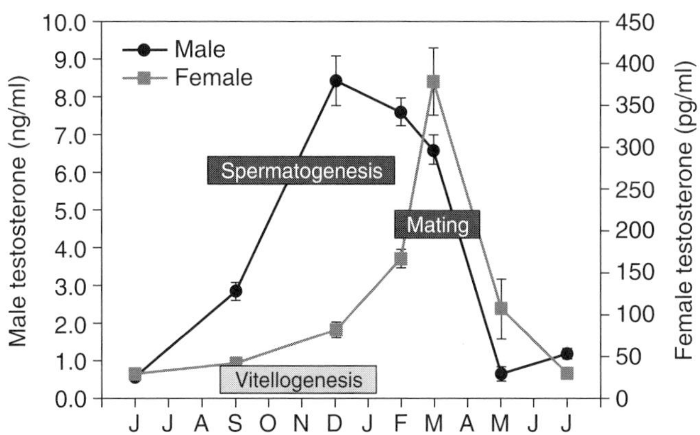

Fig. 8.1. Seasonal testosterone profiles for captive male and female *Lepidochelys kempii* maintained under seminatural conditions at the Cayman Turtle Farm, Grand Cayman, British West Indies. Values represent means ± SE (n = 10).

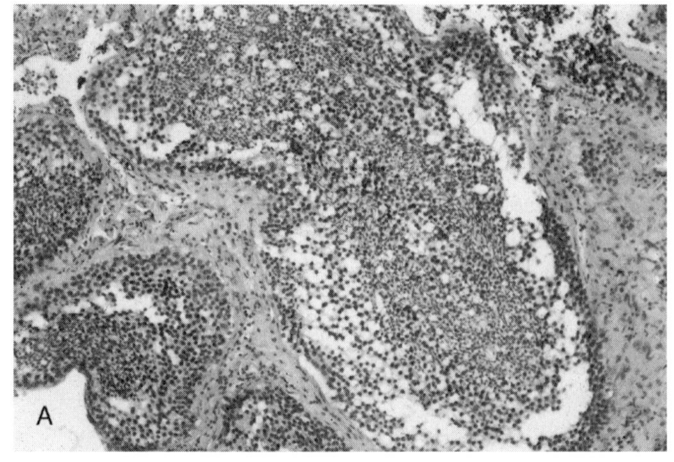

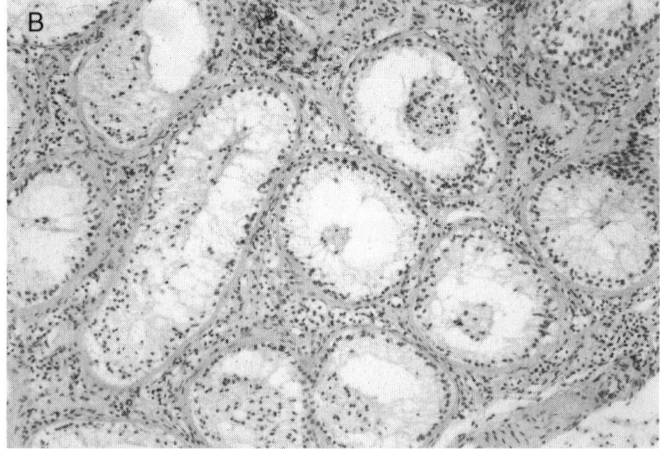

Fig. 8.2. Testis biopsies from captive male *Lepidochelys kempii* from May and November showing seasonal variation in the level of spermatogenesis. During May, the testis is regressed with little to no spermatogenesis following the mating period. In contrast, the testis is highly active in November before mating with active spermatogenesis, as can be seen by the increased lumen size of the seminiferous tubules and the increased number of developing spermatids present (both sections are at equal magnification).

Female Ridley Turtles

Female Reproductive System

The female reproductive system has been studied more thoroughly than that of the male. The basic female reproductive anatomy has paired reproductive organs located abdominally (Hamann et al., 2003). Adult females have a pair of ovaries with expanded stroma with follicles and paired oviducts at least 1.5 cm in diameter and suspended in the body cavity. Both ovaries function synchronously during reproduction and contain previtellogenic follicles, vitellogenic follicles, atretic follicles, and scars, depending on the season (Fig. 8.3). Oviducts are long (4–5 m in length in *L. olivacea*) (Owens, 1980) in both species and also function synchronously, undergoing changes in preparation for reproduction (e.g., thickening of the oviductal wall, development of the albumin gland and shell gland regions) (Rostal et al., 1990; Fig. 8.3). Ovulation occurs approximately 25 days before the first nesting of the season. Ovulation for subsequent nests occurs within 24–48 hours following a successful nesting emergence. Fertilization occurs in the upper oviduct, where sperm are stored following copulation with the male. No specialized sperm storage structures are present in sea turtles. Instead, sperm are stored in the ducts of the albumin glands in the upper region of the oviduct (Gist and Jones, 1989; Rostal et al., 1990). Following ovulation, it appears that sperm are passively forced from the gland ducts, and fertilization occurs. Following fertilization, extraembryonic membranes form, and albumin layers are laid down around the ovum as it travels through the albumin gland portion of the oviduct (Rostal et al., 1990). Once the developing ovum passes the isthmus, it enters the shell gland region of the oviduct, where the calcified shell is laid down (Rostal et al., 1990). This process is quite fast, and the earliest signs of a shelled egg can be detected with ultrasound within 48 hours. The eggs are then held in this region of the oviduct until nesting. Both ovaries function during each nesting event, and a female normally has shelled oviductal eggs in addition to large preovulatory follicles in each ovary during the nesting season (Fig. 8.3).

Vitellogenesis

Preparation for reproduction in the female is dependent on vitellogenesis. Hamann et al. (2003) note that the process of vitellogenesis is very similar in reptiles. During vitellogenesis, estradiol (E_2) is secreted by the granulosa cells of the ovary into the circulation. Circulating E_2 subsequently stimulates the liver to secret the vitellogenin protein. The vitellogenin protein then is taken up by the maturing follicles in the ovary. During vitellogenesis, ovarian and follicular growth occurs before reproduction (Hamann et al., 2003; Fig. 8.4). Vitellogenesis results in the incorporation of yolk proteins into the developing oocyte within the follicle. Vitellogenesis has been demonstrated to be estradiol-17β (E_2)–dependent in a variety of reptiles (Ho, 1987). The source of the E_2 is the granulosa cells of the previtellogenic follicles in response to gonadotropin secretion by the pituitary. Wibbels et al. (1990) measured elevated levels of E_2 in wild female *C. caretta* 1–2 months before migration. In addition, injections of E_2 in captive immature *C. mydas* and *L. kempii* stimulated vitellogenesis (Owens, 1976; Heck et al., 1990, 1997).

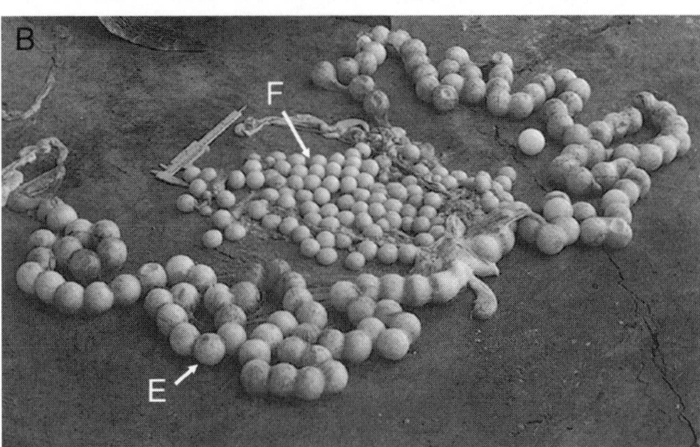

Fig. 8.3. Reproductive tract of a female *Lepidochelys olivacea* apparently killed in a shrimp trawl. (A) Note the volume of eggs (E) and ovarian follicles (F) in the carapace of the female. (B) The reproductive tract of the female showing both oviducts containing shelled eggs (left oviduct, 51 eggs; right oviduct, 65 eggs) as well as fully developed preovulatory follicles in the ovaries (88 ovarian follicles).

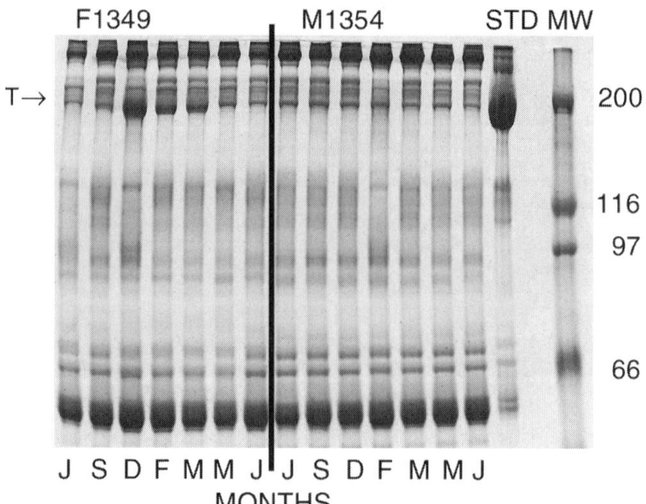

Fig. 8.4. Polyacrylamide gel showing seasonal variation in female vitellogenin protein levels corresponding with follicular development before the onset of mating and nesting. T, testosterone.

Ovarian follicle development and size were monitored in captive *L. kempii* (Rostal et al., 1998). Before mating, follicles were fully developed and consistent in size, indicating that vitellogenesis is complete (Rostal et al., 1998). Vargas et al. (2002) developed a vitellogenin ELISA to measure the levels of vitello-genin in blood. The pattern of vitellogenin in the blood of captive female *L. kempii* correlated strongly with both gel electrophoretic results and circulating levels of total calcium. Vitellogenesis is complete before mating in *L. kempii,* as indicated by the decline in vitellogenin and total calcium (Rostal et al., 1998). Wibbels et al. (1990) suggest that in the larger sea turtle species (e.g., *C. caretta* and *C. mydas*), vitellogenesis may continue into the early nesting season. However in *D. coriacea,* which nests up to 10 times or more in a single nesting season, vitellogenesis is complete before the arrival of the female at the nesting beach (Rostal et al., 1996, 2001). Hamann et al. (2002) observed that total lipids in yolk follicles were similar to those found in egg yolks during the early, middle, and late nesting season.

Female Ovarian Cycle

Most available data come from nesting female turtles because they are easily encountered on the beach. Our knowledge of the seasonal reproductive pattern is still based largely on studies of captive female *L. kempii* (Rostal, 1991; Rostal et al., 1998); however, more data have become available for female *L. olivacea* from open waters in the eastern Pacific Ocean (Plotkin et al., 1997; Kalb, 1999). Female *L. kempii* also display distinct seasonal cycles in circulating testosterone, estradiol, progesterone, total calcium, and vitellogenin (Figs. 8.1 and 8.5). The female ovary undergoes an associated or prenuptial recrudescence before the mating period. Four to six months before the mating period, increased levels of circulating estradiol, vitellogenin protein, and total calcium are observed in the female *L. kempii* (Rostal et al., 1998; Fig. 8.5).

Follicular maturation and ovarian recrudescence have been confirmed using ultrasonography in *L. kempii.* As the ovarian follicles mature and increase in size from previtellogenic to vitellogenic, the synthesis of testosterone increases. Circulating testosterone increases in association with ovarian recrudescence before the mating period. At mating, the ovary is fully developed in *L. kempii,* and the entire complement of follicles for the nesting season is present. Coincident with the onset of mating, testosterone and estradiol are also at their maximum levels. As the nesting season progresses, subsequent clutches are ovulated, and serum testosterone and estradiol levels are observed to decline to their nadir in June and July (late nesting). During the internesting period, the female may contain both fully shelled eggs in her oviducts plus multiple preovulatory follicles in her ovaries filling the

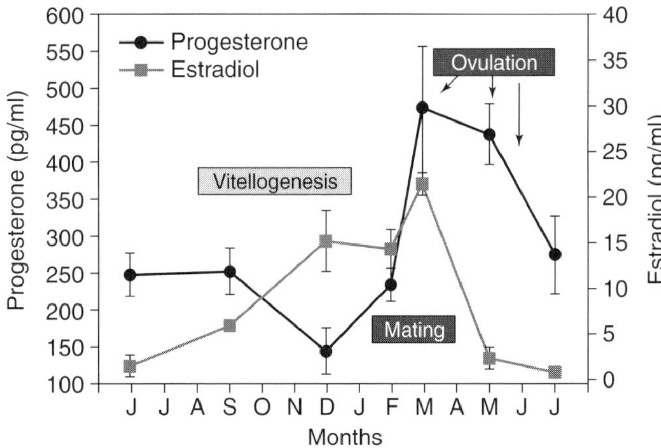

Fig. 8.5. Seasonal estradiol and progesterone profiles for captive female *Lepidochelys kempii* maintained under seminatural conditions at the Cayman Turtle Farm, Grand Cayman, British West Indies. Values represent means ± SE ($n = 10$).

majority of the body cavity (Fig. 8.3A). Both oviducts and both ovaries function as can be seen in Figure 8.3B. *Lepidochelys kempii* is capable of multiple nesting in captivity (Rostal et al., 1998). As each clutch of follicles is ovulated, a proportion of the steroid source (i.e., granulosa cells of the follicles) appears to be depleted, and serum testosterone, E_2, and progesterone decline. Observations from both wild *L. kempii* and *L. olivacea* also display a decline in circulating steroid levels as subsequent clutches of eggs are ovulated from the ovary. Similar declines in testosterone, E_2, and progesterone over the nesting season have been observed for wild *C. caretta* (Wibbels et al., 1990; Whittier et al., 1997) and *D. coriacea* (Rostal et al., 1996, 2001).

Follicle size or diameter is related to the size of the ova and the amount of energy invested by the female in reproduction. The energy stored as yolk platelets in the ova of the follicle plus other substances essential for embryogenesis will produce the hatchling plus a yolk reserve that the hatchling will be dependent on for the first few months of its life. The degree of variation in follicle size within and between females will influence hatchling size between the populations and species. Ovarian follicle size has been measured in wild populations of *L. kempii* and *L. olivacea*. Rostal et al. (1997) observed that follicle size (mean ± standard error [SE] = 2.5 ± 0.02 cm, $n = 40$) in *L. kempii* was consistent throughout the nesting season and between females. Data from wild populations of *L. olivacea* nesting at Playa Nancite, Costa Rica display similar patterns in follicle size (mean ± SE = 2.46 ± 0.01 cm, $n = 108$) (Kalb, 1999) to *L. kempii*. Follicle size was consistent among females and years in *L .olivacea* (D. C. Rostal, unpublished data). Preovulatory follicle size is similar within the ridley populations studied, suggesting a minimal-size hatchling is likely to be produced.

Courtship and Mating

Courtship and mating occur before nesting in captivity (Rostal et al., 1997). Once nesting begins, a significant decline in mating activity is observed. Mass aggregations have been observed off several of the major nesting beaches for ridley turtles. Shortly before and during the early portion of the nesting season, ridley turtles can readily be seen courting and mating in the nearshore waters (Plotkin et al., 1995, 1996; Kalb, 1999). At this time, females have fully developed ovaries with large preovulatory follicles (Rostal et al., 1997). Good accounts of courtship behavior in wild ridley turtles have been difficult to obtain because of the deep water usually present off the nesting beaches (Plotkin et al., 1995, 1996; Kalb, 1999). However, mating pairs have also been observed far offshore and separate from these aggregations as well. The environmental or physiological factors that trigger these mass aggregations still remain a mystery. Males will remain in the area of the nesting beach for an extended time but do depart the vicinity as the major nesting season begins (Plotkin et al., 1996; Kalb, 1999). This departure from the nesting beach area by males was associated with a decline in male testosterone levels. This pattern is similar to that observed in captive male *L. kempii* at the Cayman Turtle Farm (CTF) (Rostal et al., 1998; Rostal, 2005). Similar patterns are reported for wild male *C. caretta* (Wibbels et al., 1990).

Seasonal reproduction in sea turtles is controlled by hormonal function. Testosterone appears to function in regulating seasonal reproduction in both male and female sea turtles. The long-term elevation of testosterone in males appears to have primarily a physiological role but may also have a behavioral role in priming specific regions of the brain. In female *L. kempii,* however, testosterone would appear to be directly involved in triggering receptivity and the onset of mating. This associated pattern may represent the primitive pattern for tropical species. The onset of mating activity occurs following the increase in female testosterone from December to March. Mating activity was positively correlated with female testosterone levels. Mating occurred in captivity during a 3- to 4-week period preceding nesting during which certain females appeared to be receptive. In female *C. mydas,* a distinct "heat" period of 10–15 days has been observed during which a particular female is receptive (Wood and Wood, 1980; Comuzzie and Owens, 1990). Licht et al. (1979) noted that testosterone levels were high during mating in captive female *C. mydas.* Following this period, the female is no longer receptive and will avoid interactions with males. A similar pattern was observed for captive female *L. kempii* (Rostal et al., 1998). Females appeared to be receptive for only part of this mating period. The mating period for female *L. kempii* was too short to discern a "heat" period; however, the fact that certain females did avoid males later in the mating period suggests a similar period of receptivity. Following mating, female *L. kempii* were observed to avoid males by moving into the shallow portion of the enclosure (personal observation). Booth and Peters (1972) observed that wild female *C. mydas* that were unreceptive moved into shallow areas to avoid males. During May (mid-nesting), virtually no mating behavior (courtship or mounts) by males was recorded. Nesting in captivity occurred from mid April to late June. In the wild population of *L. kempii* at Rancho Nuevo, Tamaulipas, Mexico, the nesting season begins in mid-April and continues into July (Pritchard and Márquez-M., 1973). Hatchlings emerge from nests into September.

Ovulation

Following courtship and mating, the first ovulation of the nesting season is proposed to occur. Because turtles may be far from the nesting beach when this occurs, it has been difficult to document. Based on captive work with the *L. kempii* at the CTF, it is thought to occur shortly after successful mating, when the turtles appear to no longer be receptive to males. This is between 3 and 4 weeks before the first nesting event of the season. We do know that subsequent ovulations following the first nesting event occur within 48 hours following the completion of nesting and that specific hormonal events occur during this period.

Progesterone is primarily associated with ovulation in sea turtles. Progesterone levels are reported to increase sharply 24–48 hours following nesting in *L. olivacea* (Licht et al., 1982), *C. mydas* and *C. caretta* (Licht et al., 1979; Wibbels et al., 1992). A significant increase in progesterone was first observed in March in association with mating and the probable ovulation of the first clutch of eggs. As the nesting season progressed, we observed a steady decline in progesterone levels into July. Progesterone levels monitored at the time of nesting were also observed to decline with each subsequent clutch a female laid (Fig. 8.4).

Luteinizing hormone (LH) is observed to increase sharply following nesting concurrent with the observed surge in progesterone and ovulation of the ovum from the follicle into the upper oviduct, where it will be fertilized (Licht et al., 1979, 1982; Wibbels et al., 1992). Unlike the sequential decline observed in progesterone over the course of the nesting season, LH is elevated only during the brief period of ovulation and then remains low until the next ovulation. Once the follicle has left the ovary and enters the oviduct, a new series of events is observed to occur as the follicle is fertilized, the extraembryonic membranes are formed, the albumin layers are laid down, and the shell is formed as the developing egg travels down the oviduct. This sequence is reviewed in greater detail in the next section.

Oviductal Egg Development

Oviductal egg formation and development have not been well studied in sea turtles. Ultrasonography has been used successfully to track egg development in both captive and wild sea turtles (Fig. 8.6). Ovarian follicles (Fig. 8.6B) are ovulated in response to LH, and progesterone surges. Once ovulated, the ovum is first fertilized in the upper albumin gland region of the oviduct. As the fertilized ovum travels through the albumin gland region of the oviduct, it is surrounded by a nonechoic albumin layer. Once the albumin layer is completed, the ovum plus albumin enter into the shell gland region of the oviduct where the shell membrane and the eggshell form (Fig. 8.6B; day 14). In ridley turtles, fertilized eggs continue to develop further during the extended interesting period of 24–28 days. During this period, the yolk has become polarized with nonechoic fluid filling the animal pole and an echoic yolk filling the vegetal pole (Fig. 8.6C). The vitelline membrane is closer to the shell, and the albumin layer is reduced as a result of the movement of fluids from outside the vitelline membrane to the nonechoic animal pole region inside the vitelline membrane. This polarization observed with ultrasound occurs only in fertilized eggs, whereas unfertilized eggs retain the classic spherical yolk surrounded by a thick albumin layer. Following the completion of the nesting season, the remaining follicles that have not been ovulated undergo atresia (Fig. 8.6D).

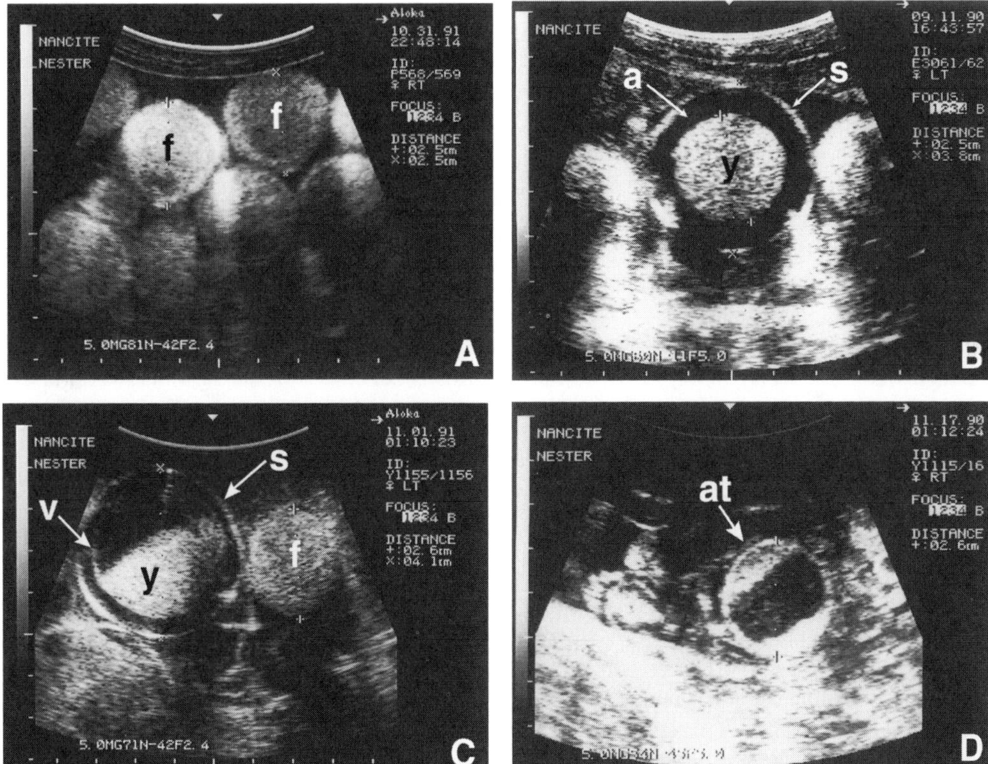

Fig. 8.6. Ultrasound scans of wild *Lepidochelys olivacea* showing various ovarian and reproductive structures. (A) Preovulatory follicles (f) in a mature ovary. (B) Fully shelled oviductal egg (approximately 14 days postovulation) showing the calcified shell (s), the nonechoic albumin layer (a), and the echoic yolk (y). (C) Developing oviductal egg (approximately 24 days postovulation) in which the yolk has become polarized with a nonechoic fluid-filled animal pole and an echoic yolk-filled vegetal pole. Note that the vitelline membrane is closer to the shell and the albumin layer is reduced because of the movement of fluids from outside the vitelline membrane to the nonechoic animal pole region inside the vitelline membrane. (D) Atretic follicle in a female following the laying of her last nest for the season. Note the "cat-eye" appearance and nonechoic fluid-filled core of the atretic follicle.

Atresia is the process by which energy originally stored as yolk platelets in the oocyte during vitellogenesis and follicular growth is reabsorbed from the oocyte and potentially reused by the fasting turtle. During this process, the follicle forms a unique "cat-eye" appearance as observed with ultrasound, in which a nonechoic line forms in the echoic yolk (Fig. 8.6D). This "cat-eye" appearance is unique to sea turtles and has not been observed in any other turtles or alligators (*Alligator mississippiensis*) (personal observation). This band of nonechoic material widens until the follicle appears to lose spherical integrity, and the follicle loses shape. The follicle then, over time, reduces in size as it eventually becomes a corpus albicans, leaving behind a small scar.

Oviposition

The process of nesting, or oviposition, is thought to be the most dangerous process for an adult female turtle. During this time, the adult female is potentially exposed to terrestrial predators and other risks for up to several hours during a single nesting event. The fact that the female repeats this response several times a nesting season and multiple times during her life increases the risk of potential injury. During this process the female must leave the safety of the sea for which she is specially adapted and must crawl to above the high-water line to lay her eggs where they will successfully incubate. During oviposition, a series of specific behavioral and endocrine events occur. A series of hormones are observed to change over the course of the nesting season. In particular, the steroid hormones testosterone, progesterone, and estradiol all decline while corticosterone levels are actually observed to increase and become more responsive as the nesting season progresses and the female emerges repeatedly from the safety of the ocean to deposit her eggs.

In addition to LH, which surges following the completion of nesting, two other hormones, arginine vasotocin (AVT), an ancestral form in the vasotocin family of hormones, and neurophysin, are known to change dramatically during the nesting process inclusive of the emergence and return to the sea. Figler et al. (1989) monitored changes in arginine vasotocin and neurophysin in *L. olivacea* nesting during an arribada at Oaxaca, Mexico and in *C. caretta* nesting at Mon Repo, Australia. AVT increases during oviposition, peaking at the time that the first egg is laid and subsequently declining as eggs are laid to near prenesting levels within a 30- to 50-minute period. The results of Figler et al. (1989) support the hypothesis that AVT has a physiological role in oviducal contraction in sea turtles.

Nesting Patterns, Arribada Behavior, and Conservation

The nesting patterns of ridley turtles are unique among the sea turtles. The pattern of mass nesting (arribada) by ridley turtles is characterized by an extended or longer than normal internesting period. Internesting intervals range from as short as 9 days observed in *D. coriacea* (Rostal et al., 2001) to as much as 66 days in *L. olivacea* depending on environmental conditions (Plotkin et al., 1995, 1997). In *L. kempii* the internesting period between arribadas averages 25 days (Rostal et al., 1997). The internesting period in *L. olivacea* is reported to be similar, averaging 28 days; however, it should be noted that at Nancite Beach it ranged from 26 to 66 days during 1990 and 1991 (Plotkin et al., 1997). It is known that some ridley turtles may nest between arribadas as solitary nesters and nest for a shorter interval (14 days) than those that nest during arribadas (Rostal et al., 1997; Kalb, 1999). However, these same individuals may have previously nested, or may later be observed nesting, during an arribada. The exact behavioral cues that stimulate arribada behavior remain a mystery. The mechanism by which ridley turtles are able to retain eggs in the oviduct for extended periods is also unknown.

Unlike other sea turtle species that nest more frequently, *L. kempii* was reported to nest only 1.3–1.5 times per season (Márquez-M. et al., 1982; Márquez-M., 1990). Kemp's ridley nesting physiology was the least studied among the sea turtles despite this species being the most critically endangered. As a result, our only estimate of the adult population is based solely on incomplete nesting data from Rancho Nuevo, Tamaulipas, Mexico. Better information is critical to the conservation of this species as well as to the estimation of the adult population. Lack of knowledge regarding the reproductive biology and true fecundity of both species of ridley turtles has contributed to overexploitation and their resultant declines.

It was previously thought, based on a fecundity index of 1.3 nests per female, that each arribada of the season was a separate population or subpopulation arriving at the nesting beach independently. This, however, does not appear to be the case. The pattern of decline in serum testosterone over the nesting season has implications regarding the nature of the nesting population. The majority of the nesting population arrives at Rancho Nuevo in mid- to late April and remains in the vicinity of Rancho Nuevo through June until they have completed nesting. This is further supported by the fact that females have been observed to nest up to four times during the season, although these were previously thought to be rare events (Márquez-M., 1994).

The pattern of nesting during the 1988, 1989, and 1990 seasons displayed multiple periods of increased nesting or arribadas. Although nesting was less synchronized during the 1988 nesting period, there were three distinct periods of nesting during 1990. Multiple recaptures of nesting females throughout the nesting season further support the ultrasonography and testosterone results of 3+ nests per female during the 1990 nesting season. The primary source of testosterone is the granulosa cells of the vitellogenic follicles. Testosterone results from the 1990 nesting season support the conclusion that the ovarian recrudescence is complete at the time of mating (Fig. 8.7). The three-stage decline observed in serum testosterone suggests that following mating, one-third of the preovulatory vitellogenic follicles are ovulated in preparation for nesting. Following nesting, ovulation is reported to occur within 48 hours postnesting in response to LH and progesterone (Licht et al., 1979). With each subsequent clutch produced, serum testosterone was observed to decline as the source of testosterone was removed by ovulation of the follicles. These observations have significant implications with regard to monitoring and managing this critically endangered species. The nesting population was estimated simply by dividing the total number of nests

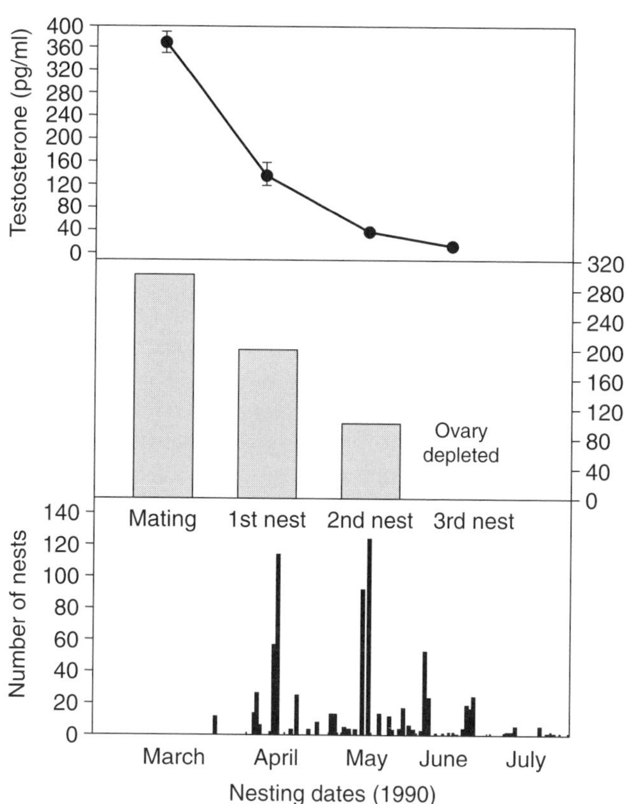

Fig. 8.7. Comparison of reproductive events in wild *Lepidochelys kempii* nesting at Rancho Nuevo, Tamaulipas, Mexico. Note the corresponding decline in testosterone levels as the number of estimated ovarian follicles declines with each subsequent ovulation and nesting event. These declines corresponded with the *arribadas* observed during the 1990 nesting season.

collected by the fecundity index of 1.3 from 1978 until early 1990s (Márquez-M. et al., 1982). The 1988 nesting season data for *L. kempii* at Rancho Nuevo, taking into account the probability of observing a given turtle on all of its nesting emergences, were compiled by Pritchard (1990). His analysis suggested that female *L. kempii* may be capable of 2.3 nests per season. Results of Rostal et al. (1997) support an estimate of 3.075 nests per female. Interestingly, Hildebrand (1963) reported that female *L. kempii* nested three times per season. On the basis of a fecundity index of 3.075, a mean egg count of 100 eggs per nest, and a short interannual interval of 1.8 years, the ridley turtles are highly fecund, similar to other sea turtle species.

Conclusions

Seasonal reproductive cycles have been well known for female turtles, but it is becoming more apparent that male turtles are also highly seasonal. Both associated and dissociated patterns have been reported for reptiles; however, most turtles show an associated or prenuptial pattern with some level of sperm storage by the female (Godwin and Crews, 2002). Male ridley turtles (*L. kempii* in particular: Rostal et al, 1998; Rostal, 2005) as well as other species in which the complete seasonal reproductive pattern is known (the desert tortoise, *Gopherus agassizi:* Rostal et al., 1994; Lance and Rostal, 2002; gopher tortoise, *Gopherus polyphemus:* Ott et al., 2000; and Galápagos tortoise, *Geochelone nigra:* Rostal et al., 1998; Schramm et al., 1999) display associated patterns. Within the sea turtles, both *C. mydas* and *C. caretta* males are known to display an associated pattern of testicular development and testosterone secretion (Licht et al., 1985; Wibbels et al., 1990). Male and female *L. kempii* both display distinct seasonal associative reproductive cycles. These cycles appear adapted to the environmental conditions that sea turtles encounter throughout their ranges. In *L. kempii,* both testicular and ovarian maturation are prenuptial, occurring in the fall and winter months when resources should be available, whereas mating and nesting occur in the spring before optimal incubation conditions at the primary nesting beach at Rancho Nuevo, Tamaulipas, Mexico. Female *L. olivacea* are equally seasonal to female *L. kempii,* but fewer data are available for male *L. olivacea*. However, based on the other parallels between *L. kempii* and *L. olivacea,* it would be unlikely that male *L. olivacea* were also not highly seasonal, displaying an associative pattern of reproduction.

Understanding reproductive biology including physiology is crucial to determining the true nesting fecundity of a species. Fecundity may be defined differently, but in the case of sea turtles, it normally represents the total number of eggs laid during a specific nesting season, which requires an accurate estimation of how many times a female nests during a specific nesting season in addition to the number of eggs laid per nesting. Although it is easy to count the eggs in a particular nest, sea turtles lay multiple clutches of eggs during the nesting season. Female sea turtles may not lay all their nests on a single beach or in the same region of a long coastline such as Rancho Nuevo, Mexico. Some species such as the island-nesting species *D. coriacea* or *E. imbricata* in the Caribbean may nest on multiple islands, making it almost impossible to record all nesting events. The determination of the true nesting fecundity of a female requires observing the female for every nesting event. This requires saturation tagging, which involves intense beach monitoring and the observation of every nesting emergence by females. The observation in recent years that females from most species will use alternate nesting beaches, which may be on different islands or coastlines, has made the determination of true nesting fecundity nearly impossible for many populations (Rostal et al., 2001). Site fidelity is often not as strong as previously reported in many populations, especially those nesting on islands. Conventional tagging studies involving several species of marine turtles have previously proven problematic (Green, 1979; Balazs, 1982; Mrosovsky, 1983; Alvarado and Figueroa, 1992). Flipper tag loss has been known to be high both within season and even more so between annual remigration to the nesting beach. Although passive integrated transponder tags (PIT) have become available in recent years, they still wait to be proven as a reliable alternative. Studies utilizing circulating testosterone levels as well as other steroids (estradiol and corticosterone, for example) have demonstrated that hormones can be measured accurately and convey important information about the reproductive status of the male as well as the female. Studies on captive *L. kempii* (Rostal, 1991) as well as in wild-nesting *C. caretta* (Wibbels et al., 1990) demonstrated that circulating testosterone levels provided a reliable index of the reproductive status of the male and female. Although serum testosterone could not be used to identify specific individuals, it could provide information as to whether a particular female had nested previously that season or was a new arrival. Rostal et al. (1990) validated the application of ultrasonography to studying the reproductive biology of female *L. kempii*. Ultrasonography is a noninvasive technique by which the ovarian status of a female can be monitored directly. The combination of ultrasonography and serum testosterone levels provides an accurate evaluation of the reproductive status of the female.

ACKNOWLEDGMENTS

I thank David W. Owens, Janice S. Grumbles, and Valentine Lance for support and assistance with many aspects of the studies presented here. The research reported here was supported by grants from the Texas A&M University Sea Grant College Program (NA85AA-D-SG128); National Science Foundation (BNS-8819940 and IBN-9124014); the Cayman Turtle Farm; the Center for Reproduction of Endangered Species, Zoological Society of San Diego; the Georgia Southern University Office of Research and Sponsored Programs; and EARTHWATCH. Research reported in this chapter was approved by the Texas A&M University Animal Care and Use Committee and the Georgia Southern University Institutional Animal Care and Use Committee.

LITERATURE CITED

Alvarado, J., and Figueroa, A. 1992. Recapturas post-anidatorias de hembras de tortuga marina negra (*Chelonia agassizii*) marcadas en Michoacan, Mexico. *Biotropica* 24:560–566.

Balazs, G. H. 1982. Factors affecting the retention of metal tags on sea turtles. *Marine Turtle Newsletter* 20:11–14.

Booth, J., and Peters, J. A. 1972. Behavioral studies on the green turtle (*Chelonia mydas*) in the sea. *Animal Behaviour* 20:808–812.

Comuzzie, D. K. C., and Owens, D. W. 1990. A quantitative analysis of courtship behavior in captive green sea turtles (*Chelonia mydas*). *Herpetologica* 46:195–202.

Eales, J. G. 1979. Thyroid function in cyclostomes and fishes. In Barrington, E. J. W. (Ed.). *Hormones and Evolution,* Volume 1. New York: Academic Press, pp. 341–436.

Figler, R. A., Mackenzie, D. S., Owens, D. W., Licht, P., and Amoss, M. S. 1989. Increased levels of arginine vasotocin and neurophysin during nesting in sea turtles. *General and Comparative Endocrinology* 73:223–232.

Gist, D. H., and Jones, J. M. 1989. Sperm storage within the oviduct of turtles. *Journal of Morphology* 199:379–384.

Godwin, J., and Crews, D. 2002. Hormones, brain and behavior in reptiles. In Pfaff, D. W., Arnold, A., Etgen, A., Fahrbach, S., Moss, R., and Rubin, R. (Eds.). *Hormones, Brain and Behavior.* New York: Academic Press, pp. 545–585.

Green, D. 1979. Double tagging of green turtles in the Galapagos Islands. *Marine Turtle Newsletter* 13:4–9.

Hamann, M., Limpus, C. J., and Whittier, J. M. 2002. Patterns of lipid storage and mobilization in female green turtles (*Chelonia mydas*). *Journal of Comparative Physiology B* 172:485–493.

Hamann, M., Limpus, C. J., and Owens, D. W. 2003. Reproductive cycles of males and females. In Lutz, P. L., Musick, J. A., and Wyneken, J. (Eds.). *The Biology of Sea Turtles.* Boca Raton, FL: CRC Press, pp. 135–161.

Heck, J., Mackenzie, D. S., and Owens, D. W. 1990. Vitellogenin from the Kemp's ridley sea turtle. *American Zoologist* 30(4):23A.

Heck, J., MacKenzie, D. S., Rostal, D., Medlar, K., and Owens, D. 1997. Estrogen induction of plasma vitellogenin in the Kemp's ridley sea turtle (*Lepidochelys kempii*). *General and Comparative Endocrinology* 107:280–288.

Hildebrand, H. H. 1963. Hallazgo del area de anidacion de la tortuga marine "lora" *Lepidochelys kempi* (Garman), en la costa occidental del Golfo de Mexico. *Ciencia* 22:105–112.

Ho, S. 1987. Endocrinology of vitellogenesis. In Norris, D., and Jones, R. (Eds.). *Hormones and Reproduction in Fishes, Amphibians, and Reptiles.* New York: Plenum Press, pp. 145–169.

Kalb, H. J. 1999. Behavior and physiology of solitary and arribada nesting olive ridley sea turtles (*Lepidochelys olivacea*) during the internesting period.

Ph.D. diss. Texas A&M University, College Station.

Lance, V. A., and Rostal, D. C. 2002. The annual reproduction cycle of the male and female desert tortoise: physiology and endocrinology. *Chelonian Conservation Biology* 4:302–312.

Licht, P., Wood, J., Owens, D. W., and Wood, F. E. 1979. Serum gonadotropins and steroids associated with breeding activities in the green sea turtle *C. mydas*. I. Captive animals. *General and Comparative Endocrinology* 39:274–289.

Licht, P., Owens, D. W., Cliffton, K., and Peñaflores, C. 1982. Changes in LH and progesterone associated with the nesting cycle and ovulation in the olive ridley sea turtle, *Lepidochelys olivacea*. *General and Comparative Endocrinology* 48:247–253.

Licht, P., Wood, J. F., and Wood, F. E. 1985. Annual and diurnal cycles in plasma testosterone and thyroxine in the male green turtle *C. mydas*. *General and Comparative Endocrinology* 57:335–344.

Márquez-M., R. 1990. *An Annotated and Illustrated Catalogue of Sea Turtle Species Known to Date. FAO Species Catalogue,* Volume 11: *Sea Turtles of the World*. FAO Fisheries Synopsis 125.

Márquez-M., R. 1994. *Synopsis of Biological Data on the Kemp's Ridley Turtle,* Lepidochelys kempii *(Garman, 1880)*. NOAA Technical Memorandum NMFS-SEFSC-343.

Márquez-M., R., Villanueva, O., and Sanchez, P. M. 1982. The population of the Kemp's ridley sea turtle in the Gulf of Mexico—*Lepidochelys kempii*. In Bjorndal, K. A. (Ed.). *Biology and Conservation of Sea Turtles*. Washington, DC: Smithsonian Institution Press, pp. 159–164.

McPherson, R. J., Boots, L. R., MacGregor, R., III, and Marion, K. R. 1982. Plasma steroids associated with seasonal reproductive changes in a multiclutched freshwater turtle, *Sternotherus odoratus*. *General and Comparative Endocrinology* 48:440–451.

Miller, J. D. 1997. Reproduction in sea turtles. In Lutz, P. L., and Musick, J. A. (Eds.). *The Biology of Sea Turtles*. Boca Raton, FL: CRC Press, pp. 51–81.

Moll, E. O. 1979. Reproductive cycles and adaptations. In Harless, M., and Morlock, H. (Eds.). *Turtles: Perspectives and Research*. New York: John Wiley & Sons, pp. 305–331.

Mrosovsky, N. 1983. *Conserving Sea Turtles*. London: The British Herpetological Society, Regents Park.

National Research Council, Committee on Sea Turtle Conservation. 1990. *Decline of the Sea Turtles: Causes and Preventions*. Washington, DC: National Academy Press.

Ott, J., Mendonça, M., Guyer, C., and Michener, W. 2000. Seasonal changes in sex and adrenal steroid hormones of gopher tortoises (*Gopherus polyphemus*). *Comparative Biochemistry and Physiology* 117:299–312.

Owens, D. W. 1976. Endocrine control of reproduction and growth in the green turtle, *C. mydas*. Ph.D. diss. University of Arizona, Tucson.

Owens, D. W. 1980. The comparative reproductive physiology of sea turtles. *American Zoologist* 20:549–563.

Owens, D. W. 1997. Hormones in the life history of sea turtles. In Lutz, P. L., and Musick, J. A. (Eds.). *The Biology of Sea Turtles*. Boca Raton, FL: CRC Press, pp. 315–341.

Plotkin, P. T., Byles, R. A., Rostal, D. C., and Owens, D. W. 1995. Independent versus socially facilitated oceanic migrations of the olive ridley, *Lepidochelys olivacea*. *Marine Biology* 122:137–143.

Plotkin, P. T., Owens, D. W., Byles, R. A., and Patterson, R. 1996. Departure of male olive ridley turtles (*Lepidochelys olivacea*) from a nearshore breeding ground. *Herpetologica* 52:1–7.

Plotkin, P. T., Rostal, D. C., Byles, R. A., and Owens, D. W. 1997. Reproductive and developmental synchrony in female *Lepidochelys olivacea*. *Journal of Herpetology* 31:17–22.

Pritchard, P. C. H. 1990. Kemp's ridleys are rarer than we thought. *Marine Turtle Newsletter* 49:1–3.

Pritchard, P. C. H., and Márquez-M., R. 1973. *Kemp's Ridley Turtle or Atlantic Ridley,* Lepidochelys kempi. IUCN Monograph No. 2: Marine Turtle Series.

Rostal, D. C. 1991. The reproductive behavior and physiology of the Kemp's ridley sea turtle, *Lepidochelys kempii* (Garman, 1880). Ph.D. diss. Texas A&M University, College Station.

Rostal, D. C. 2005. Seasonal reproductive biology of the Kemp's ridley sea turtle (*Lepidochelys kempii*): comparison of captive and wild populations. *Chelonian Conservation and Biology* 4:788–800.

Rostal, D. C., Robeck, T. R., Owens, D. W., and Kraemer, D. C. 1990. Ultrasound imaging of ovaries and eggs in Kemp's ridley sea turtles (*Lepidochelys kempii*). *Journal of Zoo and Wildlife Medicine* 21:27–35.

Rostal, D. C., Lance, V. A., Grumbles, J. S., and Alberts, A. C. 1994. Seasonal reproductive cycle of the desert tortoise (*Gopherus agassizii*) in the eastern Mojave Desert. *Herpetological Monographs* 8:72–82.

Rostal, D. C., Paladino, F. V., Patterson, R. M., and Spotila, J. R. 1996. Reproductive physiology of nesting leatherback turtles (*Dermochelys coriacea*) at Las Baulas de Guanacaste National Park,

Costa Rica. *Chelonian Conservation and Biology* 2:230–236.

Rostal, D. C., Grumbles, J. S., Byles, R. A, Márquez-M., R., and Owens, D. W. 1997. Nesting physiology of wild Kemp's ridley turtles, *Lepidochelys kempii,* at Rancho Nuevo, Tamaulipas, Mexico. *Chelonian Conservation and Biology* 2:538–547.

Rostal, D. C., Owens, D. W., Grumbles, J. S., MacKenzie, D. S., and Amoss, M. S. 1998. Seasonal reproductive cycle of the Kemp's ridley sea turtle (*Lepidochelys kempii*). *General and Comparative Endocrinology* 109:232–243.

Rostal, D. C., Grumbles, J. S., Palmer, K. S., Lance, V. A., Spotila, J. R., and Paladino, F. V. 2001. Reproductive endocrinology of the leatherback sea turtle (*Dermochelys coriacea*). *General and Comparative Endocrinology* 122:139–147.

Schramm, B. G., Casares, M., and Lance, V. A. 1999. Steroid levels and reproductive cycle of the Galapagos tortoise, *Geochelone Nigra,* living under seminatural conditions on Santa Cruz Island (Galapagos). *General and Comparative Endocrinology* 114:108–120.

Tucker, A. D. 1989. So many turtles, so little time: Underestimating fecundity and overestimating populations? In Eckert, S. A., Eckert, K. L., and Richardson, T. H. (Compilers). *Proceedings of the Ninth Annual Workshop on Sea Turtle Conservation and Biology.* NOAA Technical Memorandum NMFS-SEFC-232, pp. 81–183.

Vargas, P., Owens, D. W., and Mackenzie, D. 2002. Enzyme linked immunosorbent assay (ELISA) for sea turtle vitellogenin. In Mosier, A., Foley, A., and Brost, B. (Eds.). Proceedings of the Twentieth Annual Symposium on Sea Turtle Biology and Conservation. NOAA Technical Memorandum NMFS-SEFSC-477, p. 92.

Whittier, J. M., Corrie, F., and Limpus, C. 1997. Plasma steroid profiles in nesting loggerhead turtle (*Caretta caretta*) in Queensland, Australia: Relationship to nesting episode and season. *General and Comparative Endocrinology* 106:39–47.

Wibbels, T., Owens, D. W., Limpus, C. J., Reed, P. C., and Amoss, M. S., Jr. 1990. Seasonal changes in serum gonadal steroids associated with migration, mating, and nesting in the loggerhead sea turtle (*Caretta caretta*). *General and Comparative Endocrinology* 79:154–164.

Wibbels, T., Owens, D. W., and Rostal, D. C. 1991. Soft plastra of adult male sea turtles: an apparent secondary sex characteristic. *Herpetological Review* 22:47–49.

Wibbels, T., Owens, D. W., Licht, P., Limpus, C. J., Reed, P. C., and Amoss, M. S., Jr. 1992. Serum gonadotropins and gonadal steroids associated with ovulation and egg production in sea turtles. *General and Comparative Endocrinology* 87:71–78.

Wood, J. R., and Wood, F. E. 1980. Reproductive biology of captive green sea turtles (*Chelonia mydas*). *American Zoologist* 20:499–505.

THANE WIBBELS

9

Sex Determination and Sex Ratios in Ridley Turtles

A VARIETY OF REPTILES POSSESS temperature-dependent sex determination (TSD), including representatives in all four living orders of reptiles (Janzen and Pakustis, 1991; Lang and Andrews, 1994; Viets et al., 1994; Cree et al., 1995). The sex ratios produced by TSD are of interest for a number of reasons. From an ecological viewpoint, population sex ratios are of interest because they affect reproduction in the population. Sex ratios resulting from TSD are also of evolutionary interest because evidence suggests that many reptiles with TSD may not produce 1:1 sex ratios in a population (Wibbels, 2003).

Many reptiles with TSD are also of conservation interest because of their endangered or threatened status. The occurrence of TSD presents potential problems as well as benefits to recovery programs for endangered reptiles. Because sex ratios affect the reproductive output in a population, they can also affect the recovery and survival status of a population. Therefore, conservation programs involved in the recovery of endangered reptiles with TSD need to address TSD-related issues that could potentially benefit or hinder the recovery of a particular species. This chapter reviews what is currently known regarding sex determination and sex ratios in ridley sea turtles.

General Characteristics of TSD

Numerous studies have examined the effects of specific temperatures on sex determination in a variety of reptiles with TSD (see reviews by Janzen and Pakustis, 1991; Ewert et al., 1994; Viets et al., 1994). The period during which sex determination is sensitive to temperature has been shown to be approximately the middle third of incubation in many of the reptiles with TSD (Yntema, 1979; Bull and Vogt, 1981; Pieau and Dorizzi, 1981; Yntema and Mrosovsky, 1982; Ferguson and Joanen, 1983; Bull, 1987; Webb et al., 1987; Deeming and Ferguson, 1988; Wibbels et al., 1991). Studies of the loggerhead, *Caretta caretta* (Yntema and Mrosovsky, 1982), and the olive ridley, *Lepidochelys olivacea* (Merchant-Larios et al., 1997), indicate that the thermosensitive period also occurs during the middle third of the incubation period in sea turtles. During the thermosensitive period, most reptiles with TSD appear to conform to one of two patterns (Bull, 1980; Ewert et al., 1994): the Male:Female (MF) pattern, in which males are produced at low temperatures and females are produced at high temperatures, or the Female:Male:Female (FMF) pattern, in which males are produced at intermediate temperatures and females are produced at both low and high temperatures (Fig. 9.1). All sea turtles appear to have the MF pattern (reviewed by Mrosovsky, 1994; Wibbels, 2003). In addition to the pattern of sex determination, several other parameters should be considered in describing TSD (Mrosovsky and Pieau, 1991). The "transitional range of temperatures" (TRT) is the range of temperatures in which sex ratios gradually change from 100% male to 100% female (Fig. 9.1). In the case of sea turtles (i.e., the MF pattern), temperatures above the TRT will produce all females, and temperatures below the TRT will produce all males. The TRT can vary between species, and it has also been reported to vary among populations of leatherback sea turtles, *Dermochelys coriacea* (Chevalier et al., 1999). Within the TRT there is a "pivotal temperature" that will produce a 1:1 sex ratio (Mrosovsky and Pieau, 1991). The pivotal temperature can vary between and even within a species (Yntema, 1976; Bull et al., 1982a, 1982b; Mrosovsky, 1988; Etchberger et al., 1991; Janzen, 1992; Ewert et al., 1994; Girondot et al., 1994; Lang and Andrews, 1994; Mrosovsky, 1994; Rhen and Lang, 1998; Wibbels, 2003). Therefore, it is optimal to determine these parameters for the specific population of interest rather than extending data from one population to another. Once determined, this information can be very useful in studies or programs addressing the biology and conservation of sea turtles. In the case of conservation programs using protected egg hatcheries, for example, the choice of a specific thermal environment (i.e., a specific location on a given beach) can guarantee the production of a desired sex ratio. In the case of conservation programs protecting nests in their natural locations on the beach, incubation temperatures can be monitored to estimate hatchling sex ratios.

It is noteworthy that most previous studies examining TSD have used constant incubation temperatures to characterize sex determination. Thus, these data may not be directly applicable to natural nests, which undergo daily fluctuations in incubation temperatures. This subject has been addressed experimentally (Georges et al., 1994), and the results of that study indicate that daily temperature fluctuations can have a profound effect on sex determination, but the effect is dependent on the magnitude of the fluctuation. In species that dig deep nests (such as sea turtles), daily temperature fluctuations would be minimal. These minimal fluctuations would only alter the effect of temperature by a maximum of a few tenths of a degree Celsius in comparison to constant temperatures (Georges et al., 1994). This suggests that data from constant-temperature studies of TSD in sea turtles are applicable to studies of natural sea turtle nests. However, one should be cautious in applying constant-temperature data to natural nests. Although some studies have examined the effects of natural temperature regimens on sex ratio (Morreale et al., 1982; Standora and Spotila, 1985; Aguilar, 1987; Spotila et al., 1987), there is a distinct need for future studies addressing this subject.

Evolutionary, Ecological, and Conservation Implications of TSD

Before we review specific data regarding the ridley turtles, there are several implications of TSD

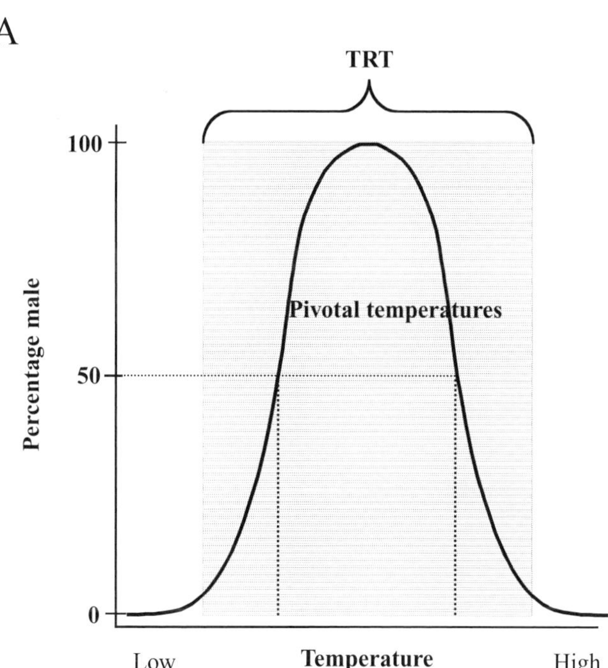

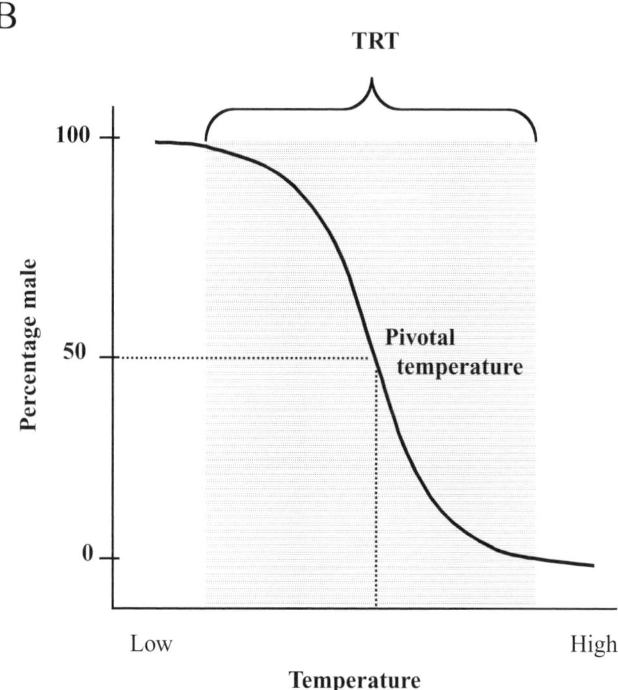

Fig. 9.1. Patterns of temperature-dependent sex determination in reptiles over the transitional range of temperatures (TRT). (A) The female: male:female, or FMF pattern, in which females are produced at relatively low and high temperatures and males are produced at intermediate temperatures. All crocodilians, some turtles, and some lizards have this pattern. (B) The male: female, or MF pattern, in which males are produced at low temperatures and females are produced at high temperatures. Many turtles, including all sea turtles, possess the MF pattern of sex determination.

that should be addressed. Such information is a prerequisite to interpreting the significance of data on sex determination and sex ratios of ridley turtles. First, evolutionary theory suggests that the primary sex ratio (e.g., hatchling sex ratio) should be approximately 1:1 if factors such as parental investment in both sexes are equal (Fisher, 1930). However, TSD has the potential of producing a wide range of sex ratios. For example, hatchling sex ratios produced from TSD in sea turtles can be highly biased (Mrosovsky, 1983, 1994; Wibbels, 2003), and such sex ratios can affect reproductive output in a population. Therefore, sex ratios from sea turtles can have profound implications for the reproductive ecology and conservation of a population. For example, a population with an adult sex ratio of 1:1 (i.e., 50% female) could potentially produce twice as many offspring as a population that is only 25% female (assuming other factors such as fertility and reproductive behavior are similar at both sex ratios). Thus, TSD can have a significant impact on the ecology and conservation of sea turtles.

As indicated above, from an evolutionary viewpoint, one might predict that TSD should produce a 1:1 sex ratio (assuming such factors as parental investment in both sexes is equal, etc.) (Fisher, 1930). However, this does not appear to be the norm in regard to hatchling sex ratios of sea turtles. A general review of hatchling sex ratios in sea turtles indicates that a wide variety of sex ratios can be produced from TSD, including many that are significantly biased (reviewed by Mrosovsky, 1994; Wibbels, 2003). Several points can be derived from a review of these hatchling sex ratio data. First, the majority of the predicted sex ratios do not conform to a 1:1 sex ratio. Although a few of the sex ratios approach 1:1, there are a predominance of beaches that produce female-biased hatchling sex ratios, and some of these biases may be extreme (greater than 90% female). Female biases appear common, but there are no reports of extreme male biases over an entire nesting season. Thus, it is possible that TSD in sea turtles may not conform to the predictions of evolutionary theory (Fisher, 1930); rather, a predominance of female biases may exist. However, other explanations are possible. It is plausible that some criteria are not fulfilled regarding Fisherian sex ratios, such as sex ratios not being at equilibrium (Mrosovsky, 1994; Godfrey et al., 1996). Additionally, the sex ratios previously reported represent only a small sampling of all sea turtle populations and nesting beaches; therefore, the available data could be affected by sampling bias.

Alternatively, TSD may produce biased sex ratios that are derived by natural selection. Although biased sex ratios may appear to be an enigma in regard to evolutionary theory (Bull and Charnov, 1989), a variety of hypotheses have been proposed that could explain how biased sex ratios could evolve (reviewed by Shine, 1999; also see Bull and Charnov, 1989; Reinhold, 1998; Girondot, 1999). For example, it has been proposed that differential fitness of males versus females in a sea turtle population could select for biased sex ratios (Bull and Charnov, 1989; Shine, 1999). Analysis of naturally occurring sex ratios combined with a better understanding of the reproductive ecology of sea turtles could provide insight on whether such hypotheses may be the underlying cause of skewed sex ratios that are often seen in sea turtle populations.

Another question regarding the ecology and evolution of sex determination and sex ratios in sea turtles is the physical or physiological means by which a specific sex ratio is derived in nature. For example, it is plausible that pivotal temperatures might evolve to match the temperatures of a given nesting beach. Alternatively, a sea turtle population may choose a specific nesting beach because its temperatures fall within those needed to produce a given range of sex ratios. A more comprehensive understanding of naturally occurring sex ratios together with data on pivotal temperatures and beach temperatures should provide insight on this question.

As indicated above, TSD can produce a wide variety of sex ratios, and those sex ratios affect the reproductive potential of a population. The example discussed above suggested that a greater proportion of females in a population may result in the production of more offspring (assuming that the bias does not decrease fertility or alter reproductive behavior, etc.). It exemplifies the implications for the conservation of endangered sea turtles with TSD; however, this is an oversimplified example. In regard to conserving an actual population of sea turtles, a wide variety of questions arise about the effects of TSD on sex ratio and survival status. For example, what effect does sex ratio have on the reproductive ecology? Are female-biased sex ratios better for the survival of a TSD species than male-biased sex ratios? The scenario becomes more complex if one decides to manipulate sex ratios in an effort to enhance the recovery of an endangered species. To manipulate sex ratios, one must address a fundamental question regarding the conservation of reptiles with TSD: What sex ratio is optimal for the recovery of a given species? Further, if sex ratios are to be manipulated, logistic questions arise such as "Which temperature ranges produce each sex, and which specific temperatures within these ranges are best for producing each sex?" Addressing these questions requires a comprehensive understanding of TSD in a species, including the effects of sex ratio on reproductive ecology. Unfortunately, comprehensive information of this sort is not available for any reptile with TSD. However, certain aspects of TSD have been studied in some species, and this information can provide insight on the conservation implications of TSD. One purpose of this chapter, then, is to review what is known about TSD in ridley turtles and discuss the conservation relevance of this information.

TSD in the Olive Ridley

Several studies have examined the effects of specific temperatures on sex determination in ridley turtles. In the olive ridley, *Lepidochelys olivacea,* the thermosensitive period of sex determination was examined, and it appears to approximate the middle third of incubation (Merchant-Larios et al., 1997). Two studies have attempted to define the pivotal temperature and TRT of olive ridleys from Playa Nancite, Costa Rica (McCoy et al., 1983; Wibbels et al., 1998b). Both of these studies were conducted in the laboratory using constant-temperature incubators. In the initial study (McCoy et al., 1983), incubation temperatures less than approximately 28.0°C produced males, and temperatures of 32.0°C or above produced females. That study estimated a pivotal temperature near 30°C. In the latter study (Wibbels et al., 1998b), results showed that an incubation temperature of 27.0°C produced all males, 29.4°C produced mostly males, and suggested a pivotal temperature nearer to 31°C, which is slightly higher than that recorded by McCoy et al. (1983). It is possible that variations between the two studies could relate to factors such as interclutch variation or experimental error (e.g., differences in the thermal characteristics of the incubators, etc.). Subsequent studies using olive ridley eggs from Mexico have used 27.0°C to consistently produce males or 32.0°C to consistently produce females in experiments examining sex determination in the olive ridley (Merchant-Larios et al., 1989; Merchant-Larios and Villalpando, 1990; Merchant-Larios et al., 1997; Merchant-Larios, 1998).

There is also a report describing sex determination in olive ridleys from the beach at Gahirmatha, India (Dimond and Mohanty-Hejmadi, 1983). Three temperature ranges were used in that study (i.e., temperatures in the laboratory incubators fluctuated within each range). Incubation temperatures of approximately 26–27°C produced males, 29–30°C produced females, and 31–32.0°C produced females.

Collectively, these studies indicate that olive ridley turtles possess an MF pattern of sex determination. Studies of olive ridleys from Nancite Beach, Costa Rica, suggest a pivotal temperature between 30.0°C and 31.0°C. In contrast, a distinctly lower pivotal temperature is suggested for olive ridleys from Gahirmatha, India (less than 29.0°C). In regard to the upper and lower limits of the TRT, the data from Nancite Beach indicate that incubation temperatures above 32.0°C consistently produce all females, and temperatures below 27.0°C consistently produce all males (McCoy et al., 1983; Wibbels et al., 1998b). The data from Gahirmatha olive ridleys suggest the TRT is shifted to slightly lower temperatures in comparison to the data from Nancite Beach because incubation temperatures of approximately 29–30°C or above produce all females and temperatures of approximately 26–27°C or below produce all males (Dimond and Mohanty-Hejmadi, 1983). Although these examples of female-producing temperatures and male-producing temperatures provide insight into the TRT of the olive ridley, the results from these previous studies do not necessarily indicate precise upper and lower limits of the TRT for the olive ridley.

TSD in the Kemp's Ridley

Sex determination has also been examined in the Kemp's ridley, *Lepidochelys kempii* (Aguilar, 1987; Shaver et al., 1988). Based on data from eggs incubated in Styrofoam boxes, Shaver et al. (1988) estimated a pivotal temperature of 30.2°C. Temperatures above 30.8°C produced all females. At temperatures between 29.0°C and 30.0°C, the sex ratios varied from all males to mostly females, and temperatures between 28.0°C and 29.0°C produced mostly males. Aguilar (1987) examined natural incubation conditions and found that temperatures of approximately 30.8°C or above were conducive to the production of females, and temperatures of 28.4°C or below were conducive to the production of males. Collectively, these data indicate that the Kemp's ridley has an MF pattern of TSD with a pivotal temperature near 30.2°C. Temperatures of approximately 31.0°C or above produce all females, and temperatures of approximately 28.0°C or below produce all males. However, as with the olive ridley, the upper and lower limits of the TRT are not precisely known.

Evaluation of Pivotal Temperatures in Ridley Turtles

The pivotal temperatures estimated for the olive ridley ranged from approximately 31°C to less than 29°C. First, it is of particular interest that two of the pivotal temperatures estimated for turtles from the same nesting beach (Nancite) varied by approximately 0.5–1.0°C. This could be attributed to a number of causes, and it exemplifies the problems associated with attempting to define a specific pivotal temperature for a sea turtle species or population. It is plausible that some or all of this variation could be caused by variation among clutches. Interclutch variation in pivotal temperatures has been noted for a variety of reptiles with TSD (Yntema, 1976; Bull et al., 1982c; Mrosovsky, 1988; Etchberger et al., 1991; Janzen, 1992; Ewert et al., 1994; Girondot et al., 1994; Lang and Andrews, 1994; Mrosovsky, 1994; Rhen and Lang, 1998). Of the two studies that examined pivotal temperatures in olive ridleys at Nancite Beach, one study used eggs from three clutches (McCoy et al., 1983), and the other used eggs from seven clutches (Wibbels et al., 1998b). Additionally, the difference in estimated pivotal temperatures could be attributed to experimental error and/or differences in experimental protocols. For example, the thermal stability of the incubators and the accuracy of the temperature-recording devices may vary between laboratories. In light of the thermal stability of most laboratory incubators and the precision and accuracy of most temperature-recording devices, it is often difficult to accurately estimate a pivotal temperature to the 0.1°C level (Wibbels, 2003). Therefore, some of the variation between the pivotal temperature estimates for olive ridleys at Nancite Beach could be the result of experimental error. In any case, the results from those studies suggest a pivotal temperature in the 30.0–31.0°C range.

Previously estimated pivotal temperatures for sea turtles range from approximately 27.7°C to 31.0°C with the majority between 29.0°C and 30.0°C. The pivotal temperatures estimated for olive ridleys at Nancite Beach are on the high end of this range (ranging from approximately 30.0°C to 31.0°C). In contrast, the pivotal temperature for Gahirmatha olive ridleys may be relatively low (less than 29.0°C). This suggests a relatively large difference between the two populations (Table 9.1). Again, some of the variation may be associated with factors such as experimental error and/or differences in experimental protocols. However, the magnitude of the variation suggests that it may well represent a true interpopulation difference in pivotal temperatures. Previous studies indicate that pivotal temperature can vary between populations, and even between egg clutches (Limpus et al., 1985; Mrosovsky, 1988; Etchberger et al., 1991; Ewert et al., 1994; Georges et al., 1994; Mrosovsky, 1994). Further, the magnitude of the variation in pivotal temperatures between these two populations suggests that the olive ridley may be an advantageous species for examining intraspecific variation in pivotal temperatures. The olive ridley inhabits tropical oceans worldwide and has widely distributed nesting beaches including nesting locations in the Pacific, Atlantic, and Indian Oceans (Ernst and Barbour, 1989), and it is plausible that the thermal characteristics of these geographically diverse nesting beaches may select for a variety of pivotal temperatures. Consequently, future studies examining and comparing pivotal temperatures of olive ridleys from nesting beaches worldwide could provide vital insight into the evolution of pivotal temperatures in sea turtles. In fact, because of its

Table 9.1 Comparison of pivotal temperatures of ridley turtles to those of other sea turtles species

Species and location	Estimated pivotal temperature (°C)	Reference
Olive ridley		
Costa Rica	~ 30	McCoy et al., 1983
Costa Rica	~ 31	Wibbels et al., 1998b
India	< 29	Dimond and Mohanty-Hejmadi, 1983
Kemp's ridley		
Mexico	30.2	Shaver et al., 1988[a]
Loggerhead		
United States	~ 30	Yntema and Mrosovsky, 1982
United States	29.0	Mrosovsky, 1988
Australia	27.7, 28.7	Limpus et al., 1985
Australia	~ 29.0	Georges et al., 1994
S. Africa	29.7	Maxwell, 1988
Brazil	29.2	Marcovaldi et al., 1997
Greece	29.3	Mrosovsky et al., 2002
Leatherback		
Suriname and French Guiana	29.5	Rimblot et al., 1985
Rimblot-Baly et al., 1987		
Costa Rica	29.4	Binckley et al., 1998
Hawksbill		
Antigua	29.2	Mrosovsky et al., 1992
Brazil	29.6	Godfrey et al., 1999
Green		
Suriname	28.8	Mrosovsky et al., 1984
Costa Rica	~28.5–30.3	Standora and Spotila, 1985[b] Spotila et al., 1987[b]
Flatback		
Australia	29.5	Hewavisenthi and Parmenter, 2000

Note: These studies are based on laboratory studies with constant temperatures except when noted. The location indicates where eggs were collected.

[a]Eggs were incubated in Styrofoam boxes with seminormal daily temperature fluctuations.

[b]Based on field studies of natural nests.

widespread distribution and relative abundance, the olive ridley may represent an optimal species for examining interspecific variation in pivotal temperatures.

In the case of the Kemp's ridley, there is only one primary nesting beach, which is located near Rancho Nuevo, Mexico. Most of the nesting for the entire species occurs on that beach (Márquez-M., 1994). Nesting beach sand temperatures and nest incubation temperatures have been examined at Rancho Nuevo, and those temperatures are relatively high (Geis et al., 2000; Wibbels et al., 2000a, 2000b; Geis et al., 2002, 2004). The estimated pivotal temperature for the Kemp's ridley (30.2°C) is also relatively high compared to most pivotal temperature estimates in sea turtles (Table 9.1). It is plausible that the Kemp's ridley has evolved a relatively high pivotal temperature to match the warm beach temperatures.

Gonadal Differentiation in Ridley Turtles

A detailed study of gonadal differentiation has been reported for the olive ridley (Merchant-Larios et al., 1989), and the structure of gonads at hatching has also been described (Wibbels et al., 1998b; Merchant-Larios, 1999). The study by Merchant-Larios et al. (1989) indicated that both males and females initially develop an undifferentiated gonad. Sexual differentiation of the gonad occurred earlier in the female than in the male. The ovaries began to show sexual differentiation toward the start of the last third of incubation, and in females, the cortical region of the gonad proliferated, and the medullary cords regressed. Although the cortex of the ovary proliferated at female-producing temperatures, follicles did not form before hatching. In contrast to the female, the gonads of males showed little sexual differentiation before hatching in the study by Merchant-Larios et al. (1989), but these gonads were clearly distinguishable from the differentiated ovary (Merchant-Larios et al., 1989). Wibbels et al. (1989) examined male and female gonads from hatchling olive ridleys histologically. A typical ovary had a well-developed cortex and regressed medullary cords (Fig. 9.2). In contrast, the testis lacked a cortex, and the medullary cords had developed into seminiferous tubules (Fig. 9.2). A study by McCoy et al. (1983) suggested that hatchling gonads were differentiated to the extent that gross morphology could be used to determine the sex of hatchlings accurately. Wibbels et al. (1989) addressed that subject by comparing the results based on gross morphology to those based on histological examination of gonadal tissue. The results indi-

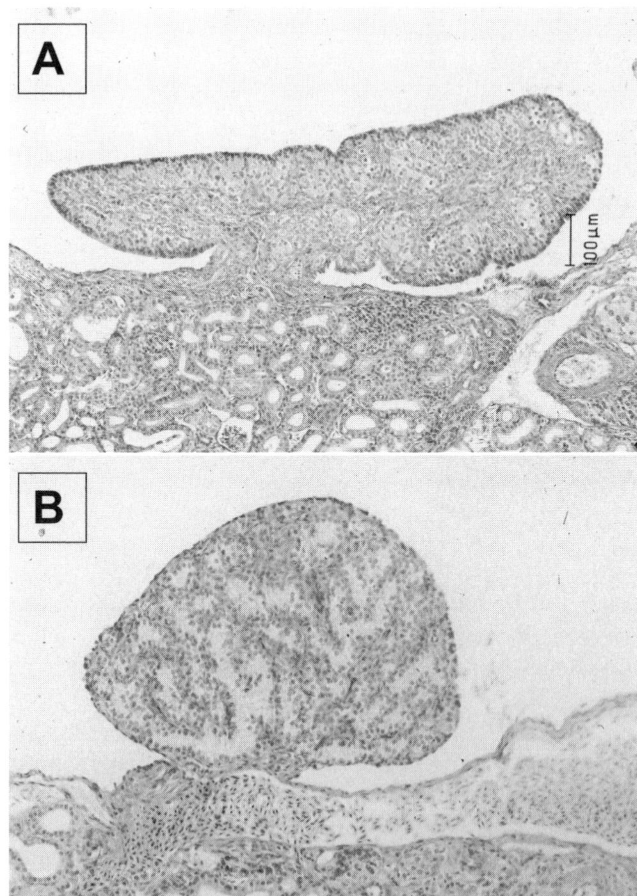

Fig. 9.2. Hatchling ovary (A) and testis (B) from olive ridleys (hematoxylin and eosin staining). The ovary has a well-developed outer layer (i.e., cortex) with dark staining. The inner region of the ovary (i.e., medulla) lacks any distinct organization. In contrast, the testis lacks a well-developed cortex, and it has organized groups of cells (i.e., sex chords) in the medulla that form seminiferous tubules.

cated that gross morphology was an accurate indicator of gender.

Several studies have utilized the morphology and/or histology of gonads to determine the sex of hatchling Kemp's ridleys (Aguilar, 1987; Shaver et al., 1988; Wibbels et al., 1989). Thus, the male and female gonads can be distinguished on the basis of characteristics such as the degree of cortical development or medullary regression. However, there have been no detailed reports showing the chronology of gonadal differentiation in Kemp's ridley.

Physiology of TSD in Ridley Turtles

Most studies of TSD in ridley turtles have investigated pivotal temperatures, TRTs, and sex ratios. The majority of studies on the physiology of TSD in turtles have been conducted with freshwater turtles, many of which have a similar MF pattern of TSD (Jeyasuria et al., 1994; Wibbels et al., 1994; Pieau, 1996; Lance, 1997; Wibbels et al., 1998a). These studies are reviewed by Wibbels (2003) relative to what is known in sea turtles. Although this subject has not been addressed in the Kemp's ridley, there have been several studies regarding the physiology of TSD in the olive ridley.

One hypothesis regarding TSD is that a female-producing temperature may induce the undifferentiated gonad to produce estrogen, which then stimulates ovarian differentiation (Jeyasuria et al., 1994; Pieau, 1996; Jeyasuria and Place, 1997; Pieau et al., 1998). The "estrogen hypothesis" was originally developed because treatment of eggs with estrogen would feminize embryos incubating at male-producing temperatures, including olive ridley embryos (Merchand-Larios et al., 1997; T. Wibbels, unpublished data). However, there are some conflicting data regarding the estrogen hypothesis because studies of some freshwater turtles and crocodilians suggest that estrogen production may be a "downstream event" that occurs after ovarian differentiation during TSD (White and Thomas, 1992; Smith and Joss, 1994a, 1994b; Smith et al., 1995; Willingham et al., 2000; Gabriel et al., 2001; Murdock and Wibbels, 2003). This includes a study of the olive ridley turtle, in which estrogen levels in the gonad were similar at both male- and female-producing temperatures during the thermosensitive period of sex determination (Salame-Mendez et al., 1998).

An alternate estrogen hypothesis is that the brain rather than the gonad may be the source of estrogen production during the thermosensitive period (Jeyasuria and Place, 1998; Merchant-Larios, 1998; Salame-Mendez et al., 1998). A study of the olive ridley suggested that the brain senses temperature and that there is an increased production of estrogen in the diencephalon/mesencephalon portion of the brain during the thermosensitive period (Merchant-Larios, 1998; Salame-Mendez et al., 1998). Consistent with the hypothesis that the brain may be involved in TSD, it has been shown that the nervous system innervates the gonads before sexual differentiation in the olive ridley (Gutierrez-Ospina et al., 1999).

In addition to examining the possible involvement of estrogen in TSD, many studies have begun to examine specific genes in reptiles that might be involved in the sex determination and/or gonadal differentiation cascade of other vertebrates (Lance, 1997; Wibbels et al., 1998a; Western and Sinclair, 2001). Genes or gene products such as steroidogenic factor-1 (*SF-1*), antimullerian hormone (*AMH*), *SOX9, DMRT1, DAX1,* and Wilms tumor 1 (*WT-1*) have been identified in reptiles and have been suggested to be involved in gonadal development and/or differentiation in other vertebrates (reviewed by Wibbels, 2003). Although most of these genes have not been studied in any sea turtles, the expression of *SOX9* has been studied in the olive ridley (Moreno-Mendoza et al., 1999; Merchant-Larios, 2001; Moreno-Mendoza et al., 2001; Torres-Maldonado et al., 2001). The results of those studies indicate that the *SOX9* gene is expressed during testicular differentiation but down-regulated during ovarian differentiation in the olive ridley. This suggests that *SOX9* may be involved in the sexual differentiation of the testis.

Thus, a few aspects of the physiology of TSD have been addressed in the olive ridleys, but in general, the physiology and genetics of TSD in ridley turtles (as well as in any species of sea turtle) are not well understood. There is, then, a distinct need for future studies addressing the physiology and genetics of TSD in ridley turtles.

Sex Ratios in Ridley Turtles

A variety of studies have examined sex ratios in various size classes within ridley populations. The following sections review data on sex ratios of hatchlings, immature turtles, and adult turtles. Studies of hatchlings have primarily focused on those produced in conservation programs. In many ridley conservation programs, nests are moved to protected areas (hatcheries) to prevent predation and poaching. As a result, many of these data are not representative of natural sex ratios; rather, they represent sex ratios produced in egg hatcheries.

Olive Ridley Hatchling Sex Ratios

In the case of the olive ridley, there are many examples of conservation programs that move nests to hatcheries (Shanker, 1994; Hoekert et al., 1996; Arauz and Naranjo, 2000; Hasbun et al., 2000; Shanker, 2000; Garcia et al., 2003). These hatcheries are generally located on the beach in areas where conservationists can protect eggs from predation and poaching. The movement of eggs to protected hatcheries has the potential of altering the natural incubation temperatures and, thus, sex ratios (Mortimer, 1999). Therefore, it is important that these programs evaluate sex ratios and/or incubation temperatures. A program in El Salvador, for example, uses shading in hatcheries in an attempt to maintain a temperature in the range of 30–31°C (Hasbun et al., 2000). This temperature range was chosen because it avoids extremely high temperatures that might increase mortality, and it should avoid highly skewed sex ratios. A detailed study examining the effects of moving olive ridley eggs to hatcheries was conducted at Cuixmala Beach in western Mexico (Garcia et al., 2003). That study included the sexing of a subset of hatchlings (based on histology of gonads) from hatchery nests as well as in nests that were left in their natural location on the nesting beach (i.e., in situ nests). Nest temperatures were monitored twice each day using thermocouples, and the results indicated that the hatchery nest temperatures were similar to those of in situ nests (30.6°C and 30.5°C, respectively). Predicted sex ratios were 1.35F:1.00M and 1.41F:1.00M for the hatchery and in situ nests, respectively. In this specific case, then, the egg hatchery had similar thermal characteristics and produced similar sex ratios to in situ nests. The female-biased sex ratio from in situ nests revealed by that study also provides insight on the natural sex ratio produced on Cuixmala Beach. However, these data are based on nine nests laid during one month of the nesting season (October, 1994), so they may not be indicative of the overall sex ratio produced from in situ nests on that beach during the entire nesting season. Another study suggested a possible male-biased hatchling sex ratio for olive ridleys from in situ nests at Gahirmatha, India (Mohanty-Hejmadi and Sahoo, 1994). That study indicates that the majority of olive ridleys nest during the first *arribada* of the year, and up to 80% of the hatchlings from that arrribada are males.

Although there are a few examples of studies examining sex ratios or hatchery sand temperatures in olive ridleys, there is a need for large-scale studies estimating overall sex ratios from specific nesting beaches. The olive ridley may also be an excellent species for examining the variability of sex ratios among different nesting beaches because of its broad geographic distribution and numerous nesting beaches. For example, it would be of interest to compare hatchling sex ratios produced from major olive ridley nesting beaches along the Pacific coasts of Mexico and Central America. That is, do all of these beaches produce similar sex ratios, or do sex ratios vary because of factors such as the specific thermal characteristics of each beach?

Kemp's Ridley Hatching Sex Ratios

Hatchling sex ratios have also received attention in the Kemp's ridley. In contrast to the olive ridley, the Kemp's ridley has a limited geographic distribution and has only one primary nesting beach (Márquez-M., 1994). Because of the endangered status of Kemp's ridley, almost all nests are moved to protected hatcheries (Márquez-M., 1994). The main hatchery (or "corral") is located on the Kemp's ridley's primary nesting beach near Rancho Nuevo, Mexico, and the great majority of nests have been moved to that hatchery over the past several decades (Fig. 9.3). Beginning about 15 years ago, an increasing number of hatcheries were established. These "satellite" corrals are located at various distances north and south of Rancho Nuevo to decrease the distance that eggs have to be transported during relocation.

There have been several studies examining sex determination and sex ratio of Kemp's ridleys at Rancho Nuevo. Initial evaluation of nest and sand temperatures in the main corral suggested that temperatures were appropriate to produce mixed sex ratios or female biases, thus avoiding strong male biases (Aguilar, 1987; P. Burchfield, personal communication). Additionally, a study examining the effects of shading eggs has been conducted in the main corral at Rancho Nuevo (Carrasco et al., 2000). In that study, 40 nests were shaded with plastic mesh screening, and 67 unshaded nests were used as controls. Hatchlings found dead in the nest or unhatched late-stage embryos were histologically examined to verify sex (a total of 53 samples from the shaded nests and 62 samples from unshaded nests). Temperatures were periodically recorded using a remote sensor at approximately midnest depth. The result of that study indicated that the shaded nests produced a slight male bias, whereas the unshaded nests produced all females (Carrasco et al., 2000). The temperature recorded in the shaded area averaged approximately 30.2°C, and the unshaded area averaged approximately 31.1°C.

Fig. 9.3. The main egg corral used for protecting Kemp's ridley eggs is located adjacent to Barra Coma, near Rancho Nuevo, Mexico. (A) Main corral during the 2002 nesting season. (B) Main corral during the 1981 nesting season. Both corrals are in the same general location. Each photo was taken from the same approximate location on a high dune that borders the beach.

Comprehensive studies have been conducted in more recent years using small temperature data loggers that can be inserted directly into nests. These data loggers record temperature at preprogrammed intervals (e.g., every hour) for the entire nesting season or nest incubation period. From 1998 through 2003, temperature data loggers have been used to monitor sand temperature in the corrals at midnest depth throughout the nesting season (Geis et al., 2000; Wibbels et al., 2000a, 2000b; Geis et al., 2002, 2003). In general, sand temperatures gradually increase during the start of the nesting season (late March and April) and are at or above pivotal temperature by mid- to late May (Fig. 9.4). Because the densest nesting occurs in May, the majority of eggs experience female-producing temperatures by the time they enter their thermosensitive period of sex determination (i.e., the second third of the incubation period). During June and July, sand temperatures remain relatively high (normally above pivotal temperature) for the remainder of the nesting season but can decrease episodically because of rain.

In addition to recording sand temperatures, data loggers have been used to directly record nest temperature in the corrals from 1998 through 2003 (Geis et al., 2000; Wibbels et al., 2000a, 2000b; Geis et al., 2002, 2003, 2004). During each season a subset of nests was sampled including nests from each of the arribadas. Average nest temperatures during the middle third of the incubation period were used to predict sex ratios (Yntema and Mrosovsky, 1982; Georges et al., 1994; Hanson and Wibbels, 1999) based on the pivotal temperature and TRT predicted for the Kemp's ridley (Aguilar, 1987; Shaver et al., 1988). The results are consistent with the sand temperature data and suggest that the corrals consistently produced overall female-biased sex ratios (Geis et al., 2000; Wibbels et al., 2000a, 2000b; Geis et al., 2002, 2003, 2004). Although the data indicate that females predominate, males were predicted to be produced early in the nesting season when sand and nest temperatures were relatively cool. Thus, the results suggest that both males and females are produced in the corrals during a typical nesting season, but fe-

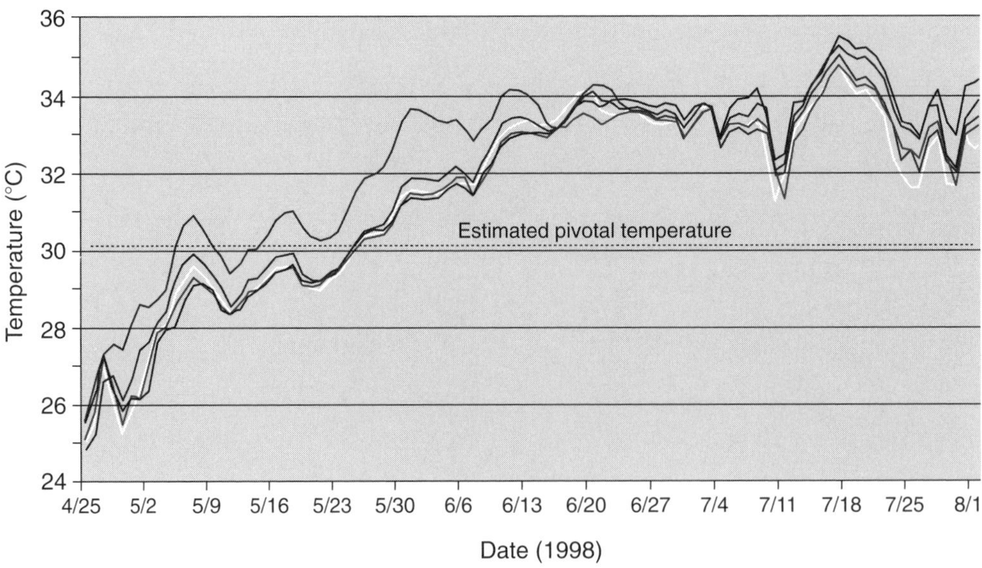

Fig. 9.4. Example of sand temperatures at midnest depth (approximately 35 cm) in the main corral near Rancho Nuevo during the 1998 nesting season. The estimated pivotal temperature (temperature producing a 1:1 sex ratio) is 30.2°C for Kemp's ridleys, and it is shown on the figure. The sand temperature gradually increases during the early portion of the nesting season and is above pivotal temperature by mid- to late May. Nesting typically begins in April and is heaviest in May, and most nesting ends by early July. Sex is determined during the middle third of incubation, which starts approximately 15–18 days after a nest is laid.

males predominate. Although these recent studies examined only 5 years of temperature data, it is of interest that the main corral has been placed in the same general position on the nesting beach for several decades (Fig. 9.3). Therefore, it is plausible that the main corral may have experienced similar temperatures in previous years. If this is the case, then a female-biased sex ratio may have occurred for many years at Rancho Nuevo.

Recent studies at Rancho Nuevo have also investigated hatchling sex ratios on the natural nesting beach (Geis et al., 2002, 2003, 2004; A. Geis and T. Wibbels, unpublished data). During the 2001 through 2003 nesting seasons, transects were set up to record sand temperature at midnest depth for an approximately 7-km stretch of beach at Rancho Nuevo. Additionally, a subset of nests were left in their natural locations to incubate (i.e., approximately 20–70 in situ nests per season with data loggers). Protective covers consisting of wide mesh fence material were placed just under the surface of the sand above the nest to prevent predation. The preliminary findings indicate that the beach shows similar temperature trends as the corrals, but on average, the beach is slightly cooler than the corrals. However, temperatures were still warm enough on the beach to produce an overall female bias, but not as strong a bias as in the corrals. These data suggest that Kemp's ridley may produce a "natural" hatchling sex ratio that is female biased (Geis et al., 2002, 2003, 2004; A. Geis and T. Wibbels, unpublished data).

Sex Ratios Produced in Head-Start Program

Finally, it should be noted that in previous years some of the eggs from Rancho Nuevo were transported to Padre Island National Seashore (PINS) for incubation and "experimental imprinting" before being captive-reared in the Head Start Program (Shaver et al., 1988; Shaver and Wibbels, Chapter 14). Data from the initial years of that program suggested a male-biased sex ratio in the 1978, 1979, 1981, and 1983 year classes (Shaver et al., 1988; Wibbels et al., 1989). The incubation facility at PINS was subsequently modified to raise incubation temperatures and increase the proportion of females (Shaver et al., 1988). As a result, the 1984 through 1992 year classes of head-started turtles were female biased (Caillouet et al., 1995). For example, 77.5% of the turtles examined from the 1985–1988 year classes were identified as females (Shaver et al., 1988).

Sex Ratio of Immature and Adult Turtles

Although hatchling sex ratio data have been examined in both species of ridleys, there is a paucity of data on the sex ratios of juvenile and adult olive ridley turtles. The following section, then, focuses on the Kemp's ridley. A variety of previous studies have investigated sex ratios of immature (i.e., juvenile and/or subadult) and adult Kemp's ridleys from a variety of locations in the Gulf of Mexico and Atlantic Ocean (Table 9.2). Adult sex ratios are noted in Table 9.2, but the data are difficult to interpret because of the possibility of biased sampling because adult sea turtles are known for sex-specific migration patterns (Henwood, 1987; Wibbels et al., 2000b). Therefore, although adult sex ratios are noted, the primary focus of the following discussion is on the immature portion of the population. Additionally, in an effort to more easily recognize any sex ratio variation among size classes of turtles, the data for turtles in Table 9.2 are arbitrarily divided into different size classes of turtles (<40 cm, 40–60 cm, and >60 cm) whenever possible.

A number of studies have examined sex ratios of immature and adult Kemp's ridleys (Table 9.2). These studies represent numerous sampling locations and sampling periods, ranging from 1977 to 1997. The sex ratios reported often vary among studies, but some trends predominate. A sex ratio of 1.4:1.0 (female:male) was reported for stranded juveniles in Cape Cod Bay from 1977 through 1987 (Danton and Prescott, 1988). A 2.0:1.0 (female:male) sex ratio was reported for 21 cold-stunned Kemp's ridleys from Long Island Sound, New York during 1987, whereas 9 turtles stranded during 1985 in that same area were all females (Morreale et al., 1992). A study of stranded Kemp's ridleys along the lower Texas coast from 1983 to 1989 found an approximate 1:1 overall sex ratio, but those results included some adult turtles and many

Table 9.2 Sex ratios reported for immature and adult Kemp's ridley sea turtles

Location	Sampling period	Sampling method	Age class / status / size class[a]	Sexing method	n	Sex ratio	Reference
Cape Cod, MA	1977–1987	Strandings	I	Necropsy	48	1.4F:1.0M	Danton and Prescott, 1988
South Texas coast	1983–1989	Strandings	I and A	Necropsy	39	1.0F:1.8M	Shaver, 1991
South Texas coast	1983–1989	Strandings	I, HS < 50 cm CCL	Necropsy	42	1.6F:1.0M	Shaver, 1991
Long Island Sound, NY	1985	Strandings	I, < 37.6 cm SCL	Histology	9	9:0F:0.0M	Morreale et al., 1992
Long Island Sound, NY	1987	Strandings	I, < 37.6 cm SCL	Histology	21	2:0F:1.0M	Morreale et al., 1992
Upper Texas coast	1986–1992	Strandings	I, < 40 cm SCL	Necropsy	78	2.9F:1.0M	Stabenau et al., 1996
Upper Texas coast	1986–1992	Strandings	I, > 40.0 and < 60.0 cm SCL	Necropsy	12	2.8F:1.0M	Stabenau et al., 1996
Upper Texas coast	1986–1992	Strandings	A	Necropsy	28	13.0F:1.0M	Stabenau et al., 1996
Cedar Keys, FL	1992	Net	I, < 40.0 cm SCL	Testosterone	12	3.0F:1.0M	Gregory and Schmid, 2001
Cedar Keys, FL	1992	Net	I, > 40.0 and < 60.0 cm SCL	Testosterone	24	1.4F:1.0M	Gregory and Schmid, 2001
Upper TX and LA coast	1993–1997	Net	I, < 40.0 cm SCL	Testosterone	149	1.2F:1.0M	Coyne, 2000
Upper TX and LA coast	1993–1997	Net	I, > 40.0 and < 60.0 cm SCL	Testosterone	41	1.4F:1.0M	Coyne, 2000
Upper TX and LA coast	1993–1997	Net	I, HS	Testosterone	20	9.0F:1.0M	Coyne, 2000
Upper TX and LA coast	1993–1997	Net	A?, > 60.0 cm SCL	Testosterone	5	5.0F:0.0M	Coyne, 2000
Upper TX and LA coast	1994	Strandings	I, < 40 cm	Necropsy	64	1.0F:1.4M	Cannon, 1998
Upper TX and LA coast	1994	Strandings	I, > 40 and < 60.0 cm SCL	Necropsy	24	1.2F:1.0M	Cannon, 1998

Note: Adult turtles are listed separately from immature turtles when possible. Immature turtles are arbitrarily divided into two size classes when possible (less than 40.0 cm versus 40.0 to 60.0 cm). Range of carapace lengths is indicated when possible. Head-start (captive-reared) turtles are listed separately from wild turtles when possible and designated as "HS" in the Status column. Sexing methods included "Necropsy" followed by examination of external gonad morphology, "Histology" of the gonad, and analysis of blood testosterone levels (see Wibbels et al., 2000b).

[a]I, immature turtles; A, adult turtles; HS, head start (captive-reared turtles); CCL, curved carapace length; SCL, straight carapace length.

captive-reared turtles from the Head Start Program (Shaver, 1991). Table 9.2 lists the head-started and wild turtles separately in that study. The wild turtles (which included some adults) had a male-biased sex ratio, whereas the head-started turtles had a female-biased sex ratio. Another study reported an overall female-biased sex ratio of 3.2:1.0 for Kemp's ridleys that stranded on the upper Texas coast from 1986 through 1992 (Stabenau et al., 1996). However, those data included adult turtles that showed a strong female bias (13.0:1.0). If the adults are excluded, the immature sex ratio from that study is still female biased, but it declines to approximately 2.9:1.0 or 2.8:1.0 (female:male), depending on the size class of turtle (i.e., <40 cm or 40–60 cm, respectively). The similarity in the two sex ratios from the different size classes of immature turtles is notable. Another study reported an overall 1:1 sex ratio for Kemp's ridleys stranded on the upper Texas and Louisiana coasts during 1994 (Cannon, 1998), with the smaller immature turtles showing a slight male bias and the larger immature turtles showing a slight female bias (Table 9.2). Testosterone levels were used to predict a 1.5:1.0 (female:male) sex ratio of juvenile Kemp's ridleys captured on the upper Texas and Louisiana coast from 1992 through 1997 (Coyne, 2000); however, excluding head-started turtles from that study results in a 1.3:1.0 (female:male) sex ratio. The head-started turtles captured in that study showed a strong female bias (9.0:1.0). Testosterone levels were also used to predict an overall sex ratio of 1.8:1.0 (female: male) for juvenile Kemp's ridleys captured off Homasassa, Florida (Gregory and Schmid, 2001). The sex ratios in that study varied from 3.0:1.0 (female:male) for the smaller immature turtles to 1.4:1.0 (female:male) for the larger immature turtles. Finally, there are two recent studies not listed in Table 9.2 that used the testosterone sexing technique to estimate sex ratios of immature Kemp's ridley captured in the Gulf of Mexico in Ten Thousand Islands, Florida, during 2000 and 2001 (W. N. Witzell, unpublished data) and captured near Steinhatchee, Florida, during 1998 through 2000 (A. Geis, unpublished data). The results from both of those studies suggest a female-biased sex ratio for immature Kemp's ridleys in this area.

Thus, a variety of sex ratios have been reported for juvenile Kemp's ridleys ranging from slightly male biased to strongly female biased. The reasons for the variation are unknown but could relate to many factors. It is possible that sex ratios could change over time in a population. For example, yearly variation in temperatures on the nesting beach could cause yearly variation in the hatchling sex ratios, which would be reflected in future years in the juvenile portion of the population. If this is the case, then the sex ratios recorded in a study may be dependent on the year(s) in which the study is conducted. Sex ratio variation over time could explain the variation in sex ratio reported in two of the studies that used the same methodology (necropsy of stranded turtles) and the same location (along the upper Texas and Louisiana coasts). A study from 1986 through 1992 (Stabenau et al., 1996) found an overall female bias, whereas the study in 1994 found an approximate 1:1 sex ratio (Cannon, 1998). A more recent study (1992–1997) along the upper Texas and Louisiana coasts found an overall sex ratio of approximately 1.5:1.0 (female:male) using tangle nets to capture juvenile ridleys (Coyne, 2000). In that study, sex ratio varied significantly over time, and the greatest female biases were recorded during the last 2 years of the study (approximately 70% female during 1996 and 1997). Thus, it is possible that sex ratio fluctuation over time could account for at least some of the variation in sex ratios reported for Kemp's ridleys.

There are also alternative factors that could account for the diversity of sex ratios reported. First, there may be a sampling bias associated with the sampling technique. For example, it may be misleading to directly compare overall sex ratios from studies using stranded turtles to studies in which turtles are captured by nets. Netting studies often exclude the smallest size range of turtles because wide-mesh nets are used to decrease bycatch. Optimally, sex ratio comparisons should be made between the same size classes of turtles.

Second, it is possible that sex ratio may vary among sampling locations. Seasonal or yearly variations of winds and currents in the Gulf of Mexico could affect the recruitment of young Kemp's ridleys into specific foraging areas (e.g., west coast of Florida versus Texas coast). Thus, cohorts from certain nesting seasons may be more likely to recruit to certain foraging areas, and sex ratios could vary between cohorts depending on the nesting beach temperature during the season that they hatched. It is also possible that certain foraging areas may be more attractive to certain size classes of juveniles because of characteristics such as depth and current.

Although a clear understanding of sex ratios and sex ratio dynamics in the immature portion of the Kemp's ridley population has not yet emerged, a few trends are suggested. In general, sex ratios reported for immature Kemp's ridleys range from slightly male biased to strongly female biased. However, reports of female biases predominate. Data from the late 1990s (Coyne, 2000) suggest the possibility of a distinct female bias during more recent years. Additionally, data on head-started turtles captured in these studies suggest a strong female-biased sex ratio in that group of turtles, and female-biased sex ratios were previously suggested for the 1984 through 1992 year classes of yearling turtles released by the Head Start Program (Caillouet et al., 1995).

The possible occurrence of a female bias in juveniles is consistent with recent studies indicating the production of female-biased hatchling sex ratios at Rancho Nuevo, Mexico (see discussion under Kemp's Ridley Hatchling Sex Ratios). The data from the nesting beach studies are too recent to be reflected in the juvenile data reviewed above. However, recent hatchling sex ratio data could be indicative of data from previous years at Rancho Nuevo (see discussion under Kemp's Ridley Hatchling Sex Ratios)

The review of previous studies of immature Kemp's ridley sex ratios (Table 9.2) suggests that female-biased sex ratios may predominate. This female bias may not represent a "natural" sex ratio for Kemp's ridleys because it reflects the intense conservation efforts that have been directed toward this endangered species over the past several decades. The current juvenile sex ratio is a result of incubation temperatures in the corrals during previous years. However, this does not preclude the possibility that the manipulated juvenile sex ratio might be similar to a "natural" sex ratio if all nests were incubated in situ on the nesting beach (see discussion above). For example, female-biased sex ratios are not unique to the Kemp's ridley. They have been reported in previous studies of other sea turtles (reviewed by Wibbels, 2003). Although nonbiased sex ratios and male-biased sex ratios have also been reported in some sea turtle populations (Chaloupka and Limpus, 2001; Wibbels, 2003), female-biased sex ratios are most commonly reported. Additionally, the data reviewed above (see Kemp's Ridley Hatchling Sex Ratios) from the nesting beach at Rancho Nuevo suggest that nests incubated in situ could produce a female bias (Geis et al., 2002, in press a, in press b).

Regardless of whether a female-biased sex ratio is similar to a "natural" sex ratio, it is of conservation and ecological significance because it will affect reproduction. The significance of a female-biased sex ratio to the reproductive ecology in sea turtles is not fully understood. In some aspects, a significantly female-biased sex ratio may be beneficial to an endangered population because of the high value of mature females in propagating the species. With a greater number of females, the production of hatchlings could increase, which may enhance the stability of the population. However, one should be cautious when considering such hypotheses without support from empirical data. For example, although one male may be able to inseminate multiple females, it is unknown at what point the percentage of males may become insufficient to facilitate maximum fertilization rates in a population. If males become a limiting factor in the reproductive ecology of the Kemp's ridley, then reproductive output in the population could decrease (Coyne, 2000). However, low fertility has not been reported to be a problem in the Kemp's ridley population. Low numbers of males could also result in the loss of genetic diversity within a population; however, there is currently no evidence that this is a problem in the Kemp's ridley population (Kichler, 1996; Kichler et al., 1999). Thus, the data suggest that a female bias may be present in the Kemp's ridley population, and it is quite possible that a female bias is advantageous to the recovery of this endangered sea turtle.

Summary and Future Research

Like all sea turtles, ridleys have been shown to possess an MF pattern of TSD in which low incubation temperatures produce males and warm temperatures produce females. Pivotal temperatures appear to vary between species and populations of ridleys and possibly even within populations. The pivotal temperatures reported for olive ridleys appear to span virtually the entire range of pivotal temperatures reported for sea turtles (Table 9.1). Although some of this variation may result from factors such as experimental error, the data suggest that the olive ridley may be an excellent model for examining variation in pivotal temperatures. There are numerous aspects of TSD in ridleys that require further investigation. In regard to characterizing TSD, future studies verifying the precise upper and lower limits of the TRT for ridleys are needed. Additionally, it is not clear if the TRT of ridleys varies between species and/or populations, as has been suggested for leatherback sea turtles (Chevalier et al., 1999). Although the thermosensitive period of sex determination has been studied in the olive ridley (Merchant-Larios et al., 1997), it has not been examined in the Kemp's ridley. Further, the physiology and genetics of TSD in ridley turtles (as well as in all sea turtles) are not well understood. For example, there are a number of genes that that have been identified in reptiles that have been hypothesized to be involved in sex determination and/or sex differentiation. However, most of these genes have not been studied in ridley turtles.

In regard to sex ratios, several studies have examined hatchling sex ratios produced in conservation programs for ridley turtles. For example, data indicate that the first few years of the Kemp's Ridley Head Start Program may have produced male-biased sex ratios, but incubation temperatures were then increased to ensure the production of female-biased sex ratios in later years. There have also been studies of sex ratios produced in the hatcheries for both olive ridleys and Kemp's ridleys. Applied aspects of those studies indicate that the shading of nests is an effective means of preventing mortality in situations where incubation temperatures become excessively warm. Shading of ridley nests was also shown to produce mixed sex ratios in situations that might normally produce all females. Data from the Kemp's ridley nesting beach near Rancho Nuevo suggest that female biases are produced in that program. The effect of a female bias on the reproductive ecology of sea turtles is unknown, but it may be advantageous to the recovery of the Kemp's ridley if males do not become a limiting factor. In view of the range of sex ratios that can result from TSD, there is a distinct need to monitor hatchling sex ratios in ridley conservation programs.

Data on hatchling sex ratios from natural (in situ) ridley nests are limited, but initial data from the Kemp's ridley's primary nesting beach suggest that in situ nests produced an overall female-biased sex ratio during recent nesting seasons. In addition to the data on the Kemp's ridley, it would be of interest from an ecological and evolutionary point of view to evaluate natural sex ratios produced from the various nesting beaches of olive ridleys. Such information combined with data on pivotal temperatures and TRTs could provide insight into the ecological and evolutionary significance of TSD and would address a variety of questions. For example, what range of sex ratios are produced on ridley nesting beaches? Do hatchling sex ratios vary significantly among different nesting beaches and populations? Do pivotal temperatures vary among populations relative to the nesting beach temperatures? These questions exemplify the need for future studies of hatchling sex ratios and TSD in ridley turtles.

This chapter also reviewed data on sex ratios in the immature and adult portion of the Kemp's ridley population. A variety of sex ratios have been reported, but there is a predominance of female biases reported for the Kemp's ridley. Again, the effect of a female bias on the reproductive ecology of the Kemp's ridley is unknown. It is easy to assume that a female bias may be advantageous to the recovery of an endangered species because it has the potential of increasing egg production. However, if males become a limiting factor in reproduction, then fertility could become a problem. Therefore, in addition to monitoring hatchling sex ratios, it would be advantageous to monitor fertility during nesting beach conservation programs for ridleys. Although sex ratios of immature and adult Kemp's ridley have been the subject of several studies, there is a need for studies of sex ratios in olive ridley populations. Because of the wide distribution of olive ridley populations, it would be of interest to examine the variability of population sex ratios in that species. Population sex ratio data, combined with data on the effects of sex ratio on reproductive output in populations, could provide insight on a major question facing sea turtle conservationists: What sex ratios are optimal for the recovery of an endangered sea turtle?

In conclusion, sex determination represents a very interesting aspect of the physiology and life history of the ridley turtles. Previous studies have provided baseline data on TSD and sex ratios in ridleys, but there are many questions that remain to be answered. It is only after such information becomes available that we will be able to comprehend the significance of TSD to the conservation, ecology, and evolution of the ridley turtles.

ACKNOWLEDGMENTS

The completion of this chapter was made possible through the support of the Mississippi-Alabama Sea Grant Consortium and the UAB Department of Biology. I thank Alyssa Geis for her assistance with Figure 9.3 and Marc Zelickson for comments on the manuscript and for his assistance with Figure 9.1.

LITERATURE CITED

Aguilar, H. R. 1987. Influencia de la temperatura de incubacion sobre la determination del sexo y la duracion del period de incubacion en la tortuga lora (*Lepidochelys kempi,* Garman, 1880). Instituto Politecnico Nacional Mexico, D.F.

Arauz, R. M., and Naranjo, I. 2000. Conservation and research of sea turtles, using coastal community organizations as the cornerstone of support—Punta Banco and the indigenous Guaymi community of Conte Burica, Costa Rica. In Abreu-Grobois, F. A., Briseno-Duenas, R., Márquez-M., R., and Sarti, L. (Compilers). *Proceedings of the Eighteenth International Symposium on Sea Turtle Biology and Conservation.* NOAA Technical Memorandum NMFS-SEFSC-436, pp. 238–240.

Binckley, C. A., Spotila, J. R., Wilson, K. S., and Paladino, F. V. 1998. Sex determination and sex ratios of Pacific leatherback turtles, *Dermochelys coriacea. Copeia* 1998:291–300.

Bull, J. J. 1980. Sex determination in reptiles. *Quarterly Review of Biology* 55:3–21.

Bull, J. J. 1987 Temperature-sensitive periods of sex determination in a lizard: comparison with turtles and crocodiles. *Journal of Experimental Zoology* 241:143–148.

Bull, J. J., and Charnov, E. L. 1989. Enigmatic reptilian sex ratios. *Evolution* 43:1561–1566.

Bull, J. J., and Vogt, R. C. 1981. Temperature-sensitive periods of sex determination in emydid turtles. *Journal of Experimental Zoology* 218:435–440.

Bull, J. J., Vogt, R. C., and Bulmer, M. G. 1982a. Heritability of sex ratio in turtles with environmental sex determination. *Evolution* 36: 326–332.

Bull, J. J., Vogt, R. C., and McCoy, C. J. 1982b. Sex determining temperatures in turtles: A geographic comparison. *Journal of Experimental Zoology* 256:339–341.

Bull, J. J., Vogt, R. C., and McCoy, C. J. 1982c. Sex determining temperatures in turtles: A geographic comparison. *Journal of Experimental Zoology* 256:339–341.

Caillouet, C., Fontain, C. T., Manzella-Turpak, S. A., and Shaver, D. J. 1995. Survival of head-started Kemp's ridleys (*Lepidochely kempi*) released into the Gulf of Mexico or adjacent bays. *Chelonian Conservation and Biology* 1:285–292.

Cannon, A. C. 1998. Gross necropsy results of sea turtles stranded on the upper Texas and western Louisiana coasts, 1 January–31 December 1994. In Zimmerman, R. (Ed.). *Characteristics and Causes of Texas Marine Strandings.* NOAA Technical Report NMFS 143, pp. 81–85.

Carrasco-A., M., Márquez-M., R., Benitez-V., V., Diaz-F., J., and Jimenez-Q., C. 2000. The effect of temperature change on the sex ratio of Kemp's ridley nests in the hatchery center at Rancho Nuevo, Tamaulipas, Mexico. In Kalb, H. J., and Wibbels, T. (Compilers). *Proceedings of the Nineteenth Annual Symposium on Sea Turtle Conservation and Biology, South Padre Island, Texas,* NOAA Technical Memorandum NMFS-SEFSC-443, pp. 128–129.

Chaloupka, M., and Limpus, C. 2001. Trends in the abundance of sea turtles resident in southern Great Barrier Reef waters. *Biological Conservation* 102:235–249.

Chevalier, J., Godfrey, M. H., and Girondot, M. 1999 Significant difference of temperature-dependent sex determination between French Guiana (Atlantic) and Playa Grande (Costa Rica, Pacific) leatherbacks (*Dermochelys coriacea*). *Annales des Sciences Naturelles*. 20:147–152.

Coyne, M. S. 2000. Population sex ratio of the Kemp's ridley sea turtle (*Lepidochelys kempii*): problems in population modeling. Ph.D. diss. Texas A&M University, College Station.

Cree, A., Thompson, M. B., and Daugherty, C. H. 1995. Tuatara sex determination. *Nature* 375:543.

Danton, C., and Prescott, R. 1988. Kemp's ridley in Cape Cod Bay, Massachusetts—1987 field research. In Schroeder, B. (Compiler). *Proceedings of the Eighth Annual Workshop on Sea Turtle Conservation and Biology, Fort Fisher, NC*. NOAA Technical Memorandum NMFS-SEFSC-214, pp. 17–18.

Deeming, D. C., and Ferguson, M. W. 1988. Environmental regulation of sex determination in reptiles. *Philosophical Transactions of the Royal Society of London* 322:19–39.

Dimond, M. T., and Mohanty-Hejmadi, P. 1983. Incubation temperature and sex differentiation in a sea turtle. *American Zoologist* 23:1017.

Ernst, C. H., and Barbour, R. W. 1989. *Turtles of the World*. Washington, DC: Smithsonian Institution Press.

Etchberger, C. R., Phillips, J. B., Ewert, M. A., Nelson, C. E., and Prange, H. D. 1991. Effects of oxygen concentration and clutch on sex determination and physiology in red-eared slider turtles (*Trachemys scripta*). *Journal of Experimental Zoology* 258:394–403.

Ewert, M. A., Jackson, D. R., and Nelson, C. E. 1994. Patterns of temperature-dependent sex determination in turtles. *Journal of Experimental Zoology* 270:3–15.

Ferguson, M. W., and Joanen, T. 1983. Temperature dependent sex determination in the *Alligator mississippiensis*. *Journal of Zoology* 200:143–177.

Fisher, R. A. 1930. *The Genetical Theory of Natural Selection*. Oxford: Clarendon Press.

Gabriel, W., Blumberg, B., Sutton, S., Place, A., and Lance, V. 2001. Alligator aromatase cDNA sequence and its expression in embryos at male and female incubation temperatures. *Journal of Experimental Zoology* 290:439–448.

Garcia, A., Ceballos, G., and Adaya, R. 2003. Intensive beach management as an improved sea turtle conservation strategy in Mexico. *Biological Conservation* 111:253–261.

Geis, A., Wibbels, T., Márquez-M., R., Garduno, M., Burchfield, P., and Peña, J. 2000. Evaluation of hatchling Kemp's ridley sex ratios using nest incubation temperatures at Rancho Nuevo, Mexico. *American Zoologist* 40:1026–1027.

Geis, A., Wibbels, T., Garduno-D., M., Márquez-M., R., Burchfield, P., and Schroeder, B. 2002. Evaluation of Kemp's ridley hatchling sex ratios within egg corral and in situ nests at the primary nesting beach. *Integrative and Comparative Biology* 42:1234.

Geis, A., Wibbels, T., Márquez-M., R., Garduno, M., Burchfield Peña, J., Schroeder, B., Quintero, A. S., Ortiz, J., and Molina, G. H. 2003. Evaluation of sex ratios in egg corral and in situ nests during the 2001 nesting season. In Seminoff, J. A. (Compiler). *Proceedings of the Twenty-Second Annual Symposium on Sea Turtle Biology and Conservation*. NOAA Technical Memorandum NMFS-SEFSC-503, pp. 188–189.

Geis, A., Wibbels, T., Márquez-M., R., Garduno-D., M., Burchfield, P., and Peña, J. 2004. Predicted sex ratios of hatchling Kemp's ridleys produced in egg corrals during the 1998, 1999, and 2000 nesting seasons. In Coyne, M. S., and Clark, R. D. (Compilers). *Proceedings of the Twenty-First International Sea Turtle Symposium*. NOAA Technical Memorandum NMFS-SEFSC-528, pp. 175–177.

Georges, A., Limpus, C. J., and Stoutjesdijk, R. 1994. Hatchling sex in the marine turtle *Caretta caretta* is determined by proportion of development at a temperature, not daily duration of exposure. *Journal of Experimental Zoology* 270:432–444.

Girondot, M. 1999. A fifth hypothesis for the evolution of TSD in reptiles. *Trends in Ecology and Environment* 14:359–360.

Girondot, M., Zaborski, P., Servan, J., and Pieau, C. 1994. Genetic contribution to sex determination in turtles with environmental sex determination. *Genetics Research* 63:117–127.

Godfrey, M. H., Barreto, R., and Mrosovsky, N. 1996. Estimating past and present sex ratios of sea turtles in Suriname. *Canadian Journal of Zoology* 74:267–277.

Godfrey, M. H., D'Amato, A. F., Marcovaldi, M. A., and Mrosovsky, N. 1999. Pivotal temperature and predicted sex ratios for hatchling hawksbill turtles from Brazil. *Canadian Journal of Zoology* 77:1465–1473.

Gregory, L. F., and Schmid, J. R. 2001. Stress responses and sexing of wild Kemp's ridley sea turtles (*Lepidochelys kempii*) in the northeastern Gulf of Mexico. *General and Comparative Endocrinology* 124:66–74.

Gutierrez-Ospina, G., Jimenez-Trejo, F. J., Favila, R., Moreno-Mendoza, N., Rojas, L. G., Barrios,

F. A., Diaz-Cintra, S., and Merchant-Larios, H. 1999. Acetylcholinesterase-positive innervation is present at undifferentiated stages of the sea turtle *Lepidochelys olivacea* embryo gonads: implications for temperature-dependent sex determination. *Journal of Comparative Neurology* 410:90–98.

Hasbun, C. R., Vasquez, M., Leon, E. A., and Thomas, C. 2000. The use of shade over olive ridley, *Lepidochelys olivacea,* hatcheries. In Abreu-Grobois, F. A., Briseno-Duenas, R., Márquez-M., R., and Sarti, L. (Compilers). *Proceedings of the Eighteenth International Symposium on Sea Turtle Biology and Conservation.* NOAA Technical Memorandum NMFS-SEFSC-436, p. 158.

Henwood, T. 1987. *Ecology of east Florida sea turtles: Movements of loggerhead turtles, Caretta caretta, in the vicinity of Cape Canaveral, Florida, as determined by tagging experiments (1978–84).* Washington, DC: U.S. Department of Commerce, p. 29.

Hewavisenthi, S., and Parmenter, C. J. 2000. Hydric environment and sex determination in the flatback turtle (*Natator depressus* Garman) (Chelonia; Cheloniidae). *Australian Journal of Zoology* 48:653–659.

Hoekert, W. E. J., Schouten, A. D., Van Tienen, L. H. G., and Weijerman, M. 1996. Is the Surinam olive ridley on the eve of extinction? First census data for olive ridleys, green turtles and leatherbacks since 1989. *Marine Turtle Newsletter* 75:1–4.

Janzen, F. 1992. Heritable variation for sex ratio under environmental sex determination in the common snapping turtle (*Chelydra serpentina*). *Genetics* 131:155–161.

Janzen, F., and Pakustis, G. L. 1991. Environmental sex determination in reptiles: ecology, evolution, and experimental design. *Quarterly Review of Biology* 66:149–179.

Jeyasuria, P., and Place, A. 1997. Temperature-dependent aromatase expression in developing diamondback terrapin (*Malaclemys terrapin*) embryos. *Journal of Steroid Biochemistry and Molecular Biology* 61:415–425.

Jeyasuria, P., and Place, A. 1998. The brain–gonadal embryonic axis in sex determination of reptile: A role for cytochrome P450 Arom. *Journal of Experimental Zoology* 281:428–449.

Jeyasuria, P., Roosenberg, W. M., and Place, A. 1994. The role of p450 aromatase in sex determination of the diamondback terrapin. *Journal of Experimental Zoology* 270:95–111.

Kichler, K. 1996. Microsatellites and marine turtle conservation: the Kemp's ridley diversity project. In Bowen, B. W., and Witzell, W. N. (Eds.). *Proceedings of the International Symposium on Sea Turtle Conservation Genetics.* NOAA Technical Memorandum NMFS-SEFSC-396, pp. 95–97.

Kichler, K., Holder, M. T., Davis, S. K., Márquez-M., R., and Owens, D. W. 1999. Detection of multiple paternity in the Kemp's ridley sea turtle with limited sampling. *Molecular Ecology* 8:819–830.

Lance, V. 1997. Sex determination in reptiles: An update. *American Zoologist* 37:504–513.

Lang, J. W., and Andrews, H. V. 1994. Temperature-dependent sex determination in crocodilians. *Journal of Experimental Zoology* 270:28–44.

Limpus, C. J., Reed, P. C., and Miller, J. D. 1985. Temperature dependent sex determination in Queensland sea turtles: intraspecific variation in *Caretta caretta.* In Grigg, G., Shine, R., and Ehmann, H. (Eds.). *Biology of Australasian Frogs and Reptiles.* Chipping Norton, NSW, Australia: Surrey Beatty and Sons and The Royal Zoological Society of New South Wales, pp. 343–351.

Marcovaldi, M. A., Godfrey, M. H., and Mrosovsky, N. 1997. Estimating sex ratios of loggerhead turtles in Brazil from pivotal incubation durations. *Canadian Journal of Zoology* 75:755–770.

Márquez-M., R. 1994. *Synopsis of Biological Data on the Kemp's Ridley Sea Turtle,* Lepidochelys kempi *(Garman, 1880).* NOAA Technical Memorandum NMFS-SEFC-343.

Maxwell, J. A., Motara, M. A., and Frank, G. H. 1988. A micro-environmental study of the effect of temperature on the sex rations of the loggerhead turtle, *Caretta caretta,* from Tongaland, Natal. *South African Journal of Zoology* 23:342–350.

McCoy, C. J., Vogt, R. C., and Censky, E. J. 1983. Temperature-controlled sex determination in the sea turtle *Lepidochelys olivacea. Journal of Herpetology* 17:404–406.

Merchant-Larios, H. 1998. The brain as a sensor of temperature during sex determination in the sea turtle *Lepidochelys olivacea. Journal of Experimental Zoology* 281:510.

Merchant-Larios, H. 1999. Determining hatchling sex. In Eckert, K. L., Bjorndal, K. A., Abreu-Grobois, F. A., and Donnelly, M. (Eds.). *Research and Management Techniques for the Conservation of Sea Turtles.* IUCN/SSC Marine Turtle Specialist Group Publication No. 4, pp. 130–135.

Merchant-Larios, H. 2001. Temperature-dependent sex determination in reptiles: the third strategy. *Journal of Reproduction and Development* 47:245–252.

Merchant-Larios, H., and Villalpando, I. 1990. Effect of temperature on gonadal sex differentiation in the sea turtle *Lepidochelys olivacea:* an organ culture study. *Journal of Experimental Zoology* 254:327–331.

Merchant-Larios, H., Fierro, I. V., and Urruiza, B. C. 1989. Gonadal morphogenesis under controlled temperature in the sea turtle *Lepidochelys olivacea. Herpetological Monographs* 1989:43–61.

Merchant-Larios, H., Ruiz-Ramirez, S., Moreno-Mendoza, N., and Marmolejo-Valencia, A. 1997. Correlation among thermosensitive period, estradiol response, and gonad differentiation in the sea turtle *Lepidochelys olivacea. General and Comparative Endocrinology* 107:373–385.

Mohanty-Hejmadi, P., and Sahoo, G. 1994. Biology of the olive ridleys of Gahirmatha, Orissa, India. In Bjorndal, K. A., Bolten, A. B., Johnson, D. A., and Eliazar, P. J. (Compilers). *Proceedings of the Fourteenth Annual Symposium on Sea Turtle Biology and Conservation.* NOAA Technical Memorandum NMFS-SEFSC-351, pp. 90–93.

Moreno-Mendoza, N., Harley, V., and Merchant-Larios, H. 1999. Differential expression of SOX9 in gonads of the sea turtle *Lepidochelys olivacea* at male- and female-promoting temperatures. *Journal of Experimental Zoology* 284:705–710.

Moreno-Mendoza, N., Harley, V. R., and Merchant-Larios, H. 2001. Temperature regulates SOX9 expression in cultured gonads of *Lepidochelys olivacea,* a species with temperature sex determination. *Developmental Biology* 229:319–326.

Morreale, S. J., Ruiz, G. J., Spotila, J. R., and Standora, E. A. 1982. Temperature dependent sex determination: current practices threaten conservation of sea turtles. *Science* 216:1245–1247.

Morreale, S. J., Meylan, A. B., Sadove, S. S., and Standora, E. A. 1992. Annual occurrence and winter mortality of marine turtles in New York waters. *Journal of Herpetology* 26:301–308.

Mortimer, J. A. 1999. Reducing threats to eggs and hatchlings: hatcheries. In Eckert, K. L., Bjorndal, K. A., Abreu-Grobois, F. A., and Donnelly, M. (Eds.). *Research and Management Techniques for the Conservation of Sea Turtles.* IUCN/SSC Marine Turtle Specialist Group Publication No. 4, pp. 175–178.

Mrosovsky, N. 1983. *Conserving Sea Turtles.* London: The Zoological Society of London.

Mrosovsky, N. 1988. Pivotal temperatures for loggerhead turtles *(Caretta caretta)* from northern and southern nesting beaches. *Canadian Journal of Zoology* 66:661–669.

Mrosovsky, N. 1994. Sex ratios of sea turtles. *Journal of Experimental Zoology* 270:16–27.

Mrosovsky, N., and Pieau, C. 1991. Transitional range of temperature, pivotal temperature and thermosensitive stages for sex determination in reptiles. *Amphibia-Reptilia* 12:169–179.

Mrosovsky, N., Dutton, P. H., and Whitmore, C. P. 1984. Sex ratios of two species of sea turtles nesting in Suriname. *Canadian Journal of Zoology* 62:2227–2239.

Mrosovsky, N., Bass, A. L., Corliss, L. A., Richardson, J. I., and Richardson, T. H. 1992. Pivotal and beach temperatures for hawksbill turtles nesting in Antigua. *Canadian Journal of Zoology* 70:1920–1925.

Mrosovsky, N., Kamel, S., Rees, A. F., and Margaritoulis, D. 2002. Pivotal temperature for loggerhead turtles (*Caretta caretta*) from Kyparissia Bay, Greece. *Canadian Journal of Zoology* 80:2118–2124.

Murdock, C. A., and Wibbels, T. 2003. Cloning and expression of aromatase in a turtle with temperature-dependent sex determination. *General and Comparative Endocrinology* 130:109–119.

Pieau, C. 1996. Temperature variation and sex determination in reptiles. *Bioessays* 18:19–26.

Pieau, C., and Dorizzi, M. 1981. Determination of temperature sensitive stages for sexual differentiation of the gonads in the embryos of the turtles, *Emys orbicularis. Journal of Morphology* 170: 373–382.

Pieau, C., Dorizzi, M., Richard-Mercier, N., and Desvages, G. 1998. Sexual differentiation of gonads as function of temperature in the turtle *Emys orbicularis:* endocrine function, intersexuality and growth. *Journal of Experimental Zoology* 281:400–408.

Reinhold, K. 1998. Nest-site philopatry and selection for environmental sex determination. *Evolutionary Ecology* 12:245–250.

Rhen, T., and Lang, J. 1998. Among-family variation for environmental sex determination in reptiles. *Evolution* 52:1514–1520.

Rimblot, F., Fretey, J., Mrosovsky, N., Lescure, J., and Pieau, C. 1985. Sexual differentiation as a function of the incubation temperature of eggs in the sea turtle *Dermochelys coriacea* (Vandelli, 1761). *Amphibia-Reptilia* 6:83–92.

Rimblot-Baly, F., Lescure, F., Fretey, J., and Pieau, C. 1987. Sensibilite a la temperature de la differenciation sexuelle chez la Tortue Luth, *Dermochelys coriacea* (Vandelli, 1971); application des donnees de l'incubation artificielle a l'etude de la sex-ratio dans la nature. *Annales des Sciences Naturelles Zoologie, Paris* 8:277–290.

Salame-Mendez, A., Herrera-Munoz, J., Moreno-Mendoza, N., and Merchant-Larios, H. 1998. Response of diencephalons but not the gonad to female-promoting temperature with elevated estradiol levels in the sea turtle *Lepidochelys olivacea. Journal of Experimental Zoology* 280:304–313.

Shanker, K. 1994. Conservation of sea turtles on the Madras coast. *Marine Turtle Newsletter* 64:3–6.
Shanker, K. 2000. Conservation and management of the olive ridley on the Madras Coast in South India. In Abreu-Grobois, F. A., Briseno-Duenas, R., Márquez-M., R., and Sarti, L. (Compilers). *Proceedings of the Eighteenth International Sea Turtle Symposium.* NOAA Technical Memorandum NMFS-SEFSC-436, p. 29.
Shaver, D. J. 1991. Feeding ecology of wild and head-started Kemp's ridley sea turtles in South Texas waters. *Journal of Herpetology* 25:327–334.
Shaver, D. J., Owens, D. W., Chaney, A. H., Caillouet, C. W., Burchfield, P., and Márquez-M., R. 1988. Styrofoam box and beach temperatures in relation to incubation and sex ratios of Kemp's ridley sea turtles. In Schroeder, B. (Compiler). *Proceedings of the Eighth Annual Workshop on Sea Turtle Biology and Conservation.* NOAA Technical Memorandum, NMFS-SEFC-214, pp. 103–108.
Shine, R. 1999. Why is sex determined by nest temperature in many reptiles? *Trends in Ecology and Environment* 14:186–189.
Smith, C., and Joss, J. 1994a. Steroidogenic enzyme activity and ovarian differentiation in the saltwater crocodile, *Crocodylus porosus. General and Comparative Endocrinology* 93:232–245.
Smith, C., and Joss, J. 1994b. Uptake of ^{3}H-estradiol by embryonic crocodile gonads during the period of sexual differentiation. *Journal of Experimental Zoology* 270:219–224.
Smith, C., Elf, P., Lang, J., and Joss, J. 1995. Aromatase enzyme activity during gonadal sex differentiation in alligator embryos. *Differentiation* 58: 281–290.
Spotila, J. R., Standora, E. A., Morreale, S. J., and Ruiz, G. J. 1987. Temperature dependent sex determination in the green turtle (*Chelonia mydas*): effects on the sex ratio on a natural nesting beach. *Herpetologica* 43:74–81.
Stabenau, E. K., Stanley, K. S., and Landry, A. M., Jr. 1996. Sex ratios from stranded sea turtles on the upper Texas coast. *Journal of Herpetology* 30:427–430.
Standora, E. A., and Spotila, J. R. 1985. Temperature dependent sex determination in sea turtles. *Copeia* 1985:711–722.
Torres-Maldonado, L., Moreno-Mendoza, N., Landa, A., and Merchant-Larios, H. 2001. Timing of SOX9 downregulation and female sex determination in gonads of the sea turtle *Lepidochelys olivacea. Journal of Experimental Zoology* 290:498–503.
Viets, B. E., Ewert, M. A., and Talent, L. G. 1994. Sex-determining mechanisms in squamate reptiles. *Journal of Experimental Zoology* 270: 45–56.
Webb, G. J. W., Beal, A. M., Manolis, S. C., and Dempsey, K. E. 1987. *Wildlife Management: Crocodiles and Alligators.* Winnelie, Australia: Surrey Beatty and Sons.
Western, P. S., and Sinclair, A. H. 2001. Sex, genes, and heat: triggers of diversity. *Journal of Experimental Zoology* 290:624–631.
White, R. B., and Thomas, P. 1992. Adrenal–kidney and gonadal steroidogenesis during sexual differentiation in a reptile with temperature-dependent sex determination. *General and Comparative Endocrinology* 88:10–19.
Wibbels, T. 2003. Critical approaches to sex determination in sea turtles. In Lutz, P. L., Musick, J. A., and Wyneken, J. (Eds.). *The Biology of Sea Turtles.* Boca Raton, FL: CRC Press, pp. 103–134.
Wibbels, T., Morris, Y. A., Owens, D. W., Dienberg, G., Noell, J., Leong, J. K., King, R. E., and Millan, R. M. 1989. Predicted sex ratios from the International Kemp's Ridley Sea Turtle Head Start Research Project. In Caillouet, C. W., Jr., and Landry, A. M., Jr. (Eds.). *Proceedings of the First International Symposium on Kemp's Ridley Sea Turtle Biology, Conservation and Management,* Galveston, TX, Sea Grant Publication, TAMU-SG-89-105, pp. 82–89.
Wibbels, T., Bull, J. J., and Crews, D. 1991. Chronology and morphology of temperature-dependent sex determination. *Journal of Experimental Zoology* 260:71–381.
Wibbels, T., Bull, J. J., and Crews, D. 1994. Temperature-dependent sex determination: a mechanistic approach. *Journal of Experimental Zoology* 270:71–78.
Wibbels, T., Cowan, J., and LeBoeuf, R. D. 1998a. Temperature-dependent sex determination in the red-eared slider turtle, *Trachemys scripta. Journal of Experimental Zoology* 281:409–416.
Wibbels, T., Rostal, D. C., and Byles, R. 1998b. High pivotal temperature in the sex determination of the olive ridley sea turtle from Playa Nancite, Costa Rica. *Copeia* 1998:1086–1088.
Wibbels, T., Márquez-M., R., Garduno-D., M., Burchfield, P., and Peña-V., J. 2000a. Incubation temperatures in Kemp's ridley nests during the 1998 nesting season. In Kalb, H. J., and Wibbels, T. (Compilers). *Proceedings of the Nineteenth Annual Symposium on Sea Turtle Conservation and Biology.* NOAA Technical Memorandum NMFS-SEFSC-443, pp. 133–134.
Wibbels, T., Owens, D. W., and Limpus, C. J. 2000b. Sexing juvenile sea turtles: Is there an accurate

and practical method? *Chelonian Conservation and Biology* 3:756–761.

Willingham, E., Baldwin, R., Skipper, J., and Crews, D. 2000. Aromatase activity during embryogenesis in the brain and adrenal-kidney-gonad of the red-eared slider turtle, a species with temperature-dependent sex determination. *General and Comparative Endocrinology* 119:202–207.

Yntema, C. L. 1976. Effects of incubation temperatures on sexual differentiation in the turtle, *Chelydra serpentina*. *Journal of Morphology* 150: 453–462.

Yntema, C. L. 1979. Temperature levels and periods of sex determination during incubation of eggs of *Chelydra serpentina*. *Journal of Morphology* 159: 17–27.

Yntema, C. L., and Mrosovsky, N. 1982. Critical periods and pivotal temperatures for sexual differentiation in loggerhead turtles. *Canadian Journal of Zoology* 60:1012–1016.

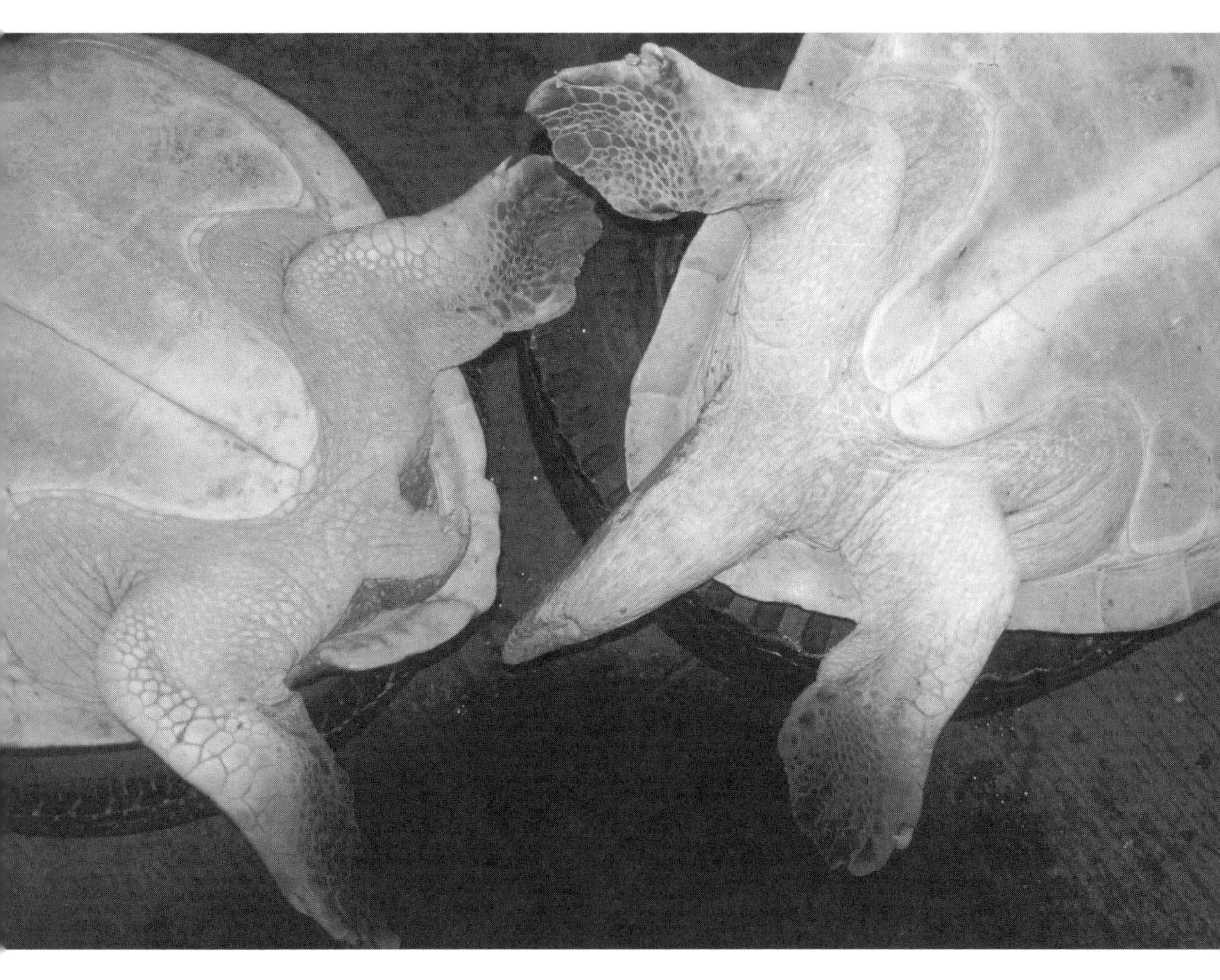

MICHAEL COYNE
ANDRE M. LANDRY, JR.

10

Population Sex Ratio and Its Impact on Population Models

HUMAN POPULATION INCREASES and habitat degradation have driven many historically abundant species, such as the Kemp's ridley (*Lepidochelys kempii*), toward extinction. Correspondingly, attempts to preserve rare or endangered taxa through proactive management have been compromised by decisions based more on ease of implementation or accessibility to particular life stages than on clear expectations of population responses to management (Crouse et al., 1987). The bulk of sea turtle conservation efforts have focused on two easily accessible life-history stages: eggs and adult females on the nesting beach.

Models that generate meaningful output for management decisions are based on input realistically describing the life history of the population. Fortunately, data related to reproduction (adult female remigration rate, number of eggs per nest, seasonal nesting frequency) and critical to developing useful population models are available because of focused efforts on the nesting beach. Unfortunately, many other variables essential to realistic sea turtle models (i.e., survivorship, age at maturity, and longevity) remain difficult to quantify (Bustard, 1979). One such variable is population sex ratio.

Accurate information on the sex ratio of wild animal populations is particularly useful, providing a baseline against which to assess conservation strategy (Mrosovsky, 1994). As an example, temperature-dependent sex determination (TSD) in sea turtles provides a potential recovery tool for these threatened and endangered species (Mrosovsky, 1980; Morreale et al., 1982; McCoy et al., 1983; Shaver et al., 1988). Resource managers may choose to masculinize, feminize, or balance a population's gender by manipulating incubation temperatures (Shaver et al., 1988). Although the ability to control sex of hatchlings is advantageous, choosing the appropriate phenotype option without knowledge of the natural sex ratio or of how subsequent alterations might impact population balance is problematic.

Determining sex in sea turtles is not trivial (Wibbels et al., 2000). No evident secondary sexual characteristics develop until maturity (Wibbels et al., 1991a; Owens, 1997). Consequently, the only reliable method of sexing immature animals is through direct observation of the gonads (Van Der Heiden et al., 1985; Mrosovsky and Benabib, 1990), which requires sacrificing the turtle, use of surgery (i.e., laparoscopy), or necropsy of stranded carcasses (Shaver, 1991; Stabenau et al., 1996). Sacrifice is not a viable alternative for an endangered species, and although laparoscopy has proven reliable and mostly harmless (Limpus et al., 1994), sample size requirements and field restrictions under endangered species regulatory framework make dependence on this procedure impractical. Finally, advanced decomposition of most stranded sea turtles often renders sex determination difficult or unreliable (Heinly, 1990).

This chapter focuses on the results of a study utilizing radioimmunoassay (RIA) determination of blood plasma testosterone (T) concentration in conjunction with limited laparoscopy to determine the sex of Kemp's ridley sea turtles (Coyne, 2000). These techniques have been used concurrently in other studies to successfully determine the sex of loggerhead (*Caretta caretta;* Wibbels et al., 1987) and green (*Chelonia mydas;* Bolten et al., 1992) sea turtles. These results and information available in the scientific literature were used to develop a population dynamics model for Kemp's ridleys (Coyne, 2000) that highlighted areas of insufficient knowledge or lack of understanding concerning Kemp's ridley population dynamics (Grant, 1986). The final model was used primarily to assess the impact of various population sex ratio values on Kemp's ridley sea turtle demography.

Background

Embryonic Sex Determination

The influence of temperature on sea turtle embryonic sex determination presents an interesting puzzle. TSD, a particular type of environmental sex determination, acts in sea turtles to produce female hatchlings at warm temperatures and males at cool temperatures. The sensitive period for sex determination appears to occur around the middle third of incubation (Yntema and Mrosovsky, 1982). The threshold or pivotal temperature that produces a sex ratio of 1:1 appears to be close to 29°C for all species for which data are available. Interestingly, olive ridleys (*L. olivacea*) from Nancite, Costa Rica appear to have a higher pivotal temperature than other sea turtles: ~31°C (Wibbels et al., 1998).

In addition to temperature, administration of exogenous estrogen can influence sea turtle sex ratios. Application of estrogen to an egg incubating at a male-producing temperature can reverse the effect of temperature and result in a female hatchling (Raynaud and Pieau, 1985; Gutzke and Bull, 1986; Bull et al., 1988; Crews et al., 1989, 1991; Wibbels et al., 1991a, 1991b; Tousignant and Crews, 1994). This provides a potential method for obtaining female offspring without regard to incubation temperature.

Conservation History

TSD and sex ratios have an interesting history with regard to Kemp's ridley recovery efforts. During 1978–1988 Kemp's ridley eggs were collected from Rancho Nuevo, Mexico in plastic bags as they were laid, packed in Styrofoam boxes containing sand from the Padre Island National Seashore (PINS) near Corpus Christi, Texas, and transferred to the PINS for incubation and hatching in the same boxes (Shaver et al., 1988; Burchfield and Foley, 1989). Before 1985, the pivotal incubation temperature of Kemp's ridley eggs was unknown (Shaver et al., 1988; Fletcher, 1989) and appears not to have been considered despite published reports of TSD in other sea turtle species at the time (Mrosovsky and Yntema, 1980; Yntema and Mrosovsky,

1980, 1982; Miller and Limpus, 1981; Ruiz et al., 1981; Morreale et al., 1982; Mrosovsky, 1982).

Concern arose regarding the Kemp's ridley program adversely affecting sex ratios (Mrosovsky, 1985), and incubation temperatures were intentionally raised in 1985 in an effort to increase the proportion of females emerging from eggs incubated at PINS (Shaver et al., 1988; Fletcher, 1989). Reviewers later concluded that most turtles from year classes through 1984 were male dominated, and those from the 1985–1992 year classes were female dominated (Shaver et al., 1988; Wibbels et al., 1989; Caillouet, 1995).

A review of incubation temperatures in relation to percentage females produced from the 1982–1987 clutches ($n = 32$) estimated that the pivotal temperature for Kemp's ridley was 30.2°C, with temperatures above 30.8°C producing 100% females (Shaver et al., 1988). However, the small sample size available for review raises concern about the accuracy of the calculated pivotal temperature. In addition, variation of the pivotal temperature within species is not well defined, and other factors (i.e., temperature variance, osmotic stress, maternally derived steroids in yolk) have been reported to affect embryonic sex determination (Ackerman, 1996; Bowden et al., 2000). In any case, one must assume that pivotal temperatures produced by relatively constant incubator conditions can be extrapolated to the nesting beach, particularly when nest temperature is used as an indicator of sex (Wibbels, 1998; Wibbels and Geis, 1999).

A problem with this assumption is that little is known about the long-term variation of beach temperature and its effect on sex ratios (Mrosovsky, 1994). A limited study of beach temperature profiles at Rancho Nuevo and PINS suggested, based on the reported pivotal temperature, that clutches undergoing the middle third of incubation early in the season should produce primarily males, a mixture of males and females at midseason, and primarily females late in the season (Standora and Spotila, 1985; Shaver et al., 1988). However, considerable temperature variation can be expected within a nesting beach over time. In addition, local variations in nesting beach temperature as a result of vegetative cover, local weather conditions, and embryonic metabolic heat in the nest should be taken into account.

Some authors have advocated the use of TSD or administration of exogenous estrogen to produce more females as a conservation tool in the recovery of endangered or threatened turtle populations (Tousignant and Crews, 1994; Vogt, 1994). This suggestion is based on the assumption that female turtles are more important than males because one male can inseminate many females. Captive breeding programs that produce and release hatchlings at a ratio of 6–20 females for every male have been suggested as a recovery strategy for rare and declining populations (Vogt, 1994).

Others have cautioned against manipulating hatchling sex ratio, suggesting that an understanding of the influence of sex ratio manipulation and natural sex ratio variation in sea turtles is needed (Mrosovsky and Godfrey, 1995; Lovich, 1996; Girondot et al., 1998). In general, adult sex ratios in turtle populations can vary according to several underlying factors (Gibbons, 1990; Lovich and Gibbons, 1990), including sex-specific mortality (Gibbons, 1968; Parker, 1984, 1990), immigration and emigration (Parker, 1984; Gibbons et al., 1990; Lovich, 1990), growth (Chaloupka and Limpus, 1997; Limpus and Chaloupka, 1997), and maturation time (Gibbons, 1990; Gibbons and Lovich, 1990; Lovich et al., 1990).

Sex Determination

Kemp's ridley exhibits sexual dimorphism in adults (Márquez-M., 1994). These differences become most evident as secondary sexual characteristics of subadults in the final phase of maturation and in adults. Mature males have a large tail that extends well beyond the carapace and a strongly curved flipper claw used to hold the female during copulation (Owens, 1997). Mature females do not exhibit secondary sexual characteristics but typically bear scratches and scars on the anterior edge of the carapace caused by the male during copulation.

Some pubescent or sexually developing male turtles can be identified by a soft plastron (Wibbels et al., 1991a). The function of the softened male plastron has not been investigated, but it may facilitate mounting during copulation (Owens, 1997). A method, both simple and reliable, for sexing immature sea turtles has yet to be identified.

Techniques that have been employed to determine sex in immature sea turtles include histological analysis of gonads, gross examination of stranded carcasses, H-Y antigen assay, laparoscopy, and RIA. More recently, attempts have been made to develop an assay to detect mullerian inhibiting hormone (Wibbels et al., 2000).

Histological examination, used most frequently to determine sex of hatchling sea turtles, involves removing the gonads and subsequent microscopic examination (Mrosovsky et al., 1984). This technique requires that subjects be euthanized, which is undesirable when one is studying threatened or endangered sea turtles. In addition, this methodology can be quite laborious (Mrosovsky and Benabib, 1990).

Necropsy and observation of gonads also have been used to determine sex of stranded Kemp's ridleys (Danton and Prescott, 1988; Shaver, 1991; Morreale et al., 1992; Stabenau et al., 1996). Although it is desirable to obtain all available data from every stranded sea turtle, sex determination can be difficult or unreliable, particularly in severely decomposed carcasses (Heinly, 1990). For example, gonadal decomposition resulted in sex being determined in only 50% of the stranded sea turtles from the upper Texas coast examined by Stabenau et al. (1996). In addition, stranded turtles may not provide a representative sample of the entire population (Epperly et al., 1996).

Detection of the H-Y antigen also has been used to sex green and loggerhead sea turtles (Wellins, 1987; Foley, 1994). This methodology uses fluorescence microscopy, x-ray film, or a cytotoxicity assay to detect the male-specific cell surface histocompatibility antigen. Results have been consistent with the pattern of H-Y-positive males found in most other vertebrates but have not been confirmed using other methods. It is worth noting that Zaborski et al. (1988) reported H-Y-negative males and H-Y-positive females in the European pond turtle (*Emys orbicularis*) when the eggs are incubated at the pivotal temperature, suggesting that H-Y antigen reflects a sexual genotype that may be overridden by temperature in TSD animals.

One reliable method for determining sex and reproductive status in sea turtles involves laparoscopy to examine the gonads and associated reproductive ducts (Wibbels, 1988; Limpus et al., 1994; Wibbels et al., 2000). Laparoscopy has been used to assess sex, maturity, and breeding status in green turtles (Limpus and Reed, 1985), loggerheads (Limpus, 1985; Wibbels et al., 1987), hawksbills (*Eretmochelys imbricata*) (Limpus, 1992; R. van Dam, personal communication), and olive ridleys (*L. olivacea*) (Plotkin et al., 1996). However, this procedure requires specialized equipment and training and is difficult to perform in the field (Wibbels et al., 1987, 1991b, 2000). Laparoscopy also has been used to validate sex determined via RIA (Wibbels, 1988; Coyne, 2000).

First described for use with sea turtles by Owens et al. (1978), RIA has been used to assess sex ratios and study reproductive cycles in green turtles (Owens et al., 1978; Bolten et al., 1992), loggerheads (Wibbels et al., 1987, 1990), and olive ridleys (Plotkin et al., 1997). Three studies that used RIA to sex Kemp's ridleys have been limited to head-started (Wibbels et al., 1989), captive (Morris, 1982), and adult turtles (Rostal, 1991), the last focusing on the reproductive cycle. The study referenced herein is the only large-scale study to characterize sex ratio of wild Kemp's ridleys (Coyne, 2000).

Population Modeling

Demographic models allow one to simulate a population's response to various factors and assess the relative importance of each model variable to model output. Simulation modeling as a management tool has provided the impetus to model animal populations either to assess their status, validate potential management scenarios for threatened populations, or develop harvest quotas for commercially valuable species (Heppell and Crowder, 1994; Heppell et al., 1995; Heppell and Crowder, 1996; Heppell, 1998; Turtle Expert Working Group, 1998). These population models are generally constructed from life table data, of which essential parameters continue to elude sea turtle investigators (Bustard, 1979).

Until recently only portions of sea turtle population models had appeared in the literature (Richardson and Richardson, 1982). For example, Hughes (1974) used annual egg production, egg survival, and the observed recruitment to an adult nesting loggerhead population to estimate juvenile survival rates. Various investigators have used remigration-interval frequency and seasonal population counts to estimate number of nesting females (Carr et al., 1978; Márquez-M., 1994; Turtle Expert Working Group, 1998). Additionally, models have been developed for density-dependent population regulation through the mechanism of intraspecific nest destruction (Bustard and Tognetti, 1969; Girondot et al., 2002).

The need to assess current and proposed management efforts has resulted in recent attempts to develop comprehensive models for threatened and endangered sea turtle populations. Models developed for loggerhead sea turtles suggest that it may be more valuable to protect older cohorts at sea, for example through the use of turtle excluder devices (TEDs), than those on the nesting beach (Crouse et al., 1987; Heppell et al., 1996a). Similar models have been developed to assess the status of loggerhead and Kemp's ridley populations (Turtle Expert Working Group, 1998). However, the paucity of data prohibits these models from providing reliable quantitative analyses of important life-history variables such as survivorship and age at maturity.

Chaloupka and Limpus (1997) presented results of robust statistical models based on long-term studies and relatively large datasets. These models represent the first opportunity to overcome shortcomings of previous modeling efforts. For the most part, few data have been available for sea turtles from the time they leave the nesting beach as hatchlings until adult females return to nest. Uncertainty surrounding age at maturity and the fact that survivorship is likely highly variable has made it difficult to estimate survivorship for interim life-history stages or to assess the impact of management efforts.

A great deal is known about Kemp's ridley nesting behavior and reproductive parameters, primarily as a result of the ongoing over 20-year joint United States–Mexico conservation program at Rancho Nuevo, Mexico. However, Kemp's ridley population models still lack reliable estimates for those critical variables associated with at-sea life-history stages (Márquez-M., 1994; Heppell et al., 1996b; Turtle Expert Working Group, 1998).

In particular, the effect of sex ratio on population dynamics in sea turtles is poorly understood. It has been suggested that hatchling sex ratios be altered through artificial incubation or application of estrogen in order to influence population sex ratios (Vogt, 1994). This suggestion relies on the assumption that female turtles are more important than males because one male can inseminate many females. However, caution should be used until the potential demographic and ecologic consequences of manipulating sex ratios in turtle populations are better understood (Lovich, 1996).

Multiple paternity has been documented in several turtle species including snapping turtles (*Chelydra serpentina*) and loggerhead turtles (Harry and Briscoe, 1988; Galbraith et al., 1989, 1993; Bollmer et al., 1999). In addition, Gist and Jones (1989) and Palmer et al. (1998) documented several turtle species that possess the ability to store sperm. Together, these characteristics raise interesting questions about their adaptive advantages (Lovich, 1996). What effect do multiple paternity and sperm storage have on population size, offspring viability, and fecundity?

Sugg and Chesser (1994) modeled gene correlations in populations with different mating strategies and have suggested that multiple paternity increases effective population size over that expected from polygyny and monogamy by maintaining genetic variation. Madsen et al. (1992) suggested that multiple copulations by female adders (*Vipera berus*) might enhance offspring viability, either because of inadequate quantities of sperm from a single mating, additional nutrients derived from the seminal fluid, or some genetic advantage. Mrosovsky and Godfrey (1995), citing Chan (1991), suggested that poor hatch rates of leatherback sea turtles (*Dermochelys coriacea*) in Malaysia may be attributable to an insufficient number of males to fertilize clutches. These potential adaptive advantages of multiple paternity and sperm storage are effective only when the number of reproductive males is sufficient to facilitate multiple insemination of reproductive females.

Sex Ratios

Sea turtles were captured in jetty and beachfront habitats immediately adjacent to Calcasieu Pass, Louisiana, and Sabine Pass, Bolivar Roads (Galveston), and at the inshore habitat of Matagorda Bay, Texas, from September 1992 through September 1997. Capture was accomplished utilizing one or more stationary entanglement nets set adjacent to one another (Coyne, 2000).

Sex Validation

The sex was verified for 79 individual Kemp's ridleys, including 4 head-started turtles (3 females, 1 male). Three of these individuals were recaptured once and another twice. In situ observation of gonads was used to verify sex of individual turtles and corroborate testosterone RIA results (Wood et al., 1983). Approximately 20% of all captured turtles (74 of 361) were subject to laparoscopic examination of the gonads. Three turtles were necropsied by National Marine Fisheries Service (NMFS–Galveston, Texas) personnel after stranding between 284 and 345 days after capture, and two others were identified as females from flipper tags attached while they were nesting in Mexico. This yielded overall positive identification of 47 female and 32 male Kemp's ridleys, with 44 and 31 wild individuals, respectively.

Sixty-three capture blood samples (including 5 recapture samples) were obtained from 58 individual turtles whose sexes were verified and included 32 female (with 3 recaptures) and 26 male (with 2 recaptures) samples. Individuals ranged in size from 21.8 to 62.1 cm straight carapace length (SCL) for females (mean = 42.8 ± 12.8 cm; ±1 SE used throughout) and from 23.9 to 59.2 cm SCL for males (mean = 37.3 ± 2.3 cm). These samples were used to set a sexing criterion for all blood samples collected at capture for Kemp's ridleys. Plasma T concentration for sex-verified turtles exhibited a mean of 4.8 ± 0.5 pg/ml for females and 186.3 ± 73.6 pg/ml for males, with respective gender ranges of 0.17–12.0 pg/ml and 18–2,063 pg/ml. There was a significant difference in plasma T between male and female turtles (F = 136.1, $P \leq 0.0001$; DF = 59). These data result in a plasma T sexing criterion for Kemp's ridley of ≤12 pg/ml for females and ≥18 pg/ml for males, with all individuals whose plasma T concentration fell between these criteria designated as indeterminate.

Blood Sampling

An attempt was made to sample blood from the dorsal cervical sinus of each turtle captured. Initial blood samples were obtained between 7 and 19 minutes (mean 12.3 ± 0.2) after turtles were first observed in the net. Plasma T concentration was determined by RIA in the laboratory of David Owens as described by Valverde (1996). Assay sensitivity was 0.5 pg/assay tube (Endocrine Sciences, 1972). Intra- and interassay variabilities were 5.64% and 9.77%, respectively.

Blood samples were obtained from 247 Kemp's ridley captures (including 7 recaptures). Plasma T ranged from undetectable to 2,063 pg/ml. The calculated sexing criteria, applied to all individual Kemp's ridleys for which capture blood was obtained, resulted in 134 females, 95 males ($\chi^2 = 5.84$, $P \leq 0.0157$), and 10 indeterminates (samples ranging between 12 and 18 pg/ml). Plasma T results of seven recapture samples, including five females and two males, were consistent with results from initial capture. Another 16 females and 6 males identified by other means (laparoscopy only, necropsy, or nesting event) increased the experiment lot to 150 females and 101 males. Twenty sexed Kemp's ridleys were part of the NMFS head-start program, including 2 males and 18 females, resulting in 132 wild females and 99 wild males.

Sex Ratio

The various sexing techniques produced a 1.3:1.0 female-to-male sex ratio for wild Kemp's ridley ($n = 231$, $\chi^2 = 4.77$, $P \leq 0.0289$) and an overall ratio of 1.5F:1.0M when head-start turtles are included ($n = 251$, $\chi^2 = 9.57$, $P \leq 0.0020$). Both ratios are significantly different from 1F:1M. Sex ratio for head-start turtles was 9.0F:1.0M ($n = 20$, $\chi^2 = 41.99$, $P \leq 0.0001$). Subsequent analyses include only wild Kemp's ridley to eliminate confounding effects related to head-start sex ratios.

Larger ridleys exhibited significant female bias, most obvious in the ≥60 cm SCL category (Fig. 10.1; $\chi^2 = 5.00$, $P \leq 0.0253$) with five females and no males. All ridleys ≥50 cm SCL exhibited a 2.3F:1.0M sex ratio ($n = 26$, $\chi^2 = 3.85$, $P \leq 0.0499$), whereas those ≥40 cm SCL exhibited a sex ratio of 1.7F:1.0M ($n = 46$, $\chi^2 = 3.13$, $P \leq 0.0768$). Ridleys less than 40 cm SCL yielded a 1.2F:1.0M ratio ($n = 149$, $\chi^2 = 1.51$, $P \leq 0.2191$). The smallest ridleys (20–29.9 cm SCL) exhibited a balanced ratio of 1.0F:1.0M ($n = 43$, $\chi^2 = 0.02$, $P \leq 0.8788$).

A similar sex distribution across size classes is evident from Kemp's ridley stranding records for the U.S. Gulf and Atlantic coasts. Data were obtained from the NMFS Sea Turtle Stranding and Salvage Network, and analysis included wild

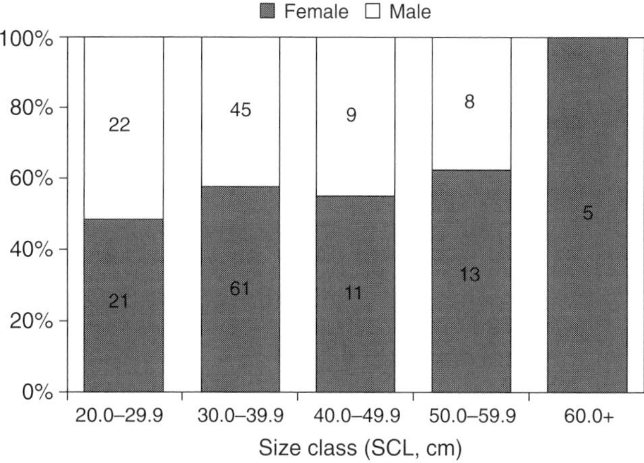

Fig. 10.1. Percentage contribution of female and male wild Kemp's ridleys to 10-cm size classes (SCL) captured between September 1992 and October 1997. Numbers within histogram bars denote sample size.

Kemp's ridleys stranded during 1986 through 1996 from Texas to Maine for which SCL and sex were recorded. Additional ridleys were included for which sex and curved carapace length (CCL) were recorded by converting CCL to SCL using a regression equation calculated from captured Kemp's ridleys ($n = 285$; $r^2 = 0.9920$; Coyne, 2000): SCL = 0.346 + 0.948 · CCL. This equation accurately predicted SCL using CCL from those stranded animals for which both SCL and CCL were recorded ($n = 1{,}344$, $r^2 = 0.9950$).

The sex ratio calculated from these stranding records was 1.4F:1.0M ($n = 584$, $\chi^2 = 5{,}001.4$, $P \leq 0.0001$). Again, gender makeup of smaller size classes was relatively uniform, with ridleys <40 cm SCL exhibiting a sex ratio of 1.2F:1.0M ($n = 403$, $\chi^2 = 2.4$, $P \leq 0.1225$). Larger size classes were predominantly female, with ≥40.0 cm SCL turtles exhibiting a sex ratio of 2.1F:1.0M ($n = 236$, $\chi^2 = 28.5$, $P < 0.0001$).

Results

Gregory and Schmid (2001) reported a sex ratio of 1.8F:1.0M ($n = 36$) for wild-captured Kemp's ridley sea turtles near Cedar Key, Florida during 1992 (Table 10.1). Stabenau et al. (1996) reported an overall sex ratio of 3.2F:1.0M ($n = 144$) for Kemp's ridley stranded along the upper Texas coast from Sabine Pass to the west end of Matagorda Island during 1986–1992. Head-started Kemp's ridleys in the latter study exhibited a sex ratio of 7.5F:1.0M ($n = 17$), whereas wild cohorts were 3.0F:1.0M ($n = 127$). Danton and Prescott (1988) reported a sex ratio of 1.4F:1.0M ($n = 48$) for stranded Kemp's ridleys from Cape Cod, Massachusetts. No mention was made of head-started individuals. Shaver (1991) reported an overall sex ratio of 1.0F:1.0M ($n = 81$) for ridleys stranded on south Texas beaches during 1983–1989. Forty-two of the observed strandings were head-started Kemp's ridleys (1.6F:1.0M), with the remaining wild cohorts exhibiting a sex ratio of 1.0F:1.8M ($n = 39$).

Coyne (2000) suggested an overall sex ratio for Kemp's ridley sea turtles of 1.5F:1.0M ($n = 251$) versus 1.3F:1.0M ($n = 231$) for wild cohorts only (Table 10.1). Head-started Kemp's ridleys account for a portion of the female bias, exhibiting a sex ratio of 9.0F:1.0M ($n = 20$). This is not surprising because approximately 81% of all head-started ridleys released since 1984, after discovery of the pivotal sex determination temperature for Kemp's ridleys, have been females (Stabenau et al., 1996). Their impact will likely decrease now that the Head Start Program is no longer in effect (100–200 ridleys are still raised each year by NMFS-Galveston for experimental purposes).

A wide range of sex ratios has been reported by various Kemp's ridley studies, and it is possible to derive an array of sex ratios within individual studies (Table 10.1). Plausible hypotheses to explain differences within and between studies include sexual bias of (1) gonadal decay rates in stranded animals; (2) annual sex ratio production; (3) stranding and/or capture rates; (4) movement or migration patterns; (5) geographic distribution; or (6) survival/mortality rates.

Evidence of sexually biased gonadal decay rates (hypothesis 1) among Kemp's ridleys has been provided by Owens (personal communica-

Table 10.1 Review of Kemp's ridley sea turtle (*Lepidochelys kempii*) sex ratios reported in the literature

Sex ratio	*n*	Time frame	Source
1.3F:1.0M overall	251	1993–1997	Coyne, 2000
1.1F:1.0M wild	231		
8.0F:1.0M head-start	20		
1.0F:1.0M overall	20	1993	Owens, personal communication
1.8F:1.0M wild	36	1992	Gregory and Schmid, 2001
3.2F:1.0M overall	144	1986–1992	Stabenau et al., 1996
3.0F:1.0M wild	127		
7.5F:1.0M head-start	17		
1.0F:1.0M overall	81	1983–1989	Shaver, 1991
1.0F:1.8M wild	39		
1.6F:1.0M head-start	42		
1.4F:1.0M	48	1977–1987	Danton and Prescott, 1988

tion), who examined gonads of 44 carcasses weighing 1–5 kg stranded in the Grand Isle, Louisiana, area during spring 1993. Owens noted that testes were more compact and solid, whereas immature ovaries were very thin and tissue-like, possibly making it more difficult to distinguish ovaries in highly decomposed individuals. This difficulty in distinguishing ovarian tissue could presumably cause an erroneous male bias in analysis of stranded animals because a greater proportion of males would be identifiable. However, two studies cited in Table 10.1 suggest a female bias in stranded Kemp's ridley (Danton and Prescott, 1988; Stabenau et al., 1996). Owens' examination occurred a year after the stranding event and involved frozen carcasses previously necropsied for other evaluations. Stabenau et al. (1996), on the other hand, necropsied only relatively fresh carcasses.

Evidence to support annual variation in hatchling sex ratio production is lacking (hypothesis 2). The fact that most Kemp's ridley nests (>90%) have been transplanted to corrals since 1978 (Márquez-M., 1994) has resulted in eggs from a given *arribada* being subjected to similar environmental conditions to dictate sex determination during incubation. However, seasonal and annual variation in factors such as temperature and rainfall should be sufficient to insure production of both sexes. Although annually sampled sex ratios of juvenile to adult Kemp's ridley appear to vary considerably (Coyne, 2000), it is unlikely that this is a function of hatchling sex ratio production but instead is related to Kemp's ridley behavior and/or distribution dynamics. In any case, four studies mentioned previously (Table 10.1) covered a period of at least 4 years. Although longer studies are desirable, this should help mitigate confounding annual effects.

Sex-specific migration patterns, distribution, and mortality rates (hypotheses 3–6) are related in that they are contingent on Kemp's ridley movement and/or behavior. It is unclear whether males and females utilize similar habitats or ranges at each life-history stage or whether adults of each sex follow the same reproductive migratory path to the nesting beach. Such differences could have profound effects on exposure to various mortality factors, subsequent stranding rates (percentage of dead animals that actually end up on the shore), and exposure to near-shore capture efforts (Coyne, 2000).

Wibbels et al. (1987, 1990) suggested that increased circulating testosterone in adult male loggerhead sea turtles may affect or coincide with migration and other reproductive events including spermatogenesis, courtship, and mating. A similar peak coinciding with spring interpond breeding migration has been suggested (Gibbons, 1968) for plasma T levels in male painted turtles (*Chrysemys picta*). Male slider turtles (*Pseudemys scripta*) and Concho water snakes (*Nerodia harteri paucimaculata*) have been reported to emigrate more frequently and farther than their female counterparts, whereas younger

P. scripta males appear to be more sedentary than larger males (Parker, 1984; Whiting et al., 1997).

Figure 10.2 presents a hypothetical testosterone model for Kemp's ridley by life-history stage adapted from data collected by Coyne (2000), Morris (1982), Rostal (1991), and Rostal et al. (1997). In general, the model suggests that adult male Kemp's ridleys maintain a relatively high level of testosterone year round, which builds to a peak coinciding with mating migration and then slowly decreases to baseline again. Adult females exhibit a similar response, with testosterone levels peaking at or near the onset of migration and then slowly decreasing to baseline during the nesting season. Seasonal analyses suggest that plasma T in juvenile and subadult male Kemp's ridley in the Gulf of Mexico is characterized by a slight increase during June through September (Coyne and Landry, 2000). Juvenile and subadult females are expected to exhibit little variation relative to their male counterparts because their baseline plasma T levels are considerably lower.

Although plasma T baseline values and ranges typical of each sex/life stage of Kemp's ridley are not well understood, the following values have been reported in the literature. Maximum plasma T detected during this study (2,063 pg/ml) was from a confirmed head-start male with a 59.7-cm SCL. Rostal et al. (1997) reported a maximum plasma T of 219 pg/ml from a nesting female at Rancho Nuevo, Mexico. Maximum plasma T observed in nonnesting, 60-cm SCL and larger females reported by Coyne (2000) was 5.07 pg/ml ($n = 6$). Plasma T in confirmed male juveniles (circa 20–40 cm SCL) ranged from 18 to 213 pg/ml ($x = 51.5 \pm 11.2$) and 68 to 2,063 pg/ml ($x = 394.6 \pm 172.5$) in subadults (circa 40–60 cm SCL). Juvenile and subadult females exhibited ranges of 0.2–12.0 pg/ml ($x = 5.3 \pm 0.9$) and 0.2–8.2 pg/ml ($x = 4.6 \pm 0.6$), respectively.

Elevated plasma T associated with the nesting season (Morris, 1982; Rostal, 1991; Rostal et al., 1997) may be partially responsible for initiating migration in females. Satellite telemetry (Renaud et al., 1996) from an adult female captured at Calcasieu Pass (August 11, 1995) and later observed nesting at Rancho Nuevo (April 23, 1996, and May 19, 1996) supports this supposition because she began a concerted southerly track at approximately the same time (December) plasma T is suggested to have begun rising (Rostal, 1991). Owens (1997) suggests that reproductive migration in female sea turtles coincides with a peak in annual plasma T concentration and that adult males experience an earlier peak and thus begin migration earlier. These observations are speculative, as direct experimentation linking testosterone to migration behavior has not been completed.

Despite a small sample size ($n = 11$), previous analysis of plasma T data suggests an exponential increase in circulating T with size in subadult male Kemp's ridleys (Coyne and Landry, 2000). The apparent plasma T increase in subadult males may result in a larger home range and/or alteration of habitat utilization, similar to the proposed elevated plasma T initiation of migration in adults. This hypothesis is supported by radio and satellite telemetry suggesting that larger Kemp's ridleys utilize deeper waters and larger home ranges (Renaud et al., 1995).

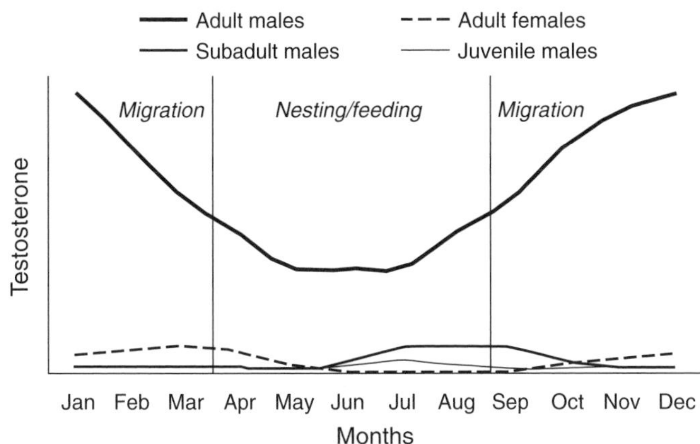

Fig. 10.2. Generalized testosterone model at each life-history stage of male and female Kemp's ridley sea turtles. Adult responses adapted from Rostal (1991) and Morris (1982).

Plasma T clearly plays a role in seasonal reproduction in both male and female Kemp's ridleys and may play a behavioral role in triggering female receptivity and onset of mating (Rostal, 1991). However, although plasma T may influence Kemp's ridley behavior and movement patterns, a myriad of factors are probably involved, including temperature, season, reproductive status, and prey availability.

Finally, there is the possibility that one or more of the aforementioned studies (Table 10.1) does not accurately reflect Kemp's ridley population demographics. Size distribution (21.8–62.1 cm SCL) of turtles for which sex was determined was representative of the entire capture lot (Coyne, 2000). It is unclear whether the same is true of previous sex ratio studies of stranded Kemp's ridleys (Danton and Prescott, 1988; Shaver, 1991; Stabenau et al., 1996). Size distribution of wild Kemp's ridleys captured at Sabine Pass correlate well (Pearson product-moment = 0.9220) with wild ridleys stranded during the same period (1993–1996) in NMFS statistical subarea 18 (Sabine Pass through Galveston County, Texas) (Coyne, 2000). However, these stranding data include smaller cohorts not encountered in netting operations and a greater proportion of subadult (40–60 cm SCL) and adult turtles (greater than 60 cm SCL). The size distribution of stranded Kemp's ridleys is consistent with those presented by Shaver (1991) and Stabenau et al. (1996).

The proposed testosterone model (Fig. 10.2) for Kemp's ridley sea turtles may well explain observed differences in the sex ratios of stranded and captured turtles. Shaver (1991) and Stabenau et al. (1996) reported similar size distribution patterns of juvenile Kemp's ridleys suspected to inhabit shallow, nearshore waters, where they become more susceptible to capture by nearshore netting operations and postmortem stranding. Stranded postjuvenile ridleys sexed by Stabenau et al. (1996) exhibited an increasing female bias with size. Perhaps increased plasma T in subadult (pubescent) males is, in part, responsible for this observation. Larger males move farther from shore and/or maintain larger home ranges, thereby (1) decreasing the chance that a dead animal will reach shore; (2) enhancing the chance the carcass will reach shore in a deteriorated state, making sex determination more difficult; (3) increasing the likelihood of being exposed to different mortality pressures; or (4) placing them out of range of nearshore capture efforts. Females, which exhibit increased plasma T only in association with the breeding season, may otherwise remain closer to shore and thus are more susceptible to nearshore capture, mortality, and/or stranding.

Similarly, the majority of female bias observed by Stabenau et al. (1996) also applies to larger size classes, particularly ridleys greater than 60 cm SCL. A review of only wild juvenile and subadult turtles (20–60 cm) examined by Stabenau et al. (1996) yielded a sex ratio of approximately 2.3F:1.0M, closer to but still considerably greater than the 1.3F:1.0M (n = 190) observed by Coyne (2000). The disparity between results from these two studies is even greater for ≤40 cm (1.2F:1.0M, n = 149) and ≤30 cm SCL (1.0F:1.0M, n = 43) juvenile turtles.

This model of testosterone–behavior interaction may help explain increasing female bias with size observed by Coyne (2000) and the exclusive capture of females among Kemp's ridleys greater than 60 cm SCL. Of these eight larger turtles, six were verified as females, and two others exhibited plasma T levels consistent with female Kemp's ridleys (5.4 and 10.3 pg/ml, respectively). The largest confirmed or predicted male captured was 59.1 cm SCL. Although it would be more difficult to verify, an alternative explanation is that maturing males suffer greater mortality and thus exhibit lower abundance at maturity than do adult females. However, the fact that both stranding and capture data exhibit similar sex ratio patterns does not support such a case.

If one assumes that male *L. kempii* are not subject to differential mortality, then the 1:1 sex ratio observed in juveniles most closely represents the primary sex ratio of the population. Implicit in this assumption is that subadult and adult males are encountered less frequently than their female counterparts because of differences in behavior and/or habitat use. However, sex ratio variations reported across size classes and in stranded animals cannot be dismissed, particularly in view of data presented by Stabenau et al. (1996).

Population Modeling

In regard to potential influences of multiple paternity, sperm storage, and multiple insemination, the model utilized in this chapter assesses effects of various sex ratios on the Kemp's ridley population under the following assumptions: (1) sex ratio does not influence fecundity or offspring viability; and (2) male bias increases fecundity or offspring viability. The effect of these assumptions was assessed by altering clutch frequency and size in relation to sex ratio. Changing hatch success in relation to sex ratio was not considered because this variable is extremely biased by the practice of protecting nests on the nesting beach. Fecundity effects were analyzed in relation to sex ratio. The Kemp's ridley population sex ratio was considered to be 1F:1M for the baseline model. Additional sex ratio scenarios were tested to assess the sensitivity of the model to sex ratio changes and potential impact on the population including 3F:1M, 2F:1M, 1F:2M, and 1F:3M.

Model Development

The strategy used to develop the model was to sacrifice precision to realism and generality (Levins, 1966). In other words, the model does not attempt to predict the population exactly, only the population's response to changing variables built into the model (Table 10.2). Primary concern with qualitative rather than quantitative results permits a flexible, graphic model to be developed that generally assumes that functions are increasing or decreasing, greater or lesser than some value, instead of specifying the mathematical form of an equation. The litany of variables needed for a robust statistical model can be reduced to manageable proportions by abstracting many functions into a reduced number of higher-level functions. For example, it is not necessary to know how many eggs are nonviable and how many succumb to predation or inundation if the proportion of eggs laid that develop into hatchlings can be estimated.

The basic process to be modeled was annual size of the Kemp's ridley population (Fig. 10.3). Number of nests was used as an indicator of the population as a whole because there is currently no way to estimate total population size or even that of a given life-history stage. Comprehensive monitoring of the Rancho Nuevo nesting beach since 1978 makes annual number of nests one of the best indices of population status for Kemp's ridley, assuming that number of nests laid each year is an appropriate index (Turtle Expert Working Group, 1998).

The Kemp's ridley life cycle was grossly simulated to estimate annual number of nests. Key components of the life cycle were evaluated, including variables related to (1) reproduction; (2) growth/age; and (3) mortality (see Coyne, 2000, for complete details).

Caveats and Assumptions

There are many assumptions related to development of this model that are important for one to consider before continuing the process (see Coyne, 2000, for complete details):

- The model utilizes deterministic versus stochastic modeling.
- The model population is not density dependent.
- Reported nest counts from the nesting beach do not represent actual total number of *L. kempii* nests.
- Effort and coverage of nesting beaches have expanded over the years.
- Number of nests may not be an accurate indicator of population status or size.
- All adult females were assumed to be equally fecund.
- Duration of the pelagic posthatchling stage remains in doubt.
- Atlantic and Gulf of Mexico cohorts were treated equally.
- Remigration rate was grossly estimated.
- No life expectancy limit is built into the model.
- The role and function of male-mediated fecundity are not well understood.

Model Results

The size and growth of any population depend on the annual number of births and deaths, the timing of maturation, and reproduction and death in each individual's life. This model was designed to be a generalized representation of the Kemp's ridley population and was intended to respond in a manner similar to the natural population. Attempting to predict exact population responses would be a difficult, if not impossible, task because many aspects of *L. kempii* life history remain unknown.

Table 10.2 Definition of variables and baseline values used in Kemp's ridley population model (Coyne, 2000)

Variable	Definition	Value
Mortality	Annual life-stage specific mortality rate.	
Posthatchling	Pool of animals, presumably in the pelagic environment, encompassing the time when hatchlings leave the beach until becoming juvenile. Typically referred to as "lost year."	Size: up to 20 cm SCL Age: 0–2 years Mortality: 0.5580
Juvenile	Postpelagic pool of animals of a size and/or age at which little or no sexual development is occurring.	Size: 20–40 cm SCL Age: 2–4 years Mortality: 0.3220
Subadult	Pool of animals of a size and/or age at which sexual development is occurring.	Size: 40–62.5 cm SCL Age: 4–10 years Mortality: 0.2260
Adult	Pool of animals that have reached sexual maturity.	Size: 62.5+ cm SCL Age: 10+ years Mortality: 0.1100
Sex ratio	Proportion of female and male turtles as determined by this study.	1F:1M
Remigration rate	Proportion of adult females that nest in a given year.	0.768
Reproductive females	Number of adult females that nest in a given year as determined by remigration rate.	
Clutch frequency	Mean number of clutches per season per reproductive female.	3
Nests	Number of nests deposited by all reproductive females as a function of clutch frequency.	
Clutch size	Mean number of eggs laid per clutch.	95
Eggs laid	Number of eggs laid each nesting season as a function of nests and clutch size.	
Hatch rate	Proportion of hatchlings from eggs laid that successfully leave the nesting beach.	0.705
Hatchlings	Total number of hatchlings that survive from eggs laid as a function of hatch rate.	

Sex Ratio Dynamics

Studies reporting *L. kempii* sex ratios other than 1F:1M have frequently detected a female bias, ranging from 1.4F:1M to 3.2F:1M (Table 10.1). On the surface, a naturally occurring female bias seems an obvious strategy that a population can utilize to increase productivity. From a management perspective, it presents a potential and very attractive tool to aid in the recovery of threatened or endangered species. However, as noted by other authors, sea turtle sex ratio dynamics may not be that simple, and their manipulation may lead to undesirable consequences (Mrosovsky and Godfrey, 1995; Lovich, 1996).

The assertion that adult sex ratios are naturally biased in some turtle species, as has been shown in numerous studies, has profound implications for any program attempting to manipulate sex ratio to manage a population (Morreale et al., 1982). A major concern would be the impact of such a program on the reproductive ecology of a species, specifically as it relates to effects of multiple paternity, sperm competition, fertility, and intraspecific competition on population persistence.

If one assumes that proportion of available males has little or no impact on production, as long as enough males are available for mating, then the current model confirms that an increased proportion of females in the population

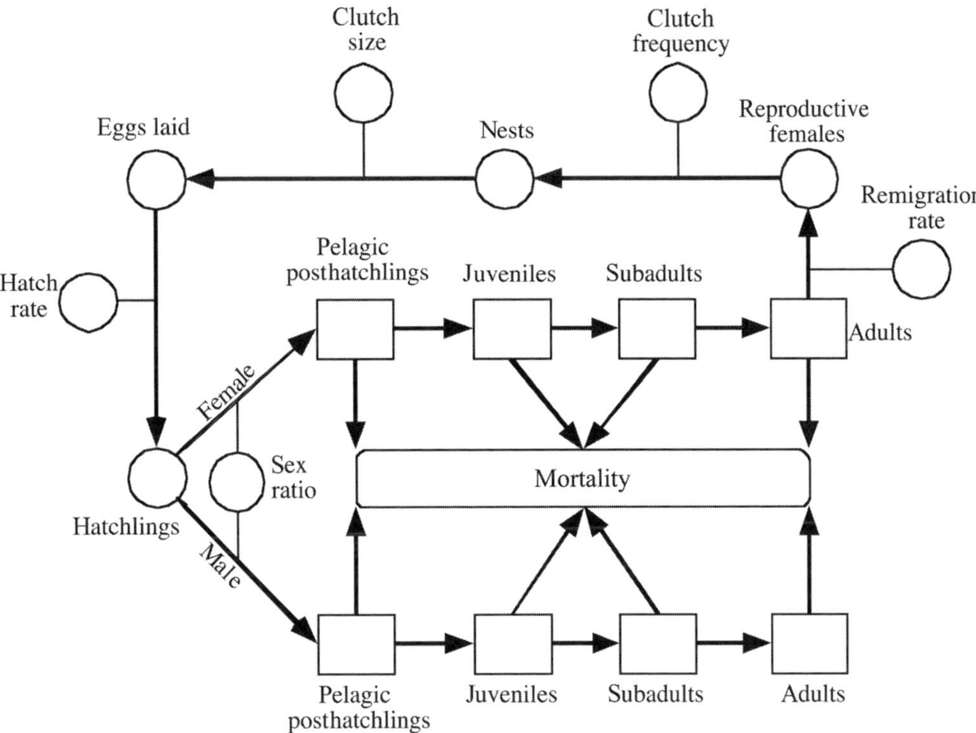

Fig. 10.3. Conceptual representation of the Kemp's ridley demographic model.

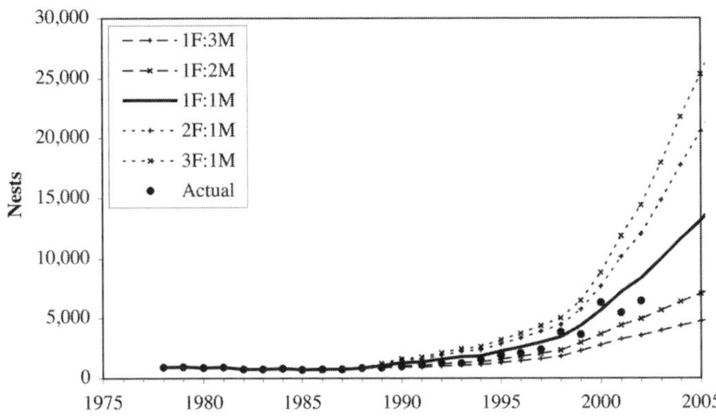

Fig. 10.4. Actual number of Kemp's ridley nests reported each year at Rancho Nuevo, Mexico, versus that predicted by the population model assuming a primary sex ratio of 1F:1M. The correlation coefficient for 1989–2002 is 0.947. Also included are model results assuming a primary sex ratio of 1F:3M, 1F:2M, 2F:1M, and 3F:1M.

will dramatically enhance hatchling production (Fig. 10.4). Under this scenario, nest production increases as a power of percentage female hatchlings produced, yielding a 271% increase in predicted nests after 50 years with a sex ratio of 3F:1M and an 81% decrease with 1F:3M:

$$\%\text{ Nest increase} = 0.3005 \cdot (\%\text{ female hatchlings produced})^{2.2615}$$

However, an increasing body of knowledge suggests that there is a benefit to having some minimum proportion of adult males in a sea turtle population. For example, multiple copulations or more time spent copulating can potentially yield a greater number of fertilized eggs, thereby increasing fecundity (Wood and Wood, 1980; Madsen et al., 1992; Mrosovsky and Godfrey, 1995). In addition, copulation with multiple

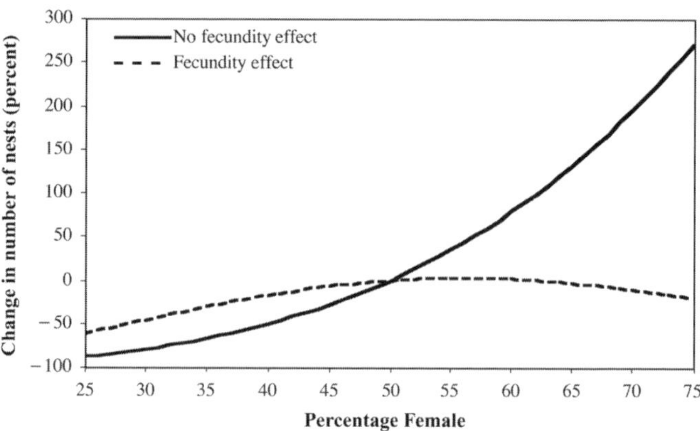

Fig. 10.5. Percentage change from baseline in predicted number of nests after a 50-year model run, with and without fecundity effects, across a range of sex ratios (1F:3M to 3F:1M). The fecundity effect curve (dashed line) suggests an optimum sex ratio of 56.2% female.

males may increase clutch viability, improving overall genetic health of a clutch or clutches (Harry and Briscoe, 1988; Sugg and Chesser, 1994; Kichler et al., 1999).

It is well established that several turtle species exhibit multiple paternity (Harry and Briscoe, 1988; Kaufmann, 1992; Galbraith et al., 1993). Coupled with the sperm storage ability possessed by many turtles (Gist and Jones, 1989), multiple paternity and sperm competition must be considered significant adaptations in their reproductive strategy. This suggests that multiple paternity and sperm competition may be very important in the persistence of populations and cautions against manipulating the sex ratio of turtle populations to produce an excess of females (Lovich, 1996).

Under this scenario, reproductive output and success are partially a function of male availability. Assuming a linear relationship between percentage increase in adult males and reproductive output (Coyne, 2000), the model suggests that strong female bias actually results in a decrease in the population (Fig. 10.5). The hatchling sex ratio producing the greatest rate of reproductive return was 56.2% female (1.28F:1M). A strong female bias (3F:1M) results in a 20% decrease in the population, and a male bias (1F:3M) yields a 62% decrease. It should be understood that the actual relationship between male availability and female reproductive output is not known. These results further stress that extreme caution should be taken when considering the manipulation of sex ratio as a sea turtle conservation strategy.

Summary and Conclusions

Summary

Observed increasing female bias with size in *L. kempii* may be a spatial artifact explained by subadult-to-adult male behavioral differences that render them more difficult to capture by methods deployed in this study as well as less likely to strand. Stranding records exhibit a pattern similar to that of captured *L. kempii* and suggest that strandings and the capture lot examined here may not be entirely representative of the at-large population.

If one assumes that the capture lot is representative of the population, the overall sex ratio reported for *L. kempii* in this study, 1.3F:1M, would apply. However, it is suggested that pubescent to adult male *L. kempii* maintain larger home ranges and remain farther from shore than do their female counterparts. Analysis of juvenile turtles captured during this study (circa 20–40 cm SCL) from which a blood sample was obtained ($n = 149$) suggests that the wild *L. kempii* population exhibits a primary sex ratio near 1F:1M.

These results and a thorough review of scientific literature were used to develop a preliminary population model for *L. kempii*. The final model was used to assess the impact of various population sex ratio values on Kemp's ridley sea turtle demography. Two scenarios were tested using the model: one in which the proportion of adult males in the population has no effect on hatchling production and another in which productivity is a function of the relative abundance of adult males.

Under the first scenario, the model suggests that a greater proportion of females in the population dramatically enhances hatchling production, yielding a 271% increase in predicted nests after 50 years with a sex ratio of 3F:1M and an 81% decrease with 1F:3M. The second scenario, in which reproductive output is a function of male availability, was described in the previous section. The greatest rate of reproductive return was achieved with a sex ratio of 1.28F:1M (56.2% female).

Conclusions

Assumptions regarding sea turtle sex ratios have serious implications for both population modelers and managers. For example, the idea of using sex ratio manipulation to "jump start" declining turtle populations has great appeal, but what would the impact be on the reproductive ecology of a species? Approximation of a fecundity effect in the model for this study attempts to simulate a scenario in which sperm storage, multiple paternity, and/or sperm competition might play a role in fertility and population persistence. Under this scenario the relative abundance of adult males is critical to reproductive output, suggesting that there is an "optimum" sex ratio.

Factors related to male-mediated fecundity would be effective only when the number of reproductive males is sufficient to facilitate multiple inseminations of individual females. Sugg and Chesser (1994) have demonstrated the importance of breeding structure on gene diversity in natural and captive populations. They reported that multiple paternity increases the effective population size above that expected from polygyny or monogamy. As the number of mating males decreases, the impact of multiple paternity also decreases.

This suggests that multiple paternity and sperm competition may be important in the persistence of populations and cautions against manipulating the sex ratio of turtle populations to produce an excess of females. Model results from this study support the need for caution and further suggest that there is a point of diminishing returns beyond which increasing the proportion of adult females in the population no longer provides increased productivity. Without male-mediated fecundity, there is little doubt that as the number of reproductive females increases, hatchling production will also increase (Fig. 10.5). However, it is not clear that this is the case, and there is much evidence to the contrary.

Under the assumptions provided, the model specifies an adult sex ratio of 1.28F:1M (56.2% female) to optimize population growth (Fig. 10.5). The exact value of this "optimum" sex ratio depends greatly on the relationship between adult sex ratio and reproductive output. Finding the actual "optimum" sex ratio would be contingent on quantifying this relationship. In addition, the fecundity effect in the model in no way addresses the potential for increased genetic fitness that might be gained from multiple paternity, only a potential increase in fecundity.

Potential problems are exacerbated when it comes to *L. kempii,* as the sex ratio present in the population today may not represent a "natural" sex ratio. The majority of Kemp's ridley nests have been transplanted to corrals since 1978 (Márquez-M., 1994). It is likely that most extant *L. kempii* came from transplanted clutches, and it is uncertain what effect nest transplantation has had on population sex ratio. Although temperature has been monitored both in the corrals and on the Mexican nesting beaches (Márquez-M., 1982, 1983, 1985; Márquez-M. et al., 1986, 1987, 1989), it is not clear these data have been analyzed to fully quantify sex ratio effects related to incubation temperature. Certainly they have received no critical review. The big question is, does the *L. kempii* sex ratio presented herein reflect a natural sex ratio?

Limited data have been presented indicating that the beach at Rancho Nuevo offers a range of incubation temperatures at, above, and below the pivotal sex ratio temperature, depending on distance from the water, season, and rainfall (Standora and Spotila, 1985; Shaver et al., 1988). Unfortunately, there is a paucity of published data comparing thermal profiles of transplanted nests versus those left in situ. The NMFS Head Start Experiment represents an extreme example of the possible effects of transplantation in the predominant production of male hatchlings before the pivotal incubation temperature was discovered and integrated into experimental protocol (Wibbels et al., 1989). It is possible that transplanted nests are exposed to less temperature variation during incubation because most found nests are grouped together into relatively small areas. Such a change in the thermal profile of a nest may have unknown consequences for the sex ratio produced by that nest (Mrosovsky and Yntema, 1980).

These factors underscore the importance not only of determining the existing population sex ratio for *L. kempii* but also of whether current population structure represents a natural state. Have ongoing conservation practices influenced population sex ratios, and if so, in what way? The population model presented here can serve as an invaluable tool in answering this question, providing a mechanism through which the population impact of in situ and natural nest sex ratios can be compared. Ongoing efforts to determine primary sex ratio production at the nesting beach via nonlethal sex determination in hatchlings is critical to these efforts (Wibbels, 1998; Wibbels and Geis, 1999).

Aside from sex ratio manipulation, the model developed in this study can be used to identify other areas of potential concern and serve as a useful tool in evaluating additional management strategies. Sensitivity analysis indicated that reduction of pelagic posthatchling mortality had the single greatest impact on the population model, increasing nest production by 301% over baseline after 50 years (Coyne, 2000). Unfortunately protecting pelagic posthatchling *L. kempii* is not a likely option in the foreseeable future as little is known about their habits or habitats. Reduction of juvenile and subadult mortality in the model also generated large increases in the number of nests and has profound implications for current TED regulations and efforts under way to establish protected areas for the species.

An important question in this regard is, if TED regulations are working to protect sea turtles at sea, why have stranding rates not declined since their inception? Subadult, and certainly juvenile, *L. kempii* should be beneficiaries of existing TED regulations, but stranding rates have continued to increase (Shaver, 1995). One answer is that stranding rates have remained the same or decreased, but, as the population recovers, there are simply more turtles and hence more strandings.

However, this question deserves a closer look, both for the long-term health of the population and to prevent increasing numbers of dead turtles from washing up on shore. A better understanding of *L. kempii* distribution in time and space and across all life-history stages is critical. Such information would be invaluable in assessing which life-history stages are most likely to be impacted by trawl fisheries, examining when and where turtles occur in relation to fishing effort.

Similarly, knowledge of *L. kempii* distribution is important for identifying and protecting a critical habitat for the species. A seasonal protected area has been established for Kemp's ridley off of the PINS during the nesting season. These efforts are aimed at protecting the recent influx of new nesters to the PINS in hopes of fostering the establishment of a second nesting beach for *L. kempii*. It is also argued that this area would provide a protected swimway for adults as they migrate to and from their Mexican nesting beaches along the south Texas coast.

Identifying life-history stage- and sex-specific temporal and spatial distribution for *L. kempii* could aid in identifying the boundaries of such protected areas as well as other candidate areas. The population model presented here and others like it provide an invaluable tool in evaluating the potential impact of these conservation strategies. Within its given working assumptions, managers have the power to review and refine existing management practices and the regulatory framework and more accurately assess the potential impact of proposed conservation measures.

Future Research Needs

An extensive review of all available information related to sea turtle population modeling and the Kemp's ridley sea turtle was completed as part of this work. As a result of these efforts the following research needs were identified:

- Further refine the primary sex ratio for use in population modeling by examining sex in *L. kempii* hatchlings on the nesting beach.
- Fully describe the in situ nesting environment for *L. kempii* relative to that of nests transplanted to protective "corrals."
- Quantify the relationship between adult sex ratio and reproductive output to quantify more accurately the "optimum" sex ratio for *L. kempii*.

- Current conservation efforts on the nesting beach should be reviewed to ensure that they are not adversely affecting primary sex ratio production with regard to the "natural" or "optimum" sex ratio.
- The distribution and habitat utilization of *L. kempii* in time and space should be investigated. In particular, effort should be focused on subadult (circa 40–60 cm SCL) male *L. kempii* to determine why they appear in lesser numbers than same-size females in both capture and stranding data. Also, an attempt should be made to determine to what extent this difference is a function of differential mortality and/or habitat utilization.
- Information regarding the distribution of *L. kempii* should be analyzed to aid in identifying potential protected areas. In particular, these data should be compared with the spatial and temporal distribution of trawl fishing to refine existing TED regulations and more accurately reflect potential conflict with sea turtles.
- Investigate the need for development of life-history stage- and/or sex-specific survivorship rates for more accurate population modeling.

ACKNOWLEDGMENTS

This work was made possible with the support and cooperation of the NMFS, U.S. Army Corps of Engineers, Texas A&M University Sea Grant Program, and U.S. Fish and Wildlife Service. We give special thanks to Dr. David Owens and his laboratory for conducting the RIA and making the laparoscopies possible. Extraordinary praise goes to the turtle crew: Stacie Arms, John Christensen, Randy Clark, David Costa, Kelly Craig, Travis Hanna, Leonard and Lisa Kenyon, Vicky Poole, Karen St. John, Katie VanDenberg, Sarah Werner, and others who lent assistance. Thanks also to Jaime Peña for timely updates from the nesting beach.

LITERATURE CITED

Ackerman, R. A. 1996. The nest environment and the embryonic development of sea turtles. In Lutz, P. L., and Musick, J. A. (Eds.). *The Biology of Sea Turtles.* Boca Raton, FL: CRC Press, pp. 83–106.

Bollmer, J. L., Irwin, M. E., Rieder, J. P., and Parker, P. G. 1999. Multiple paternity in loggerhead turtle clutches. *Copeia* 1999:475–478.

Bolten, A. B., Bjorndal, K. A., Grumbles, J. S., and Owens, D. W. 1992. Sex ratio and sex-specific growth rates of immature green turtles, *Chelonia mydas,* in the southern Bahamas. *Copeia* 1992: 1098–1103.

Bowden, R. M., Ewert, M. A., and Nelson, C. E. 2000. Environmental sex determination in a reptile varies seasonally and with yolk hormones. *Proceedings of the Royal Society of London, B* 267: 1745–1749.

Bull, J. J., Gutzke, W. H. N., and Crews, D. 1988. Sex reversal by estradiol in three reptilian orders. *General and Comparative Endocrinology* 70:425–428.

Burchfield, P. M., and Foley, F. J. 1989. Standard operating procedures for collecting Kemp's ridley sea turtle eggs for the head start project. In Caillouet, C. W., Jr., and Landry, A. M., Jr. (Eds.). *Proceedings of the First International Symposium on Kemp's Ridley Biology and Conservation.* Texas A&M University Sea Grant College Program, TAMU-SG-89-105, pp. 67–70.

Bustard, H. R. 1979. Population dynamics of sea turtles. In Harless, M., and Morlock, H. (Eds.). *Turtles: Perspectives and Research.* New York: John Wiley & Sons, pp. 523–540.

Bustard, H. R., and Tognetti, K. P. 1969. Green sea turtles: A discrete simulation of density-dependent population regulation. *Science* 163: 939–941.

Caillouet, C. W., Jr. 1995. An update of sample sex composition data for head started Kemp's ridley sea turtles. *Marine Turtle Newsletter* 69:11–14.

Carr, A., Carr, M. H., and Meylan, A. B. 1978. The ecology and migrations of sea turtles, 7. The west Caribbean green turtle colony. *Bulletin of the American Museum of Natural History* 162:1–46.

Chaloupka, M., and Limpus, C. J. 1997. Robust statistical modeling of hawksbill sea turtle growth rates (Southern Great Barrier Reef). *Marine Ecology Progress Series* 146:1–8.

Chan, E. H. 1991. Sea turtles. In Kiew, R. (Ed.). *The State of Nature Conservation in Malaysia.* Kuala Lumpur: Malayan Nature Society, pp. 120–135.

Coyne, M. S. 2000. Population sex ratio of the Kemp's ridley sea turtle (*Lepidochelys kempii*): problems in population modeling. Ph.D. diss. Texas A&M University, College Station.

Coyne, M. S., and Landry, A. M., Jr. 2000. Plasma testosterone dynamics in the Kemp's ridley sea turtle. In Abreu-Grobois, F. A., Briseño Dueñas,

R., Márquez-M., R., and Sarti, L. (Eds.). *Proceedings of the Sixteenth Annual Workshop on Sea Turtle Biology and Conservation*. NOAA Technical Memorandum NMFS-SEFSC-436, p. 286.

Crews, D., Wibbels, T., and Gutzke, W. H. N. 1989. Action of sex steroid hormones on temperature-induced sex determination in the snapping turtle (*Chelydra serpentina*). *General and Comparative Endocrinology* 76:159–166.

Crews, D., Bull, J. J., and Wibbels, T. 1991. Estrogen and sex reversal in turtles: a dose-dependent phenomenon. *General and Comparative Endocrinology* 81:357–364.

Crouse, D. T., Crowder, L. B., and Caswell, H. 1987. A stage-based population model for loggerhead sea turtles and implications for conservation. *Ecology* 68:1412–1423.

Danton, C., and Prescott, R. 1988. Kemp's Ridley Research in Cape Cod Bay, Massachusetts: 1987 Field Research. In Schroeder, B. A. (Ed.). *Proceedings of the Eighth Annual Workshop on Sea Turtle Conservation and Biology*. NOAA Technical Memorandum NMFS-SEFC-214, pp.17–18.

Endocrine Sciences. 1972. *Plasma Testosterone Radioimmunoassay Procedure Antiserum No. T3-125*. Calabasas Hills, CA: Endocrine Sciences Products.

Epperly, S. P., Braun, J., Chester, A. J., Cross, F. A., Merriner, J. V., Tester, P. A., and Churchill, J. H. 1996. Beach strandings as an indicator of at-sea mortality of sea turtles. *Bulletin of Marine Science* 59:289–297.

Fletcher, M. R. 1989. The National Park Service's role in the introduction of Kemp's ridley sea turtle. In Caillouet, C. W., Jr., and Landry, A. M., Jr. (Eds.). *Proceedings of the First International Symposium on Kemp's Ridley Biology and Conservation*. Texas A&M University Sea Grant College Program, TAMU-SG-89-105, pp. 7–9.

Foley, A. M. 1994. Detecting H-Y antigen in the white blood cells of loggerhead turtles (*Caretta caretta*): a possible sexing technique. In Schroeder, B. A., and Witherington, B. E. (Eds.). *Proceedings of the Thirteenth Annual Symposium on Sea Turtle Biology and Conservation*. NOAA Technical Memorandum NMFS-SEFSC-341, p. 228.

Galbraith, D. A., Brooks, R. J., and Obbard, M. E. 1989. The influence of growth rate on age and body size at maturity in female snapping turtles (*Chelydra serpentina*). *Copeia* 1989:896–904.

Galbraith, D. A., White, B. N., Brooks, R. J., and Hoag, P. T. 1993. Multiple paternity in clutches of snapping turtles (*Chelydra serpentina*) detected using DNA fingerprints. *Canadian Journal of Zoology* 71:318–324.

Gibbons, J. W. 1968. Population structure and survivorship in the painted turtle, *Chrysemys picta*. *Copeia* 1968:260–268.

Gibbons, J. W. 1990. Sex ratios and their significance among turtle populations. In Gibbons, J. W. (Ed.). *Life History and Ecology of the Slider Turtle*. Washington, DC: Smithsonian Institution Press, pp. 171–182.

Gibbons, J. W., and Lovich, J. E. 1990. Sexual dimorphism in turtles with emphasis on the slider turtle (*Trachemys scripta*). *Herpetological Monographs* 4:1–29.

Gibbons, J. W., Greene, J. L., and Congdon, J. D. 1990. Temporal and spatial movement patterns of sliders and other turtles. In Gibbons, J. W. (Ed.). *Life History and Ecology of the Slider Turtle*. Washington, DC: Smithsonian Institution Press, pp. 201–215.

Girondot, M., Fouillet, H., and Pieau, C. 1998. Feminizing turtle embryos as a conservation tool. *Conservation Biology* 12:353–362.

Girondot, M., Tucker, A. D., Rivalan, P., Godfrey, M. H., and Chevalier, J. 2002. Density-dependent nest destruction and population fluctuations of Guianan leatherback turtles. *Animal Conservation* 5:75–84.

Gist, D. H., and Jones, J. M. 1989. Sperm storage within the oviduct of turtles. *Journal of Morphology* 199:379–384.

Grant, W. E. 1986. *Systems Analysis and Simulation in Wildlife and Fisheries Sciences*. New York: John Wiley & Sons.

Gregory, L. F., and Schmid, J. R. 2001. Stress responses and sexing of wild Kemp's ridley sea turtles (*Lepidochelys kempii*) in the northeastern Gulf of Mexico. *General and Comparative Endocrinology* 124:66–74.

Gutzke, W. H. N., and Bull, J. J. 1986. Steroid hormones reverse sex in turtles. *General and Comparative Endocrinology* 64:368–372.

Harry, J. L., and Briscoe, D. A. 1988. Multiple paternity in the loggerhead turtle (*Caretta caretta*). *Journal of Heredity* 79:96–99.

Heinly, R. W. 1990. Seasonal occurrence and distribution of stranded sea turtles along Texas and southwestern Louisiana coasts. M.S. thesis, Texas A&M University, College Station.

Heppell, S. S. 1998. Application of life-history theory and population model analysis to turtle conservation. *Copeia* 1998:367–375.

Heppell, S. S., and Crowder, L. B. 1994. Is headstarting headed in the right direction? In Schroeder, B. A., and Witherington, B. E. (Eds.). *Proceedings of the Thirteenth Annual Symposium on Sea Turtle Biology and Conservation*. NOAA Technical Memorandum NMFS-SEFSC-341, pp. 77–81.

Heppell, S. S., and Crowder, L. B. 1996. Analysis of a fisheries model for harvest of hawksbill sea turtles (*Eretmochelys imbricata*). *Conservation Biology* 10:874–880.

Heppell, S. S., Crowder, L. B., and Priddy, J. 1995. *Evaluation of a fisheries model for the harvest of hawksbill sea turtles (*Eretmochelys imbricata*) in Cuba.* NOAA Technical Memorandum NMFS-OPR-5.

Heppell, S. S., Limpus, C. J., Crouse, D. T., Frazer, N. B., and Crowder, L. B. 1996a. Population model analysis for the loggerhead sea turtle, *Caretta caretta. Queensland. Wildlife Research* 23:143–159.

Heppell, S. S., Crowder, L. B., and Crouse, D. T. 1996b. Models to evaluate headstarting as a management tool for long-lived turtles. *Ecological Applications* 6:556–565.

Hughes, G. R. 1974. *The Sea Turtles of Southeast Africa. 1. Status, Morphology and Distribution.* Research Institute of South Africa Investigative Report, volume 35.

Kaufmann, J. H. 1992. The social behavior of wood turtles, *Clemmys insculpta,* in central Pennsylvania. *Herpetological Monographs* 6:1–25.

Kichler, K., Holder, M. T., Davis, S. K., Márquez-M., R., and Owens, D. W. 1999. Detection of multiple paternity in the Kemp's ridley sea turtle with limited sampling. *Molecular Ecology* 8:819–830.

Levins, R. 1966. The strategy of model building in population modeling. *American Scientist* 54: 421–431.

Limpus, C. J. 1985. A study of the loggerhead sea turtle, *Caretta caretta,* in eastern Australia. Ph.D. diss. University of Queensland, Brisbane.

Limpus, C. J. 1992. The hawksbill turtle, *Eretmochelys imbricata,* in Queensland: population structure within a southern Great Barrier Reef feeding ground. *Wildlife Research* 19:489–506.

Limpus, C. J., and Reed, P. 1985. Green sea turtles stranded by cyclone Kathy on the southwestern coast of the Gulf of Carpentaria. *Australian Wildlife Research* 12:523–533.

Limpus, C. J., and Chaloupka, M. 1997. Nonparametric regression modeling of green sea turtle growth rates (Southern Great Barrier Reef). *Marine Ecology Progress Series* 149:23–34.

Limpus, C. J., Couper, P. J., and Read, M. A.1994. The green turtle, *Chelonia mydas,* in Queensland: population structure in a warm temperate feeding area. *Memoirs of the Queensland Museum* 35: 139–154.

Lovich, J. E. 1990. Spring movement patterns of two radio-tagged male spotted turtles. *Brimleyana* 16: 67–71.

Lovich, J. E. 1996. Possible demographic and ecologic consequences of sex ratio manipulation in turtles. *Chelonian Conservation and Biology* 2:114–117.

Lovich, J. E., and Gibbons, J. W. 1990. Age at maturity influences adult sex ratio in the turtle *Malaclemys terrapin. Oikos* 59:126–134.

Lovich, J. E., Garstka, W. R., and Cooper, W. E. 1990. Female participation in courtship behavior of the turtle *Trachemys s. scripta. Journal of Herpetology* 24:422–424.

Madsen, T., Shine, R., Loman, J., and Håkansson, T. 1992. Why do female adders copulate so frequently? *Nature* 355:440–441.

Márquez-M., R. 1982. Atlantic Ridley Project, 1982: Preliminary account. *Marine Turtle Newsletter* 23:3–4.

Márquez-M., R. 1983. Atlantic Ridley Project, 1983: Preliminary account. *Marine Turtle Newsletter* 26:3–4.

Márquez-M., R. 1985. Atlantic Ridley Project, 1984: Preliminary report. *Marine Turtle Newsletter* 32:3–4.

Márquez-M., R. 1994. *Synopsis of Biological Data on the Kemp's Ridley Turtle,* Lepidochelys kempi *(Garman, 1880).* NOAA Technical Memorandum NMFS-SEFSC-343.

Márquez-M., R., Sanchez, M., Diaz, J., and Rios, D. 1986. Atlantic Ridley Project: 1985 Preliminary report. *Marine Turtle Newsletter* 36:4–5.

Márquez-M., R., Sanchez, M., Rios, D., Diaz, J., Villanueva, A., and Arguello, I. 1987. Rancho Nuevo operation, 1986. *Marine Turtle Newsletter* 40:12–14.

Márquez-M., R., Sanchez, M., Diaz, J., and Arguello, I. 1989. Kemp's ridley research at Rancho Nuevo, 1987. *Marine Turtle Newsletter* 44:6–7.

McCoy, C. J., Vogt, R. C., and Censky, E. J. 1983. Temperature-controlled sex determination in the sea turtle *Lepidochelys olivacea. Journal of Herpetology* 17:404–406.

Miller, J. D., and Limpus, C. J. 1981. Incubation period and sexual differentiation in the green turtle *Chelonia mydas* L. In Banks, C. B., and Martin, A. A. (Eds.). *Proceedings of the Melbourne Herpetological Symposium.* Melbourne: Dominion Press, pp. 66–73.

Morreale, S. J., Ruiz, G. J., Spotila, J. R., and Standora, E. A. 1982. Temperature dependent sex determination: Current practices threaten conservation of sea turtles *Chelonia mydas. Science* 216:1245–1247.

Morreale, S. J., Meylan, A. B., Sadove, S. S., and Standora, E. A. 1992. Annual occurrence and winter mortality of marine turtles in New York waters. *Journal of Herpetology* 26:301–308.

Morris, Y. A. 1982. Steroid dynamics in immature sea turtles. Master's thesis, Texas A&M University, College Station.

Mrosovsky, N. 1980. Thermal biology of sea turtles. *American Zoologist* 20:531–547.

Mrosovsky, N. 1982. Sex ratio bias in hatchling sea turtles from artificially incubated eggs. *Biological Conservation* 23:309–314.

Mrosovsky, N. 1985. Sex ratio of Kemp's ridley: Need for evaluation. *Marine Turtle Newsletter* 32:4–5.

Mrosovsky, N. 1994. Sex ratios of sea turtles. *Journal of Experimental Zoology* 270:16–27.

Mrosovsky, N., and Benabib, M. 1990. An assessment of two methods of sexing hatchling sea turtles. *Copeia* 1990:589–591.

Mrosovsky, N., and Godfrey, M. H. 1995. Manipulating sex ratios: turtle speed ahead! *Chelonian Conservation and Biology* 1:238–240.

Mrosovsky, N., and Yntema, C. A. 1980. Temperature dependence of sexual differentiation in sea turtles: implications for conservation practices. *Biological Conservation* 18:271–280.

Mrosovsky, N., Dutton, P. H., and Whitmore, C. P. 1984. Sex ratios of two species of sea turtles nesting in Suriname. *Canadian Journal of Zoology* 62:2227–2239.

Owens, D. W. 1997. Hormones in the life history of sea turtles. In Lutz, P. L., and Musick, J. A. (Eds.). *The Biology of Sea Turtles.* Boca Raton, FL: CRC Press, pp. 315–341.

Owens, D. W., Hendrickson, J. R., Lance, V., and Callard, I. P. 1978. A technique for determining sex of immature *Chelonia mydas* using radioimmunoassay. *Herpetologica* 34:270–273.

Palmer, K. S., Rostal, D. C., Grumbles, J. S., and Mulvey, M. 1998. Long-term sperm storage in the desert tortoise (*Gopherus agassizii*). *Copeia* 1998:702–705.

Parker, W. S. 1984. Immigration and dispersal of slider turtles *Pseudemys scripta* in Mississippi farm ponds. *American Midland Naturalist* 112:280–293.

Parker, W. S. 1990. Colonization of a newly constructed farm pond in Mississippi by slider turtles and comparison with established populations. In Gibbons, J. W. (Ed.). *Life History and Ecology of the Slider Turtle.* Washington, DC: Smithsonian Institution Press, pp. 216–222.

Plotkin, P. T., Owens, D. W., Byles, R. A., and Patterson, R. 1996. Departure of male olive ridley turtles (*Lepidochelys olivacea*) from a nearshore breeding ground. *Herpetologica* 52:1–7.

Plotkin, P. T., Rostal, D. C., Byles, R. A., and Owens, D. W. 1997. Reproductive and developmental synchrony in female *Lepidochelys olivacea*. *Journal of Herpetology* 31:17–22.

Raynaud, A., and Pieau, C. 1985. Embryonic development of the genital system. In Gans, C. (Ed.). *Biology of the Reptilia*. New York: John Wiley & Sons, pp. 149–300.

Renaud, M., Carpenter, J., Williams, J., Carter, D., and Williams, B. 1995. *Movement of Kemp's ridley sea turtles (*Lepidochelys kempii*) near Bolivar Roads Pass and Sabine Pass, Texas and Calcasieu Pass, Louisiana. May 1994–May 1995.* Final report submitted to U.S. Army Corps of Engineers (Galveston and New Orleans District).

Renaud, M. L., Carpenter, J. A., Williams, J. A., and Landry, A. M., Jr. 1996. Kemp's ridley sea turtle (*Lepidochelys kempii*) tracked by satellite telemetry from Louisiana to nesting beach at Rancho Nuevo, Tamaulipas, Mexico. *Chelonian Conservation and Biology* 2:108–109.

Richardson, J. I., and Richardson, T. H. 1982. An experimental population model for the loggerhead sea turtle (*Caretta caretta*). In Bjorndal, K. A. (Ed.). *Biology and Conservation of Sea Turtles.* Washington, DC: Smithsonian Institution Press, pp. 165–176.

Rostal, D. C. 1991. The reproductive behavior and physiology of the Kemp's ridley sea turtle, *Lepidochelys kempi* (Garman, 1880). Ph.D. diss. Texas A&M University, College Station.

Rostal, D. C., Grumbles, J. S., Byles, R. A., Márquez-M., R., and Owens, D. W. 1997. Nesting physiology of Kemp's ridley sea turtles, *Lepidochelys kempi,* at Rancho Nuevo, Tamaulipas, Mexico, with observations on population estimates. *Chelonian Conservation and Biology* 2:538–547.

Ruiz, G. J., Standora, E. A., Spotila, J. R., Morreale, S. J., Camhi, M., and Ehrenfeld, D. 1981. Artificial incubation of sea turtle eggs affects sex ratio of hatchlings. In *Proceedings of the Annual Meeting of the Herpetologist League.* Memphis, TN, p. 68.

Shaver, D. J. 1991. Feeding ecology of wild and head-started Kemp's ridley sea turtles in south Texas waters. *Journal of Herpetology* 25:327–334.

Shaver, D. J. 1995. Sea turtle stranding along the Texas coast again cause for concern. *Marine Turtle Newsletter* 70:2–4.

Shaver, D. J., Owens, D. J., Chaney, A. H., Caillouet, C. W., Jr., Burchfield, P., and Márquez-M., R. 1988. Styrofoam box and beach temperatures in relation to incubation and sex ratios of Kemp's ridley sea turtles. In Schroeder, B. A. (Ed.). *Proceedings of the Eighth Annual Workshop on Sea Turtle Conservation and Biology.* NOAA Technical Memorandum NMFS-SEFSC-214, pp. 103–108.

Stabenau, E. K., Stanley, K. S., and Landry, A. M., Jr. 1996. Sex ratios from stranded sea turtles on the upper Texas coast. *Journal of Herpetology* 30:427–430.

Standora, E. A., and Spotila, J. R. 1985. Temperature dependent sex determination in sea turtles. *Copeia* 1985:711–722.

Sugg, D. W., and Chesser, R. K. 1994. Effective population size with multiple paternity. *Genetics* 137:1147–1155.

Tousignant, A., and Crews, D. 1994. Effect of exogenous estradiol applied at different embryonic stages on sex determination, growth, and mortality in the leopard gecko (*Eublepharis macularius*). *Journal of Experimental Zoology* 268:17–21.

Turtle Expert Working Group. 1998. *An assessment of the Kemp's ridley (*Lepidochelys kempii*) and loggerhead (*Caretta caretta*) sea turtle populations in the Western North Atlantic.* NOAA Technical Memorandum NMFS-SEFSC-409.

Valverde, R. A. 1996. Corticosteroid dynamics in a free-ranging population of olive ridley sea turtles (*Lepidochelys olivacea* Eschscholtz, 1829) at Playa Nancite, Costa Rica as a function of their reproductive behavior. Ph.D. diss. Texas A&M University, College Station.

Van Der Heiden, A. M., Briseno-Duenas, R., and Rios-Olmeda, D. 1985. A simplified method for determining sex in hatchling sea turtles. *Copeia* 1985:779–782.

Vogt, R. C. 1994. Temperature controlled sex determination as a tool for turtle conservation. *Chelonian Conservation and Biology* 1:159–162.

Wellins, D. J. 1987. Use of an H-Y antigen assay for sex determination in sea turtles. *Copeia* 1987: 46–52.

Whiting, M. J., Dixon, J. R., and Greene, B. D. 1997. Spatial ecology of the Concho water snake (*Nerodia harteri paucimaculata*) in a large lake system. *Journal of Herpetology* 31:327–335.

Wibbels, T. R. 1988. Gonadal steroid endocrinology of sea turtle reproduction. Ph.D. diss. Texas A&M University, College Station.

Wibbels, T. R. 1998. *1998 Kemp's Ridley Sex Ratio Project Rancho Nuevo, Mexico.* Progress Report to Instituto Nacional de la Pesca, Manzanilla, Mexico.

Wibbels, T. R., and Geis, A. 1999. *1999 Kemp's Ridley Sex Ratio Project Rancho Nuevo, Mexico.* Progress report to the Instituto Nacional de la Pesca, Manzanilla, Mexico.

Wibbels, T. R., Morris, Y. A., and Amoss, M. S., Jr. 1987. Sexing techniques and sex ratios for immature loggerhead sea turtles captured along the Atlantic coast of the United States. In Witzell, W. N. (Ed.). *Ecology of East Florida Sea Turtles.* NOAA Technical Report NMFS-53, pp. 65–74.

Wibbels, T. R., Morris, Y. A., Owens, D. W., Dienberg, G. A., Noell, J., Leong, J. K., King, R. E., and Márquez-M., R. 1989. Predicted sex ratios from the International Kemp's Ridley Sea Turtle Head Start Research Program. In Caillouet, C. W., Jr., and Landry, A. M., Jr. (Eds.). *Proceedings of the First International Symposium on Kemp's Ridley Sea Turtle Biology, Conservation, and Management.* Texas A&M University Sea Grant College Program Publication TAMU-SG-89-105, pp. 77–81.

Wibbels, T. R., Owens, D. W., Limpus, C. J., Reed, P. C., and Amoss, M. S., Jr. 1990. Seasonal changes in serum gonadal steroids associated with migration, mating, and nesting in the loggerhead sea turtle (*Caretta caretta*). *General and Comparative Endocrinology* 79:154–164.

Wibbels, T. R., Owens, D. W., and Rostal, D. 1991a. Soft plastra of adult male sea turtles: an apparent secondary sexual characteristic. *Herpetological Review* 22:47–49.

Wibbels, T., Martin, R. E., Owens, D. W., and Amoss, M. S., Jr. 1991b. Female-biased sex ratio of immature loggerhead sea turtles inhabiting the Atlantic coastal waters of Florida. *Canadian Journal of Zoology* 69:2973–2977.

Wibbels, T. R., Rostal, D. C., and Byles, R. 1998. High pivotal temperature in the sex determination of the olive ridley sea turtle, *Lepidochelys olivacea,* from Playa Nancite, Costa Rica. *Copeia* 1998:1086–1088.

Wibbels, T. R., Owens, D., and Limpus, C. 2000. Sexing juvenile sea turtles: is there an accurate and practical method? *Chelonian Conservation and Biology* 3:756–761.

Wood, J. R., and Wood, F. E. 1980. Reproductive biology of captive green sea turtles (*Chelonia mydas*). *American Zoologist* 20:499–505.

Wood, J. R., Wood, F. E., Critchley, K. H., Wildt, D. E., and Bush, M. 1983. Laparoscopy of the green sea turtle, *Chelonia mydas. British Journal of Herpetology* 6:323–327.

Yntema, C. L., and Mrosovsky, N. 1980. Sexual differentiation in hatchling loggerheads (*Caretta caretta*) incubated at different controlled temperatures. *Herpetologica* 36:33–36.

Yntema, C. L., and Mrosovsky, N. 1982. Critical periods and pivotal temperatures for sexual differentiation in loggerhead sea turtles. *Canadian Journal of Zoology* 60:1012–1016.

Zaborski, P., Dorzzi, M., and Pieau, C. 1988. Temperature-dependent gonadal differentiation in the turtle *Emys orbicularis:* concordance between sexual phenotype and serological H-Y antigen expression at threshold temperature. *Differentiation* 38:17–20.

STEPHEN J. MORREALE
PAMELA T. PLOTKIN
DONNA J. SHAVER
HEATHER J. KALB

11

Adult Migration and Habitat Utilization

Ridley Turtles in Their Element

IN FOLKLORE, THE TURTLE has long held a prominent position as a symbol of steadfast plodding and unhurried tranquility. This mythical allusion may contribute to a common misconception that turtles are sedentary animals. However, turtles in general migrate frequently, over relatively great distances, and for many different reasons throughout their long lives. Extreme among this group are sea turtles, which exhibit impressive skills at long-distance movement, especially when one considers the adults' ponderous movements on land. In their natural element, however, sea turtles are adept travelers, often displaying remarkable abilities for homing and orientation that rank them alongside such well-known long-distance migrants as geese, butterflies, and salmon. In fact, because sea turtles spend virtually their entire life at sea, they clearly have adopted many strategies that allow them to move with relative ease across vast expanses of ocean and to transit many different environments. Beginning with the initial dispersal of hatchlings, the propensity for travel of these creatures becomes established at an early stage, persists into adulthood, and indeed appears to be woven into their very evolutionary fabric.

Although all sea turtle species appear capable of extensive migrations, variability exists among species, among populations, and even among individual turtles, making any generalization necessarily imprecise. Nevertheless, for adults in general, it appears that the leatherback turtle (*Dermochelys coriacea*) migrates the farthest, the flatback turtle (*Natator depressus*) exhibits the shortest migrations, and the two species of ridley turtles (*Lepidochelys* spp.) are somewhat intermediate in this regard.

It is important to note that as more information on sea turtle movements becomes available, our understanding of their migratory behavior evolves. The flatback turtle, which is found in shallow, turbid tropical waters and bays of the Australian continental shelf (Limpus et al., 1981, 1983, 1989), once was characterized as nonmigratory (Hendrickson, 1980). Indeed they may have the most restricted migratory range of all sea turtles but nevertheless can undergo long-distance migrations of more than 1,000 km between foraging and breeding areas (Limpus et al., 1981, 1983). Similarly, the hawksbill turtle (*Eretmochelys imbricata*), which is globally distributed in warmer regions, once was believed to be a nonmigratory resident of reefs adjacent to its nesting beach (Hendrickson, 1980; Witzell, 1983; Frazier, 1984). However, tagging, telemetry, and genetic studies have revealed that hawksbill turtles can be highly migratory, traveling hundreds and thousands of kilometers between nesting beaches and foraging areas (Meylan, 1982; Parmenter, 1983; Broderick et al., 1994; Byles and Swimmer, 1994; Groshens and Vaughan, 1994; Miller et al., 1998; Meylan, 1999a; Horrocks et al., 2001).

Not much was known about the migrations of ridley turtles until the more recent incorporation of satellite telemetry into migration studies. It was assumed that Kemp's ridley (*Lepidochelys kempii*) was confined to the Gulf of Mexico for its entire life (Hildebrand, 1982), and hence, adults presumably did not undergo extensive migrations. Further support of this came from tag recoveries and observations in presumed feeding areas in Campeche, Mexico, and Louisiana (Carr, 1980; Hildebrand, 1982) and from recaptures of tagged nesting females in other areas of the Gulf of Mexico (Zwinenberg, 1977; Mysing and Vanselous, 1989) but not often from areas outside the Gulf. Thus, from early on, Kemp's ridley adult migration was thought to occur within continental shelf waters, mainly between nesting sites in Tamaulipas, Mexico, and coastal feeding grounds in the Gulf of Mexico. Migration distances were thought to range from roughly 900 km to the south, to as far as 2,000 km northward and eastward along the coast of the U.S. Gulf states. Such a migration pattern also fit nicely with the notion of the Kemp's ridley as an organism evolved to feed primarily on neritic benthic crustaceans (Hendrickson, 1980). Further conforming to the presumed strategy of moderate-distance migrations was the observed high frequency of annual return of nesting females to their main nesting site, Rancho Nuevo, Mexico.

A similar short interval of annual return to nesting beaches for olive ridleys (*Lepidochelys olivacea*) was thought to indicate shorter migrations between nesting and feeding areas for this species too (Hendrickson, 1980), but surprisingly little was recorded of the migrations of this extremely abundant and widespread turtle. Early tagging records from Central and South American nesting olive ridleys combined to depict a migration pattern that seemed similar to Kemp's ridley (Pritchard, 1973; Meylan, 1982). Olive ridleys also appeared to undergo moderate-distance migrations to coastal areas, extending for hundreds of kilometers north and south of nesting beaches. But there were indications of greater distances traveled too, such as movements of 1,900 km along the Atlantic coast and more than 2,000 km from Mexico to Colombia and Ecuador in the Pacific.

With subsequent tagging records of nesting olive ridleys from Costa Rica (Cornelius and Robinson, 1986), a slightly different pattern emerged. As in previous studies, turtles migrated both northward and southward from the two Pacific coast nesting beaches. The travel distances also ranged from tens of kilometers to over 2,000 km. However, after tagging more than 45,000 turtles, only 189 recaptures were reported, which seems like a low return rate for a presumed coastal animal. Furthermore, another indication that all olive ridleys may not migrate to coastal feeding areas emerged when three recaptures were reported from offshore waters. One turtle was captured more than 2,100 km from the beach, and all three were in water with depths greater than 3,000 m. These three records perhaps foretold what was only suspected at the time (Hendrickson, 1980; Cornelius and Robinson, 1986) and would later be confirmed by satellite tracking studies: that the adult olive ridley is an epipelagic feeder, and its migrations convey it from breeding and nesting areas to open-water feeding areas.

Undoubtedly, much pattern variation is introduced into movement studies that depend solely on sporadic tag returns. Also contributing to apparent discrepancies in generalized reports of movements is the individual variability that has been observed in adult migratory behavior. Widely diverse behavior in migration has been observed, often within species and even within individuals sharing the same nesting or feeding areas (Meylan et al., 1990; Plotkin et al., 1995; Eckert, 1998; Horrocks et al., 2001; Limpus and Limpus, 2001; Hatase et al., 2002). In these and several other studies, individuals that all converged on a common location exhibited tremendous variability in direction and magnitude of travel after departure. In extreme cases, individual turtles travel great distances, and their close associates appear to make only minimal movements between feeding and nesting grounds (Mortimer and Balazs, 2000; Limpus and Limpus, 2001). Ridley turtles seem to be no exception, and there appears to be a great deal of individual variability of movement, at least in their patterns of postnesting migration. To understand both similarities and differences in migration patterns of these turtles, it is first necessary to examine the ecological factors that motivate them.

Ecological Influences on Migration Patterns

A question one might ask is: Why do adult sea turtles make such extensive migrations? Presumably, we might speculate, to get to necessary resources. Migration is distinct from less-directed movements such as dispersal, the implied difference being one of motivational state of the organism. Migratory movements generally can be categorized as an organism's purposeful travel toward a specific goal (Papi, 1992). This is unlike random dispersal, which often is the result of many stochastic processes. Therefore, one of the principal implications of goal orientation is that some direct benefit is derived as a result of migration. The traditional example is of birds moving from one site to another to benefit from improved feeding or breeding conditions at a distant site. A less commonly acknowledged purpose of migration is to track resources that also are migrating or shifting spatially.

Although direct migration and tracking of resources are not exclusive types of behavior, there are some important differences that are reflected in the directness of travel and the fixity and predictability of migratory routes. An adult caribou migrating from winter feeding sites to spring calving grounds tends to move directly between endpoints with minimal deviation (Craighead and Craighead, 1987). The path of a nectivorous bat, however, may be less than straight as it migrates along a corridor of a specific type of cactus that fulfills its specialized dietary needs (Fleming et al., 1993). On the other extreme are migrations that have no predetermined endpoint, such as roaming schools of tuna tracking after smaller baitfish (Arocha, 1997; Menard and Marchal, 2003), or wandering albatrosses foraging for patchily distributed prey (Weimerskirch et al., 1994). For adults, such a nomadic strategy is often superimposed on circannual or seasonal migration patterns (Block et al., 2001; Mauritzen et al., 2001). Thus, many animals similarly migrate from a point of origin to a distant area and are guided by similar motivations, but their movement patterns along the way can be quite different. Interestingly, all of these types of migration appear to be manifested by sea turtles, and the two species of ridleys seem to exhibit at least two different modes of migratory behavior.

Migration is a crucial element in the life cycles and natural history of most turtles, as adults of many aquatic and terrestrial species regularly migrate to nesting areas (Cagle, 1950; Moll and Legler, 1971; Obbard and Brooks, 1980; Congdon et al., 1983), to overwintering sites (Sexton, 1959; Burke and Gibbons, 1995), and to search for mates or resources (Cagle, 1944; Ernst and Barbour, 1972; Gibbons et al., 1983; Morreale et al., 1984). Similar movements by sea turtles, on the order of hundreds and even thousands of kilometers, carry them to and from feeding areas (Carr, 1967; Balazs, 1976; Keinath et al., 1987; Ogren, 1989; Meylan et al., 1990; Epperly et al., 1995; Morreale and Standora, 1998), overwintering sites (Mendonça and Ehrhart, 1982; Henwood, 1987; Henwood and Ogren, 1987;

Witherington and Ehrhart, 1989; Morreale et al., 1992; Schmid, 1995), and mating and nesting areas (Carr, 1967; Plotkin et al., 1995; Morreale et al., 1996; Miller et al., 1998; Meylan, 1999a, 1999b; Limpus and Limpus, 2001).

When a turtle embarks on a long-distance movement, it implies that a necessary resource is not provided at its present site or that there is some cost associated with remaining at its current location. Simply stated, if sea turtles are traveling great distances, there must be benefits derived at the remote destination or along the path of travel. For adult sea turtles, the primary motivations for movement appear to be improved nesting, mating, foraging, and temperatures. It is tempting to speculate that if sea turtles migrate to obtain necessary resources at a distant location, then differences in migration patterns may reflect different needs among species or individuals. Furthermore, different migratory behaviors may indicate different reproductive or foraging strategies among species, among individuals, or between sexes, or even between such closely related species as Kemp's ridley and olive ridley.

Nesting Constraints on Migration

One of the most obvious reasons for sea turtles to travel great distances is to transport their eggs to a beach that will ensure an adequate environment for incubation and subsequent survival of offspring. For ridleys and other sea turtles, of key importance to their successful nesting is the thermal regime of the beach. Sea turtle eggs develop successfully at temperatures between 24°C and 35°C (Bustard and Greenham, 1968; Yntema and Mrosovsky, 1980; Miller and Limpus, 1981; Ackerman, 1997; Matsuzawa et al., 2002; Wibbels, Chapter 9), and it has long been recognized that temperature may limit the nesting distribution of sea turtles. Furthermore, climatic fluctuations that can affect incubation, even on tropical beaches, may contribute to a distinct seasonality to the nesting cycle (Owens, 1980; Godley et al., 2002). Thus, there is a distinct season for the Kemp's ridley, between May and July, during which virtually all of the nesting for this species takes place. Similarly, on the densest olive ridley nesting beach in the world, Gahirmatha, Orissa, India, nesting takes place between January and March (Kar and Bashkar, 1982; Pandav et al., 1995, 1998; Shanker et al., 2003b), rather than year-round. In the other months, apparently olive ridleys do not remain in nearby waters.

In addition to thermal tolerance of eggs, there are two factors related to incubation temperature that constrain sea turtles to nesting in warm climates: rate of embryonic development and sex determination. As nest temperatures increase within the thermal limits, the rate of development increases in a predictable manner (Yntema and Mrosovsky, 1982; Ackerman, 1997; Matsuzawa et al., 2002). Thus, warm climates yield more rapid incubation, which presumably contributes to short-term benefits for nesting females, enabling them to produce several successive clutches in a single nesting season. Warmer nesting beaches also may render long-term benefits by providing an environment that is more likely to produce a mixed sex ratio among offspring. For all sea turtles (and many other reptiles), the sex of the developing embryo is influenced by nest temperature, with the dividing line between males and females at approximately 29°C (for reviews see Janzen and Pakustis, 1991; Ackerman, 1997; and Wibbels, Chapter 9). Hence, beaches that provide sand temperatures around this range would tend to be favorable in the long run to nesting females. This is especially important for Kemp's ridley, which nests primarily in a single climate.

Also of potentially great importance for nesting beach selection is the immediate environment in which the newly emerged hatchlings find themselves on entering the sea. Presumably there are areas that would be more suitable than others for young offspring in terms of currents, food, temperature, and safety. Kemp's ridley hatchlings emerging from the beaches in Tamaulipas, Mexico are transported to new pelagic habitats by the predictable currents and existing gyres in the Gulf of Mexico (Collard and Ogren, 1990). Indeed, there may be comparably important currents for olive ridleys associated with the East Indian Coastal Current and the Monsoon Currents in the Bay of Bengal. The positions and the seasonal directional flow of these currents in the vicinity of East Indian nesting beaches (Vinayachandran and Yamagata,

2003) may help explain the intense aggregations of olive ridleys nesting there during a part of the year.

The many limiting factors, including current systems, temperature regime, amount of moisture, and stability of beaches, would greatly reduce the number of places in the world suitable for sea turtle nesting. Such potential constraints on nesting go a long way toward explaining why female sea turtles are inevitably linked to warmer beaches for at least the duration of a nesting season and may partially explain why tens or hundreds of thousands of olive ridley turtles vie tenaciously for nesting spaces on extremely crowded beaches in western Costa Rica (Hughes and Richard, 1974; Cornelius and Robinson, 1986; National Marine Fisheries Service and U.S. Fish and Wildlife Service, 1998), or eastern India (Pandav et al., 1995, 1998; Shanker et al., 2003b).

Postreproductive Migrations

The tie that binds adult females to certain beaches during the nesting season ostensibly also would make the nearby waters highly attractive places for turtles to establish residency. Indeed, some sea turtles do not travel far between nesting and foraging areas, but most sea turtles appear to migrate to and from nesting beaches. However, there is no simple summary that encompasses the nesting migrations of all sea turtles. Their propensity to travel seems to be variable on an individual basis rather than on the sole basis of species. As a good example, it has long been acknowledged that many adult green turtles (*Chelonia mydas*) from a nesting population spend their nonbreeding seasons in neighboring waters, whereas others from the same nesting beach travel up to several thousands of kilometers to and from nesting beaches (Carr, 1967; Meylan, 1982; Green, 1984; Limpus et al., 1992). On the whole, however, most green turtles appear to travel at least moderate distances to nest. A particularly reasonable explanation for this was proffered by Carr (1980), who pointed out that suitable nesting beaches likely are created by physical processes that are somewhat conducive to culturing good foraging habitats for green turtles. Such is probably the case for nesting beaches of other species as well. Hence, it may be just as reasonable to label these postreproductive movements as foraging migrations. But for an adult sea turtle, foraging migrations must be balanced with effective reproductive strategies, and the solutions may be different for females and males.

Females

From numerous studies in which turtles were tagged on nesting beaches, more is known about postnesting migrations of adult females than any other kind of sea turtle movement.

Piecing together the mark–recapture data established that female sea turtles of every species can move great distances after nesting. The basic model presented by Carr (1967) was that many females migrate directly to identifiable feeding areas, where they likely remain until they return to nest again. From early studies, this behavioral model seemed to describe the postnesting activities of Kemp's ridley, as many tagged females were found in crab-rich coastal areas at various distances from the nesting beach (Pritchard and Márquez-M., 1973; Zwinenberg, 1977; Carr, 1980; Hildebrand, 1982). The rapid rate of movement of individuals captured within 1–2 months after nesting, ranging from 32 to 50 km/day, also indicated a certain directness of travel to foraging areas (Zwinenberg, 1977).

The more recent introduction of satellite telemetry to study Kemp's ridleys has provided both detail and confirmation to the previous notions of postnesting movements and habitat use for this species (Fig. 11.1). The first groups of postnesting females were tracked from Rancho Nuevo, Mexico in 1987 and 1988 (Byles, 1989). In this study, 11 turtles were successful tracked various distances, lasting up to 127 days. Four turtles swam northward from the nesting beach, and seven turtles swam southward. Two of the northward-bound turtles were tracked as far as Corpus Christi, Texas, a movement of more than 500 km. The remaining seven turtles migrated southward to Veracruz, Campeche, and Yucatan, Mexico. The two longest migrations, to the northern tip of the Yucatan Peninsula, Mexico, covered more than 1,500 km. The turtles' movements along shore in water depths of less than 50 m also reinforced previous con-

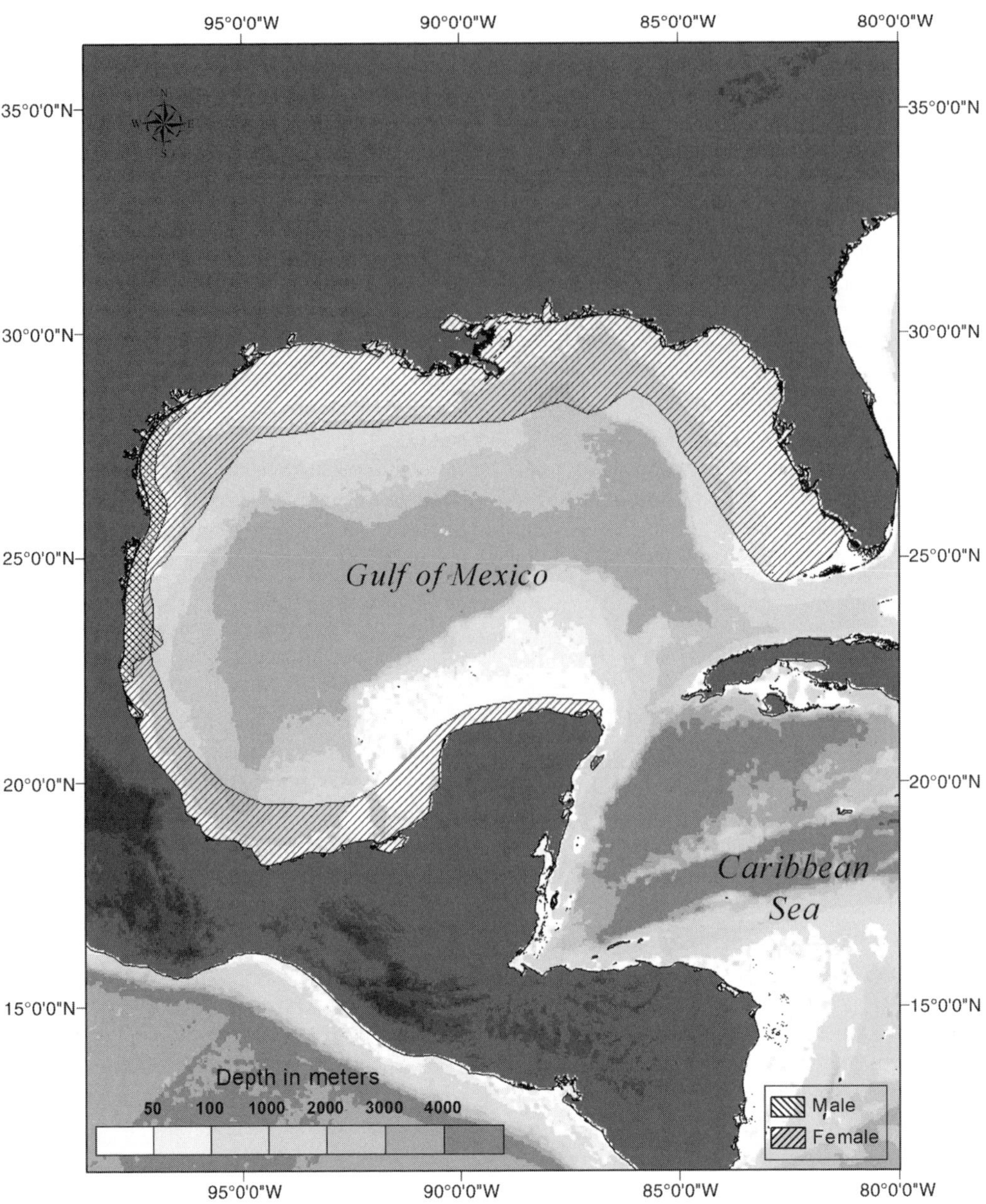

Fig. 11.1. Generalized patterns of movement of postreproductive Kemp's ridley turtles from several different studies in the Gulf of Mexico. Postnesting females leave the nesting area in Tamaulipas, Mexico, and travel along a shallow coastal corridor northward to feeding areas as far as Florida and southward to as far as the Yucatan Peninsula. Movements of males from the same area are a small subset of those of females. One of 11 males tracked moved northward to Texas waters; the remaining 10 males remained in the vicinity of the nesting beach.

tentions that adult Kemp's ridleys primarily occupied the neritic zone.

Subsequent satellite tracking of adult female Kemp's ridleys has added more detail, longer-distance migrations, and a much better view of the associated biology (Byles and Plotkin, 1994; Renaud, 1995; Renaud et al., 1996; Shaver, 1999, 2001a, 2001b). The picture that emerges for postnesting Kemp's ridleys is that these turtles travel along coastal corridors generally shallower than 50 m in depth, extending the length of the Gulf of Mexico from the Yucatan Peninsula to southern Florida (Fig. 11.1). Furthermore, several of these turtles were observed to settle in what appeared to be resident feeding areas for up to several months after migrating (Byles and Plotkin, 1994), demonstrating that Kemp's ridley postnesting migrations also might be considered foraging migrations to fixed destinations.

Such a basic model does not appear to adequately depict all of the postnesting movements of olive ridleys. Similar to the Kemp's ridley, early tagging studies reported postnesting movements ranging from nearby waters to nearly 2,000 km away (summarized in Meylan, 1982). Additionally, in Costa Rica some females appeared to remain in nearby waters for at least 2 years, whereas others moved more than 2,000 km away (Cornelius and Robinson, 1986). Some of the more rapid rates of travel for this species ranged from 37 km / day to 87 km / day (Pritchard, 1973; Meylan, 1982; Cornelius and Robinson, 1986), which were comparable to Kemp's ridley. Remarkably, however, only a small number of the tagged females were later encountered, and only a subset was found nearby. The low recapture rates could indicate long-distance migration away from nesting beaches or, alternatively, movements to offshore areas where encounters with humans would be minimized. In the case of eastern India's nesting olive ridleys, the lack of turtles captured or found stranded in months outside the nesting season was thought to indicate postnesting departure from the area (Pandav et al., 1995, 1998). Certainly, the stark contrast from hundreds of thousands of turtles to almost none does not reinforce an image of a local coastal resident.

It was not until satellite telemetry studies were undertaken that detailed movements for Pacific olive ridleys were revealed (Plotkin, 1994; Byles and Plotkin, 1994; Plotkin et al., 1994, 1995, 1996; Beavers and Cassano, 1996). The immediate pattern that emerged was one of a highly migratory turtle that existed mainly in the oceanic zone and was widely dispersed in its movements. After nesting, adult females tracked from Costa Rica moved to oceanic foraging areas (Plotkin, 1994; Plotkin et al., 1995). During their movements, they mainly resided in waters with depths greater than 3,000 m. In addition, there appeared to be no association among individuals' movements during migration, as their paths were widely separated (Fig. 11.2). The observed movements of females are almost entertaining in their diversity, each individual seeming to come up with its own innovative approach to a foraging migration. Their migratory routes also varied annually.

Although some postnesting movements may overlap in space, the dispersal behavior of olive ridleys into broad expanses of ocean suggests that foraging individuals are solitary. Furthermore, the relatively indirect tracks of individuals indicate that olive ridleys travel to general areas of open ocean, not with a lack of purpose but likely with a lack of specific predetermined location. These observations meshed with the notable lack of olive ridleys in nearshore areas beyond the breeding seasons. Furthermore, the tracking data provided a direct connection to the observations of olive ridleys widely distributed in the surface waters of the eastern Pacific Ocean, to distances greater than 2,000 km from shore (Pitman, 1990, 1993; Arenas and Hall, 1992).

Recent satellite telemetry studies of postnesting olive ridleys in eastern India also have successfully outlined their migratory behavior in the Bay of Bengal (Shanker et al., 2003a). Interestingly, it was noted that the turtles appeared to be moving randomly offshore in large circles before one turtle began a directed movement southward. Such behavior was similar to the nondirected movements of female olive ridleys in the eastern Pacific. The tracks of the eastern Indian turtles may reflect the complexity of the currents in that region (Vinayachandran and Yamagata, 2003) and reinforce tagging records that indicated that at least some of the females nesting in Orissa migrate southward to Tamil Nadu, India, and Sri Lanka. Incidentally, the longest track, which extended for more than 2,900 km, reached one of the most productive areas in the region, in the vicinity of the nutrient-rich Sri Lanka Dome (Vinayachandran et al., 2004). Thus, the movements of females away from the nesting beach, even though apparently nondirect, probably could be characterized as foraging migrations.

Males

Traditional thought is that male sea turtles finish mating and depart from the breeding grounds by

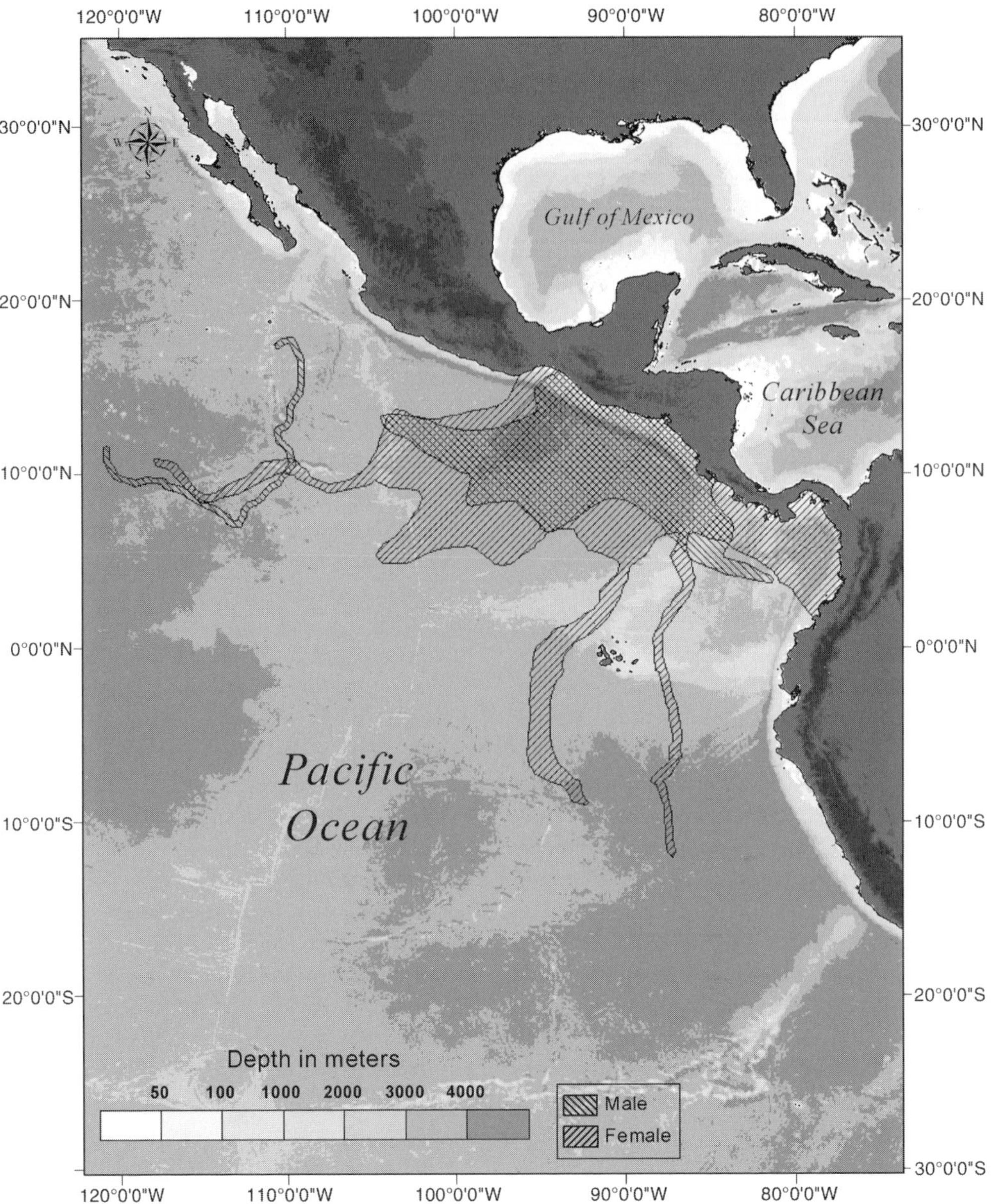

Fig. 11.2. Generalized patterns of movement of postreproductive olive ridley turtles from several different studies in the eastern tropical Pacific. Postnesting females leaving from Costa Rican beaches display widely divergent movement patterns and directions of travel, during which they remain mainly in oceanic waters greater than 3,000 m in depth. Movements of postmating males from the same nesting areas are a subset of those of females. A male mating in offshore waters also moved in areas similar to at least one female. Observed movements indicate that olive ridleys are nomadic epipelagic foragers that prey on patchily distributed food.

the time the greatest number of females emerge to lay eggs (Hendrickson, 1958; Booth and Peters, 1972; Ehrhart, 1982; Frazier, 1985; Owens, 1997). This, however, assumes a simple behavioral pattern in which males migrate to and from breeding grounds, as do nesting females. But, the selective advantage for males to return to a nesting beach may not be the same as for females, which ultimately are constrained by the quality of environment for eggs and hatchlings. Thus, it may be important to distinguish between males in general and males that are present at nesting beaches.

There is very little information on the movements of male Kemp's ridleys after breeding. Although principal courtship and mating areas for Kemp's ridleys are not well known, at least some proportion of males are present in the vicinity of Rancho Nuevo during the breeding season. Anecdotal information supplied by fishermen has indicated that mating presumably occurs before or during the nesting season in the vicinity of the nesting beach (Chavez, 1969; Pritchard, 1969; Márquez-M., 1970, 1990). Indeed, mating may take place about 30 days before the first clutch of eggs for the season is laid (Rostal et al., 1998). What these males do after mating has only recently been revealed.

To address the general lack of information on the distribution, movements, and habitat utilization of adult male Kemp's ridleys, 11 males captured near Rancho Nuevo, Tamaulipas, were monitored using satellite telemetry between 1999 and 2001 (Shaver et al., 2005). Transmitters were attached to turtles that local fishermen captured by net during the fall (September–November; $n = 3$), winter (December–February; $n = 4$), spring (March–May; $n = 2$), and summer (June–August; $n = 2$). Turtles were tracked from 73 days to 233 days, and the majority of locations were deemed to be in nearshore waters shallower than 37 m. One of the 11 males traveled northward and was last located offshore of Galveston, Texas, whereas the other 10 males remained nearby in Tamaulipas waters for the remainder of the study (Fig. 11.1). Thus, it is appears that a proportion of the adult male Kemp's ridleys reside in the vicinity of the nesting beach year-round, in sharp contrast to the pattern of female postnesting migration. Beyond the ecological aspects, a resident population of adult males also underscored the need for protection of the marine habitat adjacent to the nesting beach year-round.

The movements of these turtles may represent one of several strategies that a male ridley could adopt to mate successfully. In fact, the postbreeding movements of male olive ridleys after mating in coastal Pacific waters seem to exemplify foraging movements of olive ridleys in general (Plotkin et al., 1996). Six male olive ridleys monitored by satellite all departed from the vicinity of the nesting beach and headed to offshore waters (Fig. 11.2). Their movements were similar to those of postnesting females from the same area in that they were highly disassociated from each other (Plotkin et al., 1995), as individuals fanned out into the open Pacific, presumably to resume their wandering nomadic foraging behavior.

Breeding Migrations

Because most tagging studies on adult turtles originate on nesting beaches, precise knowledge of sea turtle migrations toward the nesting beach are relatively rare. Mostly it has been presumed that migrations to nesting beaches mirror, or at least resemble, the observed migrations back to foraging areas. This seems to be the case for some flatback turtles (Limpus et al., 1983), green turtles (Green, 1984; Limpus et al., 1992), and loggerhead turtles (*Caretta caretta*) (Limpus et al., 1992; Limpus and Limpus, 2001), all of which have been recorded making complete migrations to and from nesting and foraging areas. Satellite telemetry studies that have followed turtles through a complete migratory cycle are extremely informative in this regard. Limpus and Limpus (2001) reported some excellent examples of loggerhead turtles migrating from feeding areas to nesting grounds, using virtually the same routes on both legs of the journey. However, as with other facets of migratory behavior in sea turtles, there probably is variability among individuals and especially between the sexes.

Simply stated, wherever adult females forage, they make return migrations to nesting beaches to which they usually show strong fidelity over time. In the case of the Kemp's ridley, most females return to the primary nesting beach in Tamaulipas, Mexico. Satellite telemetry studies have shown that after nesting, adult female Kemp's ridleys typically migrate in a directed manner away from nesting to foraging areas where they take up residence as nearshore, shallow water inhabitants, (Byles, 1989; Shaver, 2001b). To date, there is one example of a female Kemp's ridley captured in a coastal foraging area and tracked by satellite as it migrated several months later to the nesting beach (Renaud et al., 1996). Indeed, the movements of this turtle, as it migrated more than 800 km southward in shallow water, looked very much like those of post-nesting females, only in reverse direction.

Thus, in spring, reproductively mature female Kemp's ridleys undertake annual migrations, probably following along the coastline from northern and southern portions of the Gulf of Mexico, where most of the adults of this species forage. Roughly a month before the nesting season begins, females and males aggregate to mate in nearshore waters near the beach at Rancho Nuevo (Pritchard, 1969; Mendonça and Pritchard, 1986; Rostal, 1991).

Similarly, many female olive ridleys return to nest on beaches to which they can be faithful for many years. At Gahirmatha, India, individual females have been recorded returning after as long as 21 years (Pandav and Kar, 2000). Some female olive ridleys, however, do not nest on just one beach. Kalb (1999) discovered that in contrast to the *arribada* nesters that exhibit a high degree of site fidelity to Nancite Beach, Costa Rica, the solitary nesters also nesting at this site do not; solitary nesters used multiple beaches during a nesting season, traveling hundreds of kilometers north and south of Costa Rica to nest on beaches in other countries.

Nearshore breeding is a common strategy for olive ridleys. As adult females gear up almost annually for reproduction, they migrate into the neritic zone, where they reside near the nesting beach for the duration of the breeding season (Kalb, 1999). There they meet with other reproductively active olive ridley females and males that similarly have migrated toward the coast, presumably from scattered oceanic foraging locations (Plotkin, 1994; Plotkin et al., 1994, 1995, 1996; Kalb et al., 1995; Pandav and Kar, 2000; Pandav et al., 2000). Within these aggregations near the nesting beaches, breeding takes place mostly before the first mass nesting emergence (Pritchard, 1969; Hughes and Richard, 1974; Cornelius and Robinson, 1986; Dash and Kar, 1990; Plotkin et al., 1991, 1996, 1997; Kalb et al., 1995; Pandav et al., 2000).

Because the mass nesting emergences of ridleys occur on a predictable annual basis at predictable sites, this undoubtedly contributes to making the nearby waters a prime mating area. Both the predictability and the potential payoffs in terms of finding reproductively active females would tend to attract males to these sites too. Thus, a simple mating strategy for males would be to migrate from foraging areas to nesting areas before the breeding season.

Inasmuch as most female Kemp's ridleys migrate each year to one nesting beach to reproduce, it could be presumed that at least a portion of the adult male Kemp's ridleys in the population make similar annual mating migrations. However, breeding migrations for male sea turtles are very poorly understood. There are many different strategies that can lead to successful mating. Nevertheless, mating probably is most successful within a relatively short time preceding oviposition by the female (Owens, 1980; Miller, 1997). Thus, a logical place for male Kemp's ridleys to increase their chance of reproductive success would be near the nesting beach, and such behavior seems to fit the generalized pattern that has been described for male sea turtles (Rostal, 1991).

But, even in the simple biological model of the Kemp's ridley, it is probably not that straightforward. Males foraging in distant feeding areas cannot be at the nesting beach simultaneously. Also, to shuttle back and forth annually over great distances may not be as productive as other strategies. In fact, male Kemp's ridleys probably do exhibit other mating strategies. At the Rancho Nuevo nesting beach, 10 of the 11 adult males tracked by satellite transmitter remained resident in coastal waters near Rancho Nuevo after the nesting season (Shaver et al., 2005). Hence, at least some males may not ever stray far from the mating areas, even outside the breeding season. Inversely, some male turtles may not make the annual migratory journey from foraging areas to the nesting beach but instead may mate opportunistically with females they encounter. In fact, it appears that mating activities for this species are more widespread than the nesting areas, occurring in coastal and inshore waters extending from south Texas to Veracruz, Mexico (Shaver, 1992). It would be expected that male Kemp's ridleys adopting other strategies may increase their mating opportunities by frequenting predictable feeding areas or even predictable pathways of female migration.

Olive ridley males probably exhibit a similar array of mating strategies. The numerous reports of mating in nearshore waters off nesting beaches obviously implicate this as a strategy employed by males. Nevertheless, a great deal of breeding also takes place far from shore (Pitman, 1990, 1993; Kopitsky et al., 2000; Parker et al., 2003), which indicates that some individuals do not participate in nearshore breeding aggregations and suggests viable alternate mating strategies for olive ridleys too. In offshore waters of the eastern Pacific, the olive ridley is by far the most abundant species of turtle (Pitman, 1990, 1993). Shipboard observations of thousands of individuals indicate that both male and female olive ridleys are distributed throughout tropical latitudes, in waters extending to more than 4,000 km offshore. Therefore, there may be ample opportunity to encounter a mate, especially in areas of potential convergences (Plotkin, 1994) or in the region of nesting.

The various mating behaviors exhibited by male ridley turtles are representative of the suite of strategies used by males in response to a variety of selective pressures characteristic of sea turtle mating systems. High competition and even competitive interference among males at popular mating areas (Booth and Peters, 1972; Limpus, 1993) opens up the possibility for alternate strategies, such as mating in waters offshore (Pitman, 1990; Plotkin, 1994; Plotkin et al., 1996; Kopitsky et al., 2000; Parker et al., 2003), mating in common courtship areas (Booth and Peters, 1972; Green, 1984; Henwood, 1987; Limpus, 1993; FitzSimmons et al., 1997), or mating along migration routes as females pass by on their way to the nesting beach (Meylan et al., 1992; FitzSimmons et al., 1997; FitzSimmons, 2000).

However, females may exert much more control over the relative success of different mating strategies than males simply by being receptive for a very short time before nesting (FitzSimmons, 2000). This greatly increases the chance of encountering a receptive female closer in time and in space to the nesting beach. Males depending on chance encounters with females in the vast oceanic realm may not fare as well as males that go to known sites where breeding occurs. This would tend to strongly favor the migration of males to breeding areas, many of which likely are not that far from nesting areas, and would go a long way to explain why adult males would undertake migrations from foraging areas to breeding sites.

Conclusions

The basic essence of a sea turtle's existence as an adult seems to be that when it is not engaged in breeding or breeding migrations, it is engaged in foraging or foraging migrations. Foraging migrations, like breeding migrations, can be of any magnitude imaginable, either minuscule or oceanic in scale, and just about every distance in between. In adult foraging behavior, there is a great deal of variability among species, ranging from the herbivorous grazing of green turtles to the coastal foraging on benthic invertebrates of Kemp's ridleys, to the open-water epipelagic feeding of olive ridleys. The inherent differences in foraging behavior undoubtedly can be manifested in noticeable differences in migratory behavior among sea turtle species.

Such differences are starkly obvious for olive ridleys, which are nomadic migrants feeding on tunicates and other epipelagic fauna in open ocean waters, and Kemp's ridleys, which are coastal migrants that travel along relatively narrow coastal corridors and feed on shallow-water benthic invertebrates. The differences in foraging strategies and destinations result in different migratory routes and obvious differences in directness of travel. The foraging migrations of the Kemp's ridley appear more goal-oriented and directed toward a specific destination. The less direct movements of olive ridleys do not reflect an inability to navigate but rather represent an effective means of searching for patchily distributed prey. These characteristic migration and foraging behaviors form the basis of the distinction between the two ridley species.

The two species of ridley turtles, for the most part, breed on a yearly basis. This apparently is true for females (Márquez-M., 1994; Pandav and Kar, 2000) and males (Pandav and Kar, 2000; Shaver, 2005). Hence, their annual cycle seems to be composed of migrating to foraging areas, migrating to breeding areas, and migrating back to foraging areas. This pattern can be repeated yearly and can persist for decades (Pandav and Kar, 2000). Such a condensed migratory cycle can be considered the most basic behavioral model common to sea turtles. However, males and females might have different selective forces shaping the patterns of their foraging migrations, especially in the magnitude of distance away from mating areas they can migrate. The different selective forces arise from the interplay between foraging success and reproductive success.

Ridley mating migrations appear to be representative of sea turtles in general. The basic outline is for both males and females to show up in areas where there is a reasonable chance for successful reproduction. Often, but not always, this entails migration. Ultimately, all female sea turtles need to migrate to nesting beaches, either from nearby or distant foraging areas. Hence, males can virtually ensure encounters with females by meeting them near nesting beaches. Many males of different species appear to strongly favor this strategy, some taking it to extremes by remaining near the breeding area year-round (Booth and Peters, 1972; Henwood, 1987; FitzSimmons et al., 1997; Shaver et al., 2005). Some males emulate females by migrating to and from breeding and foraging areas (Henwood, 1987; Limpus, 1993; FitzSimmons et al., 2000; Pandav et al., 2000).

Just as males and females may adopt different strategies, so too can different species, depending a lot on their breeding and foraging movements. Male Kemp's ridleys may visit regular coastal feeding areas where females predictably occur, intercept females in predictable migratory corridors, migrate to the nesting beach before the breeding season, or remain resident in nearby waters all year. Male olive ridleys may migrate to the nesting beach to breed; they may intercept females in offshore waters as they coalesce toward the nesting beaches; or they may opportunistically mate on chance encounters at sea. Notwithstanding gender differences, interestingly, in the satellite telemetry studies summarized in Figures 11.1 and 11.2, the movement and habitat utilization patterns of males overlapped greatly with those of females for both species of ridleys.

Undoubtedly, much remains to be learned about sea turtle migrations. For years, sea turtle studies were concentrated on turtles at their nesting beaches and nearshore waters. The inherent difficulties of open water research posed a major obstacle for studying sea turtles in their natural element. Now with new techniques in telemetry and remote sensing, the daunting chore of studying the ecology and behavior of turtles migrating over great distances during oceanic travel has been alleviated. It is a relatively easy task for sea turtles of any species to be monitored over long periods, in all weather conditions, and whether the individual is at or below the surface.

As studies on oceanic movements proliferate, it is becoming even more evident that the ability to migrate is ingrained in all of the sea turtle species. A concise statement made for loggerhead and green turtles, and probably broadly applicable to all sea turtles, is that "there are probably no 'non-migratory' members . . . just turtles that migrate very short distances" (Limpus et al., 1992). Indeed, a finely honed ability to migrate over sometimes vast distances seems to be an inherent feature of sea turtle ecology. More important, any conservation plans involving turtles must include serious consideration of their potential and propensity for long-distance migration.

LITERATURE CITED

Ackerman, R. A. 1997. The nest environment and the embryonic development of sea turtles. In Lutz, P. L., and Musick, J. A. (Eds.). *The Biology of Sea Turtles*. Boca Raton, FL: CRC Press, pp. 83–106.

Arenas, P., and Hall, M. 1992. The association of sea turtles and other pelagic fauna with floating objects in the eastern tropical Pacific Ocean. In Salmon, M., and Wyneken, J. (Compilers). *Proceedings of the Eleventh Annual Workshop on Sea Turtle Biology and Conservation*. NOAA Technical Memorandum NMFS-SEFC-302, pp. 7–10.

Arocha, F. 1997. The reproductive dynamics of swordfish *Xiphias gladius* L. and management implications in the northwestern Atlantic. Ph.D. diss. University of Miami, Coral Gables, FL.

Balazs, G. H. 1976. Green turtle migrations in the Hawaiian Archipelago. *Biological Conservation* 9:125–140.

Beavers, S. C., and Cassano, E. R. 1996. Movements and dive behavior of a male sea turtle (*Lepidochelys olivacea*) in the eastern tropical Pacific. *Journal of Herpetology* 30:97–104.

Block, B. A., Dewar, H., Blackwell, S. B., Williams, T. D., Prince, E. D., Farwell, C. J., Boustany, A., Teo, S. L. H., Seitz, A., Walli, A., and Fudge, D. 2001. Migratory movements, depth preferences, and thermal biology of Atlantic bluefin tuna. *Science* 293:1310–1314.

Booth, J., and Peters, J. 1972. Behavioural studies on the green turtle (*Chelonia mydas*) in the sea. *Animal Behaviour* 20:808–812.

Broderick, D., Moritz, C., Miller, J. D., Guinea, M., Prince, R. I. T., and Limpus, C. J. 1994. Genetic studies of the hawksbill turtle *Eretmochelys imbricata:* evidence for multiple stocks in Australian waters. *Pacific Conservation Biology* 1:123–131.

Burke, V. J., and Gibbons, J. W. 1995. Terrestrial buffer zones and wetland conservation: a case study of freshwater turtles in a Carolina bay. *Conservation Biology* 9:1365–1369.

Bustard, H. R., and Greenham, P. 1968. Physical and chemical factors affecting hatching in the green sea turtles, *Chelonia mydas. Ecology* 49:269–276.

Byles, R. A. 1989. Satellite telemetry of Kemp's ridley sea turtle, *Lepidochelys kempi,* in the Gulf of Mexico. In Eckert, S. A., Eckert, K. L., and Richardson, T. H. (Compilers). *Proceedings of the Ninth Annual Workshop on Sea Turtle Conservation and Biology.* NOAA Technical Memorandum NMFS-SEFC-232, pp. 25–26.

Byles, R. A., and Plotkin, P. T. 1994. Comparison of the migratory behavior of the congeneric sea turtles *Lepidochelys olivacea* and *L. kempii.* In Schroeder, B. A., and Witherington, B. E. (Compilers). *Proceedings of the Thirteenth Annual Symposium on Sea Turtle Biology and Conservation.* NOAA Technical Memorandum NMFS-SEFSC-341, p. 39.

Byles, R. A., and Swimmer, Y. B. 1994. Post-nesting migration of *Eretmochelys imbricata* in the Yucatán Península. In Bjorndal, K. A., Bolten, A. B., Johnson, D. A., and Eliazar, P. J. (Compilers). *Proceedings of the Fourteenth Annual Symposium on Sea Turtle Biology and Conservation.* NOAA Technical Memorandum NMFS-SEFSC-351, p. 202.

Cagle, F. R. 1944. Home range, homing behavior, and migration in turtles. *Miscellaneous Publications of the Museum of Zoology University of Michigan* 61:1–34.

Cagle, F. R. 1950. The life history of the slider turtle, *Pseudemys scripta troosti* (Holbrook). *Ecological Monographs* 20:32–34.

Carr, A. F. 1967. *So Excellent a Fishe: A Natural History of Sea Turtles.* New York: Scribner, 1984 revised edition.

Carr, A. 1980. Some problems of sea turtle ecology. *American Zoologist* 20:489–498.

Chavez, H. 1969. Tagging and recapture of the lora turtle (*Lepidochelys kempi*). *International Turtle and Tortoise Society Journal* 3:14–19, 32–36.

Collard, S. B., and Ogren, L. H. 1990. Dispersal scenarios for pelagic post-hatchling sea turtles. *Bulletin of Marine Science* 47:23–243.

Congdon, J. D., Tinkle, D. W., Britenbach, G. L., and van Loben Sels, R. 1983. Nesting ecology and hatching success in the turtle, *Emydoidea blandingi. Herpetelogica* 39:417–429.

Cornelius, S. E., and Robinson, D. C. 1986. Post-nesting movements of female olive ridley turtles tagged in Costa Rica. *Vida Silvestre Neotropical* 1:12–23.

Craighead, D. J., and Craighead, J. J. 1987. Tracking caribou using satellite telemetry. *National Geographic Research* 3:462–479.

Dash, M., and Kar, C. S. 1990. *The Turtle Paradise: Gahirmatha (An Ecological Analysis and Conservation Strategy).* New Dehli: Interprint.

Eckert, S. A. 1998. Perspectives on the use of satellite telemetry and other electronic technologies for the study of marine turtles, with reference to the first year long tracking of leatherback sea turtles. NOAA Technical Memorandum, NMFS-SEFSC-415, pp. 44–46.

Ehrhart, L. M. 1982. A review of sea turtle reproduction. In Bjorndal, K. A. (Ed.). *Biology and Conservation of Sea Turtles.* Washington, DC: Smithsonian Institution Press, pp. 29–38.

Epperly, S. P., Braun, J., and Chester, A. J. 1995. Aerial surveys for sea turtles in North Carolina waters. *Fishery Bulletin* 93:254–261.

Ernst, C. H., and Barbour, R. W. 1972. *Turtles of the United States.* Lexington: University Press of Kentucky.

FitzSimmons, N. N., Goldizen, A. R., Norman, J. A., Moritz, C., Miller, J. D., and Limpus, C. J. 1997. Philopatry of male marine turtles inferred from mitochondrial markers. *Proceedings of the National Academy of Sciences USA* 94:8912–8917.

FitzSimmons, N. N., Limpus, C. J., Miller, J. D., Prince, R. I. T., and Moritz, C. 2000. Male marine turtles: questions beyond gene flow. In Kalb, H. J., and Wibbels, T. (Compilers). *Proceedings of theNineteenth Annual Sea Turtle Symposium on Sea Turtle Biology and Conservation.* NOAA Technical Memorandum NMFS-SEFSC-443, pp. 11–13.

Fleming, T. H., Nunez, R. A., and Sternberg, L. 1993. Seasonal changes in the diets of migrant and non-migrant nectarivorous bats as revealed by carbon stable isotope analysis. *Oecologia* 94:72–75.

Frazier, J. 1984. Marine turtles in the Seychelles and adjacent territories. In Stoddard, D. R. (Ed.). *Biogeography and Ecology of the Seychelles Islands.* The Hague: Dr. W. Junk Publishers, pp. 417–468.

Frazier, J. 1985. *Marine Turtles in the Comoro Archipelago.* Amsterdam: Royal Netherlands Academy of Arts and Sciences.

Gibbons, J. W., Greene, J. L., and Congdon, J. D. 1983. Drought-related responses of aquatic turtle populations. *Journal of Herpetology* 17:242–246.

Godley, B. J., Richardson, S., Broderick, A. C., Coyne, M. S., Glen, F., and Hays, G. 2002. Long-term satellite telemetry of the movements and habitat utilisation by green turtles in the Mediterranean. *Ecography* 25:352–362.

Green, D. 1984. Long-distance movements of Galapagos green turtles. *Journal of Herpetology* 18: 121–130.

Groshens, E. B., and Vaughan, M. R. 1994. Post-nesting movements of hawksbill sea turtles from Buck Island Reef National Monument, St. Croix, USVI. In Schroeder, B. A., and Witherington, B. E. (Compilers). *Proceedings of the Thirteenth Annual Symposium on Sea Turtle Biology and Conservation.* NOAA Technical Memorandum NMFS-SEFSC-341, pp. 69–71.

Hatase, H., Takai, N., Matsuzawa, Y., Sakamoto, W., Omuta, K., Goto, K., Arai, N., and Fujiwara, T. 2002. Size-related differences in feeding habitat use of adult female loggerhead turtles (*Caretta caretta*) around Japan determined by stable isotope analyses and satellite telemetry. *Marine Ecology Progress Series* 233:273–281.

Hendrickson, J. R. 1958. The green sea turtles *Chelonia mydas* (Linn.) in Malaya and Sarawak. *Proceedings of the Zoological Society of London* 130:455–535.

Hendrickson, J. R. 1980. The ecological strategies of sea turtles. *American Zoologist* 20:597–608.

Henwood, T. A. 1987. Movements and seasonal changes in loggerhead turtle *Caretta caretta* aggregations in the vicinity of Cape Canaveral, Florida (1978–1984). *Biological Conservation* 40:191–202.

Henwood, T. A., and Ogren, L. H. 1987. Distribution and migrations of immature Kemp's ridley turtles (*Lepidochelys kempi*) and green turtles (*Chelonia mydas*) off Florida, Georgia, and South Carolina. *Northeast Gulf Science* 2:153–159.

Hildebrand, H. 1982. A historical review of the status of sea turtle populations in the western Gulf of Mexico, In Bjorndal, K. (Ed.). *Biology and Conservation of Sea Turtles.* Washington, DC: Smithsonian Institution Press, pp. 447–453.

Horrocks, J. A., Vermeer, L. A., Krueger, B., Coyne, M., Schroeder, B. A., and Balazs, G. H. 2001. Migration routes and destination characteristics of post-nesting hawksbill turtles satellite-tracked from Barbados, West Indies. *Chelonian Conservation and Biology* 4:107–114.

Hughes, D. A., and Richard, J. D. 1974. The nesting of the Pacific ridley turtle *Lepidochelys olivacea* on Playa Nancite, Costa Rica. *Marine Biology* 24:97–107.

Janzen, F. J., and Pakustis, G. L. 1991. Environmental sex determination in reptiles: ecology, evolution, and experimental design. *Quarterly Review of Biology* 66:149–179.

Kalb, H. J. 1999. Behavior and physiology of solitary and arribada nesting olive ridley sea turtles (*Lepidochelys olivacea*) during the internesting period. Ph.D. diss. Texas A&M University, College Station.

Kalb, H., Valverde, R. A., and Owens, D. W. 1995. What is the reproductive patch of the olive ridley sea turtle? In Richardson, J. I., and Richardson, T. H. (Compilers). *Proceedings of the Twelfth Annual Workshop on Sea Turtle Biology and Conservation.* NOAA Technical Memorandum NMFS-SEFSC-361, pp. 57–60.

Kar, C. S., and Bhaskar, S. 1982. The status of sea turtles in the Eastern Indian Ocean. In Bjorndal, K. A. (Ed.). *Biology and Conservation of Sea Turtles.* Washington, DC: Smithsonian Institution Press, pp. 365–372.

Keinath, J. A., Musick, J. A., and Byles, R. A. 1987. Aspects of the biology of Virginia's sea turtles: 1979–1986. *Virginia Journal of Science* 38:329–336.

Kopitsky, K., Pitman, R. L., and Plotkin, P. 2000. Investigations on at-sea mating and reproductive status of olive ridleys, *Lepidochelys olivacea,* captured in the eastern tropical Pacific. In Kalb, H. J., and Wibbels, T. (Compilers). *Proceedings of the Nineteenth Annual Symposium on Sea Turtle Biology and Conservation.* NOAA Technical Memorandum NMFS-SEFSC-443, pp. 160–162.

Limpus, C. J. 1993. The green turtle, *Chelonia mydas,* in Queensland: breeding males in the Southern Great Barrier Reef. *Australian Wildlife Research* 20:513–523.

Limpus, C. J., and Limpus, D. J. 2001. The loggerhead turtle, *Caretta caretta,* in Queensland: breeding migrations and fidelity to a warm temperate feeding area. *Chelonian Conservation and Biology* 4:142–153.

Limpus, C. J., Parmenter, C. J., Parker, R., and Ford, N. 1981. The flatback turtle *Chelonia depressa* in Queensland: the Peak Island rookery. *Herpetofauna* 13:14–18.

Limpus, C. J., Parmenter, C. J., Baker, V., and Fleay, A. 1983. The Crab Island sea turtle rookery in north-eastern Gulf of Carpentaria. *Australian Wildlife Research* 10:173–184.

Limpus, C. J., Zeller, D., Kwan, D., and MacFarlane, W. 1989. Sea turtle rookeries in the northwestern Torres Strait. *Australian Wildlife Research* 16:517–525.

Limpus, C. J., Miller, J. D., Parmenter, C. J., Reimer, D., Mclachlan, N., and Webb, R. 1992. Migration of green (*Chelonia mydas*) and loggerhead (*Caretta caretta*) turtles to and from eastern Australian rookeries. *Australian Wildlife Research* 19:347–358.

Márquez-M., R. 1970. Las tortugas marinas de Mexico. Thesis, I. P. N., Escuela Nacional de Ciencias Biologias.

Márquez-M., R. 1990. *FAO Species Catalogue,* Volume 11: *Sea Turtles of the World. An Annotated and Illustrated Catalogue of Sea Turtle Species Known to Date.* FAO Fisheries Synopsis No. 125. Rome: FAO.

Márquez-M., R. 1994. *Synopsis of Biological Data on the Kemp's Ridley Turtle,* Lepidochelys kempii *(Garman, 1880).* NOAA Technical Memorandum NMFS-SEFSC-343.

Matsuzawa, Y., Sato, K., Sakamoto, W., and Bjorndal, K. A. 2002. Seasonal fluctuations in sand temperature: effects on the incubation period and mortality of loggerhead sea turtle (*Caretta caretta*) pre-emergent hatchlings in Minabe, Japan. *Marine Biology* 140:639–646.

Mauritzen, M., Derocher, A. E., and Wiig, Ø. 2001. Space-use strategies of female polar bears in a dynamic sea ice habitat. *Canadian Journal of Zoology* 79:1704–1713.

Menard, F., and Marchal, E. 2003. Foraging behaviour of tuna feeding on small schooling *Vinciguerria nimbaria* in the surface layer of the equatorial Atlantic Ocean. *Aquatic Living Resources* 16:231–238.

Mendonça, M. T., and Ehrhart, L. M. 1982. Activity, population size and structure of immature *Chelonia mydas* and *Caretta caretta* in Mosquito Lagoon, Florida. *Copeia* 1982:161–167.

Mendonça, M. T., and Pritchard, P. C. H. 1986. Offshore movements of post-nesting Kemp's ridley sea turtles (*Lepidochelys kempii*). *Herpetologica* 42:373–380.

Meylan, A. 1982. Sea turtle migration—evidence from tag returns. In Bjorndal, K. A. (Ed.). *Biology and Conservation of Sea Turtles.* Washington, DC: Smithsonian Institution Press, pp. 91–100.

Meylan, A. B. 1999a. International movements of immature and adult hawksbill turtles (*Eretmochelys imbricata*) in the Caribbean region. *Chelonian Conservation and Biology* 3:189–194.

Meylan, A. B. 1999b. Status of the hawksbill turtle (*Eretmochelys imbricata*) in the Caribbean region. *Chelonian Conservation and Biology* 3:177–184.

Meylan, S. B., Bowen, B. W., and Avise, J. C. 1990. A genetic test of the natal homing versus social facilitation models for green turtle migrations. *Science* 248:724–727.

Meylan, P. A., Meylan, A. B., and Yeomans, R. Interception of Tortuguero-bound green turtles at Bocas Del Totor Province, Panama. In Salmon, M., and Wyneken, J. (Compilers). *Proceedings of the Eleventh Annual Workshop on Sea Turtle Biology and Conservation.* NOAA Technical Memorandum NMFS-SEFSC-302, p. 74.

Miller, J. D. 1997. Reproduction in sea turtles. In Lutz, P. L., and Musick, J. A. (Eds.). *The Biology of Sea Turtles.* Boca Raton, FL: CRC Press, pp. 51–81.

Miller, J. D., and Limpus, C. J. 1981. Incubation period and sexual differentiation in the green turtle, *Chelonia mydas* L. In Banks, C., and Martin, A. (Eds.). *Proceedings of the Melbourne Herpetological Symposium.* Melbourne: The Royal Melbourne Zoological Gardens, pp. 66–77.

Miller, J. D., Dobbs, K. A., Limpus, C. J., Mattocks, N., and Landry, A. M. 1998. Long-distance migrations by the hawksbill turtle, *Eretmochelys imbricata,* from north-eastern Australia. *Australian Wildlife Research* 25:89–95.

Moll, E. O., and Legler, J. M. 1971. The life history of a neotropical slider turtle, *Pseudemys scripta* (Schoepff), in Panama. *Bulletin of the Los Angeles County Museum of Natural History and Science* 11:1–102.

Morreale, S. J., and Standora, E. A. 1998. *Early Life Stage Ecology of Sea Turtles in Northeastern U.S. Waters.* NOAA Technical Memorandum NMFS-SEFSC-413.

Morreale, S. J., Gibbons, J. W., and Congdon, J. D. 1984. Significance of activity and movement in the yellow-bellied slider turtle (*Pseudemys scripta*). *Canadian Journal of Zoology* 62:1038–1042.

Morreale S. J., Meylan, A. B., Sadove, S. S., and Standora, E. A. 1992. Annual occurrence and winter mortality of marine turtles in New York waters. *Journal of Herpetology* 26:301–308.

Morreale, S. J., Standora, E. A., Spotila, J. R., and Paladino, F. V. 1996. Migration corridor for sea turtles. *Nature* 384:319–320.

Mortimer, J. A., and Balazs, G. H. 2000. Post-nesting migrations of hawksbill turtles in the granitic Seychelles and implications for conservation. In Kalb, H. J., and Wibbels, T. (Compilers). *Proceedings of the Nineteenth Annual Symposium on Sea Turtle Biology and Conservation.* NOAA Technical Memorandum NMFS-SEFSC-443, pp. 22–26.

Mysing, J. O., and Vanselous, T. M. 1989. Status of satellite tracking of Kemp's ridley sea turtles. In

Caillouet, C. W., Jr., and Landry, A. M., Jr. (Eds.). *Proceedings of the First International Symposium on Kemp's Ridley Sea Turtle Biology, Conservation, and Management.* Texas A&M University Sea Grant College Publication TAMU-SG-89-105, pp. 112–115.

National Marine Fisheries Service and U.S. Fish and Wildlife Service. 1998. *Recovery Plan for U.S. Pacific Populations of the Olive Ridley Turtle (*Lepidochelys olivacea*).* Silver Spring, MD: National Marine Fisheries Service.

Obbard, M. E., and Brooks, R. J. 1980. Nesting migrations of the snapping turtle (*Chelydra serpentina*). *Herpetologica* 36:158–162.

Ogren, L. H. 1989. Distribution of juvenile and subadult Kemp's ridley turtles: preliminary results from 1984–1987 surveys. In Caillouet, C. W., Jr., and Landry, A. M., Jr. (Eds.). *Proceedings of the First International Symposium on Kemp's Ridley Sea Turtle Biology, Conservation and Management.* Texas A&M Publication TAMU-SG-89-105, pp. 116–123.

Owens, D. 1980. The comparative reproductive physiology of sea turtles. *American Zoologist* 20: 549–523.

Owens, D. W. 1997. Hormones in the life history of sea turtles. In Lutz, P., and Musick, J. A. (Eds.). *The Biology of Sea Turtles.* Boca Raton, FL: CRC Marine Science Series, CRC Press, pp. 315–341.

Pandav, B., and Kar, C. S. 2000. Reproductive span of olive Ridley turtles at Gahirmatha Rookery, Orissa, India. *Marine Turtle Newsletter* 87:8–9.

Pandav, B., Choudhury, B. C., and Kar, C. S. 1995. A note on the occurrence of sub-adult olive ridley turtles along the Gahirmatha coast. *Marine Turtle Newsletter* 71:15–17.

Pandav, B., Choudhury, B. C., and Shanker, K. 1998. The Olive Ridley sea turtle *(Lepidochelys olivacea)* in Orissa: an urgent call for an intensive and integrated conservation programme. *Current Science* 75:1323–1328.

Pandav, B., Banugopan, K., Sutaria, D., and Choudhury, B. C. 2000. Fidelity of male olive ridley sea turtles to a breeding ground. *Marine Turtle Newsletter* 87:9–10.

Papi, F. 1992. General aspects. In Papi, F. (Ed.). *Animal Homing.* London: Chapman and Hall, pp. 1–18.

Parker, D. M., Dutton, P. H., Kopitsky, K., and Pitman, R. L. 2003. Movement and dive behavior determined by satellite telemetry for male and female olive ridley turtles in the eastern tropical Pacific. In Seminoff, J. A. (Compiler). *Proceedings of the Twenty-Second Annual Symposium on Sea Turtle Biology and Conservation.* NOAA Technical Memorandum NMFS-SEFSC-503, pp. 48–49.

Parmenter, C. J. 1983. Reproductive migration in the hawksbill turtle (*Eretmochelys imbricata*). *Copeia* 1983:271–273.

Pitman, R. L. 1990. Pelagic distribution and biology of sea turtles in the eastern tropical Pacific. In Richardson, T. H., Richardson, J. I., and Donnelly, M. (Compilers). *Proceedings of the Tenth Annual Workshop on Sea Turtle Biology and Conservation.* NOAA Technical Memorandum NMFS-SEFC-278, pp. 143–148.

Pitman, R. L. 1993. Seabird associations with marine turtles in the eastern Pacific Ocean. *Colonial Waterbirds* 16:194–201.

Plotkin, P. T. 1994. Migratory and reproductive behavior of the olive ridley turtle, *Lepidochelys olivacea* (Eschscholtz, 1829), in the eastern Pacific Ocean. Ph.D. diss. Texas A&M University, College Station.

Plotkin, P., Polak, M., and Owens, D. W. 1991. Observations on olive ridley sea turtle behavior prior to an arribada at Playa Nancite, Costa Rica. *Marine Turtle Newsletter* 53:9–10.

Plotkin, P. T., Byles, R. A., and Owens, D. W. 1994. Post-breeding movements of male olive ridley sea turtles *Lepidochelys olivacea* from a nearshore breeding area. In Bjorndal, K. A., Bolten, A. B., Johnson, D. A., and Eliazar, P. J. (Compilers). *Proceedings of the Fourteenth Annual Symposium on Sea Turtle Biology and Conservation.* NOAA Technical Memorandum NMFS-SEFSC-351, p. 119.

Plotkin, P. T., Byles, R. A., Rostal, D. C., and Owens, D. W. 1995. Independent vs. socially facilitated migrations of the olive ridley, *Lepidochelys olivacea. Marine Biology* 122:137–143.

Plotkin, P. T., Owens, D. W., Byles, R. A., and Patterson, R. 1996. Departure of male olive ridley turtles (*Lepidochelys olivacea*) from a nearshore breeding area. *Herpetologica* 52:1–7.

Plotkin, P. T., Rostal, D. C., Byles, R. A., and Owens, D. W. 1997. Reproductive and developmental synchrony in female *Lepidochelys olivacea. Journal of Herpetology* 31:17–22.

Pritchard, P. C. H. 1969. Studies of the systematics and reproductive cycles of the genus *Lepidochelys,* Ph.D. diss. University of Florida, Gainesville.

Pritchard, P. C. H. 1973. International migrations of South American sea turtles (Cheloniidae and Dermochelidae). *Animal Behaviour* 21:18–27.

Pritchard, P. C. H., and Márquez-M., R. 1973. Kemp's ridley or Atlantic ridley, *Lepidochelys kempii.* IUCN Monograph No. 2 (Marine Turtle Series), Morges, Switzerland.

Renaud, M. L. 1995. Movements and submergence patterns of Kemp's ridley turtles (*Lepidochelys kempii*). *Journal of Herpetology* 29:370–374.

Renaud, M. L., Carpenter, J. A., Williams, J. A., and Landry, A. M., Jr. 1996. Kemp's ridley sea turtle (*Lepidochelys kempii*) tracked by satellite telemetry from Louisiana to nesting beach at Rancho Nuevo, Tamaulipas, Mexico. *Chelonian Conservation and Biology* 2:108–109.

Rostal, D. C. 1991. The reproductive behavior and physiology of the Kemp's ridley sea turtle, *Lepidochelys kempi* (Garman, 1880). Ph.D. diss. Texas A&M University, College Station.

Rostal, D. C., Owens, D. W., Grumbles, J. S., MacKenzie, D. S., and Amoss, M. S., Jr. 1998. Seasonal reproductive cycle of the Kemp's ridley sea turtle (*Lepidochelys kempi*). *General and Comparative Endocrinology* 109:232–243.

Schmid, J. R. 1995. Marine turtle populations on the east-central coast of Florida: results of tagging studies at Cape Canaveral, Florida, 1986–1991. *Fishery Bulletin* 93:139–151.

Sexton, O. J. 1959. Spatial and temporal movements of a population of the painted turtle *Chrysemys picta marginata* (Agassiz). *Ecological Monographs* 29:113–140.

Shanker, K., Choudbury, B. C., Pandav, B., Tripathy, B., Kar, C. S., and Kar, S. K. 2003a. Tracking olive ridley turtles from Orissa. In Seminoff, J. A. (Compiler). *Proceedings of the Twenty-Second Annual Symposium on Sea Turtle Biology and Conservation.* NOAA Technical Memorandum NMFS-SEFSC-503, pp. 50–51.

Shanker, K., Pandav, B., and Choudhury, B. C. 2003b. An assessment of the olive ridley turtle nesting population in Orissa, India. *Biological Conservation* 15:149–160.

Shaver, D. J. 1992. Kemp's ridley sea turtle reproduction. *Herpetological Review* 23:59.

Shaver, D. J. 1999. Kemp's ridley sea turtle project at Padre Island National Seashore, Texas. In McKay, M., and Nides, J. (Eds.). *Proceedings from the Seventeenth Annual Gulf of Mexico Information Transfer Meeting,* U.S. Department of the Interior, Minerals Management Service, Gulf of Mexico OCS Region, MMS 99-0042, p. 342.

Shaver, D. J. 2001a. *Padre Island National Seashore Kemp's ridley Sea Turtle Project and Texas Sea Turtle Nesting and Stranding 2000 Report.* Washington, DC: U.S. Geological Survey, Department of the Interior.

Shaver, D. J. 2001b. U.S. Geological Survey/National Park Service Kemp's ridley sea turtle research and monitoring programs in Texas. In McKay, M., Nides, J., Lang, W., and Vigil, D. (Eds.). *Proceedings of the Gulf of Mexico Marine Protected Species Workshop.* U.S. Department of the Interior, Minerals Management Service, Gulf of Mexico OCS Region, MMS-2001-039, pp. 121–124.

Shaver, D. J., Schroeder, B. A., Byles, R. A., Burchfield, P. M., Peña, J., Márquez-M., R., and Martinez, H. J. 2005. Movements of adult male Kemp's ridley sea turtles (*Lepidochelys kempii*) in the Gulf of Mexico investigated by satellite telemetry. *Chelonian Conservation and Biology* 4:817–827.

Vinayachandran, P. N., and Yamagata, T. 2003. Comment on "Indian Ocean: Validation of the Miami Isopynic Coordinate Ocean Model and ENSO Events During 1958–1998" by V. E. Haugen et al., *Journal of Geophysical Research* 108:C6, doi:10.1029/2002JC001624.

Vinayachandran, P. N., Chauhan, P., Mohan, M., and Nayak, S. 2004. Biological response of the sea around Sri Lanka to summer monsoon. *Geophysical Research Letters* 31:L01302, doi:10.1029/2003GL018533.

Weimerskirch, H., Doncaster, C. P., and Cuenotchaillet, F. 1994. Pelagic seabirds and the marine environment: foraging of wandering albatrosses in relation to the availability and distribution of their prey. *Proceedings of the Royal Society of London Series B-Biological Sciences* 255:91–97.

Witherington, B. E., and Ehrhart, L. M. 1989. Hypothermic stunning and mortality of marine turtles in the Indian River lagoon system, Florida. *Copeia* 1989:696–703.

Witzell, W. N. 1983. Synopsis of biological data on the hawksbill turtle, *Eretmochelys imbricata* (Linnaeus 1766). *FAO Fisheries Synopsis* 137:1–78.

Yntema, C. L., and Mrosovsky, N. 1980. Sexual differentiation in hatchling loggerheads (*Caretta caretta*) incubated at different controlled temperatures. *Herpetologica* 36:33–36.

Yntema, C. L., and Mrosovsky, N. 1982. Critical periods and pivotal temperatures for sexual differentiation in loggerhead sea turtles. *Canadian Journal of Zoology* 60:1012–1016.

Zwinenberg, A. J. 1977. Kemp's ridley, *Lepidochelys kempi* (Garman, 1880), undoubtedly the most endangered marine turtle today (with notes on the current status of *Lepidochelys olivacea*). *Bulletin of the Maryland Herpetological Society* 13:170–192.

STEPHEN E. CORNELIUS
RANDALL ARAUZ
JACQUES FRETEY
MATTHEW H. GODFREY
RENE MÁRQUEZ-M.
KARTIK SHANKER

12

Effect of Land-Based Harvest of *Lepidochelys*

THOUGH GENERALLY NOT PRIZED for food or shell as are most of the other sea turtles, *Lepidochelys* nonetheless has been severely impacted by humans. The olive ridley's *(Lepidochelys olivacea)* widespread solitary nesting distribution and mass nesting (known as *arribazones* in Mexico and *arribadas* elsewhere) at over a dozen locales have supported extensive subsistence use and facilitated commercial harvests in much of its range. The nesting of Kemp's ridley (*Lepidochelys kempi*), historically at a single site in the western Gulf of Mexico, has likewise contributed to its vulnerability to overuse. As both olive ridley and Kemp's ridley emerge from the sea to lay eggs, land-based harvesting of adult females and their eggs occurs at nesting beaches. This chapter reviews the nature of exploitation at ridley nesting beaches and explores its impact on regional populations and, by extension, on the global survival of the species.

Nature of the Harvest of Adults and Eggs

The first encounters between humans and sea turtles most likely occurred at nesting beaches. As precivilization peoples expanded from inland to coastal environments, the discovery of relatively large, slow moving, egg-laying animals emerging from a mysterious and probably intimidating ocean must have been welcome. Collecting eggs and / or slaughtering an adult sea turtle were certainly easier than capturing fleet-footed, winged, or fierce land animals. Subsistence hunting of sea turtles and collection of eggs thus evolved at many places and has continued to the modern age (Frazier, 1980; Mosseri-Marlio, 1998; Bird et al., 2001). Eggs are nutritious, easy to transport, and often are in demand beyond the immediate vicinity of the harvest area. Over time, commercial markets developed in some locales, expanding on or supplanting the original subsistence use (Caldwell, 1963; Nietschmann, 1982). The arribada behavior of ridley turtles facilitated efficient extraction and conveyance of eggs to local, national, and international markets and contributed to industrial-scale harvest of eggs on certain nesting beaches.

It is generally assumed that slaughter of female ridleys on land historically was less common in rural coastal communities than egg harvesting. In some Asian and African societies, sea turtles are considered sacred, and their killing is forbidden (Hendrickson, 1958). A more secular view offers that it is foolish to kill the provider of highly prized eggs. In Central America, some coastal communities are composed of displaced farmers who had little tradition or experience in harvesting food from the sea (Cornelius, 1986) and thus were culturally unprepared to eat sea turtles, though they prized eggs. The direct take of adult turtles in fisheries at sea, particularly where they concentrate for mating, feeding, or nesting, generally has had a more severe impact on population survival than the killing of females on land and is discussed in Chapter 13. Nevertheless, an assessment of land-based harvests throughout the range of *Lepidochelys* makes it clear that unmanaged removal of nests and, to a lesser extent, the killing of adults are implicated in the decline of many ridley populations.

Western Atlantic Ocean

Both casual and organized harvests of adults and eggs of all nesting sea turtle species historically were widespread in the Guianas and northeastern Brazil. In some locations it has diminished recently as a result of the erosion of nesting beaches, disappearance of populations, improved protection efforts, and possibly increased awareness of the plight of sea turtles.

Historical accounts of olive ridley adult and egg harvests in Suriname come mostly from the former arribada nesting site at Eilanti Beach. According to Geijskes (1945, cited in Reichart and Fretey, 1993), about 1,500 nesting females were killed annually during most of the 1930s. The direct take of adults has apparently diminished over time, but egg harvesting was intense and reached nearly 100% in the late 1960s (Schulz, 1975). Although a federal law in 1970 officially banned the collection of olive ridley eggs, tension between the government and an Amerindian community over control of natural resources, including turtles (Kloos, 1971), resulted in uncontrolled egg harvests occurring in the late 1980s and early 1990s both at Eilanti Beach and elsewhere (Reichart, 1993; Reichart and Fretey, 1993). Despite the current ban on harvesting olive ridley eggs, more than 40% of olive ridley nests were taken during the peak season in 1995 (Hoekert et al., 1996). Very little information is available from French Guiana and Guyana, although it seems likely that at least some low-level egg harvesting occurred in the past (Pritchard, 1969). In Brazil, initial surveys of sea turtle nesting activity in the early 1980s revealed unorganized but widespread harvest of adults and eggs of all species nesting along the Sergipe coast (Marcovaldi and Marcovaldi, 1999).

The first evidence of land-based exploitation of Kemp's ridley was from an amateur film made in 1947 of dozens of villagers collecting large quantities of eggs on 2 km of beach near Rancho Nuevo, Tamaulipas, Mexico (Hildebrand, 1982; Márquez, 1994), which for years was the only known nesting beach for the species. Subsequent interviews suggested that a minimum of 80% of the nests, approximately 33,000, were transported by pack animal and vehicle to nearby population centers (Hildebrand, 1963). Although commercial fishing of green turtles *(Chelonia mydas)* was carried out extensively along the Texas and Louisiana coasts in the early 1900s and resulted in incidental catch of an occasional Kemp's ridley, there was never an organized fishery targeting the species anywhere in the Gulf of Mexico or on the nesting beach (Hildebrand, 1982). It is probable that adult ridleys were occasionally killed in small numbers at Rancho Nuevo, Mexico, for food or medicinal purposes (Hildebrand, 1963).

Eastern Atlantic Ocean

With some exceptions, nesting olive ridleys are systematically captured along the entire West African coastline and sold in local and regional markets. The meat is a highly desired food, the bones and fat are used in the traditional pharmacopoeia, and the carapaces are sold as ornaments and fetishes in tourist markets and serve local religious purposes. Olive ridleys are the target species in some locales, but more often eggs and adults are taken in association with harvests of more abundant sea turtle species.

A survey of 27 West African countries (including Macaronesia) indicated that females were killed on land in 14 of them (Fretey, 2001). Ridleys are part of a very large trafficking in sea turtle meat on the island of São Thomé, by far the most important market in Africa for the species. Females are seized on the beach, and adult males and females are captured at sea. They are kept alive for a few days until they are sold to itinerant buyers who transport them to urban markets. In Togo and Benin, the olive ridley is the most frequently killed turtle. In Cameroon and Angola, the carapaces are crafted into large masks lavishly decorated with bronze (Fretey, 1998). Though eggs are traditionally preferred over meat on Bioko Island in Equatorial Guinea (Tomas et al., 2001), ridleys are occasionally transported live to the capital, Malabo, and mixed with the more abundant green turtles to satisfy a demand for this specialty meat at Christmas. In the Ningo ethnic group east of Accra, Ghana, and in the Gulf of Benin, sea turtles are totemic and are not eaten, although nests are widely dug up by dogs and pigs. Paris and Agardy (1993) note that residents of the Bijagos Archipelago in Guinea-Bissau prize sea turtle eggs and meat and systematically search for nests and females on the beaches during the nesting season. Egg gathering is particularly widespread in Orango National Park, Guinea-Bissau, where theoretically they should be better protected than elsewhere in the country (Barbosa et al., 1998).

Eastern Pacific Ocean

The consumption of turtle eggs and adults turtles in the eastern Pacific predates the arrival of Europeans (Coe and Flannery, 1967; Nabhan, 2003). Márquez et al. (1990) note the importance turtles played in the culture and diet of Mexican indigenous groups, including the Seris of Sonora, the Pomaros in Michoacan, and the Huaves in Oaxaca. No turtle taboos are known from either native or Hispanic societies, and traditional demand in part is based on the consumption of eggs as a presumed aphrodisiac.

Before the 1960s, coastal residents from Jalisco, Mexico, to Panama harvested olive ridley eggs for subsistence and small-scale local or regional commerce only (Cliffton et al., 1982; Cornelius, 1986). In Honduras, eggs supplied markets in neighboring countries (Lagueux, 1991). In Mexico, eggs were dried, packed in sacks, and transported to regional markets as a substitute for hard-to-acquire meat (Márquez, 2000). Although historical data of egg harvests are few, there is no doubt that egg collection reached nearly 100% at many of the region's solitary ridley nesting beaches and continues at this level in many places today. Improved access to remote sites, an abundance of ridleys, a heightened demand from city dwellers, and stagnant or declining economic conditions, as well as political instability in rural coastal areas, have all contributed to increased turtle egg commerce, especially in Central America (Cornelius, 1982; Hásbun and Vásquez, 1992; Muccio, 1999).

At solitary nesting beaches in Panama, El Salvador, and Guatemala, egg collectors are paid from $2 to $10 for a clutch of 100 eggs (Arauz, 2000b). Very large arribadas of 100,000–200,000 nesting females at Ostional, Costa Rica, stimulated massive egg harvests since at least the early 1970s. From 1988 to 1997, 5.4–38.6% of the total estimated nests were transported from the beach (Ballestero et al., 2000), supplying a large portion of the national demand (Fig. 12.1). In 2001, the harvest generated over $45,000 income for the local community (Morera, 2001, cited in Araya et al., 2002). And although eggs continue to be a supplemental wild food for many residents of solitary and arribada nesting beaches in the eastern Pacific, they are now more impor-

Fig. 12.1. Members of the Integrated Development Association of Ostional, Costa Rica, collecting and packing olive ridley eggs during an arribada in 1998. Photo by Randall Arauz.

tant as a creator of income for needy residents than as a source of protein (Campbell, 1998).

An estimated 2,000,000 turtles, the great majority olive ridleys, were fished from Mexican waters between 1965 and 1970 (Cliffton et al., 1982; Márquez et al., 1990) to supply a rapidly expanding international trade in turtle skin and leather. Although most of the turtles were captured at sea, females often were slaughtered as they emerged to nest, their skin removed and carcasses initially discarded on the beach. Slaughterhouses later recognized the commercial value of eggs and meat and developed a capacity and market for these "byproducts" of the turtle fishery. At the same time, egg collection from Mexican arribada beaches increased steadily, eventually producing 150 metric tons of eggs annually, or the equivalent of approximately 45,000 nests. By 1966, excessive harvest had reduced the tonnage collected by 70%, and subsequently motivated the government to prohibit all egg trade. Olive ridleys were the major source of this egg harvest (Márquez, 1976).

Indian Ocean

Killing of solitary nesting adults and egg harvesting for subsistence or small-scale commerce has been widespread and significant in the Indian Ocean (Frazier, 1982a). Eggs are sought for their supposed aphrodisiac properties and as feed for livestock (Nareshwar, 1998; Qureshi, 2001a), skin for rough leather, calipee and blood for health remedies (Murthy and Menon, 1976; Murthy, 1981; Priyadarshini, 1998), and meat for use as fishing bait (Rajasekhar and Subba Rao, 1988).

In many parts of the Indian subcontinent and Southeast Asia, adult turtles are venerated as an incarnation of one of the gods of the Hindu trinity and hence not consumed (Rajagopalan, 1984). Muslims likewise generally do not eat turtles because Islamic custom considers turtle as *haram* or unclean (Shockley, 1949; Hatt, 1957; Groombridge and Luxmoore, 1989). Nevertheless, some Bangladeshi communities trade in turtle products (Islam, 2002b), and eggs are extensively collected in many Muslim countries in the Indian Ocean and Southeast Asia (Frazier, 1982a, 1982b; Limpus, 1993). Eating turtles and their eggs may increasingly be a matter of personal interpretation by Muslims (Clark and Khatib, 1993). Christian and tribal communities on the Indian subcontinent, Andaman and Nicobar Islands, and Southeast Asia are not averse to killing and eating turtles and consuming eggs. Theobald (1868) stated that the "Burmese do not greatly care for" *Lepidochelys* and in fact consider it inedible. In much of South India, turtles are considered bad omens and cannot be brought into the house. There is a saying in coastal Tamil Nadu (K. Shanker, personal communication) and Lakshadweep (Tripathy et al., 2002) that "the *aamai* [turtle] is closely followed by the *ameen* [tax collector]," a belief that, were it widely held, would no doubt work to sea turtles' advantage everywhere.

Harvesting turtle eggs for human and domestic animal consumption historically was widespread in the Indian Ocean (Murray, 1884; Maxwell, 1911; Burton, 1918; Deraniyagala, 1939; Hendrickson and Alfred, 1961) and continues today largely wherever ridleys nest (Valliapan and Whitaker, 1975; Siow and Moll, 1982; Kar and Bhaskar, 1982; Banerjee, 1985; Das, 1989; Choudhury, 2001; Islam, 2002a). Barring a few protected areas such as Point Calimere Wildlife Sanctuary and the Gulf of Mannar Biosphere Reserve, more than 90% of the turtle nests in Tamil Nadu, India, are eaten by humans and animals (Bhupathy and Saravanan, 2002). A similar situation occurs in Madras, India, except where local conservationists relocate eggs to a protected area or hatchery (Shanker, 2003). Rajasekhar and Subba Rao (1988) report that most nests are dug up by humans (55%) and animals (45%) in Andhra Pradesh, India, and Tripathy et al. (2003) report that almost all solitary nests are destroyed there, mostly by animals. Dattatri and Samarajiva (1982) reported that olive ridley eggs are exploited throughout Sri Lanka where nesting occurs, and it has been estimated that some 20,000 eggs were consumed annually (Hoffman, 1975, cited in Wickremasinghe, 1981). Wild pigs, feral dogs, and humans take most nests in the Andaman Islands (Andrews et al., 2001).

A major olive ridley commercial egg harvest occurred in coastal Myanmar at the turn of the last century (Maxwell, 1911). Nesting beaches were leased to local businessmen, who collected and sold 1,500,000 eggs annually. Beach leasing to the highest bidder continued until 1986, and virtually all eggs were collected. Thereafter, and until 1996 when all nesting beaches were protected, local residents collected eggs and were required to leave one third to hatch, although this regulation was rarely complied with.

Another important commercial harvest of sea turtle eggs has occurred since colonial days at the arribada beach at Gahirmatha, Orissa, India (Kar, 2001). After independence, licensed collectors sold eggs in poor riverside communities (Kar, 1988, 2001), where they were consumed immediately, preserved by sun drying, or transported in large quantities to markets in Calcutta (Kar, 2001). An estimated 1,500,000 eggs were legally harvested in 1973 (FAO, 1974), although the illegal take was probably much more (Kar, 1988, 2001). In 1974, 800,000 eggs were collected (FAO, 1974), after which all licensed harvesting was halted (Kar, 2001).

Turtles are fished and caught opportunistically on beaches in parts of the region. Bhaskar (1979) reported that nesting turtles were killed in the Nicobar and Andaman Islands (Fig. 12.2). Turtle meat is still consumed by settlers and aboriginal tribes, who are exempt from laws protecting turtles. In Great Nicobar Island, most turtles, many of which are ridleys, are caught as they emerge to nest near beach villages (Sivakumar, 2002). Salm (1976) estimated that about 50,000 people depended on sea turtles for subsistence in northern Sri Lanka. It is estimated that 1,500 turtles were caught annually in Jaffna, and about 2,000–3,000 on the entire island (Frazier, 1980). Many of these animals were green turtles; however, olive ridleys also were captured. Killing of nesting females has increased along these beaches; today not a single nesting female is spared along the western coast of Sri Lanka (Hewavisenthi, 1990). For the last 10 years, residents at Kandakuliya have butchered at least 20 turtles weekly (over 1,000 annually), although the majority of these turtles likely were caught at sea (Anonymous, 1999).

The largest direct fishery of olive ridleys in the Indian Ocean occurred during the 1970s in Baluchistan, Pakistan, and eastern India for the leather and skin trade (Frazier, 1980; Qureshi, 2001b). Das (1985) reported that before 1981, 80,000 ridleys were transported to Howrah, Orissa per season. Many accounts report an annual catch of 50,000 ridleys or more from the Orissa and West Bengal coast until the 1980s (Silas et al., 1983a, 1983b; Kar and Dash, 1984). In addition, more than 90,000 dead ridleys have stranded on the Orissa coast since 1994, representing only a part of the actual mortality. Here

Fig. 12.2. Nesting ridleys have been heavily harvested on the eastern coast of India and Sri Lanka in the past. They are still taken in the Andaman and Nicobar islands, where aboriginal people are exempt from India's wildlife protection laws. Photo by Kartik Shanker.

and elsewhere, the lack of precise data makes it difficult to distinguish between the relative contribution of at-sea and land-based harvests. In a rare study, Rajasekhar and Subba Rao (1988) reported that of 800 ridleys commercially caught in Andrah Pradesh, India, 648 were captured during mating (offshore), and 170 while nesting (1 turtle in 4 taken on the beach). Thus, it seems probable that a minor portion of the tens of thousands of ridleys transported from Gahirmatha, Orissa, were slaughtered on the beaches.

Impact of Land-Based Harvest on Global Populations

Many ridley populations have experienced drastic declines in the past 30 years, although the precise reasons are diverse and not always clear. In some cases, declines may be blamed in part on naturally occurring phenomena. For example, the coastline in Suriname and French Guiana is characterized by constant erosion and accretion of sand caused by oceanic currents (Schulz, 1975). Thus, a perceived decrease in one nesting population may be the result of females simply shifting among different beaches depending on accessibility and suitability of beach habitat (Schulz, 1971).

On certain arribada beaches, nest survival is often markedly low, so low that the value of these beaches for sustained recruitment to populations has been questioned (Robinson, 1983, cited in Cornelius et al., 1991). The high egg mortality is attributed to very dense nesting during arribadas, where the turtles themselves physically disturb from 20% to 50% of the deposited egg clutches. If a subsequent arribada occurs before completion of the incubation period, this results in disturbance of previously deposited nests and further mortality. Furthermore, contamination of the nest may occur when it is disturbed, resulting in the proliferation of fungi and bacteria or the depletion of oxygen, which would prevent embryo development (Cornelius et al., 1991; Valverde et al., 1998). At the completely protected Nancite Beach in Santa Rosa National Park, Costa Rica, low hatchling recruitment over several decades may explain a population decline (Valverde et al., 1998), although female mortality away from the nesting beach possibly contributed. In general, however, the downward trend at many solitary and some arribada beaches can be attributed to uncontrolled egg harvests and/or the slaughter of females on land.

Western Atlantic Ocean

Overall, the harvest of olive ridley adults and eggs likely affected the size of the major nesting populations in the Western Atlantic. In Suriname, the annual killing of thousands of females in the 1930s and the near complete collection of eggs in different years have contributed to an overall decline the past 65–70 years. The low numbers of nesting ridleys observed in Suriname in the last decade have prompted suggestions that the species is on the verge of extinction because of egg poaching and accidental captures at sea (Hoekert et al., 1996). However, this is probably an exaggeration, as greater monitoring effort in recent years has revealed several hundred nests in French Guiana (Talvy and Vie, 2000; Talvy, 2001). The lack of historical information in French Guiana makes it difficult to analyze population trends there (Marcovaldi, 2001). In Guyana, olive ridleys at Shell Beach and elsewhere in the country are nearly extirpated following decades of unrestricted slaughter of nesting females (Schulz, 1982). In Sergipe, Brazil, low-level egg harvest may have perpetuated the small size of the population until recent times, when better protection was instituted (Marcovaldi, 2001).

Other factors are likely to have been important in shaping trends in the size of the nesting aggregations, including incidental capture in off-shore fisheries (Tambiah, 1994; Guéguen, 2000). It is interesting to note that although the shrimp fisheries in the region were developed in the 1960s, it appears that trawling only became intense in the 1970s (Dintheer et al., 1989), by which time the decline in olive ridley nests at several beaches, notably in Guyana and Suriname, was already evident. In any case, the total size of these nesting populations is relatively small and warrants further protection from threats including egg harvests and fisheries-related mortality.

Hildebrand's (1963) discovery of the filmed evidence of a large arribada of an estimated 40,000+ Kemp's ridleys at Rancho Nuevo has served as an undeniable benchmark for documenting the species' plummet. Within 20 years, this species' single nesting aggregation had diminished to 5,000 females (Hildebrand, 1982; Márquez et al., 2001), and between 1978 and 1991 only 200 emerged to nest each season (INP-FWS unpublished data in U.S. Fish and Wildlife Service and National Marine Fisheries Service, 1992). The astonishing decline in the number of nesting females is largely attributed to two concurrent factors. First, the highly organized commercial egg harvest that operated unfettered through 1966, when the area was given official protection, drastically dampened recruitment. The harvest was so widespread and intensive that Hildebrand (1963) predicted the eventual demise of the Rancho Nuevo population. He offered anecdotal information to suggest that similar arribada beaches on the Tamaulipas coast already had disappeared. Second, the species' survival prospects were significantly worsened by the expansion of shrimp trawling in the western Gulf of Mexico in the late 1940s and 1950s, which incidentally captured adult and subadult ridleys (Magnuson et al., 1990; U.S. Fish and Wildlife Service and National Marine Fisheries Service, 1992).

Eastern Atlantic Ocean

Even though it appears clear that egg harvest and slaughter of nesting olive ridleys are widespread in West Africa, the lack of basic data on these activities and the status and trend of nesting aggregations at specific beaches all make it difficult to evaluate the relative impact of land-based harvest. Compounding the assessment often is the inability to distinguish between turtles captured in offshore fisheries and those taken on the beach. For example, Bijagos Archipelago residents attribute a decline in nesting adults to industrial trawlers and not to their own egg-harvesting activities (Paris and Agardy, 1993). Until systematic monitoring of egg harvest and female slaughter is established at the principal olive ridley nesting beaches, it is not possible to evaluate the evident decline in these populations. Given that villagers on the entire West African coastline methodically capture nearly all nesting females, except where research stations are established, their impact is certainly significant and perhaps enormous (J. Fretey, personal communication).

Eastern Pacific Ocean

The high rate of egg harvest, harvest in turtle fisheries, and incidental capture in other fisheries during the past four decades have greatly affected olive ridleys in the eastern Pacific Ocean. In fact, probably no other region has seen a greater overall decline in total number of nesting olive ridleys than the coastline from Central Mexico to Panama.

Egg use in Mexico was largely clandestine, unrestrained, and subsistence-based except between 1964 and 1967, when large-scale commercial harvesting was permitted. The overall effect of historic subsistence egg use on recruitment to the population was probably minimal, however. Before 1950, the number of olive ridleys nesting in Mexico was conservatively estimated to be 10,000,000 females (Cliffton et al., 1982). If this number is accurate, the nesting populations were likely large enough to withstand the relatively modest subsistence use by local communities. Beginning in the mid-1960s, egg harvests, land-based slaughter of nesting adults, and at-sea fisheries that targeted olive ridleys intensified in Mexico. The concurrence of these stressors, together with an expanding fishery for olive ridleys in Ecuador during the 1970s (Green and Ortiz-Crespo, 1982), devastated major Mexican nesting populations at Piedra de Tlacoyunque, Guerrero; Chacahua, Oaxaca; and Mismaloya, Jalisco, and caused the near collapse of the La Escobilla, Oaxaca arribada nesting population. Additionally, the number of nesting turtles at over 15 other important arribada and solitary nesting beaches in Mexico were severely depleted for the same reasons (Peñaflores et al., 2001). Today, La Escobilla Beach remains the sole surviving original olive ridley arribada nesting population in Mexico. The others of 40 years ago, thus far, have not responded to increased protection on the beach and in nearshore waters (Márquez et al., 1998).

Central America's large numbers of olive ridleys never were fished intentionally in the vicinity of their nesting beaches, nor were they affected by subsistence egg harvests (Cornelius, 1982). The first report of commercial egg harvesting leading to disappearance of an arribada occurred in the early 1970s at Masachapa and Pochomil, two sites in Nicaragua that currently host only solitary nesting ridleys (Nietschmann, 1975). By the early 1980s, Central American coastal residents and government officials acknowledged that olive ridleys were declining, with the exception of Ostional Beach. This decline was attributed to widespread egg harvesting augmented by the incidental capture of turtles in shrimp trawls and by the commercial turtle fishery in Ecuador (Cornelius, 1982).

Though direct take of ridleys in Ecuador and Mexico turtle fisheries ended by 1990 (Valverde et al., 1998), incidental capture in Central American shrimp trawls continues to be high because of the reluctance of operators to install turtle excluder devices that reduce mortality of captured turtles (Arauz, 1995, 2000b). Additionally, egg loss now approaches 100% at most unmanaged or unprotected beaches in Central America (Lagueux, 1991; Hasbún and Vásquez, 1992). In Guatemala, solitary nesting turtles are estimated to have declined 34% between 1981 and 1997 (Muccio, 1999). Coastal residents in El Salvador are convinced that sea turtle populations are steadily declining (Arauz, 2000a), even with nest protection programs in place. There is some evidence that arribadas at Isla Cañas, Panama are decreasing in number (J. A. Cordoba, personal communication).

Indian Ocean

The number of nesting olive ridleys has declined throughout the Indian Ocean, with the significance of land-based harvests varying among sites and with population size. These harvests are implicated in declines at solitary nesting beaches but are of secondary importance at the major arribada beach in Orissa, India, which was affected by an olive ridley fishery in the 1970s and more recently by incidental capture in shrimp trawls. Even though subsistence hunting remains widespread in many parts of the Indian Ocean and is responsible for some population declines (Frazier, 1982a), market-driven harvesting of eggs or females on nesting beaches is of greatest conservation concern now.

The consumption of sea turtle meat and eggs has actually lessened in many parts of the region in recent years as a result of either the parallel decline of the nesting populations or the implementation of wildlife protection laws. Declines of solitary nesting olive ridleys have been recorded in Bangladesh (Islam, 2002a), Myanmar (Thorbjarnarson et al., 2000), Malaysia (Limpus, 1995), and Pakistan (Asrar, 1999). In many of these cases, excessive harvest of eggs is suspected to be the cause of the decline. Thorbjarnarson et al. (2000) blamed long-term noncompliance with egg harvest regulations for Myanmar's ridley decline. Nesting appeared to have stabilized between 1996 and 2001 on the Bangladeshi island of St. Martin's (Islam, 2002a). However, reports that 35 ridleys nested in a single night at St. Martin's during the late 1980s (S. M. A. Rashid, unpublished data) indicate far greater nesting densities in the past, which is consistent with the high level of egg harvest reported on this island and in the region (Islam et al., 1999; Islam, 2002a, 2002b). The olive ridley population at Terengganu, Malaysia, declined precipitously from thousands of annual nests to just a few dozen. This decrease and the decimation of Thailand's Andaman Sea populations are thought to have resulted from long-term overharvest of eggs (Limpus, 1995).

In several locales, both excessive egg harvest and killing of nesting adults are suspected. Surveys of the large diffuse solitary nesting populations on the Madras coast in northern Tamil Nadu in 1973 and 1974 reported high egg predation and large-scale killing of nesting females (Valliapan and Whitaker, 1975; Whitaker, 1977). Comparison of nest densities in the region then and more recently (Shanker, 1995) show no clear evidence of a decrease. Bhupathy and Saravanan (2002), however, report that questionnaire-based interviews suggested that the nesting population was declining over the last 10 years. In the early 1980s several authors reported that the most common species in northern Sri Lanka was the olive ridley (Wickremasinghe, 1979, 1981; Dattatri and Samarajiva, 1982). Within 20 years, however, surveys indicated that the green turtle has become the most abundant sea turtle (Anonymous, 1999). Although there is indirect evidence of a significant decline in Sri Lanka's olive ridleys, it is difficult to attribute this to land-based harvest only because of the long history of at-sea fishing of turtles (Frazier, 1982a).

Although most estimates of the number of turtles nesting during arribadas at Gahirmatha, Orissa, India are unreliable, Pandav and Choudhury (1999) concluded that the nesting population has definitely declined. The failure of arribadas to appear in three of the past five years and a decrease in the body size of nesting females between 1996 and 2002 suggest a potential or imminent decline (Shanker et al., 2003). However, the relative importance of land-based mortality in Orissa is secondary to the offshore capture of adult ridleys in the fishery of the 1970s and current mortality as bycatch in shrimp trawls and finfish gill nets (Pandav, 2000; Wright and Mohanty, 2002). Although there are few records of olive ridley nesting in the western Indian Ocean (Frazier 1982b; Ross and Barwani, 1982), most accounts suggest that the populations there are declining as well. In all likelihood, this is a result of heavy egg collection, the purposeful capture of adults, and bycatch in other fisheries.

Other Land-Based Stressors

Sand mining, aquaculture development, beach armoring, and fishing harbor and tourism facilities collectively contribute to habitat loss, light pollution, and overall heightened human contact with turtles. These development activities have increased in many parts of the olive ridley's range and today pose threats to major nesting sites along the east coast of India (Pandav and Choudhury, 1999) and at certain beaches in Central America (Cornelius, 1982). However, there is no evidence that these actions have contributed significantly to observed population declines. As existing coastal villages grow and new population centers are established for tourism and general habitation, the significance of these land-based threats to ridleys, and indeed to all sea turtles, will certainly rise.

Stable or Increasing Populations

Some ridley populations appear to be stable or may even be increasing, although the causal factors are not well understood. Information on status and trends is often anecdotal and difficult to evaluate. In addition, techniques to estimate population sizes vary, are inconsistently applied sometimes, and thus compromise statistically valid analyses.

It is difficult to ascertain what the size of the nesting population at Sergipe, Brazil, was before 1980. The fact that olive ridleys did not have a local common name when the original surveys were conducted (Marcovaldi, 2001) suggests that they were never very abundant. Low-level egg harvests in the 1980s may have constrained the growth of this population, and strict nest protection afforded for the past 20 years is now translating into an increase in Sergipe's nesting population size (Marcovaldi, 2001; da Silva et al., 2003).

Certain aspects of the arribadas in Ostional, Costa Rica, suggest that this olive ridley nesting population, perhaps the largest in the world, is increasing in size. For example, the frequency of occurrence and duration of arribadas have increased, and the amount of beach used for nesting also has expanded (Chaves, 2002, 2004). However, the size of the nesting population remains uncertain. An 11-year review of the impact of egg harvesting concluded that weaknesses in the quadrant population census method made it difficult to assess the degree of the upward trend (Ballestero et al., 2000). A transect census method (Gates et al., 1996), in use since 1999, has resulted in a dramatic increase in the estimated nesting population size, often surpassing 500,000 turtles in a single arribada (Chaves, 2002, 2004). Clusella et al. (2000) found that both quadrant and transect methods overestimated the total number of nesting turtles counted during an arribada when compared to direct counts.

A spectacular case of how heightened protection of eggs and females at the nesting beach can result in relatively rapid recovery of a nesting population is illustrated by La Escobilla, Oaxaca, Mexico. In the five years leading up to 1969, over 3,100,000 adult male and female olive ridleys were killed on or near arribada nesting beaches in Mexico (Cliffton et al., 1982). By the middle of the 1970s, a lethal combination of intensive egg poaching, slaughter of nesting females, and nearshore turtle fisheries pushed the largest olive ridley nesting population in Mexico toward imminent collapse (Cahill, 1978). The establishment of research camps and hatcheries in 1968 and subsequent enhancement of beach protection, assisted by a complete halt in the nearshore turtle fishery in 1990, are believed to have stimulated recovery of the nesting population, perhaps exceeding preharvest levels (Márquez et al., 1996b, 1998; Márquez, 2001; Peñaflores et al., 2001). Olive ridley reproduction at La Escobilla rebounded from approximately 50,000 nests in 1988 to over 700,000 nests in 1994 (Márquez et al., 1996b) and to more than a million nests by 2000 (Márquez et al., 2002). Because ridleys may mature in less than 15 years (Chaloupka and Zug, 1997; Schmid and Witzell, 1997; Snover, 2002), the initial increase in the nesting population that occurred within 6–7 years probably was a response to the closure of the nearshore fishery. A combination of egg protection and the restriction on harvesting adult turtles at sea is contributing to the continued population recovery. In any case, this beach provides evidence that depleted nesting populations have the potential to recover rapidly if timely protection is afforded on land and in nearshore waters and if juveniles and adults are not overexploited elsewhere.

An innovative and potent recovery plan for the Kemp's ridley, now in its third decade of implementation, is showing promising results (Shaver and Wibbels, Chapter 14). The nesting population at Rancho Nuevo, Mexico, has increased steadily from a low point of 200 turtles in the 1980s to approximately 2,000 turtles in the 2000 season (Márquez et al., 2001). A single arribada of slightly more than 1,000 turtles was reported in 2001 (P. Burchfield, personal communication), and an estimated 3,600 turtles produced over 9,000 nests in 2003 (C. Caillouet, personal communication). Although the trend is encouraging, the species continues to be, as Carr (1967) noted, "by all odds the most precariously ensconced marine turtle in the world."

Egg Harvests and Hatchery Production

Most countries regulate the collection and commerce of ridley eggs, with total or temporary bans being the most popular measures. However, conservation programs that combine the relocation of natural nests to protective enclosures, or egg hatcheries, with controlled commercial harvests have been attempted in many areas. Results of these approaches have been mixed.

Controlled Egg Harvests

Some Central American countries permit a legal commerce in eggs but proclaim closures at specific nesting beaches during certain times. Unfortunately, enforcement of regulations is uneven and, in many places, totally ignored. Human and economic resources for effective enforcement are scarce or nonexistent. The more difficult problem is that the practice of taking eggs is deeply rooted in rural societies, and restrictions are broadly unpopular in light of the extreme poverty of coastal communities. As a result, olive ridleys continue to retain great value and commercial acceptance and are an object of clandestine egg and meat harvests. A poacher's relatively small effort results in a high profit margin, and despite the risk, turtle capture and egg collecting represent economically attractive endeavors. At many beaches it will be impossible to completely eliminate harmful poaching of eggs and turtles until basic human needs are met in nearby poor coastal communities, effective campaigns are instituted to support alternative fisheries or other economic activities, and strict enforcement of existing conservation laws occurs.

Nicaragua prohibits egg harvesting from October 1 to January 31 and year round in protected areas (Valle, 1997). However, enforcement of this closed period is poor, and very few eggs are left to incubate anywhere in the country (Camacho and Cáceres, 1995). To address this, managed use and monitoring have been fostered at Isla Juan Venado, Chacocente, and La Flor Wildlife Refuges. Coastal communities are granted permits to take eggs that are at high risk of being destroyed by nesting turtles, beach erosion, predation, or illegal harvesting. Under this scheme, residents harvested over 600,000 eggs annually between 1993 and 1999. Another 74,000 nests were protected each year (in situ at Chacocente and La Flor; in hatcheries at Isla Juan Venado) and produced over 2,000,000 hatchlings. However, egg-harvesting quotas seem to be based more on the demands of the surrounding coastal communities than on the conservation needs of the turtles, which results in chaotic, illegal egg commerce (Hope, 2000; Smith, 2002). At La Flor Refuge, enforcement of harvesting controls has improved since 1997 with the involvement of a nongovernmental organization. Nevertheless, after more than a decade of protection and active management of these arribada beaches, there has been no increase in the nesting population.

In a few arribada populations, particularly at Ostional, Costa Rica, there is evidence that controlled harvest of eggs provides local benefits without adversely affecting recruitment (Cornelius et al., 1991). Unfortunately, the distribution of legally harvested eggs is determined almost entirely by demand in the capital city, San José, and not at markets near other nesting beaches. Thus, the potential conservation value of swamping the entire country with relatively inexpensive eggs and reducing poaching pressure at hard-to-protect solitary nesting beaches has not been realized (Drake, 1996).

Hatcheries

The various egg-harvesting schemes all anticipate an increase in overall hatchling production from practically nothing to levels that will sustain natural recruitment and perpetuate economic and social values to coastal communities. Although no one has yet determined what that level needs to be, protecting a significant portion of eggs is critical. Where respect for conservation regulations and adequate enforcement exist, this may be accomplished by protecting eggs in situ. Where these conditions are not present, or if nonhuman predation is significant, conservationists often relocate eggs to fenced enclosures to provide protection. Hatcheries that contribute significantly to the maintenance or recovery of depleted populations normally have local residents who understand the importance of harvest regulations, assist in the collection of eggs for the hatcheries, and participate in other conservation programs. Where these conditions are absent, hatcheries seem less effective.

A hatchery and egg donation system has been used widely in Guatemala and El Salvador since the late 1970s. Coastal communities, academic institutions, government entities, and nongovernmental organizations collaborate in the operation of hatcheries to protect some of the eggs. Participating egg collectors donate a portion of each clutch harvested to their local hatchery as a condition for permission to sell eggs. A total of 188,699 sea turtle eggs (92–95% corresponded to olive ridleys) were protected in Guatemalan hatcheries during 1999–2001, producing 159,843 hatchlings (Jolón et al., 2002). This represents the recruitment of nearly 1,600 clutches that likely would have been lost otherwise. Decades of operating hatcheries at Punta Raton, Honduras, have resulted in no observed population increase. However, these efforts may have at least maintained a nesting population that suffered near 100% loss of eggs for 40 years (Lagueux, 1991). The establishment and maintenance of hatcheries during the past 30 years in Madras (Shanker, 2003) and Gujarat (Sunderraj et al., 2002), India, may have helped sustain olive ridley populations that were under severe human and feral animal predation. A hatchery program in Jalisco, Mexico, apparently has produced positive results (Garcia et al., 2003).

In other instances, hatcheries have not been instrumental either in maintaining populations or in contributing to their recovery. Many solitary nesting beaches do not have local hatcheries, and resident egg collectors do not contribute significantly to the protection of eggs (Muccio, 1999; Arauz, 2000a). Muccio (1999) estimated that only 2% of the clutches laid in Guatemala were protected in hatcheries and that, so far, no increase in the nesting population has been observed. At a small arribada beach at Isla Cañas, Panama, only a few clutches are relocated to a hatchery each night, leaving the vast majority for commercial harvest. Many hatcheries set up in Sri Lanka for tourism have been poorly managed with low hatching success of translocated eggs, suggesting that these hatcheries actually may be detrimental (Amarasooriya, 1996; Richardson, 1996; Hewavisenthi, 2001). In Hawkes Bay, Pakistan, dramatic declines have occurred in the olive ridley nesting population despite the existence of a hatchery program (Asrar, 1999). In summary, even when hatcheries are well managed, they rarely afford protection to a significant portion of the total eggs deposited.

An exception to the presumed small contribution of hatcheries to total nesting success of ridleys has been at Rancho Nuevo, Mexico. All Kemp's ridley nests found there are transported to protective hatcheries at this beach. In the past, eggs were also relocated to a hatchery in North Padre Island, Texas (Márquez, 1994). The slow but steady increase in the nesting population at Rancho Nuevo (Márquez et al., 1996a; Heppell, 1997) and the increase in number of females nesting in Texas have been aided by egg protection in hatcheries (Shaver and Caillouet, 1998). It has been suggested that the recovery of the La Escobilla, Mexico, olive ridley nesting population is in great measure a result of beach protection efforts, including hatcheries that began in 1967 (Peñaflores et al., 2001). The enforced ban on turtle fishing instituted in 1990 may have been as significant, however (A. Abreu, personal communication). Declines in ridley nesting populations in Madras, India, may have been arrested by local conservation programs, where eggs have been collected by conservation volunteers and incubated in hatcheries since 1974 (Shanker, 2003). It is worthwhile noting that well-run hatchery programs that incorporate public outreach and environmental education activities indirectly benefit sea turtle populations by building a constituency for their conservation, even if the hatcheries themselves have no or insignificant recruitment value (Juarez and Muccio, 1997; Shanker and Pilcher, 2003).

Conclusion

Limpus (1995) noted that the olive ridley's exceptionally large arribadas, as well as cosmopolitan range, might have bestowed a false sense of security on its conservation status. Unfortunately, it is now clear that neither its wide distribution nor its great abundance has buffered it from the same perils faced by other sea turtle species. Neither can one be sanguine about the future of Kemp's ridley despite recent evidence that it is beginning to recover.

With few exceptions, intensive harvesting of ridley eggs has resulted in population declines (Table 12.1). This is most evident with Kemp's ridley in Mexico, the olive ridley arribada beaches in Suriname, Nicaragua, and possibly Panama, as well as many solitary nesting beaches in Central America, Myanmar, Malaysia, and elsewhere in the Indian Ocean. Lack of information makes it difficult to assess cause and effect in West African nesting beaches. Egg harvests certainly contributed to the collapse of the olive ridley arribada populations on the Pacific coast of Mexico and to the current downward trend in Orissa, India, although direct take and incidental capture in shrimp trawls and gill nets are the principal causes.

Assessing the impact of land-based harvests on the olive ridley's global status is complicated by individual population anomalies. In the early 1970s, the two massive arribada beaches in Costa Rica, Nancite and Ostional, were estimated to host 200,000 and 100,000 nesting females, respectively (Hughes and Richard, 1974). These were likely underestimates, as subsequent work estimated Nancite's total nests in 1980 in excess of 300,000 (Cornelius et al., 1991), and even larger arribadas were estimated at Ostional (Cornelius et al., 1991; Ballestero et al., 2000). Absolute protection of Nancite Beach in Costa Rica for nearly 35 years has done nothing to forestall a major population decline over this period (Valverde et al., 1998). In contrast, Ostional Beach has been subjected to constant subsistence use and huge commercial harvests of eggs for perhaps 50 years. Yet today the nesting population at Ostional Beach seems stable and may be increasing. In Mexico, three major arribada beaches and many smaller ones were devastated by less than a decade of intense commercial adult and egg harvests, yet 15 years after a comprehensive halt in fishing of ridleys in the eastern Pacific, and 13 years after strict protection was instituted on the nesting beaches, only La Escobilla has shown a measurable, indeed a very remarkable, recovery.

Today, the number of female olive and Kemp's ridleys captured on land is far surpassed by that lost incidentally in bottom trawling, gill nets, long lines, and other fishing operations. This appears to be the case especially in the Gulf of Guinea and West Africa, the Guianas, the Indian subcontinent, Central America, and Mexico for the olive ridley and the western Gulf of Mexico for Kemp's ridley and is likely the primary force reducing nesting populations or obstructing their global recovery.

Many countries have designed sea turtle conservation programs that attempt to respond to cultural and ethnic connections to turtles, status of local populations, and capacity for enforcement. Absolute protection is normally prompted when populations are devastated, as in the case of the Kemp's ridley in Mexico. Recent research and protection projects in West and Central Africa are making progress and will, we hope, lead to a significant reduction in land-based harvests of the olive ridley there, although much better monitoring of population trends is needed. Sustained utilization models based on collaboration among coastal communities, academic institutions, nongovernmental organizations, and regulatory authorities have been attempted in less critical situations, largely in the eastern Pacific. Certain populations have shown a resilience and ability to rebound from long-term and intensive overharvests, perhaps in part because recovery efforts have been timely. Escobilla's recovery from near collapse may portend well for some depleted nesting beaches currently under active protection and management, particularly in the West Atlantic and Indian Ocean for olive ridleys and the Gulf of Mexico for Kemp's ridley. The managed egg harvest at Ostional, which thus far has resulted in no apparent adverse impact on population recruitment, suggests that certain very large arribadas may be capable of providing significant economic benefits through the controlled use of eggs while building a conservation constituency in coastal communities. This may be especially relevant for the future management of beaches at Escobilla, Mexico, and Gahirmatha, India.

Unfortunately, most experiences at management of solitary nesting beaches and small arribada beaches have yet to prove the legitimacy

Table 12.1 Relative extent and trend of land-based and other stressors and collective impact on selected *Lepidochelys* populations

Population	Subsistence egg harvest	Commerical egg harvest	Land-based subsistence adult killing	Land-based commercial adult killing	Other population stressors and trend	Collective impact on population	Reference
Lepidochelys kempii							
Rancho Nuevo, Mexico	High ⇒ zero	High ⇒ zero	Low ⇒ zero		Bycatch	Severe ⇒ Slow recovery	Hildebrand, 1963; Magnuson et al., 1990
Lepidochelys olivacea							
Shell Beach, Guyana	High ⇒ zero	High ⇒ zero				Extirpated	Schulz, 1982
Eilanti, Suriname	High ⇒ low	High ⇒ low		High ⇒ zero	Bycatch	Severe decline	Hoekert et al., 1996; Tambiah, 1994
Sergipe, Brazil	Mod ⇒ low	Mod ⇒ low	Mod ⇒ low			Steady increase	Da Silva et al., 2003
Escobilla, Mexico	Mod ⇒ low	High ⇒ zero	Low ⇒ zero	High ⇒ zero	Past fishery	Severe decline Steady recovery	Peñaflores et al., 2001
Mismaloya, Piedra de Tlacoyunque, Chacahua, Mexico	Mod ⇒ low	High ⇒ zero	Low ⇒ zero	High ⇒ zero	Past fishery	Severe decline	Peñaflores et al., 2001
Bijagos Archipelago, Guinea-Bissau	? ⇒ high		? ⇒ high		Bycatch	In decline	This chapter
Nancite, Costa Rica	Low ⇒ zero				Bycatch Hyperdense nesting	Severe decline	Valverde et al., 1998
Ostional, Costa Rica	High ⇒ high	High ⇒ high			Bycatch	Stable or increasing	This chapter
Masachapa/Pochomil, Nicaragua	High ⇒ low	High ⇒ zero				Extirpated	Nietschmann, 1975
Orissa, India	Mod ⇒ low	High ⇒ zero	Low ⇒ zero	Low ⇒ zero	Past fishery; bycatch	In decline	Shanker et al., 2003
Baluchistan, Pakistan				Mod ⇒ zero	Past fishery	Severe decline	Asrar, 1999; Qureshi, 2001b
Myanmar	High ⇒ low	High ⇒ zero				Extirpated	Thorbjarnarson et al., 2000
Terengganu, Malaysia	High ⇒ low	High ⇒ zero				Extirpated	Limpus, 1995
St. Martin, Bangladesh	? ⇒ high	? ⇒ high				Severe decline	Islam, 2002a

of the nexus of conservation and sustainable use. The speed at which these programs have been instituted largely has not been matched by technical skills to monitor the population or political and enforcement capacity to control their use. Evidence suggests that not enough eggs are protected to allow the recruitment of a sufficient number of hatchlings to sustain regional populations, especially in light of an ever-increasing human population and its associated pressures, increased demand for eggs, alteration of nesting habitat by beach mining and construction, and the paramount issue of needless capture and death of juveniles and adults in other fishing operations. The last has replaced land-based harvest as the principal stressor of ridley populations globally. Clearly there is much to be learned about the parameters contributing to the decline and occasional recovery of regional populations of ridley turtles.

LITERATURE CITED

Amarasooriya, D. 1996. Some observations of marine turtle hatcheries in Sri Lanka. In da Silva, A. (Ed.). *Proceedings of the International Conference on the Biology and Conservation of the Amphibians and Reptiles of South Asia.* Kandy, Sri Lanka: Amphibia and Reptile Research Organization of Sri Lanka, pp. 19–20.

Andrews, H. V., Krishnan, S., and Biswas, P. 2001. The status and distribution of marine turtles around the Andaman and Nicobar archipelago. Government of India/UN Development Program Sea Turtle Project Report. Tamil Nadu, India: Madras Crocodile Bank Trust.

Anonymous. 1999. Unpublished beach survey report to IUCN. Turtle Conservation Programme, Sri Lanka.

Arauz, R. 1995. A description of the Central American shrimp fisheries with estimates of incidental capture and mortality of sea turtles. In Keinath, J., and Keinath, D. (Eds.). *Proceedings of the Fifteenth Annual Symposium on Sea Turtle Biology and Conservation.* NOAA Technical Memorandum NMFS-SEFSC-387, pp. 5–9.

Arauz, R. 2000a. *Diagnóstico de la Situación Actual de las Tortugas Marinas en El Salvador.* Comité Nacional para la Conservación de la Tortuga Marina en El Salvador. Estrategia Nacional Para la Conservación de las Tortugas Marinas de El Salvador. San Salvador: Comisión Centroamericana de Ambiente y Desarrollo.

Arauz, R. 2000b. Equitable fisheries and conservation of marine resources: An evaluation of Public Law 101-162, Section 609. Central American Report. Washington, DC: Smithsonian Institution Research and Conservation Center.

Araya, A., Rojas, M., and Villalobos, J. 2002. La explotación de los huevos de tortuga en la región de Guanacaste, Refugio Nacional de Vida Silvestre Ostional. Universidad Nacional, Escuela de Ciencias Biológicas, Facultad de Ciencias Exactas y Naturales.

Asrar, F. F. 1999. Decline of marine turtle nesting populations in Pakistan. *Marine Turtle Newsletter* 83:13–14.

Ballestero, J., Arauz, R. M., and Rojas, R. 2000. Management, conservation, and sustained use of olive ridley sea turtle eggs (*Lepidochelys olivacea*) in the Ostional Wildlife Refuge, Costa Rica: an eleven year review. In Abreu-Grobois, F. A., Briseño Dueñas, R., Márquez-Milán, R., and Sarti-Martínez, A. L. (Eds.). *Proceedings of the Eighteenth International Sea Turtle Symposium.* NOAA Technical Memorandum NMFS-SEFSC-436, pp. 4–5.

Banerjee, R. 1985. The marine turtle *Lepidochlys olivacea* Eschscholtz, its occurrence and captive rearing in Sunderbans. Paper 26. Presented at the Symposium on Endangered Marine Animals and Marine Parks, Cochin.

Barbosa, C., Broderick, A., and Catry, P. 1998. Marine turtles in the Orango National Park (Bijago Archipelago, Guinea-Bissau). *Marine Turtle Newsletter* 81:6–7.

Bhaskar, S. 1979. Sea turtle survey in the Andaman and Nicobars. *Hamadryad* 4(3):2–19.

Bhupathy, S., and Saravanan, S. 2002. Status survey of sea turtles along the Tamil Nadu coast. A GOI-UNDP Sea Turtle Project Report. Coimbatore, India: Salim Ali Centre for Ornithology and Natural History.

Bird, R. B., Smith, E. A., and Bird, D. W. 2001. The hunting handicap: costly signaling in human foraging strategies. *Behavioral Ecology and Sociobiology* 50:9–19.

Burton, R. W. 1918. Habits of the green turtle (*Chelonia mydas*). *Journal of the Bombay Natural History Society* 25(3):508.

Cahill, T. 1978. The shame of Escobilla. *Outside (San Francisco and New York)* February:22–27, 62–64.

Caldwell, D. K. 1963. The sea turtle fishery of Baja

California, Mexico. *California Fish and Game* 49:140–151.

Camacho, M. G., and Cáceres, J. G. 1995. Importancia de la Protección y Conservación de las Tortugas Marinas en Nicaragua y el Aprovechamiento de Huevos en el RVS Chacocente. In *Proceedings 1er Encuentro Ciclo de Conferencias Biológicas: Un Día con la Tortuga Marina. Parque Nacional Volcán Masaya, Septiembre, 1994,* pp. 40–65.

Campbell, L. M. 1998. Use them or lose them? Conservation and the consumptive use of marine turtles at Ostional, Costa Rica. *Environmental Conservation* 25:305–319.

Carr, A. 1967. *The Sea Turtle—So Excellent a Fishe.* Garden City, NY: Natural History Press.

Chaloupka, M., and Zug, G. R. 1997. A polyphasic growth function for the endangered Kemp's ridley sea turtle, *Lepidochelys kempii. Fishery Bulletin* 95:849–856.

Chaves, G. A. 2002. Plan de aprovechamiento para la utilizacion racional, manejo y conservacion de los huevos de la tortuga marina lora (*Lepidochelys olivacea*) en el Refugio de Vida Silvestre de Ostional, Santa Cruz, Guanacaste, Costa Rica. Ministerio de Ambiente y Energia, Area Conservacion Tempisque.

Chaves, G. A. 2004. Nesting activity of sea turtles in Ostional Beach, Costa Rica: 30 years of research. In Coyne, M. S., and Clark, R. D. (Compilers). *Proceedings of the Twenty-First Annual Symposium on Sea Turtle Biology and Conservation.* NOAA Technical Memorandum NMFS-SEFSC-528, pp. 144–145.

Choudhury, B. 2001. A short survey on sea turtles in West Bengal. A GOI-UNDP Sea Turtle Project Report. Calcutta. India: Nature, Environment and Wildlife Society.

Clark, F., and Khatib, A. A. 1993. *Sea Turtles in Zanzibar: a Preliminary Study.* Zanzibar Environmental Study Series No. 15a, 1993. The Commission for Lands and Environment Zanzibar.

Cliffton, K., Cornejo, D. O., and Felger, R. S. 1982. Sea turtles of the Pacific Coast of Mexico. In Bjorndal, K. (Ed.). *Biology and Conservation of Sea Turtles.* Washington, DC: Smithsonian Institution Press, pp. 199–209.

Clusella, S., Saenz, J., and Fernandez, M. 2000. Comparison of three methods for estimating the size of olive ridley *(Lepidochelys olivacea)* arribadas at Nancite Beach, Santa Rosa National Park, Costa Rica. In Abreu-Grobois, F. A., Briseño-Dueñas, R., Márquez, R., Sarti, L. (Compilers). *Proceedings of the Eighteenth International Sea Turtle Symposium.* NOAA Technical Memorandum NMFS-SEFSC-436, pp. 58–59.

Coe, M. D., and Flannery, K. V. 1967. Early cultures and human ecology in south coastal Guatemala. *Smithsonian Contributions to Anthropology* 3.

Cornelius, S. 1982. Status of sea turtles along the Pacific coast of middle America. In Bjorndal, K. (Ed.). *Biology and Conservation of Sea Turtles.* Washington, DC: Smithsonian Institution Press, pp. 211–219.

Cornelius, S. 1986. *The Sea Turtles of Santa Rosa National Park.* Fundación de Parques Nacionales, Costa Rica. Programa Educación Ambiental UNED.

Cornelius, S. E., Alvarado-Ulloa, M., Castro, J. C., Mata del Valle, M., and Robinson, D. C. 1991. Management of olive ridley sea turtles (*Lepidochelys olivacea*) nesting at playas Nancite and Ostional, Costa Rica. In Robinson, J. R., and Redford, K. H. (Eds.). *Neotropical Wildlife Use and Conservation.* Chicago: The University of Chicago Press, pp. 111–135.

Das, I. 1985. Marine Turtle Drain. *Hamadryad* 10:17.

Das, I. 1989. Sea turtles and coastal habitats in southeastern Bangladesh. Project report to the Sea turtle Rescue Fund/Center for Marine Conservation, Washington DC.

da Silva, A. C., de Castilhos, J. C., Rocha, D. A. S., Oliveira, F. L. O., Weber, M. I., and Barata, P. C. R. 2003. Nesting biology and conservation of the olive ridley sea turtle (*Lepidochelys olivacea*) in the state of Sergipe, Brazil. In Seminoff, J. (Ed.). *Proceedings of the Twenty-Second Annual Symposium on Sea Turtle Biology and Conservation.* NOAA Technical Memorandum NMFS-SEFSC-503, p. 89.

Dattatri, S., and Samarajiva, D. 1982. The status and conservation of sea turtles in Sri Lanka. A project of the Sea Turtle Rescue Fund, Center for Environmental Education, Washington, DC.

Deraniyagala, P. E. P. 1939. *The Tetrapod Reptiles of Ceylon,* Volume 1. Colombo: Ceylon Government Press.

Dintheer, C., Gilly, B. J. Y., Le Gall, M., Lemoine, M., and Rosé, J. 1989. La recherche et la gestion de la pêcherie de crevettes pénéides en Guyane française de 1958 à 1988: trente années de surf. *Equinoxe* 28:21–33.

Drake, D. L. 1996. Marine turtle nesting, nest predation, hatch frequency, and nesting seasonality on the Osa Peninsula, Costa Rica. *Chelonian Conservation and Biology* 2:89–92.

FAO. 1974. *India: A Preliminary Survey of the Prospects for Crocodile Farming (Based on the Works of Dr. H. R. Bustard).* Rome: FAO, pp. 1–50.

Frazier, J. G. 1980. Exploitation of marine turtles in the Indian Ocean. *Human Ecology* 8:329–370.

Frazier, J. G. 1982a. Subsistence hunting in the Indian Ocean. In Bjorndal, K. (Ed.). *Biology and Conservation of Sea Turtles.* Washington, DC: Smithsonian Institution Press, pp. 391–396.

Frazier, J. G. 1982b. Status of sea turtles in the Central Western Indian Ocean. In Bjorndal, K. (Ed.). *Biology and Conservation of Sea Turtles.* Washington, DC: Smithsonian Institution Press, pp. 385–389.

Fretey, J. 1998. Les carapaces-masques du Royaume des Bamum (ouest Cameroun). *La Tortue* 43: 7–10.

Fretey, J. 2001. *Biogeography and Conservation of Marine Turtles of the Atlantic Coast of Africa/ Biogéographie et conservation des tortues marines de la côte atlantique de l'Afrique.* CMS Technical Series Publication 6, UNEP/CMS Secretariat, Bonn, Germany.

Garcia, A., Ceballos, G., and Adaya, R. 2003. Intensive beach management as an improved sea turtle conservation strategy in Mexico. *Biological Conservation* 111:253–261.

Gates, C. E., Valverde, R. A., and Mo, C. L. 1996. Estimating arribada size using a modified instantaneous count procedure. *Journal of Agricultural, Biological, and Environmental Statistics* 1:275–287.

Green, D., and Ortiz-Crespo, F. 1982. Status of sea turtle populations in the Central Eastern Pacific. In Bjorndal, K. (Ed.). *Biology and Conservation of Sea Turtles.* Washington, DC: Smithsonian Institution Press, pp. 221–233.

Groombridge, B., and Luxmoore, R. 1989. Costa Rica. In Groombridge, B., and Luxmoore, R. (Eds.). *The Green Turtle and Hawksbill (Reptilia Cheloniidae): World Status, Exploitation and Trade.* Lausanne: CITES (Convention on International Trade in Endangered Species).

Guéguen, F. 2000. Captures accidentelles de tortues marines par la flotille crevettière de Guyane Française. *Bulletin Societé Herpetologie Française* 93:27–93.

Hasbún, C. R., and Vásquez, M. 1992. *Plan de Acción Para el Inventario y Diagnóstico de las Playas de Arribo de la Tortuga Marina y de la Población que Explota Dicho Recurso.* San Salvador: Asociación Ambientalista Amigos del Arbol (AMAR).

Hatt, R. T. 1957. Turtling at Hawks Bay, a beach on the Arabian Sea. *Newsletter Cranbrook Institute of Science* 26(5):53–58.

Hendrickson, J. R. 1958. The green sea turtle *Chelonia mydas* (Linn.) in Malaya and Sarawak. *Proceedings of the Zoological Society of London* 130: 455–535.

Hendrickson, J. R., and Alfred, E. R. 1961. Nesting populations of sea turtles on the east coast of Malaya. *Bulletin of the Raffles Museum Singapore* 26:59–69.

Heppell. S. 1997. On the importance of eggs. *Marine Turtle Newsletter* 76:6–8.

Hewavisenthi, S. 1990. Exploitation of marine turtles in Sri Lanka: historic background and the present status. *Marine Turtle Newsletter* 48:14–19.

Hewavisenthi, S. 2001. Sri Lanka's hatcheries: boon or ban. *Marine Turtle Newsletter* 60:19–22.

Hildebrand, H. H. 1963. Hallazgo del área de anidación de la tortuga lora *Lepidochelys kempii* (Garman), en la costa occidental de Golfo de México (Rept., Chel.). *Ciencia, México* 22: 105–112.

Hildebrand, H. H. 1982. A historical review of the status of sea turtle populations in the Western Gulf of Mexico. In Bjorndal, K. (Ed.). *Biology and Conservation of Sea Turtles.* Washington, DC: Smithsonian Institution Press, pp. 447–453.

Hoekert, W. E. J., Schouten, A. D., van Tienen, L. H. G., and Weijerman, M. 1996. Is the Surinam olive ridley on the eve of extinction? First census data for olive ridleys, green turtles, and leatherbacks since 1989. *Marine Turtle Newsletter* 75:1–4.

Hope, R. 2000. Egg harvesting of the olive ridley marine turtle (*Lepidochelys olivacea*) along the Pacific coast of Nicaragua and Costa Rica: an arribada sustainability analysis. Masters thesis, University of Manchester.

Hughes, D. A., and Richard, J. D. 1974. Nesting of the Pacific ridley *Lepidochelys olivacea* on Playa Nancite, Costa Rica. *Marine Biology* 24: 97–107.

Islam, M. Z. 2002a. Marine turtle nesting at St. Martin's Island, Bangladesh. *Marine Turtle Newsletter* 96:19–21.

Islam, M. Z. 2002b. Threats to sea turtles in St. Martin's Island, Bangladesh. *Kachhapa* 6:6–10.

Islam, M. Z., Islam, S. Z., and Rashid, S. M. A. 1999. Sea turtle conservation program in St. Martin's Island by CARINAM: a brief review. *Tigerpaper* 26:17–22.

Jolón, M., Sánchez, R., España, P., Andrade, H., Carballo R., and Ruíz, R. 2002. Informe Nacional de Acciones de Protección y Conservación de Tortugas Marinas en Guatemala, 1999–2002. Unidad Nacional de Manejo de la Pesca y Acuicultura (UNIPESCA), Consejo Nacional de Areas Protegidas (CONAP) y Escuela de Biología, Universidad de San Carlos.

Juarez, R., and Muccio, C. 1997. Sea turtle conservation in Guatemala. *Marine Turtle Newsletter* 77:15–17.

Kar, C. S. 1988. Ecological studies on the Pacific Ridley Sea Turtle, *Lepidochelys olivacea (Eschscholtz, 1829)* in the Orissa Coast. Ph.D. diss. Sambalpur University, Sambalpur.

Kar, C. S. 2001. Review of threats to sea turtles in Orissa. In Shanker, K., and Choudhury, B. C. (Eds.). *Proceedings of the National Workshop for the Development of a National Sea Turtle Conservation Action Plan, Bhubaneshwar, Orissa.* Dehradun, India: Wildlife Institute of India, pp. 15–19.

Kar, C. S., and Bhaskar, S. 1982. The status of sea turtles in the eastern Indian Ocean. In Bjorndal, K. (Ed.). *The Biology and Conservation of Sea Turtles.* Washington, DC: Smithsonian Institution Press, pp. 365–372.

Kar, C. S., and Dash, M. C. 1984. Conservation and status of sea turtles in Orissa. *CMFRI Special Publication 18,* pp. 93–107.

Kloos, P. 1971. *The Maroini River Caribs of Surinam.* Assen: K. Van Gorcum.

Lagueux, C. 1991. Economic analysis of sea turtle eggs in a coastal community on the Pacific coast of Honduras. In Robinson, J. G., and Redford, K. H. (Eds.). *Neotropical Wildlife Use and Conservation.* Chicago: The University of Chicago Press, pp. 136–144.

Limpus, C. J. 1993. The worldwide status of marine turtle conservation. In Nacu, A., Trono, R., Palma, J. A., Torres, D., and Agas, F. (Eds.). *Proceedings of the First ASEAN Symposium-Workshop on Marine Turtle Conservation, Manila, Philippines,* pp. 43–61.

Limpus, C. J. 1995. Global overview of the status of marine turtles: a 1995 viewpoint. In Bjorndal, K. (Ed.). *Biology and Conservation of Sea Turtles, Revised Edition.* Washington, DC: Smithsonian Institution Press.

Magnuson, J. J., Bjorndal, K. A., DuPaul, W. D., Graham, G. L., Owens, D. W., Pritchard, P. C. H., Richardson, J. I., Saul, G. E., and West, C. W. 1990. *Decline of the Sea Turtles, Causes and Prevention.* Washington, DC: National Academy Press.

Marcovaldi, M. Â. 2001. Status and distribution of the olive ridley turtle, *Lepidochelys olivacea,* in the western Atlantic Ocean. In Eckert, K. L., and Albreu-Grobois, F. A. (Eds.). *Proceedings of the Regional Meeting "Marine Turtle Conservation in the Wider Caribbean Region: a Dialogue for Effective Regional Management." Santo Domingo, 16–18 November 1999.*: WIDECAST, IUCN-MTSG, WWF, and UNEP-CEP, pp. 52–56.

Marcovaldi, M. Â., and Marcovaldi, G. G. 1999. Marine turtles of Brazil: the history and structure of Projeto TAMAR-IBAMA. *Biological Conservation* 91:35–41.

Márquez-M., R. 1976. Estado actual de la pesquería de Tortugas Marinas en México, 1974. Instituto Nacional de Pesca, México. Serie Inf. INP/SI, 46:27.

Márquez-M., R. 1994. *Synopsis of biological data on the Kemp's ridley sea turtle,* Lepidochelys kempi *(Garman, 1880).* NOAA Technical Memorandum, NMFS-SEFSC-343, p. 91.

Márquez-M., R. 2000. *Las Tortugas Marinas y nuestro tiempo. La Ciencia para todos 144.* 2nd ed. México: Fondo de Cultura Económica.

Márquez-M., R. 2001. *Las tortugas marinas en México. estrategias y resultados—Tortuga Lora.* Cd. Victoria, Tamaulipas: INP-PNITM.

Márquez-M., R., Vasconcelos, J., and Peñaflores, C. 1990. *XXV Años de Investigacion, conservacion y proteccion de la tortuga marina.* Manzanillo: Secretaria de Pesca, Instituto Nacional de la Pesca.

Márquez-M., R., Byles, R. A., Burchfield, P., Sanchez, M., Diaz, J., Carrazco, M. A., Leo, A. S., and Jiménez, C. 1996a. Good news! Rising numbers of Kemp's ridleys nest at Rancho Nuevo, Tamaulipas, Mexico. *Marine Turtle Newsletter* 73:2–5.

Márquez-M., R., Peñaflores, C., Vasconcelos, J., and Albavera, E. 1996b. Olive ridley turtles (*Lepidochelys olivacea*) show signs of recovery at La Escobilla, Oaxaca. *Marine Turtle Newsletter* 73: 5–7.

Márquez-M., R., Jiménez, M. C., Carrasco, M. A., and Villanueva, N. A. 1998. Comentarios acerca de las tendencias poblacionales de las tortugas marinas del género *Lepidochelys* después de la veda total de 1990. *Oceánides* 13:41–62.

Márquez-M., R., Burchfield, P., Carrazco, M. A., Jiménez, C. Diaz, J., Garduño, M., Leo, A., Peña, J., Bravo, R., and Gonzalez, E. 2001. Update on the Kemp's ridley turtle nesting in Mexico. *Marine Turtle Newsletter* 92:2–4.

Márquez-M., R., Carrasco, M. A., and Jiménez, M. C. 2002. The marine turtles of Mexico: An update. In Kinan, I. (Ed.). *Proceedings of the Western Pacific Sea Turtle Cooperative Research and Management Workshop, Hawaii, Feb. 5–8, 2002.* App. IV. pp. 281–285. Honolulu: Western Pacific Regional Fishery Management Council.

Maxwell, F. D. 1911. Government report on the turtle banks of the Irrawaddy Division Rangoon. Cited in Smith, M. A. 1931. *Reptilia and Amphibia,* Volume 1, in Stephenson, J. (Ed.). *The*

Fauna of British India, including Ceylon and Burma. London: Taylor and Francis.

Mosseri-Marlio, C. 1998. Marine turtle exploitation in Bronze Age Oman. *Marine Turtle Newsletter* 81:7–9.

Muccio, C. 1999. *Hacia Una Estrategia Nacional Para la Recuperación de Tortugas Marinas en Guatemala.* Guatemala: Asociación Rescate y Conservación de Vida Silvestre.

Murray, J. A. 1884. *The Vertebrate Zoology of Sind.* London: Richardson and Company.

Murthy, T. S. N. 1981. Turtles: their natural history, economic importance and conservation. *Zoologiana* 4:57–65.

Murthy, T. S. N., and Menon, G. K. 1976. The turtle resources of India. *Seafood Export Journal* 8:1–12.

Nabhan, G. P. 2003. Singing the turtles to sea: the Comcáac (Seri) art and science of reptiles. Berkeley and Los Angeles: University of California Press, p. 317.

Nareshwar, E. K. 1998. Evaluation of sea turtle nesting beaches for promoting participatory conservation at Sundarvan Beyt Dwarka, India. *Hamadryad* 22:121–122.

Nietschmann, B. 1975. Of turtles, arribadas, and people. *Chelonia* 2:6–9.

Nietschmann, B. 1982. The cultural context of sea turtle subsistence hunting in the Caribbean and problems caused by commercial exploitation. In Bjorndal, K. (Ed.). *Biology and Conservation of Sea Turtles.* Washington, DC: Smithsonian Institution Press, pp. 439–445.

Pandav, B. 2000. Conservation and management of olive ridley sea turtles on the Orissa coast. Ph.D. diss. Utkal University, Bhubaneshwar, India.

Pandav, B., and Choudhury, B. C. 1999. An update on the mortality of the olive ridley sea turtles in Orissa, India. *Marine Turtle Newsletter* 83: 10–12.

Paris, B., and Agardy, T. 1993. La tortue verte et la tortue olive de ridley de l'Archipel des Bijagos: Identification de leur importance dans le contexte mondial et contribution à la proposition de zonage d'une réserve de la biosphère.

Peñaflores, C., Vasconcelos, J., Albavera, E., and Jiménez, M. C. 2001. Especies sujetas a proteccion especial. Tortuga golfina. In Cisneros, M. A., Belendez, L. F., Zarate, E., Gaspar, M. T., Lopez, L. C., Saucedo, C., and Tovar, J. (Eds.). *Sustentabilidad y Pesca Responsible en Mexico. Evaluacion y Manejo. 1999–2000.* Publicado en CD. México: Instituto Nacional de la Pesca / SEMARNAT, pp. 1001–1021.

Pritchard, P. C. H. 1969. Sea turtles of the Guianas. *Bulletin of the Florida State Museum* 13:85–140.

Priyadarshini, K. V. R. 1998. Status, ecology and management of olive ridley sea turtles and their nesting habitats along north coastal Andhra Pradesh. Annual Report (January 1997 to June 1998). New Delhi: WWF-India.

Qureshi, T. 2001a. Information sheet on Ramsar Wetlands: Jiwani Coastal Wetlands, Pakistan. Unpublished report. Karachi: IUCN Pakistan.

Qureshi, T. 2001b. Information sheet on Ramsar Wetlands: Ormara Turtle Beaches, Pakistan. Unpublished Report. Karachi: IUCN Pakistan.

Rajagopalan, M. 1984. Value of sea turtles to India. In Silas, E. G. (Ed.). *Proceedings of the Workshop on Sea Turtle Conservation,* Special Publication 18. Cochin, India: Central Marine Fisheries Research Institute, pp. 49–58.

Rajasekhar, P. S., and Subba Rao, M. V. 1988. Conservation and management of the endangered olive ridley sea turtle *Lepidochelys olivacea* (Eschscholtz) along the northern Andhra Pradesh coastline, India. *B.C.G. Testudo* 3(5): 35–53.

Reichart, H. A. 1993. *Synopsis of Biological Data on the Olive Ridley Sea Turtle* Lepidochelys olivacea *(Eschscholtz 1829) in the Western Atlantic.* NOAA Technical Memorandum NMFS-SEFSC-336.

Reichart, H. A., and Fretey, J. 1993. *WIDECAST Sea Turtle Recovery Action Plan for Suriname* In Eckert, K. L. (Ed.). CEP Technical Report No. 2. Kingston, Jamaica: UNEP Caribbean Environment Programme.

Richardson, P. 1996. The marine turtle hatcheries of Sri Lanka: A turtle conservation programme review and assessment of current hatchery practices and recommendations for their improvements. Unpublished TCP report submitted for DWLC and NARA, Colombo, Sri Lanka.

Ross, J. P., and Barwani, M. A. 1982. Review of sea turtles in the Arabian area. In Bjorndal, K. (Ed.). *The Biology and Conservation of Sea Turtles.* Washington, DC: Smithsonian Institution Press, pp. 365–372.

Salm, R. V. 1976. Critical marine turtle habitats of the Northern Indian Ocean. Contract report to the IUCN, Morges, Switzerland.

Schmid, J. R., and Witzell, W. N. 1997. Age and growth of wild Kemp's ridley turtles (*Lepidochelys kempi*): cumulative results from tagging studies in Florida. *Chelonian Conservation and Biology* 2:532–537.

Schulz, J. P. 1971. Nesting beaches of sea turtles in west French Guiana. *Koninklijke Nederlandse Akademie van Wetenschappen* 74:396–404.

Schulz, J. P. 1975. Sea turtles nesting in Surinam. *Zoologische Verhandelingen* 143:1–143.

Schulz, J. P. 1982. Status of sea turtle populations nesting in Surinam with notes on sea turtles nesting in Guyana and French Guiana. In Bjorndal, K. (Ed.). *The Biology and Conservation of Sea Turtles.* Washington, DC: Smithsonian Institution Press, pp. 435–437.

Shanker, K. 1995. Conservation of sea turtles on the Madras coast. *Marine Turtle Newsletter* 64:3–6.

Shanker, K. 2003. Thirty years of sea turtle conservation on the Madras coast: a review. *Kachhapa* 8:16–19.

Shanker, K., and Pilcher, N. 2003. Marine turtle conservation in south and southeast Asia: Hopeless cause or cause for hope? *Marine Turtle Newsletter* 100:43–51.

Shanker, K., Pandav, B., and Choudhury, B. C. 2003. An assessment of olive ridley (*Lepidochelys olivacea*) nesting population in Orissa on the east coast of India. *Biological Conservation* 115: 149–160.

Shaver, D. J., and Caillouet, C. W. 1998. More Kemp's ridley turtles return to South Padre Island to nest. *Marine Turtle Newsletter* 82:1–5.

Shockley, C. H. 1949. Herpetological notes for Ras Jiunfri, Baluchistan. *Herpetologica* 5(6):121–123.

Silas, E. G., Rajagopalan, M., Bastion Fernando, A., and Dan, S. S. 1983a. Marine turtle conservation and management. A survey of the situation in Orissa during 1981–82 and 1982–83. *Marine Fisheries Information Service Technical and Extension Service* 50:13–23.

Silas, E. G., Rajagopalan, M., and Dan, S. S. 1983b. Marine turtle conservation and management: a survey of the situation in West Bengal 1981/82 and 1982/83. *Marine Fisheries Information Service Technical and Extension Service* 50:24–33.

Siow, K. T., and Moll, O. M. 1982. Status and conservation of estuarine and sea turtles in west Malaysia. In Bjorndal, K. A. (Ed.). *Biology and Conservation of Sea Turtles.* Washington, DC: Smithsonian Institution Press, pp. 339–347.

Sivakumar, K. 2002. Sea turtles nesting in the South Bay of Great Nicobar Island. *Marine Turtle Newsletter* 96:17–18.

Smith, P. 2002. Turtle Egg Trafficking. Sister City Update. *Richland Center-Santa Teresa Sister City Project Newsletter.* Richland Center, WI. Issue 17.

Snover, M. L. 2002. Growth and ontogeny of sea turtles using skeletochronology: methods, validation and application to conservation. Ph.D. diss. Duke University, Durham, NC.

Sunderraj, W. S. F., Vijay Kumar, V., Joshua, J., Serebiah, S., Patel, I. L., and Saravana Kumar, A. 2002. *Status of the Breeding Population of Sea Turtles along the Gujarat Coast.* A GOI-UNDP Sea Turtle Project Report. Gujarat Institute of Desert Ecology, Bhuj, India.

Talvy, G. 2001. Marine turtle program Kwata. In Schouten, A. K., Mohadin, K., Adhin, S., and McClintock, E. (Eds.). *Proceedings of the V Regional Marine Turtle Symposium for the Guianas, 25–27 September 2001, Paramaribo, Suriname.* WWF Technical Report No. GFECP-9, pp. 12–14.

Talvy, G., and Vie, J. C. 2000. Evaluation of the importance of French Guiana eastern beaches as nesting sites for turtles. In Kelle, L., Lochon, S., Thérèse, J., and Desbois, X. (Eds.). *Proceedings of the Third Meeting on the Sea Turtles of the Guianas, 15–16 July 1999, Mana, French Guiana,* pp. 26–29.

Tambiah, C. R. 1994. Saving sea turtles or killing them: the case of U.S. regulated TEDs in Guyana and Suriname. In Bjorndal, K. A., Bolten, A. B., Johnson, D. A., and Eliazar, P. J. (Eds.). *Proceedings of the Fourteenth Annual Symposium on Sea Turtle Biology and Conservation.* NOAA Technical Memorandum NMFS-SEFSC-351, pp. 149–151.

Theobald, W. 1868. 1867 Catalogue of the Reptiles of British Birma, embracing the provinces of Pegu, Martaban, and Tenasserim: with descriptions of new or little-known species. *Journal of the Linnean Society—Zoology* 10:4–20.

Thorbjarnarson, J. B., Platt, S. G., and Khaing, S. T. 2000. Sea turtles in Myanmar: past and present. *Marine Turtle Newsletter* 88:10–11.

Tomas, J., Fretey, J., Ragr, J., and Castroviejo, J. 2001. Tortues marines de la façade Atlantique de l'Afrique. Genre *Lepidochelys.* Quelques données concernant la présence de *L. olivacea* (Eschscholtz, 1929) dans l'île de Bioko (Guinée Equatoriale). *Bulletin de la Société Herpétologique de France* 98:31–42.

Tripathy, B., Choudhury, B. C., and Shanker, K. 2002. *A Survey of Marine Turtles and Their Nesting Habitats in the Lakshadweep Islands, India.* A GOI UNDP Sea Turtle Project. Wildlife Insititute of India, Dehradun, India.

Tripathy, B., Shanker, K., and Choudhury, B. C. 2003. Important nesting habitats of olive ridley turtles (*Lepidochelys olivacea*) and their nesting habitats along the Andhra Pradesh of eastern India. *Oryx* 37(4):454–463.

U.S. Fish and Wildlife Service and National Marine Fisheries Service. 1992. *Recovery Plan for the Kemp's Ridley Sea Turtle* Lepidochelys kempii. St. Petersburg, FL: National Marine Fisheries Service.

Valle, E. 1997. Marco Jurídico de Protección a las Tortugas Marinas. In *Memoria del 1 Taller Problemática de las Tortugas Marinas en Nicaragua, 18–19 Setiembre, El Crucero, Managua, Nicaragua.*

Valliapan, S., and Whitaker, R. 1975. Olive ridleys on the Coromandel Coast of India. *Herpetological Review* 6:42–43.

Valverde, R. A., Cornelius, S. E., and Mo, C. L. 1998. Decline of the olive ridley sea turtle (*Lepidochelys olivacea*) nesting assemblage at Nancite Beach, Santa Rosa National Park, Costa Rica. *Chelonian Conservation and Biology* 3(1):58–63.

Whitaker, R. 1977. A note on the sea turtles of Madras. *Indian Forester* 103:733–734.

Wickremasinghe, S. 1979. Turtle tales. *Loris* 15: 78–86.

Wickremasinghe, S. 1981. Turtles and their conservation. *Loris* 15:313–315.

Wright, B., and Mohanty, B. 2002. Olive ridley mortality in gill nets in Orissa. *Kachhapa* 6:18.

JACK FRAZIER
RANDALL ARAUZ
JOHAN CHEVALIER
ANGELA FORMIA
JACQUES FRETEY
MATTHEW H. GODFREY
RENE MÁRQUEZ-M.
BIVASH PANDAV
KARTIK SHANKER

13

Human–Turtle Interactions at Sea

THE TWO LIVING SPECIES OF ridley sea turtles have very different geographic distributions with virtually no sympatry. Kemp's ridley (*Lepidochelys kempii*) inhabits the Gulf of Mexico and the North Atlantic, primarily on the western side (Márquez-M., 1990, 1994a, 1994b; Weber, 1995; Frazier, 2000), but it also occurs on the eastern or European side (Brongersma, 1972; USFWS-NMFS, 1992). The olive ridley (*L. olivacea*) is found regularly in the Indian and Pacific Oceans (Márquez-M., 1990; Pritchard and Plotkin, 1995) and also in the South Atlantic (Reichart, 1993; Fretey, 2001). Sympatry could occur in the Caribbean Sea as well as along the interface between the North and South Atlantic basins, particularly at the western extreme of Africa, but although there are scattered records of *L. olivacea* from the northwestern coasts of Africa and the North Atlantic, the species is evidently not common in these waters (Fretey, 2001; Foley et al., 2003), and records of *L. kempii* from outside the North Atlantic are questionable (e.g., Fretey, 2001). With the exception of a few very rare records, neither species of *Lepidochelys* is found in the Mediterranean.

In addition to the differences in geographic distributions, the two species also seem to be separated by differences in habitat preference, at least once the pelagic hatchling phase has been completed. *L. kempii* normally stays near to the coast or at least over the continental shelf (Byles, 1988), whereas at least in the eastern tropical Pacific, *L. olivacea* routinely occurs in open ocean or high seas (Pitman, 1990, 1992; Plotkin et al., 1993). Moreover, *L. kempii,* considered a distinct species, is the rarest of all sea turtles, whereas *L. olivacea* is regarded as the most abundant (Márquez-M., 1990; Pritchard, 1997). As a result, there are considerable differences between these two species in the types and frequencies of human–turtle interactions that occur at sea.

Unfortunately, many treatises of marine turtle biology and conservation combine discussions of different species, leaving the reader with the impression that there are no substantive differences among species' biology, ecology, conservation status, conservation threats, and conservation actions. Although there are many similarities between the two species of *Lepidochelys,* this chapter separates the discussions by species in an effort to emphasize that not only the species but the different populations need to be evaluated separately when one is considering both biological and conservation issues.

In addition, it is not uncommon for marine turtles to be discussed in terms of continental distributions: for example, "marine turtles in Africa" or "marine turtles in South America." However, these are marine animals, so meaningful geographic groupings and categories need to be based on ocean basins, not on continental land masses (see, for example, Frazier, 1998). As a result, some parts of the discussions herein may be repetitive, as certain situations in some ocean basins are repeated in others.

Furthermore, distinctions between human–turtle interactions on land and human–turtle interactions at sea are often arbitrary and ambiguous: some human activities on land have profound effects on turtles at sea, and vice versa. Hence, certain parts of the discussion in the present chapter touch on, or even repeat, discussions in Chapter 12 of this volume.

Finally, a number of terms are commonly used when referring to human–turtle interactions, and although many of these have become conventional or even trendy and politically correct, they often carry implicit values through positive or negative connotations that are unjustified. For example, there have been recent critiques of fashionable expressions such as "sustainable use" referring to the exploitation of living marine resources (Jackson, 2001), a concept paramount for the dogma of "sustainable development" (Frazier, 1997). Likewise, the term "harvest" is also commonly used in referring to the exploitation of living marine resources, but it would normally be more appropriate to refer to "extraction" or "mining" rather than "harvesting." On the other hand, descriptions of "poaching" often connote socially negative actions and attitudes that are inappropriate, or at least unsubstantiated. As a result, this chapter attempts to use terms that are as neutral as possible, such as "exploitation," "interactions," and "take," expressions that are meant to be without ethical or social values and neutral in connotation.

When evaluating older records of ridley turtles, it is important to bear in mind the fact that for many years there was considerable confusion about the identity of these species. Despite the careful studies of P. E. P. Deraniyagala (1939), ridleys (*Lepidochelys* spp.) and loggerheads (*Caretta caretta*) not only were confused but were often thought to be the same species. There were also commonly held opinions that *L. kempii,* known as the "bastard" turtle, was a hybrid between loggerhead and green turtles (Carr, 1955). These misunderstandings led to repeated problems of misidentification of specimens, with consequent confusion and errors in understanding geographic ranges and other aspects of their biology (Frazier, 1985).

Prehistoric information on human–turtle interactions involving *Lepidochelys* spp. is scanty. There are archaeological records of *L. kempii* from relatively few sites, most of which are in Florida, with rarely more than one or two individuals per site (Frazier, 2003, 2005a). *L. olivacea* seems to have been reported from just one archaeological site: Anuradhapura, Sri Lanka (800–250 BC) (Chandraratne, 1997).

Assuming that synchronous, massed nesting (known by a variety of names in Latin America including *arribazón, arribada,* and *morrinda*) occurred in prehistoric times, as happens today for both species, one would expect the dense concentrations of nesting turtles and their eggs to have been heavily exploited during prehistoric times. However, the limited data do not indicate any significant prehistoric interaction between humans and ridley turtles, unlike the case with *Chelonia mydas* and *Eretmochelys imbricata* (Frazier, 2003, 2004, 2005a). This lack of information, however, should be interpreted with care:

no evidence does not justify the conclusion that the phenomenon did not exist.

Finally, there is a general lack of detailed, systematic information on contemporary human–turtle interactions, and routinely one is obliged to make use of out-of-date, incomplete, sketchy, anecdotal, and/or hearsay accounts. It is essentially a matter of personal choice whether one errs on the side of emphasizing that the lack of information is best interpreted as no evidence for human–turtle interactions or takes the alternate route and underscores that what little is known, in the face of major shortcomings in the data, indicates that we can see but the tip of the iceberg. Not infrequently, those trained in occidental, reductionist, or positivist science choose the first option. The second approach is in accordance with the precautionary principle, now an integral part of fisheries agreements such as the FAO Code of Conduct for Responsible Fisheries, that stipulates that "the absence of adequate scientific information should not be used as a reason for postponing or failing to take measures to conserve target species, associated or dependent species, and nontarget species and their environment" (Article 6.5; www.fao.org/fi/agreem/codecond/ficonde.asp).

Lepidochelys kempii: Kemp's Ridley Turtle

Human–turtle interactions with *L. kempii* are limited by the relatively restricted geographic distribution of this species when compared with *L. olivacea*. The majority of nesting occurs in the State of Tamaulipas, Mexico, particularly at Rancho Nuevo (Márquez-M., 1994a, 1994b; Burchfield et al., 1997). Hence, virtually all forms of direct exploitation, and most other interactions, involving breeding turtles have taken place in the western extreme of the Gulf of Mexico. This has included directed takes of reproductive animals immediately off the nesting beaches as well as incidental capture in shrimp trawls, particularly in U.S. waters (Márquez-M., 1994a, 1994b). For example, a take of 5,000 *L. kempii* from the Rancho Nuevo area was authorized by the Mexican government in 1970, although in the end there was little if any directed exploitation that year (Márquez-M., 1994a, 1994b).

This species seems to prefer coastal, or even inshore, waters that are less than 50 m deep and rich in crabs. Immature *L. kempii* are known to concentrate in shallow waters, bays, and sounds of the northern Gulf of Mexico, particularly western Texas, Louisiana, and northwestern Florida, as well as along the western north Atlantic, particularly North Carolina, Chesapeake Bay, Long Island Sound, and Cape Cod Bay. There are regular seasonal migrations, with most of the animals leaving the cooler northern waters as autumn and winter set in and migrating south and sometimes farther offshore (Weber, 1995). These areas of higher density and seasonal movements clearly relate to places and seasons in which the chances of marine interactions are higher (Frazier, 2000).

L. kempii have long been known to occur in European Atlantic waters (Brongersma, 1972), and they have been documented twice from the Azores, but there are no confirmed records from the Atlantic coast of Africa (Fretey, 2001). Thus, the numbers that occur in the eastern Atlantic are relatively small, and there is little evidence that human activities, such as incidental capture in fisheries, are a significant cause of mortality in these areas.

Directed Take

Commercial fisheries for marine turtles were active before 1950 along most of the U.S. coast of the Gulf of Mexico, especially in Texas, Louisiana, and Florida, and three species were commonly taken, canned, and marketed widely (Hildebrand, 1963; Witzell, 1994a, 1994b). Unfortunately, records are incomplete, and, moreover, at the time of the fishery, *L. kempii* were either confused with loggerheads (*C. caretta*) or commonly regarded to be hybrids between loggerheads and green turtles (*C. mydas*) (Carr, 1955). As a consequence, few useful records of commercial fisheries for *L. kempii* are available. Nonetheless, it has been estimated that the number of *L. kempii* taken in these commercial fisheries was never large; Hildebrand (1982) stated that "there was no organized fishery for the species anywhere except as a by-catch of a green turtle fishery near Cedar Key, Florida." In 1970 a permit from Mexican fisheries authorities was provided for exploiting turtles at the only massed nesting beach at Rancho Nuevo, but because the arribada did not arrive on schedule, only five turtles were taken, according to official records (Hildebrand, 1982).

Legal exploitation of *L. kempii* ended when the species was totally protected; legislation banning direct exploitation was enacted at different times in different states and countries: 1963, Texas, state; 1970, United States, federal; 1973, Mexico, federal; 1974, Florida, state (Frazier, 2000). However, illegal trade has continued. Records from the U.S. Customs Department show that between 1983 and 1989, when the species had been fully protected in both Mexico and the United States for at least a decade, *L. kempii* products were supplied to a very active market for exotic animal skins. It is estimated that the U.S. Customs Department may apprehend less than 10% of the illegal trade, and furthermore, the available figures on confiscations by U.S. Customs only consider illegal imports into one country. Hence, the volume of the black market is clearly much greater than the official number of confiscations (Teyeliz, 2000).

Incidental Capture

Incidental capture of *L. kempii* during commercial shrimping operations, particularly in the northern Gulf of Mexico, began to be documented in the biological literature by 1973, when the shrimp fishery in the U.S. Gulf states was becoming more highly mechanized (Frazier, 2000; J. Frazier, unpublished data). By the late 1970s, there was tremendous concern that the only major nesting population of this species, at Rancho Nuevo, had declined dramatically, was literally on the brink of extinction, and a major source of mortality was incidental capture in shrimp trawls. This led to the design of turtle excluder devices (TEDs) followed by trials and tribulations in their validation and implementation, with nearly a decade of intense social conflict between and among shrimpers, conservationists, and government agencies in the shrimping communities of the United States (Weber et al., 1995; Margavio and Forsyth, 1996; Márquez-M., 1996; Weber, 1996; Frazier, 2000). More concern about the issue was raised by a study organized by the National Academy of Sciences in which it was concluded that the "incidental capture of sea turtles in shrimp trawls was identified by this committee as the major cause of [turtle] mortality associated with human activities; it kills more sea turtles than all other human activities combined" (NRC, 1990: 13).

Between 1986 and 1997, a total of 3,476 *L. kempii* were recorded stranded in the United States (this does not include turtles that were smaller than 20 cm in carapace length, cold-stunned, or head started). Of these, 6% were from the Northeast (Virginia through Maine), 25% from the Southeast (Atlantic coast of Florida through North Carolina), and 69% from the Gulf of Mexico (Texas through western Florida). Over this period, there had been a marked increase in strandings in both the Southeast and Gulf states. The western Gulf (Louisiana and Texas), which consistently has the greatest shrimp-trawling effort, also had the largest number of carcasses washed up, accounting for nearly 45% of recorded strandings. There has been a significant increase in size of turtles stranded between the 1980s and the 1990s (Turtle Expert Working Group, 2000).

Washed-up carcasses are recorded in all four seasons of the year, with most of the winter records coming from the eastern Gulf and the Atlantic seaboard (Teas, 1993). Although there are no systematic stranding networks in Mexico, it is known that there are regular strandings of adult-sized turtles at Rancho Nuevo and nearby beaches during the nesting season (Turtle Expert Working Group, 2000).

Although the most likely source of mortality for these strandings is incidental capture and drowning in shrimp trawl nets, the trends cannot be interpreted simply as an index of fishing activity (Caillouet et al., 1996). Over the same period when stranding rates increased, there was a dramatic increase in the number of nesting *L. kempii*. It has been concluded that the population is growing and in the early stages of recovery; hence, increases in strandings may be nothing more than an effect of more turtles being available to be caught and killed. The tendencies in stranding data have been taken as indications that TEDs have significantly decreased mortality of incidentally caught *L. kempii* (Turtle Expert Working Group, 2000): it is argued that the increase in strandings is a result of an increase in the number of turtles.

When one is interpreting records of strandings, it is fundamental to keep in mind that the number of carcasses documented represents an unknown proportion of the total number of animals caught and drowned at sea, and the number of recorded strandings will be related to a variety of variables, such as surface currents, winds, the time elapsed since the animals were drowned, and search effort (Epperly et al., 1996). Hence, records of strandings in just the United States leave no doubt that significant numbers of *L. kempii* have been killed in shrimp trawls.

In addition to incidental captures in shrimp trawls, there are records of *L. kempii* being taken accidentally by other fishing gear, including gillnet, hook and line, beach seine, purse seine, bag seine, cast net, butterfly net, and crab trap (Manzella et al., 1988; Márquez-M., 1994b). However, records from all of these gear types collectively represent less than a third of what was reported for just shrimp trawls, so the biggest threat to this species has been identified as shrimp trawling (NRC, 1990).

Interactions Other Than Capture

Marine pollution may take on many forms, including chemical, photic, physical, and thermal, and *L. kempii* are liable to all of these. The majority of studies indicate that after passing the pelagic phase, immature turtles return to shallow inshore waters and take up neritic habits, and at that point the turtle becomes a bottom feeder (Bjorndal, 1997). Nonetheless, there is evidence that at least some immatures may feed from the surface (Dobie, 1996), and if this habit is widespread, it would expose the animals to various forms of floating debris and contamination.

The Gulf of Mexico, through the outflow of the Mississippi River, has a high level of discharge of numerous anthropogenic compounds, many of which are known to be toxic and are likely to affect a major part of the geographic range of this turtle (Frazier, 1980a; Weber, 1995). Even though oil spills are relatively rare events, there are serious concerns about the effect of these tragedies occurring in the Gulf of Mexico (where there are scores of offshore platforms and activities related to the petrochemical industry), and there have been national initiatives at preparing contingency plans (Shigenaka, 2003). Furthermore, oil spills are not the only threat posed to marine turtles by exploration and extraction of oil; however, there is relatively little information on the effects of marine pollution on *L. kempii* (Weber, 1995; Lutcavage et al., 1997; Milton et al., 2003). Other sources of mortality include dredges, power plant intakes, and boat strikes in nearshore and estuarine areas, which are thought to be increasing, but there is little systematic information on any of these (USFWS-NMFS, 1992; Weber, 1995).

A much-ignored threat is sonic pollution resulting from various sources. These include seismic testing during oil exploration, explosive removal of abandoned oil rigs, weapons testing, and submarine low-frequency communications by the U.S. Navy. Although the results are varied, indications are that under certain conditions low-frequency pollution can pose a threat to marine turtles (Moein Bartol and Musick, 2003). Because these activities are concentrated in the Gulf of Mexico, the primary area of *L. kempii,* their relevance to this species is probably substantial.

Conclusions

It was the tremendous and growing concern for the dramatically depleted nesting populations of *L. kempii* at Rancho Nuevo (Carr, 1977), along with the growing evidence that a major source of mortality for adult-sized turtles was the drowning in commercial shrimp trawls (Frazier, 2000; Turtle Expert Working Group, 2000) that prompted the development of TEDs by the National Marine Fisheries Service of the U.S. Department of Commerce and the subsequent lobbying by conservation nongovernmental organizations (NGOs) in the late 1970s and early 1980s for mandatory use of TEDs. Combined with numerous international events and issues, including Arab-Israeli wars, increased gas prices, and increased competition for the U.S. shrimp market, the TED initiatives led to the "TED wars" in the Gulf and southeastern shrimping states of the United States (Frazier, 2000). From this emanated U.S. Public Law 101-162, section 609, which called for the United States to embargo shrimp not caught with TEDs. This one U.S. law prompted a series of international events of great importance not only for marine turtle conservation but also for the debate about environment and trade. Of particular importance was the World Trade Organization "shrimp-turtle dispute," which provided the political stimuli for the development of several international instruments. The Inter-American Convention for the Protection and Conservation of Sea Turtles (IAC) and the Memorandum of Understanding on the Conservation and Management of Marine Turtles and Their Habitats of the Indian Ocean and South-East Asia (IOSEA) are products of the TED wars (Frazier, 2002; Frazier and Bache, 2002; Bache and Frazier, 2006), even though *L. kempii* does not occur within the IOSEA region. Hence, in the last few decades, more than any other species of marine turtle, *L. kempii* has been the raison d'être for a number of high-level policy decisions, with enormous implications at the global level, despite the limited geographic range of this species. For a discussion of conflicting issues involving the conservation of *L. kempii*, see Márquez-M. (1994a, 1994b), Frazier (2000), and Turtle Expert Working Group (2000).

Although the species is well documented along the eastern seaboard of the United States, where seasonal north–south migrations have been shown to occur in several locations from New York to the Carolinas, some authors question that the individuals that have "escaped" from the Gulf of Mexico are significant for the maintenance of the population (C. W. Caillouet, Jr., in litt., July 22, 2004). For decades it has been questioned if *L. kempii* in European waters are waifs, lost from the population (Hendrickson, 1980; Brongersma, 1982). Hence, some workers consider that human–turtle interactions involving *L. kempii* in the North Atlantic, particularly European waters, are a low conservation priority. Until it can be shown that these individuals have importance to population maintenance and/or genetic diversity of the species, their conservation relevance will be in doubt.

Lepidochelys olivacea: The Olive Ridley Turtle

Although there are no commonly recognized subspecies of this wide-ranging turtle, there are thought to be some subtle morphological and color differences between certain geographic areas (Pritchard and Plotkin, 1995). Broadly, *L. olivacea* shows a population genetic structure with differences between and within the ocean basins (Bowen et al., 1998; Shanker et al., 2004; Kichler Holder and Holder, Chapter 6). Although the genetics and demography of *L. olivacea* need to be resolved, environmental characteristics, risks, and management decisions vary from country to country and from region to region. Hence, the summary of human–turtle interactions involving this turtle is organized by sectors of ocean basins.

Southwestern Atlantic Ocean

Despite the fact that *L. olivacea* is the most abundant marine turtle, it is the least commonly observed species in the western Atlantic (Marcovaldi, 2001). *L. olivacea* have been reported at sea as far north as Florida (Foley et al., 2003) and as far south as Uruguay (Frazier, 1991), encompassing a range between 25°N and 21°S (Fretey, 1999). However, on the western side of the Atlantic, they are common only in the Guianas (Guyana, Suriname, and French Guiana) and northern Brazil; elsewhere they are rarely documented.

Unlike in the eastern Pacific, where the species is commonly recorded on the high seas (Pitman, 1990, 1992; Plotkin et al., 1993), there are very few records from pelagic waters of the Atlantic, and there is no evidence that *L. olivacea* cross the Atlantic Ocean. Hence, it is likely that a large majority (and perhaps even all) of the turtles caught, intentionally or accidentally, in western Atlantic waters originate from the two principal nesting sites on the coast of South America: one in Sergipe, Brazil, and the other in French Guiana and Suriname (because the nesting beaches in these two neighboring countries are relatively close and there are numerous records of female *L. olivacea* that have nested in both countries, this can be considered as a single nesting location). Based on the available data, it appears that after nesting, turtles from these two different reproductive populations disperse to different zones and thus experience different interactions with humans, particularly problems of incidental capture at sea. Tag returns from females that nested in French Guiana/Suriname indicate that postnesting dispersion is made to one of two areas: from eastern Guyana to Amapa (Brazil) or north, from the mouth of the Orinoco River to the islands of Tobago, Trinidad, and Margarita (Pritchard, 1973; Schulz, 1975). Nesting females tagged in Sergipe have been recovered in either Sergipe or farther south in Brazil (Marcovaldi et al., 2000), indicating that *L. olivacea* from the Guianas does not share the same foraging areas with *L. olivacea* from Sergipe. Although data gathered from incidental captures in shrimp fisheries may be representative of only a part of the turtle population, it is clear that there are differences between the two nesting areas, and hence, the situation in French Guiana / Suriname ("Guiana") is treated separately from that in Brazil.

The estimated number of clutches laid by the French Guiana / Suriname population is 1,000–2,000 per year (extrapolated from Godfrey and Chevalier, 2004). This population has undergone a decline during the last century: although arribadas used to be common in Suriname, they have been absent since the 1970s (Fretey, 1989; Reichart, 1993; Pritchard and Plotkin, 1995). Remarkably, annual takes from just Suriname during the first half of the 1900s were nearly of the same size as the estimated annual number of all reproductively active females in the entire French Guiana / Suriname population today. In contrast, the estimated number of clutches laid per year by the Sergipe population is 1,000–15,000 (extrapolated from Godfrey and Chevalier, 2004). There is no evidence that this population was much larger in the past, and the trend in the annual number of nests over the past 10 years shows a clear increase (Marcovaldi, 2001; Godfrey and Chevalier, 2004).

DIRECTED TAKE. Reichart, who has considerable experience in the Guianas, stated that "probably because the *L. olivacea* is relatively rare in the western Atlantic, there is no direct fisheries effort on the species in this region" (Reichart, 1993). He did, however, explain that there is a direct, albeit spontaneous, take of nesting animals, particularly in Guyana. In fact, although there may not be an organized, or commercial, fishery, there have been significant levels of directed take, the vast majority of which has concentrated on nesting females. During the 8-year period between 1933 and 1940, an estimated 1,500 *L. olivacea* were killed annually in Suriname (Geijskes, 1945, cited in Reichart and Fretey, 1993). Yet, there is no evidence of a systematic, directed take of *L. olivacea* from the sea in this region.

Directed take in Sergipe, Brazil, was also casual and focused on nesting females. Since the inception of Projecto TAMAR in 1982, there have been numerous initiatives to stop human exploitation of marine turtles and replace it with other livelihood activities (Marcovaldi, 2001).

INCIDENTAL CAPTURE. The importance of shrimp trawling on incidental capture of turtles in waters of the Guianas was pointed out more than 30 years ago (Pritchard, 1969), less than a decade after this fishery became well developed in the region (Dintheer et al., 1989). Some of the earliest information on tag returns from French Guiana and Suriname reported that nearly all the recaptures of turtles came from shrimp trawlers that fished along the coast (Pritchard, 1969, 1973), making it clear that this fishing method has a major impact on *L. olivacea* in the region. Postnesting females from the Guianas have been recaptured as far north as Isla Margarita, Venezuela, and even into the Lesser Antilles and as far south as Amapa, Brazil (Pritchard, 1973; Schulz, 1975), all locations where there is also active shrimp trawling. Since the 1990s, incidental take in shrimp trawls in the Guianas has become better understood (e.g., see Tambiah, 1994; Hoekert et al., 1996; Laurent et al., 1999; Guéguen, 2000). The limited information indicates that *L. olivacea* are by far the most commonly caught marine turtles in shrimp trawls in the region, and they are caught in all months of the year. Different studies report that there are peaks in incidental capture from July to September (Pritchard, 1973) or from January to March (Tambiah, 1994).

Preliminary estimates of the annual mortality from shrimp trawling are some 1,600 marine turtles in Suriname (Tambiah, 1994) and another 1,000 in French Guiana (Guéguen, 2000) (most of these turtles would be *L. olivacea*). Because of overfishing, there has been increasing trawling activity closer to the coast (Dintheer et al., 1989), which is likely to result in even more incidental captures and mortality. In Guéguen's (2000) pilot study, a third of the turtles caught in shrimp trawls were landed dead in French

Guiana, whereas "direct mortality" (turtles dead on landing) in Venezuela was reported as 19% for all species (Marcano and Alio, 2000). Not surprisingly, postrelease mortality has not been estimated, but it is likely to be significant, so the figures for "direct mortality" will be underestimates of mortality caused by capture in shrimp trawls.

Although change over a period of three decades should not be surprising, it is remarkable that 30 years ago nearly a quarter of the tag recaptures were from Venezuela (Pritchard, 1973), and in recent years there have been few, if any, reports of the species captured in these waters (Guada, 2001). It is relevant that in 1979 the U.S. government passed a law (PL 101-162, section 609) that required shrimp-exporting countries to implement turtle conservation programs, namely, the mandatory use of TEDs (Frazier and Bache, 2002). Thus, it is likely that the chances of reporting tag returns from shrimp trawlers has declined, as it would be perceived as increasing the risk of having a shrimp import embargo imposed by the main importing country, the United States.

Gillnets are known to be an additional source of mortality, but annual captures are thought to be well below those for shrimp trawling (Fretey, 1999). Evidence from other fishing methods in this region indicates only occasional captures of marine turtles (e.g., Hoekert et al., 1996; Chevalier et al., 1998; Laurent, 1999a, 1999b; Chevalier, 2001; Guada, 2001).

Reichart and Fretey (1993) suggested that accidental capture of *L. olivacea* at sea was the largest unresolved problem facing *L. olivacea* in the Guianas, and Pritchard and Plotkin (1995) concluded that the dramatic decline in numbers nesting in Suriname was the result of high mortality from incidental capture in shrimp trawls throughout the region where these turtles are known to disperse, from Venezuela to Brazil.

There is a paucity of systematic information on incidental capture, and data for at-sea interactions are very limited; quantitative data are based on results of 39 recaptures of tagged nesting females, compiled before the mid-1970s (Pritchard, 1973; Schulz, 1975). Nonetheless, more detailed research on the impacts of fishing and other marine activities are likely to identify other forms of human–turtle interaction at sea.

Nesting females tagged in Sergipe, Brazil, are known to move south (Marcovaldi et al., 2000). Most marine turtles recorded as stranding dead on the coast of Sergipe have been adult-sized *L. olivacea;* from 1999 to 2002 there were 201 records, of which 71% were *L. olivacea* (J. Castilhos, personal communication). As in many parts of the world, an active shrimping fleet in Sergipe trawls closer to the coast than is legally allowed, and it is known that shrimpers catch turtles, particularly *L. olivacea,* with estimates that they may drown at least dozens in a year (Thomé et al., 2003).

Other fishing methods in Brazil seem to have little impact on *L. olivacea.* In one study of turtles captured accidentally in Almofala, Ceará, Brazil, during the 8 months between January 1 and August 31, 2001, in currais, a traditional type of fishing weir, only 1 of 75 marine turtles was *L. olivacea* (Lima et al., 2002). A larger study in Brazil found that only 3 of 207 turtles caught in currais were *L. olivacea,* and there was not one record of the species in more than 2,300 records of accidental capture in the state of São Paulo (Marcovaldi et al., 2001). Although no *L. olivacea* were reported from a recent study of by-catch on longlines in southern Brazil (Kotas et al., 2004), there are at least one confirmed record (Serafina et al., 2002) and one "highly probable" record (Pinedo and Polacheck, 2004) of incidental capture on long lines as well as several records of strandings in southern Brazil attributed to catch with hooks or nets (Soto and Beheregaray, 1997). A total of 237 dead stranded turtles were recorded between 1996 and 2000, from 125 km of the 161 km of beach in Sergipe, and of 154 carcasses that were identified as to species, 88 (57.4%) were *L. olivacea,* the vast majority of which were of adult size. The increase in shrimp trawling, particularly within 3 nautical miles of the coast, is uncontrolled and is thought to be a major factor in the mortality of turtles (Da Silva et al., 2002).

In addition, at least three of the four records of *L. olivacea* from Uruguay were caught accidentally in gillnets (A. Estrades, in litt., August 4, 2003). Recently initiated coastal monitoring and on-board monitoring programs in Brazil (Marcovaldi et al., 2002) and Uruguay (López-Mendilaharsu et al., 2003; Estrades et al., in press) will address these problems of the paucity of basic information, especially reliable data on incidental capture and impacts caused by other human activities in the marine environment.

INTERACTIONS OTHER THAN CAPTURE. There seems to be little, if any, information on various forms of marine pollution in the southwestern Atlantic and how they affect marine turtles. Given the situations in other regions, lost and discarded fishing gear, plastics, and oil spills are likely threats.

CONCLUSIONS. The available evidence indicates that there are two breeding populations of *L. olivacea* in the western Atlantic. The French Guiana–Suriname population has declined dramatically (Fretey, 1989; Reichart and Fretey, 1993) and has been named as a conservation priority (Mast et al., 2004). Mortality just from shrimp trawling appears to be a major threat to the recovery of this population. The Sergipe population, although fewer than 1,000 nesting females per year, is increasing despite whatever mortality is inflicted by fishing and other human activities. An ongoing program to monitor and mitigate fishing in Brazil (Marcovaldi et al., 2002) will help provide much-needed basic data for the design of conservation and management activities.

Shrimp trawling is active from Venezuela to Brazil, and all countries in this region have legislation requiring TEDs, with the notable exception of French Guiana (France) (Laurent et al., 1999). This legislation is to comply with requirements to have legal access to the U.S. shrimp import market, not primarily to protect marine turtles (Frazier and Bache, 2002), and in many cases there is poor, if any, implementation. The future of the nesting population in the Guianas may depend on resolving the problem of incidental capture in various fisheries and instigating responsible fisheries in the region.

At the same time, incidental captures present a remarkable paradox to marine turtle conservationists. Although this is a major source of mortality on reproductively active turtles, it has also been a unique source of information on postnesting females, particularly on their dispersion, distributions, and potential foraging areas. This information is fundamental for any conservation program.

Southeastern Atlantic Ocean

Information on marine turtles from the west coast of Africa is limited, but it is clear that *L. olivacea* is common throughout this region. The species has been confirmed, or is thought to occur, in all countries along the Atlantic coast of Africa between Mauritania and South Africa, and the highest densities are documented, or suspected, in the Gulf of Guinea between Côte d'Ivoire and Gabon, including the volcanic island chain that bisects the gulf. Juveniles have been reported in the waters of Côte d'Ivoire, Cameroon, and São Thomé (J. Gómez, H. Angoni, and J.-F. Dontaine, personal communication), and there are indications that, as well as nesting beaches, the region provides important foraging areas and migratory corridors for this species. The environmental conditions of the Cameroon estuary as well as the Niger Delta, mouth of the Upper Volta, and other estuarine and deltaic areas indicate that these could be important feeding and developmental areas for this turtle (Fretey, 2001).

As is the case on the western side of the Atlantic, there are few records from the eastern Atlantic of *L. olivacea* on the high seas. Because the coastline of West Africa is some 12,000 km long, and there is considerable marine and coastal diversity in this region, it is likely that several stocks, or management units, of *L. olivacea* occur, as seems to be the case with *C. mydas* in this region (Formia, 2002).

DIRECTED TAKE. The intentional capture of marine turtles, both on nesting beaches and at sea, is documented in 15 of 27 countries along the west coast of Africa, from Morocco to South Africa, and although *C. mydas* is preferred, *L. olivacea* is commonly captured to be eaten. In general, the turtles are killed whenever they are encountered. Most directed take is on nesting beaches, and directed fisheries for marine turtles are not generally common. However, in areas of high turtle density, some fishermen use specialized techniques, such as large mesh turtle nets set at strategic sites. Although not the most commonly captured species, *L. olivacea* is occasionally taken in these nets. Capture of *L. olivacea* at sea has been recorded in Senegal, Gambia, Sierra Leone, Liberia, Côte d'Ivoire, Ghana,

Togo, Benin, Nigeria, Cameroon, Equatorial Guinea, São Thomé and Principe, Gabon, Congo, and Angola, and the species is suspected to occur in Mauritania, Cape Verde, Guinea-Bissau, and Guinea (Fretey, 2001; Fretey et al., 2001; Formia, 2002; Barnett et al., 2004). The most important locality for *L. olivacea* in the region is São Thomé, where there is a thriving black market in turtle meat, and this is the most frequently encountered species in local wildlife markets in Togo and Benin (Fretey, 2001). Although there is strict protective legislation in most of these countries, the capture, transport, holding, marketing, killing, and consumption (all actions that are illegal) of marine turtles, including *L. olivacea,* are commonplace.

Although there are few systematic data on directed take, consumptive use of marine turtles is regarded to be a major source of their mortality throughout most of West Africa. Although *C. mydas* is the preferred species, often considered a delicacy, marine turtles of all species and sizes are consumed when available, whether at home, in bars, hotels, or restaurants. The market for meat and eggs of marine turtles is fueled by a high demand for animal food products by inhabitants of coastal villages and particularly larger towns and cities (especially in São Thomé, where *L. olivacea* is the most frequently consumed marine turtle). These turtles are often commercialized within established market systems (albeit illegal), providing a much-needed source of cash for economically marginalized villagers but especially benefiting intermediate traders and middlemen. This species is a particularly prized commodity in remote villages because the turtles can be maintained alive for several days, eliminating the need for refrigeration or preservation, and they can be more easily transported because of their smaller size and weight (compared to *C. mydas* and *Dermochelys coriacea*).

Other traditional uses of marine turtle products include medicinal oil, produced from the body fat, and powder from crushed skull bones, used to cure aches and migraines (Fretey et al., in press). In addition, carapaces of *L. olivacea* are often utilized for decoration in private dwellings, bars, and restaurants; they are also used as containers and to shield items against the rain (such as religious sculptures): this last-named function is particularly common in Cameroon, Togo, and Benin. Carapaces are often polished and painted with marine scenes or made into masks and sold in markets, souvenir shops, or alongside major roads, particularly in areas most frequented by national and international tourists. For instance, carapaces decorated with bronze ornaments and made into masks by Bamum artists in Cameroon appeal to the wealthy, local elite. Although it is often impossible to determine the origin of these turtles, it is likely that at least some result from intentional captures at sea.

INCIDENTAL CAPTURE. Despite the widespread occurrence of directed take of *L. olivacea* along the coast of West Africa, the pelagic habits of these turtles means that they are not usually available for targeted or directed coastal fisheries. Instead, they are frequent victims of incidental capture in a wide variety of fishing activities. Artisanal fisheries using gillnets, hook and line, beach seines, and large-mesh gillnets are known to catch these turtles. In virtually all instances of incidental capture by artisanal fishermen, whether the turtles are retrieved dead or alive, *L. olivacea* are retained as part of the catch and either consumed locally or sold.

In addition to small-scale, artisanal fishing, eastern Atlantic waters, namely the territorial seas and exclusive economic zones (EEZs) of West African states, are exploited by industrial fishing fleets, particularly trawlers and pelagic long liners. Pelagic fishing fleets, including those from countries in the European Union (mainly Spain), operate throughout the region. Industrial trawlers of various nationalities operate mainly out of ports in Nigeria, Cameroon, Gabon, Guinea, Senegal, and Congo-Brazzaville, concentrating their activities in the nutrient-rich waters of the Gulf of Guinea and the upwelling zone between Senegal and Mauritania. There is little, if any, regulation of bottom trawling, such as net and mesh sizes, maximum trawl times, target species, minimum sizes, catch quotas, or damage to marine environments. Moreover, it is common for these foreign trawlers to fish illegally in coastal zones reserved for artisanal fishers and also to stray into waters of states where they have no permits.

Although many countries in West Africa have adequate fisheries regulations on the books, the activities of the foreign fleets are largely unregulated. For example, Asian companies often operate the most destructive trawlers in the region, and intergovernmental agreements for funding of infrastructure (such as roads, hospitals, and government buildings, not to mention "kickbacks" involved in accepting the contracts) often guarantee them immunity from regulations and sanctions. In many areas, the protected status of marine turtles, including *L. olivacea,* is not widely known, and it is likely that fishermen retain the majority of incidentally caught sea turtles, either for sale at port or for consumption on board.

INTERACTIONS OTHER THAN CAPTURE. The Gulf of Guinea and adjacent areas are the focus of intense oil exploitation, which has extended over an increasingly wider region during the last few decades. This represents a serious potential threat to all marine turtles, but especially to *L. olivacea;* for these turtles can be found in relatively high densities in this area, which is evidently where they concentrate for feeding. Threats related to the oil industry include low-frequency, high-energy sonic perturbations from explosives used during exploration activities; chemical and physical contamination from drilling, construction of wells and platforms at sea, oil refineries, and pipelines; light pollution from gas flares and other developments; and of course oil slicks.

Turtles covered in tar are sometimes found stranded on the beaches of northwestern Cameroon; the marine terminus of the Chad/Cameroon pipeline represents a serious threat for nearby marine ecosystems and nesting beaches. Waste and plastic debris, as well as oil drums and other materials, from offshore oil platforms are routinely discarded at sea, compounding the problems of urban and industrial pollution. Other threats to both the turtles and their habitats include sewage discharge, agricultural runoff and siltation, eutrophication, discarded fishing gear, drifting logs (from logging activities), and shifting currents and erosion caused by coastal construction. However, there is little if any systematic information on any of these threats.

As mentioned above, industrial fishing, especially by foreign fleets, is intense and virtually unregulated. The resultant overfishing and environmental degradation in certain areas of the eastern Atlantic are likely to be influencing the trophic structure of prey items on which *L. olivacea* and other marine turtles feed. Certainly, they are depleting fish stocks and other marine resources on which small-scale coastal fishermen depend, forcing greater reliance on alternative sources of protein such as marine turtles.

CONCLUSIONS. Although directed fisheries for marine turtles in West Africa are generally small scale and focused on local subsistence needs, their cumulative effects over some 12,000 km of coastline, much of which is densely populated, may represent a significant source of mortality.

Mechanized trawling has increased greatly along the Atlantic coast of Africa, and there is widespread evidence of *L. olivacea* being caught, sometimes in relatively large numbers. At Bijagos Archipelago, Guinea-Bissau, local residents attribute the decline in numbers of nesting ridleys to the mortality caused by offshore trawlers (J. Fretey, personal observation). Coastal strandings of marine turtles are common and widespread, particularly in Cameroon, Nigeria, Benin, Togo, Senegal, and Gabon, and the most commonly reported species is *L. olivacea,* although *C. mydas* and *D. coriacea* are occasionally documented. The causes of strandings can seldom be identified with certainty unless the turtle washes up accompanied by fishing gear. In Nigeria, for example, several strandings have been reported entangled in pieces of thick plastic netting, showing evidence of having been cut away from the rest of the net. Mass strandings have been reported, indicating possible seasonal peaks, notably during the nesting season. For example, in February 2000, a total of 18 *L. olivacea* strandings were observed along 15 km of shore east of Accra, Ghana, and in November of the same year, 15 strandings were reported in the same area. Similar incidences have been reported in recent years in Gabon and Congo-Brazzaville (Renatura and A. Billes, personal communication).

Interaction with industrial fisheries is thought to be the gravest indirect threat to *L. olivacea* in the region. Nonetheless, as is usual for other regions, there are very little systematic data on incidental capture of marine turtles in West

Africa. This is compounded by the fact that commercial fishing vessels often do not unload or dock at African ports, and there are no on-board observer programs. Hence, inferences must be drawn from incomplete sources of information.

Many countries in West Africa lack the means (trained, motivated, and supported personnel; boats; equipment; fuel; administrative and legal support; etc.) to carry out effective coastal patrols in their territorial waters and, even worse, in their EEZs. Moreover, prosecutions for law breaking are rarely given administrative, legal, or political priority. Along the entire Atlantic coast of Africa, the only country where TEDs are legally required is Nigeria, and even there, adequate implementation of the TED laws is uncommon despite Section 609 certification by the U.S. Department of State (see Frazier and Bache, 2002).

Diverse and heavy impacts by commercial fisheries and the petroleum industry on fish stocks and other marine resources affect not only marine turtles and their habitats but also countless small-scale fisheries, coastal communities, and their livelihoods. The development of accountable and transparent coast guard brigades will be essential to reduce by-catch and environmental destruction from industrial fisheries and other development activities in the region.

Western Indian Ocean

Although the eastern Indian Ocean has some of the largest breeding concentrations of *L. olivacea* known, the species is not generally common in the western Indian Ocean, which extends from the southernmost tip of India, west to South Africa. The species nests in small numbers along the east coast of Africa, from Mozambique to Kenya, as well as on Madagascar (FAO, 2006); with the exception of some beaches in Oman, it has been rarely reported from the Arabian Peninsula (Ross and Barwani, 1982; Baldwin and Al-Kiyumi, 1999), and there are regular but relatively small numbers nesting in Sind, Pakistan (Firdous, 2000; Qureshi, 2006) and along the western coast of India, from Gujarat to Kerala (Kar and Bhaskar, 1982; Sharath, 2002, 2006; Giri, 2001; Sunderraj et al., 2001, 2006; Dileepkumar and Jayakumar, 2002; Giri and Chaturvedi, 2003, 2006), and also in the Lakshadweep Archipelago (Tripathy et al., 2006, in press). Relatively little is recorded about either the biology or conservation status of *L. olivacea* in this vast region. Some cultures of the western Indian Ocean have long histories of interacting with marine turtles, with special traditions linking their societies with these reptiles; the Bajun of southern Somalia and northern Kenya (Gudger, 1919a, 1919b; Grottanelli, 1955) and the Vezo of southwestern Madagascar (Astuti, 1995) are clear examples.

In contrast, on many parts of the Indian subcontinent, adult sea turtles have not been harmed directly because of long-established Hindu religious beliefs that turtles are an incarnation of Vishnu, one of the Gods of the "Hindu trinity." Likewise, many Muslims do not eat turtles or turtle products because their Islamic customs forbid it; in many Islamic societies turtles are considered as *haram* or unclean because they have an amphibious life (Dileepkumar and Jayakumar, 2006; Qureshi, 2006; Tripathy et al., 2006). More recently, nationally and internationally recognized, grassroots, community-based conservation programs have developed to protect these turtles in this region (e.g., Shanker and Kutty, 2005; Kutty, 2006).

DIRECTED TAKE. In general, directed take of *L. olivacea* from the sea in the western Indian Ocean is uncommon and exists at a very low level. Possible exceptions could occur where the species is especially abundant, or at least common, and there is also a custom of catching turtles at sea. However, most traditional turtle hunters, such as the Bajun and Vezo, live in areas where *C. mydas* is common, and *L. olivacea* is generally uncommon.

It was reported that marine turtles in Mozambique are killed, accidentally and intentionally, by fishing activities (Magane et al., 1998), and *L. olivacea* can be common in the north of the country. Directed take is thought to occur in Kenya (FAO, 2006: App. I), but there is little information. Fishermen from the west coast of Madagascar take *L. olivacea* for meat (Rakotonirina and Cooke, 1994). There also may have been a directed take in Baluchistan, the westernmost province of Pakistan, where for years, there have been reports of a fishery for marine turtles that may include ridleys (Frazier, 1980b; Qureshi, 2001, 2006), but few details are available, and the characteristics and magnitude of the situation are unclear.

In general, the coastal population of the shores of the western Indian Ocean from Tanzania to the Red Sea and around the Arabian Peninsula to Pakistan is Muslim (Qureshi, 2006), and most of the western coast of India is Hindu (Giri and Chaturvedi, 2006), although the Lakshadweep Islands are predominantly Muslim (Tripathy et al., 2006). In both cases there are religious and cultural taboos against consuming turtles (Dileepkumar and Jayakumar, 2006), so there would be fewer motives for people to be involved in directed take. However, there is some evidence of directed take in the Indian state of Goa, which is mainly Christian (Giri and Chaturvedi, 2006).

INCIDENTAL CAPTURE. Along the coasts of eastern Africa, shrimp trawling is known to catch significant numbers of marine turtles, and many, if not most, of these are likely to be *L. olivacea*. However, few detailed data are available.

Formerly it was assumed that there was no significant incidental catch in shrimp trawls in Mozambique but that the beach seine fishery was taking some 20 turtles (species not determined) per month (Magane et al., 1998). However, more recently a study of by-catch was conducted, and it was estimated that between 6 and 8, or perhaps as many as 12, turtles are caught by "semiindustrial" trawlers per month, yielding an annual estimate of between nearly 2,000 and more than 5,000 turtles captured annually on just the Sofala Bank, Mozambique. Although there were no specific records of *L. olivacea* (Gove et al., 2004), it is likely that this species is affected.

It was reported that, despite protective legislation, almost every turtle captured in nets in Tanzania is killed, and that the animals are caught in both artisanal and commercial shrimp fisheries. There is some indication that "large numbers" may be caught, and although there is no confirmation that *L. olivacea* is commonly taken (Haule et al., 1998), it is likely that this species is common off the Rufiji delta, where shrimp trawling is also concentrated. Although relatively few trawlers have been licensed to work in Kenya, there have been significant numbers of strandings in Ungwana and Malindi Bays for years, and *L. olivacea* is one of the species most affected. It has been estimated that at least 100–500 marine turtles (species not specified) are caught annually in this fishery (Wamukoya et al., 1998). It is also thought that this species is also caught in a variety of other inshore fishing gear in Kenya (FAO, 2006: App. I), but detailed data are scarce.

Studies from Eritrea indicate that although significant numbers of marine turtles are caught incidentally in shrimp trawls, *L. olivacea* are not included (Gebremariam et al., 1998), most likely because the species is uncommon in this region. No records of captures in shrimp trawls were available from Madagascar, but this is probably because there was no effort to document incidental capture of marine turtles (Randriamiarana et al., 1998). There is no evidence that *L. olivacea* is taken in fisheries from any of the other island territories of the western Indian Ocean, for the species is generally rare in these oceanic waters.

An on-board observer program was conducted on up to eight industrial trawlers (mainly from South Korea) during 1989 in Oman. Between May and December at least 201 turtles were recorded in the catches, about half of which were estimated to have died. However, although *L. olivacea* occurs in Omani waters, no information is available on the species of turtles caught by the trawlers (Hare, 1991).

Strandings have been reported on the coast of Sind, Pakistan (Stevens, 1998), but there seem to be no data on incidental capture. There is evidence of fisheries-related mortality in the Indian state of Gujarat, which lies just to the south, and where fisheries activity has increased since the late 1970s (Sunderraj et al., 2001).

Pair trawling off the coast of Gujarat between December 1983 and March 1984 was reported to have caught 70 *C. mydas* (Siraimeetan, 1988). However, this form of marine exploitation is one of the least selective, aside from use of dynamite and poison, so it is likely that *L. olivacea* would also have been caught. There has been a marked increase in fishing effort in the state of Gujarat, as indicated by the number of fish landing centers; these increased from 477 in 1977 to 854 in 1992. Yet, relatively few stranded *L. olivacea* have been found during recent coastal surveys: just nine carcasses were reported from a survey conducted during part of the 2000–2001 season (Sunderraj et al., 2001, 2006). Early surveys in Gujarat reported that turtle meat was often sold and that the flippers could be hacked off to make rough shoes for walking on coral (Bhaskar, 1979, 1984). However, more recent surveys in this state found no evidence for the sale of turtle meat (Sunderraj et al., 2001). Hence, at least in recent years, there is little evidence for incidental capture.

In Maharashtra state there are reports of incidental capture of *L. olivacea* by trawlers, but evidently numbers are relatively small. For religious reasons most fishermen in Maharashtra do not harm turtles but instead release them from their nets; however, meat consumption is reported from two of the five coastal districts, where it is carried out in an opportunistic manner and probably based mainly on incidental captures (Giri, 2001; Giri and Chaturvedi, 2003, 2006). It is believed that consumption of turtle meat in Goa, predominantly a Christian state, was widespread in earlier times, although this seems to be much reduced now (Giri, 2001; Giri and Chaturvedi, 2006). In contrast, meat consumption in Karnataka state is reported to be rare (Madhyastha et al., 1986), whereas in Kerala, particularly in the south where the population is mainly Christian, meat is consumed (Dileepkumar and Jayakumar, 2002, 2006). Hence, even though *L. olivacea* is common along the western coast of India, in those areas where there seems to be little demand for marine turtle meat, there appear to be few motives to catch and keep turtles.

Unlike other island territories, the Lakshadweep Archipelago has moderate numbers of *L. olivacea* nesting (Tripathy et al., 2002, 2006, in press). Here, turtles, although not usually eaten, are killed for the oil used to treat wooden boats, for bait, and for making stuffed curios; however, *L. olivacea* is not common in the inshore areas where most turtles are captured. There are only 22 records of the species in the Maldives, although it has been suggested that Maldivian offshore waters may be a significant foraging area for juveniles (Anderson et al., 2003). Remarkably, half of all *L. olivacea* reported from Maldives were either entangled in discarded fishing gear (41%) or caught incidentally in oceanic driftnets or long lines (9%). Although the sample sizes are admittedly small, it was concluded that entanglement in discarded fishing gear is a significant source of mortality for this turtle in the central Indian Ocean (Anderson et al., 2003).

INTERACTIONS OTHER THAN CAPTURE. Various forms of pollution, such as plastics, agrochemicals, urban wastes, and particularly lost and discarded fishing gear, are likely to have important impacts on marine turtles throughout the region, but there is no systematic information available. Major threats to marine turtles on the coast of Gujarat are petrochemical industries, sand mining, and harbor activities (Sunderraj et al., 2001, 2006), and similar concerns have been raised for the coast of Maharashtra (Giri and Chaturvedi, 2006).

CONCLUSIONS. Clearly, there is a tremendous paucity of basic information on both the biology and conservation status of *L. olivacea* from the western Indian Ocean. This turtle is generally uncommon in this region, and other species, namely *C. mydas* and *E. imbricata,* have been given higher priority because they are (or were) abundant and have been the focus of directed fisheries in many countries (Frazier, 1980b, 1982). If the relatively small, dispersed nesting populations throughout the western Indian Ocean are found to have unique genetic characteristics, more attention may be warranted.

Eastern Indian Ocean

Marine turtles have been important for nutritional, economic, and cultural reasons in several places in the eastern Indian Ocean. However, toward the end of the twentieth century there have been repeated signs that turtle populations in various countries had declined, and they were no longer economically viable or as culturally significant as before.

In contrast to the situation in the western Indian Ocean, *L. olivacea* is the most abundant marine turtle in the eastern Indian Ocean, with three sites in Orissa, India, where massed nesting occurs: Gahirmatha (in Bhitarkanika National Park), Devi, and Rushikulya. For decades it has been alleged that Gahirmatha hosts the largest concentration in the world of *L. olivacea* (or any other marine turtle), and although this claim recently has been challenged (Tripathy, 2002), the number of turtles that mass each year is extraordinarily large: hundreds of thousands. On the eastern shores of the Bay of Bengal, however, this species is uncommon (Tow and Moll, 1982), and in some places, such as Myanmar, there has been confusion between *C. caretta* and *L. olivacea,* which is complicated by the fact that few data are available (DF-GUM, 1999; Thorbjarnarson et al., 2000).

Remarkably, with the enormous numbers that occur in Orissa, there is no clear idea where the adult turtles live once they have left the nesting area. Out of nearly 10,000 turtles tagged at Gahirmatha, Devi, and Rushikulya, 24 have been recaptured in the Gulf of Mannar, between India and Sri Lanka, and elsewhere on the coast of Sri Lanka, leading to suggestions that this may be an important foraging area for *L. olivacea* in the eastern Indian Ocean (Kapurusinghe and Cooray, 2002; Pandav and Choudhury, 2006; see also Bhupathy and Saravanan, 2006). In addition, there are several reports of concentrations of these turtles in Sri Lankan waters, suspected to have been migrating to nesting beaches in India (Kar and Bhaskar, 1982). However, a satellite telemetry study found that one of four post-nesting females from Devi went as far as southeastern Sri Lanka, turning east before transmissions stopped, which headed the animal away from the Gulf of Mannar (J. Frazier et al., unpublished data). There is a higher probability for turtles to be observed, caught, and tags returned from active fishing areas such as the Gulf of Mannar than from areas with lower fishing effort or from the high seas. Hence, although Sri Lankan waters, particularly the Gulf of Mannar, may be an important area for *L. olivacea,* there are likely to be other locations, perhaps with even larger numbers of turtles.

DIRECTED TAKE. Directed fisheries involving *L. olivacea* are known to occur in Sri Lanka as well as in the four Indian states that border the Bay of Bengal: Tamil Nadu, Andhra Pradesh, Orissa, and West Bengal. In general, these are locations where marine turtles, including this species, are singularly abundant, at least during certain times of the year. Elsewhere in the eastern Indian Ocean basin, there appear to have been few directed fisheries for this turtle.

Marine turtles of several species have been taken at sea by Sri Lankan fishermen from various parts of the island. The Tamils and Sinhalese Christians in northern Sri Lanka have been involved in turtle exploitation for generations, and the Tamils in Jaffna were renowned as accomplished turtle catchers, using a variety of nets to capture sea turtles (Frazier, 1980b; Hewavisenthi, 1990; de Silva, 2006). Decades ago it was estimated that 1,500 turtles were caught yearly in Jaffna alone, and about 2,000–3,000 animals on the entire island (Hoffmann, cited in Frazier, 1980b, 1982). There are reports of relatively high rates of catch by Jaffna fishermen: 100 turtles taken in one fisherman's nets over a 4- to 5-day period (Somander 1963), and 78 turtles, almost all of which were *L. olivacea,* being caught in a single net (Hoffmann, cited in Kar and Bhaskar, 1982). Over 3 days, 16 turtles were butchered at Negombo village (Perera, 1986), and a daily average of 10 turtles were landed and butchered at Kandakkuliya village (Gunawardane, 1986); in both cases with these west coast villages, the majority of the turtles were *L. olivacea.* At Kandakkuliya, 13 *L. olivacea* were butchered in one morning, and it was estimated that an average of at least 20 per week were killed, yielding an estimate of over 1,000 turtles killed annually in just Kandakkuliya, a village with an estimated 1,000 fishermen (Kapurusinghe and Cooray, 2002).

Dattatri and Samarajiva (1982) reported carapaces of this turtle in nearly every coastal village of Sri Lanka (except on the east coast, which was not surveyed); in one village north of Puttalam on the west coast, they counted 250 carapaces during a single visit, and they considered *L. olivacea* to be a common and heavily exploited species. Several other reports from the early 1980s (Wickremasinghe, 1981) concurred that this was the most common species. Remarkably, surveys in the 1990s indicate that *C. mydas* is the most common species of marine turtle in Sri Lanka, and this has led to suggestions that there may have been a significant decline in *L. olivacea* in Sri Lankan waters (Kapurusinghe and Cooray, 2002). However, because there are no specific studies or long-term monitoring, the situation is unclear, and a variety of estimates have been offered for different years and coastal areas. Finally, it is not often discernible if figures for turtle exploitation in Sri Lanka deal with directed take, stemming from a specific fishery, or from incidental take, supplemental to another fishery, but with the turtles eagerly kept for consumption (Kapurusinghe and Cooray, 2002). De Silva (2006) provides a summary of historic information from Sri Lanka, indicating that at least during the later part of the nineteenth century and early twentieth century thousands of *L. olivacea* were captured each year in directed fisheries.

Directed take of turtles is not usual in India, particularly at sea. Clear exceptions seem to have occurred in certain places in the states of Tamil Nadu, Andhra Pradesh, Orissa, and West Bengal, which together form the western shore to the Bay of Bengal. The Tamils of southern India were, like the Tamils of northern Sri Lanka, accomplished turtle fishermen. In the Gulf of Mannar and Tuticorin, marine turtles have been caught for generations in special turtle nets (Kuriyan, 1950; Valliapan and Pushparaj, 1973). It has been estimated that during the middle of the twentieth century, some 5,000 turtles, of which approximately 75% were *C. mydas,* were landed annually in southern Tamil Nadu for both local consumption and export (Jones and Fernando, 1968). For the period between 1971 and 1974, it has been estimated that annual exports of turtle products from this fishery, mainly to Japan, the United States, and Europe, ranged between 1,000 and 2,500 kg (Murthy, 1981). There seem to be no detailed figures on the species composition, but *L. olivacea* should have comprised at least some of the catch; recent years appear to have experienced a relative increase in exploitation of this turtle, as the preferred *C. mydas* has become less abundant (Bhupathy and Saravanan, 2003, 2006). However, with the exception of the extreme south of the state, where turtles are (or were) intentionally captured for local consumption of both meat and blood as well as for export, most people in Tamil Nadu do not consume turtles, and intentional capture is currently relatively uncommon (Bhupathy and Saravanan, 2003). Nevertheless, illegal meat and blood are still sold in "underground" markets (K. Shanker, personal observation).

In Andhra Pradesh, the state immediately to the north, there has not been an organized turtle fishery. Nonetheless, it was reported that some 800 *L. olivacea* were caught for local trade, of which 648 were captured offshore during mating (Raja Sekhar and Subba Rao, 1988). This, however, seems to have been an isolated occurrence for this state, for there is a religious taboo on eating turtle in the fishing communities in Andhra Pradesh (Tripathy and Choudhury, 2001; Tripathy et al., 2006).

Orissa, the next state to the north, is world famous for the massed nesting beaches of *L. olivacea.* Although the inhabitants of the state are not known for eating marine turtles, the large massed nesting populations, particularly off of Gahirmatha, have attracted fishermen from West Bengal and Bangladesh, where seafood and other aquatic, freshwater animals are openly consumed, even by some Brahmins (the Hindu caste in which people typically have the most restrictions to their diet, including a ban on animal matter). As a result, there have been intense fisheries for *L. olivacea* in coastal Orissa (Biswas, 1982; Silas et al., 1983a, 1983b; Kar and Dash, 1984).

Kar and Dash (1984), Das (1985), Dash and Kar (1990), and Chada and Kar (1999) summarize the directed take of *L. olivacea* in Orissa. At first there was no organized fishery, although turtles caught incidentally in fishing nets were brought to shore and sent overland to markets. The massed nesting beach at Gahirmatha was first described in 1974, and by the mid-1970s, a turtle fishery quickly became organized. In 1975 the government of Orissa prohibited the collection of eggs and the capture of adult turtles, but at-sea captures continued (Kar, 1980). Turtles in the waters offshore of Devi and Astaranga, which is south of Gahirmatha, were caught at sea, landed, and sent by boat, truck, bus, and train to Howrah, the main fish market of Calcutta. Most Oriyas do not purposefully kill turtles because of religious reasons, and this fishery was operated mainly by immigrants or temporary inhabitants from West Bengal and Bangladesh. It was estimated that between 1975 and 1983 about 1,000 fishermen were involved in the fishery, which lasted 5–6 months a year while the turtles were massed in the offshore waters of Orissa. Turtles were simply captured directly from the surface of the sea. Although there is no detailed study, the numbers involved obviously were very large; for example, during just the 3 months from November 1974 through January 1975, more than 6,000 turtles are known to have been booked from various train stations in Orissa for delivery to Howrah. Because turtles were often booked as "fishery products" or with false documentation (to avoid prosecution for trading in legally protected species), it is impossible to know the precise numbers involved. But it is estimated that between 50,000 and 80,000 turtles of adult size and of both sexes were taken from Orissa each year until the 1982–1983 nesting season, when the authorities finally were able to patrol the offshore areas, and only about 10,000 turtles are thought to have been taken that year. However, an illegal fishery continued for a number of years later (for more details see Dash and Kar, 1990; Chada and Kar, 1999). A major demographic change that is thought to have promoted this boom in turtle captures is the mechanization of fisheries beginning in the early 1970s.

Directed fisheries for these turtles in West Bengal waters have probably been active for decades, and people from this state routinely went to Orissa to exploit eggs and turtles. A well-organized fishery, transport, and marketing network was described in 1982, and it was reasoned that this was promoted by a lack of success in other fishing activities (Silas et al., 1983b); instead of simply keeping turtles incidentally caught in nets, fishermen would specifically look for *L. olivacea* floating at the surface and catch the animals by hand. Raut and Nandi (1988) provided a summary of the fishery in West Bengal, explaining that it was a "supplementary fishery," but nonetheless, the practice had become quite common by the mid-1980s; they estimated that nearly 20,000 were captured between November 1983 and February 1984. The West Bengal and Orissa fisheries were probably based on the same population of turtles, and because the two states are neighbors, it is difficult to interpret figures and estimates of turtle capture in simple terms of state fisheries; in the main, the figures are most likely to apply to captures made in Orissa's waters.

Marine turtles are consumed by aboriginal ethnic groups as well as most settlers in the Andamans and Nicobars. There are descriptions of turtle-hunting methods as well as rituals for Great Andamanese (Man, 1883), and turtle spearing has been reported more recently (Bhaskar, 1979). Although most of the turtles taken from these two island groups are likely to have been *C. mydas,* there are significant numbers of *L. olivacea* in these waters (Bhaskar, 1993; Andrews et al., 2001, 2006), so it is also likely that they have been subjected to directed take.

Very few communities in Bangladesh consume turtle meat, but some turtles are taken for the souvenir and curio trade, which seems to be growing (Rashid and Islam, 2006). There appears to be no information on directed take of *L. olivacea* at sea from countries on the eastern side of the Bay of Bengal: Bangladesh, Myanmar (Burma), Thailand, and Malaysia. This indicates that if there is a fishery, the numbers taken annually are not large. In those areas where the human population is largely Muslim (Bangladesh, western Myanmar, southern Thailand, and Malaysia), there would be relatively little attraction in catching turtles, for they are usually not eaten.

INCIDENTAL CAPTURE. Over the past decade it has become clear that, although not well understood or studied, by-catch is a significant source of mortality to marine turtles in this region. As with information on directed take in Sri Lanka, there are various estimates on incidental capture, ranging from a total of 400 turtles annually for the entire island (Jinadasa, 1984) to 16 turtles in a three-day period at Kandakuliya fishing village (Gunawardene, 1986) to 10 turtles per day for just one village (Perera, 1986). In the last two cases, *L. olivacea* was observed to be the species most commonly captured.

An attempt to systematically estimate by-catch in Sri Lanka was conducted through a network of fishermen interviews for 12 months, between November 1999 and October 2000 (Kapurusinghe and Cooray, 2002). The largest figure comes from Galle village, in the southwest, where it was estimated that more than 2,000 turtles have been caught annually by the many vessels that harbor there and that this level of incidental capture probably has been occurring for at least a decade. On the basis of the study, it was estimated that 5,241 turtles were caught incidentally, of which 1,626 (31%) were reported to be *L. olivacea.* A wide variety of gear is used, including at least six types of nets and seines as well as hooks of various sorts. Most turtles are captured in nets, particularly bottom drift gillnets (e.g., ray and shark nets), gillnets (e.g., those that target flying fish), and tuna nets.

Although Kapurusinghe and Cooray's (2002) study is one of the few investigations in the region specifically focused on incidental capture, the results must be interpreted carefully. The 12-month survey was restricted to the southern and about half the western coasts, less than 30% of total coast. Clearly, much of the northern area, renowned for its well-established turtle fishery, has been dangerous to access because of civil war. In those areas where access was less dangerous, it was clear that in many cases the numbers reported during the survey were underestimates because fishermen were afraid of being persecuted for catching and consuming legally protected turtles. Despite laws and religious beliefs, turtles are commonly killed and consumed, and there is a widespread, if small-scale, directed fishery; in many cases incidental catches are eagerly exploited, sometimes with the explanation that this compensates for damage to gear caused by the turtles (Kapurusinghe and Cooray, 2002). Other studies have also concluded that incidental catch in Sri Lankan waters is an important source of turtle mortality (de Silva, 2006).

A survey of the Tamil Nadu coast between November 2000 and April 2001 yielded a record of 377 dead, stranded *L. olivacea,* more than 60% of which came from Nagapatinam, just north of Point Calimere and Sri Lanka (Bhupathy and Saravanan, 2002, 2003). There has been a market in the southern part of the state, where there was once a specialized turtle fishery, and sale of meat and blood persists even though it is illegal. Hence, turtles caught accidentally in the south would likely be channeled to an underground market, and it would be difficult to get reliable figures on incidental take. Although there was once a turtle fishery in Tamil Nadu, the majority of turtles consumed (illegally) today are from incidental capture, which is widespread along the coast (Bhupathy and Saravanan, 2003, 2006; K. Shanker, personal observation).

Between October and December 2000, a research trawler that operated in the northern part of Andhra Pradesh caught only *L. olivacea,* with an average rate of one turtle per 90-minute trawl. In November most were males, and in December the proportion of females increased; this was taken as evidence that the turtles were migrating through Andhra Pradesh en route to the massed nesting areas in Orissa. There were also reports of turtles being caught in gillnets and beach seines. Between May 2000 and March 2001, a total of 806 dead turtles, only 2 of which were not *L. olivacea,* were recorded, mainly along the northern coast; more than 70% of these carcasses were female (Tripathy et al., 2003, 2006). In comparison, a 3-year study beginning in 1984 produced an estimate of 577 stranded turtles along the northern coast (Raja Sekhar and Subba Rao, 1988), the majority of which were likely to have been *L. olivacea*. The apparent upsurge in strandings in recent years is likely to be from increased fishing effort with gear that is accidentally catching and drowning turtles in the waters of Andhra Pradesh, and incidental catch is regarded as a major problem in this state (Tripathy et al., 2006).

By far the most remarkable figures on incidental capture of *L. olivacea* anywhere are from Orissa. In the mid-1970s, mechanized shrimp trawling began to increase, and even then there were indications that trawlers were taking turtles at sea (Kar, 1980). Beginning in the 1982–1983 season, extraordinary numbers of turtles were documented washing up dead, with from 55 to 150 carcasses per 100 m of beach, and an estimated 7,000–7,500 carcasses over a 15-km stretch of beach (e.g., Silas et al., 1983a). During the 1983–1984 nesting season alone, it was estimated that more than 3,000 carcasses of *L. olivacea* stranded along a 10-km stretch of beach in the Gahirmatha study area, and an additional 1,000 carcasses were on beaches to the north (Dash and Kar, 1990). Every nesting season since then, for the past two decades, thousands (if not tens of thousands) of turtles have stranded dead on the beaches of Orissa. More intensive beach surveys were begun in the 1993–1994 season (Pandav et al., 1994, 1997), and these continued for a period of six seasons, until 1998–1999. A peak in stranded carcasses was reached in the 1997–1998 season, when 13,575 *L. olivacea* were counted on the beaches from Gahirmatha south some 350 km to the border with Andhra Pradesh, and over the six seasons more than 46,200 carcasses were counted. There was a clear relationship between number of trawlers active in a zone and the numbers of carcasses that were counted in that zone (Pandav, 2000). Since then,

10,000–15,000 dead turtles have been counted on the Orissa coast each year (Wright and Mohanty, 2006).

However, trawlers are not the only source of incidental capture and mortality. In the same areas of coastal Orissa where an estimated 900 trawlers operate, there are also an estimated 5,000 vessels using gillnets. On February 17, 2002, a portion of gillnet washed ashore at Gundalba Beach with 205 dead, rotting *L. olivacea* entangled. By this date, about two-thirds of the way through the season, more than 10,000 carcasses had been tallied, and it was reported that some 75,000 carcasses had been counted in the six seasons from 1996–1997 to 2001–2002 (Wright and Mohanty, 2002).

There is evidence that the fishing effort in Orissa has more than doubled in the last two decades, with substantive increases in incidental take. For example, it is estimated that there has been a threefold increase in mechanized fishing craft in India, from 19,210 in 1980 to 47,706 in 1994, while the numbers of nonmechanized craft remained relatively constant during the same period, at about 150,000, although "traditional boats" with outboard motors increased from 0 to 36,000 (Rajagopalan et al., 1996). These increases in mechanized fishing vessels, gillnets, and trawl nets, and port facilities (Dash and Kar, 1990) have promoted substantial increases in the overall fishing effort in the area where the turtles concentrate, so it is to be expected that large numbers will be caught incidental to fishing operations, even though the animals are legally protected and much of the area where they are caught is within a marine reserve. Clearly, incidental capture is a major source of ridley mortality in Orissa (Pandav et al., 2006).

During a recent coastal survey in West Bengal, 514 dead *L. olivacea* were recorded, all of which were attributed to trawlers (Roychoudhury, 2001). With an estimated 2,000 trawlers in the state, and a "rough estimate" of 20 turtles (nearly all *L. olivacea*) caught per trawler per year (Chowdhury et al., 2006), it is clear that incidental capture is a major source of turtle mortality in West Bengal. There is little information from the Andaman and Nicobar Islands, but it has been estimated that 2,000–3,000 turtles (of all species) are caught in shark nets each year, just in the Andaman Islands; turtle entanglements in discarded net is reported from both Andaman and Nicobar Islands (Andrews et al., 2006). There could also be significant illegal fishing within the EEZ by vessels from neighboring nations, particularly from Southeast Asia, and this could have a significant impact on turtles in pelagic waters, notably *L. olivacea*.

Evidently, the modernization of the fishing fleet in Bangladesh has resulted in increased mortality to marine turtles, which is greatest during the nesting season (Rashid and Islam, 2006). Various types of gillnets and drift nets used around St. Martin's Island, Bangladesh, are reported to catch marine turtles, and mechanized trawlers also operate in these waters. More than 51 carcasses of *L. olivacea* stranded during the 2000–2001 season. In 2001, fishermen reported seeing "numerous dead turtles" floating in oceanic concentrations, and it was suggested that, given the size of the phenomenon, there could have been hundreds or thousands of turtles (Islam, 2002). However, there are no details on rates of catch or to what extent *L. olivacea* are affected. Additionally, there is evidence that fishermen will kill turtles found entangled in nets, for they are regarded as bad omens (Islam, 2002). However, conflicting evidence indicates that some fishermen will not harm marine turtles in Bangladesh, as it is thought to bring bad luck (Das, 1989).

There seems to be little systematic information on incidental capture of *L. olivacea* from the eastern shores of the Bay of Bengal: Myanmar, Thailand, or Malaysia. However, given the fact that there are many fishing activities, and fishing effort generally has increased throughout the region, it is to be expected that incidental catch is a significant issue.

INTERACTIONS OTHER THAN CAPTURE. The usual threats, including marine and land-based pollution as well as lost and discarded fishing gear, are likely to be most problematic where both fishing and coastal development are most intense. In Sri Lanka, mining and construction industries have destroyed large areas of reef, and tourism and beach armoring have disturbed important nesting areas (Kapurusinghe and Cooray, 2002; de Silva, 2006). Various types of land-based pollution, stemming from increased coastal development, are of concern in Andhra Pradesh

(Tripathy et al., 2006), and the same is true in Bangladesh (Rashid and Islam, 2006). A fertilizer plant in Paradeep, Orissa, discharges effluents into the Mahanandi River, which wash directly out to sea in the vicinity of Gahirmatha. The discharges include phosphogypsum together with radioactive radium-226, which releases radon, fluorine, sulfuric acid, and sulfur dust. Preliminary surveys revealed evidence of marked environmental and trophic perturbations, and there is concern that the turtles downstream of the plant will be affected, not to mention the thousands of people who have been injured (Anon, 2002; Mohanty, 2002a, 2002b). Developing and planned port facilities, such as Dharma, which is within kilometers of the Gahirmatha nesting beach, have raised concerns among turtle conservationists (Sekhsaria, 2004a; Pandav et al., 2006), as have offshore oil exploration and extraction activities on the Orissa coast (Sekhsaria, 2004b); it is argued that this can only have serious impacts on the large seasonal concentrations of *L. olivacea*. In the Andaman Islands, plastics and other debris are abundant in several channels where turtles also concentrate and where boat strikes also occur; dumping from shipping and bilge wastes also may be a significant problem (Singh et al., 2003). In general, however, there appears to be relatively little specific information from this region on what are known from other regions to be serious threats to marine turtles.

CONCLUSIONS. General reports from the eastern Indian Ocean on directed take and the eating of turtle meat are not all directly relevant to *L. olivacea*. The general information on meat consumption, however, does indicate that at least some people in these places are not averse to killing these turtles if they come upon them, despite generalizations about religious and cultural taboos. Northern Sri Lanka, southern Tamil Nadu, and West Bengal clearly have the strongest customs of catching and consuming marine turtles, and there is some evidence that populations of *L. olivacea* may have been affected.

Nevertheless, all indications are that incidental capture not only is more widespread but results in considerably more mortality: routinely, the source of turtle meat is incidentally caught animals. Obtaining reliable estimates of rates, or even numbers, of incidentally caught *L. olivacea* is not easy. The recent estimates from Sri Lanka, although based on one of the few studies that have specifically addressed by-catch, apply to less than half the national coastline; in addition, there were clearly cases of underreporting in an attempt to avoid prosecution (Kapurusinghe and Cooray, 2002). In an estimate of the mortality of marine turtles along the entire coast of India, Rajagopalan et al. (2001) suggested that between 1997 and 1999 from about 2,000 to 3,000 turtles were caught or stranded yearly (excluding the coast of Gahirmatha) and that about half of these were dead turtles that had washed ashore. However, on the basis of the estimates from detailed state surveys, involving on-the-ground field work for just the four states of eastern India (e.g., Pandav, 2000; Roychoudhury, 2001; Bhupathy and Saravanan, 2002, 2003; Tripathy et al., 2003), the number of turtles stranded appears to be much greater than that estimated by Rajagopalan et al. (2001). Moreover, because a dead turtle at sea will be driven according to winds and currents, not all carcasses will wash ashore (Epperly et al., 1996). It is not known what proportion of the total number of dead turtles were counted on the beaches, possibly half or less.

For years it has been known that trawlers catch and kill large numbers of *L. olivacea* in the waters of Orissa, but recent studies have shown that a variety of gear, including several kinds of nets and hooks, can catch and kill substantial numbers of turtles (Kapurusinghe and Cooray, 2002). In India it was estimated that gillnets catch nearly four times as many turtles as trawlers (Rajagopalan et al., 2001), but because the overall estimates may be questionable (see above), it is unclear how to interpret the findings on relative effects of different types of gear.

There has been a growing awareness of the pressing need to develop gear modifications and strive for less destructive fishing practices, including the use of TEDs, in various countries of the region. In India this has been the source of tremendous conflict (Choudhury, 2003; Behera, 2006), but there is hope that, at least in Andhra Pradesh, there will be greater responsibility by the fisheries sector (Sankar and Raju, 2003, 2006). The situation in Orissa, where the turtle mortality is greatest, is less clear. These boat owners are politically powerful and aggressive (Pandav, 2003), and there has been at least one incident of a guard of the Orissa Forest Department being killed by a boat crew (WPSI, 2003).

The eastern Indian Ocean is of tremendous importance for *L. olivacea,* with some of the largest breeding concentrations in the world. Logically, this region must also have some of the largest concentrations of migratory turtles, but relatively little is documented. Tag returns from southern Tamil Nadu (Bhupthy and Saravanan, 2002) as well as Sri Lankan waters in the Gulf of Mannar (Kapurusinghe and Cooray, 2002) indicate that these may be important foraging and migration areas for the turtles that nest in Orissa. Further evidence comes from opportunistic reports of large concentrations of "migrating" *L. olivacea* in these areas (Hoffmann, cited in Kar and Bhaskar, 1982; Silas et al., 1984; Bhupthy and Saravanan, 2002).

Recent evidence indicates that numbers of *L. olivacea* have declined throughout the eastern Indian Ocean. In Sri Lanka it is common for fishermen to kill and sell turtles that become entangled as a means of compensating for damage to nets and loss of fish catch (Kapurusinghe and Cooray, 2002). Because *L. olivacea* evidently had been one of the most commonly caught species until the 1990s (Wickremasinghe, 1981; Dattatri and Samarajiva, 1982), significant numbers of these turtles have been caught and killed in Sri Lanka for many years. It has been suggested that incidental capture may be one of the most important sources of mortality in Sri Lanka and that this has provoked a decline in numbers (Kapurusinghe and Cooray, 2002). There are indications that the enormous nesting population in Orissa has been affected by both directed take and incidental capture (Shanker et al., 2003). Much smaller nesting populations in the region also have declined, including Bangladesh, notably St. Martin's Island (Islam, 2002); Myanmar (Thorbjarnarson et al., 2000); and Thailand, notably Phra Thong Island (Limpus, 1995; Aureggi and Chantrapornsyl, 2003).

The eastern Indian Ocean not only has some of the largest concentrations of *L. olivacea* in the world but also has some of the largest, and most impoverished, populations of people. Despite religious and cultural beliefs that protect turtles, and the acknowledged need to control destructive fishing practices, it seems unlikely that present threats to these animals are going to subside in the immediate future (Shanker and Pilcher, 2003). Numerous public education and community-based programs have made great advances in increasing the awareness of the need for protecting turtles from the diverse and large sources of mortality, and these initiatives should continue to have important effects (e.g., Kutty, 2001; Dharani, 2003; Dongre, 2003; Shanker, 2003). There also have been recommendations to allow for a controlled, and much reduced, annual take of mainly injured and deformed turtles in Orissa, as a means to provide a source of protein and give local people a direct benefit from the resource (Dash and Kar, 1990); however, this proposal has not been explored, much less implemented.

Southeast Asia: "Greater Sunda Sea"

The waters of Southeast Asia (excluding the Andaman Sea), including the Gulf of Thailand, Gulf of Tonkin, South China Sea, Celebes Sea, Java Sea, Banda Sea, Timor Sea, Arafura Sea (much of which is taken together as the Sunda Sea), and extending to the Gulf of Carpentaria, have variously been considered to be part of either the eastern Indian Ocean or the western Pacific. In fact, this area has some of the greatest diversity in marine species in the world. Although the surface area is relatively small, there is an enormous, complex coastline with large continental stretches, countless islands, and vast areas of tropical, shallow seas. This provides large areas of nesting and feeding habitat for marine turtles, including *L. olivacea.*

The species is recorded nesting throughout the region, and in most cases there have been clear, even dramatic, declines in annual numbers nesting (Chantrapornsyl, 1994; Kamarruddin, 1994; Limpus, 1994, 1997; Kalim and Dermawan, 1999; Taha, 1999; Shanker and Pilcher, 2003). In general, throughout this region much more attention has been given to *C. mydas, D. coriacea,* and *E. imbricata,* so that relatively little is documented on *L. olivacea* (Limpus, 1997). Apparently, the two largest nesting concentrations are in Terengganu, Peninsular Malaysia, and northern Australia; the former has declined in about 50 years from an estimated 1,000 to a few dozen annual nesters, and the latter hosts an estimated 500–1,000 annual nesters (Kamarruddin, 1994; Limpus, 1994; WMS-EA, 1998). The most commonly discussed problems are the overexploitation of eggs and increases in fishing activities, carried out over decades (Tow and Moll, 1982; Limpus, 1997).

One common problem is the lack of systematic, long-term data, a problem that was underscored in nearly all country reports at the Southeast Asian Fisheries Development Center (SEAFDEC)–ASEAN (Association of Southeast Asian Nations) Regional Workshop on Sea Turtle Conservation and Management, held July 26–28, 1999, in Kuala Terengganu, Malaysia. The numbers of turtles nesting in Malaysia, derived from egg collection data, have declined drastically (Ramli and Hiew, 1999), but few reports provide any information for the Malaysian states in Borneo: Sabah and Sarawak. The turtle populations along Thai coasts on the Gulf of Thailand are thought to have declined, but little detailed information is available (Charuchinda and Chantrapornsyl, 1999). Although there are no detailed data, *L. olivacea* is reported to be the most abundant marine turtle in Brunei Darussalam, but there has been a decline in nesting in the east, especially as a result of coastal development (Taha, 1999). In Cambodia little is known, but marine turtle populations are said to be decreasing, and fishing activities seem to be related to the decline (Try, 1999); little seems to be known from Vietnam (Vinh and Thuoc, 1999). Although the species seems to be relatively uncommon in the Philippines, records are widespread (Cruz, 1999). In Indonesia, it was reported that there has been an increase in nesting at two national parks in East Java, Alas Purwo and Jamursba-Medi (Kalim and Dermawan, 1999); however, the increased numbers reported may result from increased beach-monitoring efforts.

DIRECTED TAKE. In many places throughout the "Greater Sunda Sea" marine turtles are not eaten for religious and cultural reasons. However, in several places in Indonesia turtles are taken for both consumption and trade, and they have been caught at sea with nets, harpoons, and "Hawaiian slings" (Polunin and Sumertha Nuitja, 1982). It was reported that there is no "significant" take in countries neighboring Australia (Limpus, 1997); however, it is believed that the species is severely affected by directed take of nesting females along much of the coast of Vietnam (TRAFFIC, 2004). What directed take does occur is poorly understood and rarely documented, so it may be prudent to reserve judgment about whether or not directed take has been significant until more information is available.

INCIDENTAL CAPTURE. Over the past decades fishing effort has increased enormously throughout the region, and although it has been known for decades that *L. olivacea* are affected, there is little systematic information on their incidental capture and mortality. For example, between 1956 and 1978, marine fisheries landings in Peninsular Malaysia increased more than five times, with much greater effort from gillnets and trawlers, and it was generally known that this had "taken a heavy toll on turtles." It was reported that in 1973 "a bottom long line for rays caused a massive kill of ridleys at Setiu, Terengganu" (Tow and Moll, 1982). Some of the earliest and most detailed information on incidental catch is from the waters off Terengganu, Malaysia, where it was estimated that more than 440 *L. olivacea* were caught annually by trawlers, drift, and gillnets (Chan et al., 1988). Gillnets, long lines, and particularly trawlers were all identified as important sources of mortality for marine turtles in Thai waters (Chantrapornsyl, 1994; Charuchinda and Chantrapornsyl, 1999). Various fishing activities, namely trawling and long-lining, have been known to catch turtles incidentally in Indonesian waters (Polunin and Sumertha Nuitja, 1982). Although in the Philippines *L. olivacea* is not commonly sighted, and numbers are thought to be low, there seem to be records of capture throughout the archipelago (Palma, 1997). There are general comments that *L. olivacea* is "incidentally caught in fisheries operating in the Gulf of Papua," Papua New Guinea (WMS-EA, 1998), and two records of capture by trawlers in this Gulf (Ulaiwi, 1997). In Vietnam there are records of incidental capture, but because turtles are symbols of longevity, if they are still alive when incidentally caught, they are generally released back to the sea (Vinh and Thuoc, 1999). Nonetheless, it is believed that the species is severely affected by incidental capture along much of the country's coast (TRAFFIC, 2004).

In an attempt to quantify anthropogenic sources of mortality in Australia, the Marine Recovery Team had to conclude that the species is "poorly documented" and that the "lack of certain knowledge about this species within Australia is a cause of concern." Both trawls and gillnets were known to cause mortality (WMS-EA, 1998). There are reported to be "low numbers as by-catch in the Northern Prawn Trawl" (Harris, 1994, cited in WMS-EA, 1998), but the conclusion that this was not of immediate concern has been criticized, and it was argued that trawling and other gear do pose real threats to turtles, particularly *L. olivacea,* about which so little is known in these waters (Guinea and Whiting, 1997). For the population that occurs in north Australian waters, Limpus (1997) reported that "the species is drowned sufficiently commonly in fishing gear (prawn trawl and gillnets) to raise the concern that this species is probably as threatened as the loggerhead turtle in Australia." There is at least one record of an estimated 300 turtles being drowned in just one 2,000-m shark gillnet over a two-week period in northern Australia, 85% of which were *L. olivacea* (Guinea and Chatto, 1992; Limpus, 1994; WMS-EA, 1998).

INTERACTIONS OTHER THAN CAPTURE. Several locations in the region have active petrochemical ventures, including offshore exploration and extraction as well as onshore refineries and other activities. Water pollution and destruction of feeding habitat by irresponsible fishing activities were recognized as problems in Thailand (Charuchinda and Chantrapornsyl, 1999). Chan and Liew (1988) drew attention to the fact that these effects represent serious threats to marine turtles in the area, but because of a lack of fundamental information, they were compelled to recommend that basic studies be carried out. It is unclear if the institutions responsible have advanced much in this regard.

It also has been suggested that plastics may be a major source of pollution and mortality in this region (Limpus, 1994), and discarded and lost nets are known to kill turtles in Australian waters (WMS-EA, 1998), but little systematic information is available. In the mid-1990s, scattered accounts from northern Australia began to document stranded turtle carcasses, including *L. olivacea,* entangled in marine debris, including net webbing (Chatto et al., 1995). In 1996, in response to mounting quantities of marine debris on the coast of the Grove Peninsula, Northeast Arnhem Land, Northern Territory, Australia, yearly coastal surveys were begun, and a 70-km stretch of beach has been monitored between April and August/September every year. By the end of 2003, a total of 194 entangled turtles have been recorded, of which 25 were *L. olivacea.* Debris appears to concentrate in certain areas, such as in the vicinity of Cape Arnhem and Port Bradshaw, and it is mainly foreign trawl and drift net webbing, especially of Asian origin. It is during southeasterly (onshore) winds when most of the debris is pushed ashore, and stranded, entangled marine turtles are found primarily between late April and early August every year when the Southeast winds are at their strongest. There has been considerable variation in numbers of stranded animals from year to year (Roeger, 2004; S. Roeger, in litt., June 3, 2004). There are also reports of *L. olivacea* entangled in ghost nets in the Timor and Arafura Seas, north of Darwin (D. White, in litt., June 2, 2004). If the figures from these studies are extrapolated over the vast area of Austral-Asia in which marine debris are likely to pose serious threats, it is clear that the numbers of turtles negatively affected must be very large.

CONCLUSIONS. The fact that *L. olivacea* is not generally common in the "Greater Sunda Sea," and other species of marine turtles have represented important sources of food and income, means that relatively little attention has been paid to this species in this region. For example, a recent report on trade in marine turtle products in Vietnam (TRAFFIC, 2004) focused mainly on *Eretmochelys imbricata* and to a lesser extend *Chelonia mydas,* with only a few lines about *L. olivacea.* Coupled with the generalized lack of systematic information on human impacts, including directed take, incidental capture, and interactions other than capture, the result is a tremendous lack of basic information on *L. olivacea*. Ironically, Manila Bay, Philippines, is the type locality for this species.

Western Pacific Ocean

It is generally assumed that *L. olivacea* is uncommon throughout most of the western Pacific, there being neither large nesting populations nor feeding aggregations. For example, although there have been decades of careful study of marine turtles in Japan, this species has been documented only rarely and is thought to be astray in these waters (Abe, 1999). Nonetheless, the species does occur widely throughout this vast area (Eckert, 1993; Pritchard and Plotkin, 1995; Limpus, 1997; NMFS-USFWS, 1998). If, however, areas important for feeding are found, research and conservation priorities will need to be revised accordingly: it is not unlikely that coastal waters of southern China host significant numbers of these turtles, at least at certain times of the year.

DIRECTED TAKE. Despite the lack of basic information, it is generally assumed that there is little significant directed take of *L. olivacea* in the western Pacific (e.g., Limpus, 1997). Other species, however, have been part of directed fisheries, particularly *C. mydas* and *E. imbricata*.

INCIDENTAL CAPTURE. It is known that *L. olivacea* are caught incidentally by trawlers and gillnets along the east coast of Australia, particularly Queensland, but no details seem to be available (WMS-EA, 1998). There is believed to be no threat to this turtle from incidental take in U.S. territories in the eastern Pacific (NMFS-USFWS, 1998); however, there are some records of capture on long lines and plankton nets around Guam (Eckert, 1993). Many of the carapaces found in coastal villages of southern China in 1985 had come from incidental captures in fishing activities that were being modernized in the early 1980s (Frazier et al., 1988).

INTERACTIONS OTHER THAN CAPTURE. The usual problems from marine and land-based pollution, plastics, debris, and discarded fishing gear will be generally relevant to *L. olivacea* in the western Pacific. From the past history of industrial pollution in Japan, mercury for example, and the rate at which China is modernizing, it is to be assumed that there will be many different anthropogenic assaults on all species of marine turtles in this region.

CONCLUSIONS. Even if the numbers of *L. olivacea* in the western Pacific really are as small as they are assumed to be, there could be arguments for strengthening research and conservation. However, it will be necessary to determine if the western Pacific serves as a sink for turtles dispersing from the very large populations in the eastern Pacific or if the *L. olivacea* that are found in the region have unique genetic qualities.

Eastern Pacific Ocean

Lepidochelys olivacea is by far the most abundant marine turtle in the eastern Pacific, with nesting recorded from northern Mexico, evidently as far north as Baja California (Márquez-M. et al., 1976; Fritts et al., 1982) to as far south as northern Peru (Hays Brown and Brown, 1982) and massed nesting (arribada) sites in Oaxaca, Mexico (Escobilla and Morro Ayuta), Nicaragua (La Flor and Chacocente), and Costa Rica (Nancite and Ostional) (Márquez-M., 1990; Eckert, 1993; Pritchard and Plotkin, 1995; NMFS-USFWS, 1998). There is abundant evidence that large numbers of *L. olivacea* in the eastern Pacific Ocean live on the high seas, hundreds and even thousands of kilometers from a continental shore (Pitman, 1990, 1992; Arenas and Hall, 1992; IATTC, 2003a). In certain places there are remarkable concentrations at sea, such as along a band at about 5°N extending west from the South American continent to about 125°W (IATTC, 2003a).

These high densities, with tens to hundreds of thousands of turtles massing to nest on a kilometer or so of beach, attracted one of the, if not the, most intense directed fisheries for marine turtles ever documented. At the same time, the extraordinary densities of turtles aggregated in east Pacific waters presented other singularly attractive areas for at-sea captures, distant from coastal waters. In more recent times, these same concentrations of turtles present high probabilities of negative effects from various fisheries and other human activities in marine environments. In the waters around Pacific islands, however, the species is considered to be rare (Pritchard and Plotkin, 1995). It was concluded that "lack of knowledge concerning the abundance and distribution of *L. olivacea* in the northeastern Pacific constitutes a threat, particularly since important foraging grounds have not been identified" (NMFS-USFWS, 1998), an observation reflected by other authors who have reviewed the status of this turtle (e.g., Eckert, 1993).

DIRECTED TAKE. There is no solid evidence for regular, directed exploitation on *L. olivacea* in the eastern Pacific by preindustrialized societies. Traditional fisheries by the Seri (Comcáac) of northern Mexico were focused on *C. mydas* (McGee, 1898; De Grazia and Smith, 1970; Smith, 1974; Felger and Moser, 1991; Nabhan, 2003), and there is no evidence that *L. olivacea* were taken in significant numbers by these people. However, in the early 1960s, with the worldwide decimation of crocodilian populations but insatiable markets for luxury reptile skins, wildlife traffickers turned to alternate sources, especially skins of marine turtles. Because *L. olivacea* occurs in extraordinarily dense concentrations in the eastern Pacific, particularly during the reproductive season, when hundreds of thousands of reproductively active turtles aggregate near nesting beaches, this turtle was an easy target. The massed nesting populations in three different Mexican states, Mismaloya, Jalisco; Piedra del Talcoyunque, Guerrero; and Escobilla and Morro Ayuta, Oaxaca, supported major fisheries in which the animals were taken either from immediately offshore or from the nesting beaches, although the latter was banned a few years after the large-scale fishery began (Márquez-M., 1996). In the early years, some 40 kg of live turtle was reduced to about 5 kg of flipper skins, and the rest of the carcass, including some 15 kg of edible muscle and other tissues (Márquez-M. et al., 1976), was discarded. By 1968 a peak was reached with a take of more than 300,000 turtles reported by the Instituto Nacional de la Pesca (National Institute of Fisheries) (Márquez-M. et al., 1976; Frazier, 1980c).

In 1968 there were attempts to regulate the fishery, and in mid-1971 the government declared a "total ban," with a "moratorium" that officially lasted for about 18 months. A complex regulatory system was put in place, including quotas and closed seasons, with authorizations in which fishing cooperatives were to have exclusive rights to the fishery. However, "franchises" and special concessions were made as political favors, and other special interests took precedence over technical decisions (Márquez-M., 1996; see also Riding, 1985), totally undermining regulatory recommendations made by Mexican marine turtle specialists. In addition, reporting and implementation left much to be desired. The regularization of the fishery made it illegal to capture turtles on nesting beaches, and processors were obliged to make efficient use of all the body parts, even the bones. Official figures indicate that after 1969 the annual take was consistently below 100,000 turtles. However, estimates by the owner of the processing plants indicated that annual takes of more than 100,000 turtles continued until the late 1970s (see also Mack et al., 1982). Based on both official and unofficial reports, it was estimated that by 1979 more than 1.6 million turtles had been taken in the Mexican Pacific coast fishery, and nearly all of these would have been *L. olivacea* in reproductive condition (Frazier, 1980c). Indeed, one rough estimate is that "at least 2,000,000 *L. olivacea* would have been landed during the 5 years prior to 1969" (Cliffton et al., 1982).

In 1990 all marine turtle exploitation in Mexico was banned (Aridjis, 1990; DOF, 1990). However, an illegal fishery has continued along most of the Pacific coast of Mexico: there are recurrent reports from all coastal states (Baja California, Baja California Sur, Sonora, Sinaloa, Nayarit, Jalisco, Colima, Michoacán, Guerrero, Oaxaca, and Chiapas) of very large annual takes of marine turtles. For example, in just Baja California, it was estimated that some 35,000 turtles are killed yearly either by illegal hunting or as bycatch (Nichols, 2003a). South of Baja California, *L. olivacea* is one of the most common species taken by the illegal fishery. Moreover, there are reports of organized endeavors (some of which are evidently linked to illegal drug-trafficking operations) in Cacalotepec, Palmito, La Tuza, and Nuevo Ayuta, all in Oaxaca, as well as several coastal villages in Baja California Sur, Sonora, and Chiapas. Both meat and skins are distributed through well-organized networks to border regions of Baja California, Sonora, and Cuidad Juárez, Chihuahua, as well as in León, Guanajuato. There is evidence that thousands of *L. olivacea* are still taken each year along the Pacific coast of Mexico (Cantú and Sánchez, 1999; Gardner and Nichols, 2001; Nichols, 2003b; PROFEPA, 2003; also see the following section on incidental capture).

In recent years there have been reports from Guatemala of *L. olivacea* captured at sea, well off the continental shelf, to be used for shark bait (Higginson, 1989). It is not known how long this practice has been going on, how frequently it happens, or the rates of take. The small amount of information available could be interpreted as evidence that this illegal fishery is erratic, casual, and may involve relatively small numbers of turtles per year; however, there is no adequate information, and because the fishery is against the law, it will be difficult to obtain reliable estimates.

There appears to be no other evidence of directed at-sea take of *L. olivacea* elsewhere on the Pacific coast of Central America, where in general there seems to have been little demand for turtle meat (Cornelius, 1982). According to fishermen in the north of Ecuador, an organized fishery in southern Colombia was supplying skins and other products to traders in Ecuador during the period of that fishery, between 1970 and 1981 (see below). Like the Ecuadorian fishery, much of the activity included catching and skinning the turtles and then jettisoning the carcasses at sea, and the vast majority of the animals are thought to have been *L. olivacea* (Hurtado, 1982). However, even after the large Ecuadorian fishery was officially closed, organized directed takes in Pacific Colombia apparently continued between 1980 and 1985, with some boats taking as many as 50 turtles during an outing (Amorocho et al., 1992). In Ecuador, near Santa Rosa and La Libertad, Guayas Province, there was a fishery for local consumption (Green and Ortiz-Crespo, 1982).

A second legal "industrial" fishery that focused on *L. olivacea* began in Ecuador in 1970, just two years after the Mexican fishery reached its peak, and began a steep decline (Frazier, 1980c; Green and Ortiz-Crespo, 1982; Mack et al., 1982). In contrast to the Mexican fishery, the Ecuadorian turtles were all captured at sea, commonly as much as 50 km offshore. Also in contrast with the Mexican fishery, the turtles in Ecuador were first captured for exporting turtle meat, not skins, although there are accounts that skins from Ecuador were sent to Mexico in the early years of the fishery. In 1973 the documented production in turtle skins began in Ecuador, and by 1975 the quantity of exported skins represented far more turtle equivalents than did the exported meat. Remarkably, there was little domestic demand for turtle meat; it was often added to sausages without informing customers that turtle meat was included (Frazier, 1980c). The estimated annual take rose steadily from fewer than 600 turtles in 1970 to some 150,000 turtles in 1979 (Frazier, 1980c; Green and Ortiz-Crespo, 1982). Ecuador had ratified CITES in 1975, so export of the skins was in contravention to this treaty, and in 1981, the international trade in skins was stopped, thereby complying with CITES regulations (Frazier and Salas, 1982; Hurtado, 1982).

Farther south in Peru, marine turtles have been taken offshore in small-scale fisheries for decades. In the late 1970s it was reported that 7–10 small, uncovered boats from Pisco specialized in turtle fishing, landing an average of 10–30 turtles per day (Hays Brown and Brown, 1982). Nearly 20 years later, a brief survey reported that 50–80 tangle nets were being used in ports south of Lima and that estimated catches could be 30–40 turtles in two days (Arauz, 1999). The vast majority of the turtles caught in Peru are normally *C. mydas;* a city dump at San Andres had remains of 25 recently killed turtles, 17 of which were *C. mydas* and 8 were *L. olivacea* (Arauz, 1999). From limited information on species composition of several landings, it is assumed that fewer than a quarter of the turtles captured in the Peruvian fishery have been ridleys, and one estimate is that only about 1% of the turtles captured in Peruvian waters are *L. olivacea* (J. Alfaro, in litt., June 3, 2004). In March 1995, a total ban was declared on catching all marine turtles in Peru, but this did not stop the fishery. In 2001 another law provided for the legal consumption of turtles caught accidentally, which opened up endless possibilities for the interpretation "accidental catch."

L. olivacea does occur in Chilean waters, but it is not common (Frazier and Salas, 1984), and there is no directed take of any marine turtle. Although formerly Polynesian islanders hunted turtles, today there is no known directed take of marine turtles on oceanic islands of the eastern Pacific, and furthermore, there is no evidence that *L. olivacea* was ever taken in former times.

INCIDENTAL CAPTURE. It has been known for some time that *L. olivacea* in the eastern Pacific is subjected to substantial mortality from incidental capture in shrimp trawls and long lines, but the information available is incomplete (e.g., Pritchard and Plotkin, 1995; NMFS-USFWS, 1998). Hence, estimates of catch numbers and rates often vary considerably, and information on mortality is even more elusive.

The estimated annual take of *L. olivacea* on long lines in Hawaiian waters was 152, and although most of these were released alive, there is no information on postcapture mortality (NMFS-USFWS, 1998). It is estimated that an average of at least 140 turtles in the eastern tropical Pacific die each year in the tuna purse seine fleet, and most of these interactions occur when nets are set around fish-aggregating devices (FADs); the great majority of the turtles affected are *L. olivacea.* Estimated annual mortalities from this fishery between 1993 and 2002 have varied from 30 turtles to 109 turtles identified as *L. olivacea,* and if the figures for "unidentified" turtles (most of which are thought to be *L. olivacea*) are added, the values increase by at least 50% (IATTC, 2003b). It is known that long lines present a substantial risk to marine turtles in the eastern tropical Pacific, but even the scientific staff of the Inter-American Tropical Tuna Commission (IATTC), who have been working with by-catch issues for over a decade, do not have adequate data for making reliable estimates (IATTC, 2003c).

There are occasional records of *L. olivacea* being killed in gillnets and from boat strikes along the west coast of the continental United States and also records of captures in shrimp trawls in the Gulf of California as well as gillnets, traps, pound nets, haul seines, and beach seines in coastal waters of Baja California, but no numbers seem to be available (Alvarado and Figueroa, 1990; NMFS-USFWS, 1998). Data from onboard observers pertaining to incidental takes from shrimp trawlers and shark netters along the Pacific coast of Mexico are carefully guarded as confidential by government fisheries officers, so no detailed information is generally available. However, besides the obvious generality that there must be significant incidental capture in shrimp trawls, there is evidence that large mesh shark nets set along the Pacific coast of Mexico are catching significant numbers of turtles. For example, on July 22, 2004, it was reported that 125 *L. olivacea* were drowned in just one shark net about 50–70 m long, set along the coast near Santa María Tonameca, Oaxaca. Although this incident was reported to have been a case of incidental capture, it is likely that the turtles caught in shark nets are destined for the well-organized black market (see above), so this could also be considered to be part of a directed fishery, masquerading as incidental capture.

As in other parts of the tropics, the rapid development of shrimp trawling along the Pacific coast of Central America, beginning in the 1950s, led to incidental capture of *L. olivacea* as well as other species, and by the late 1970s it has been known that shrimp trawling along the Pacific coast of Central America, particularly Guatemala, Salvador, and Costa Rica, was responsible for the incidental capture and mortality of marine turtles, particularly *L. olivacea,* but there were only rough estimates of the numbers involved (Cornelius, 1982). In Guatemala 150–200 carcasses were estimated to wash ashore yearly, with no information on species composition; for El Salvador there was evidence of a direct relationship between shrimp trawl activity and numbers of carcasses that washed up, most of which were *L. olivacea,* but no numbers were provided; Honduras was thought to have had occasional takes of turtles in cast-net and set-net fisheries; Nicaragua was thought to have had few turtles killed in shrimp trawls; in contrast, estimates from Costa Rican shrimp trawlers varied from 600 to 2,000 turtles caught annually, and up to 45 in one haul or 200 taken per day, most of which were *L. olivacea;* it was claimed that the Panamanian shrimp fleet released the majority of the turtles that they caught, but it was suspected that in reality there was significant mortality. Overall, it was concluded at the end of the 1970s that the incidental capture and mortality in shrimp trawls was one of the major factors causing the decline of marine turtles along the Pacific coast of Central America (Cornelius, 1982).

More recent studies carried out in the 1990s provided estimates that bore out the earlier concerns for incidental mortality in shrimp trawls in Pacific Central America (Arauz, 1996). The estimated annual captures for all species of marine turtle were: Guatemala, 10,000; El Salvador, 21,280; Honduras, 0 (no Pacific shrimp fleet); Nicaragua, 8,000; Costa Rica, 20,762, totaling more than 60,000 turtles. Later studies indicated that catch per unit effort for the Costa Rican fleet varied depending on the type of shrimp being targeted (0.0188–0.0684 turtles/hour drag per 100-foot head rope for white shrimp, and 0.0881–0.238 turtles/hour drag per 100-foot head rope for pink shrimp). Although overall catch rates for turtles may be less than was originally estimated, based on the finding that more than 90% of the turtles caught were *L. olivacea,* and nearly 40% were landed dead (Arauz et al., 1998a), it is clear that tens of thousands of these turtles are caught and killed each year in the Central American shrimp fishery, and this is not considering the Panamanian fishery.

Strandings of dead turtles provide further evidence of incidental capture and mortality at sea. Hundreds of *L. olivacea* carcasses have been counted over a period of only a few months at Chacocente, Nicaragua, and Ostional, Costa Rica, both sites of mass nesting. Although the offshore areas in front of these two beaches are legally closed to fishing, small-scale gillnets, long lines, and also trawlers operate with apparent impunity (Orrego and Arauz, 2005); on rare occasions when authorities have inspected the offshore areas of Chacocente and Ostional Wildlife Refuges, they have found gillnets in no-fishing areas, often with entangled *L. olivacea* (Arauz, 2002). In addition, surveys along the north Pacific coast of Costa Rica over the 6-month period between August 2001 and January 2001 revealed 423 stranded carcasses, 99% of which were *L. olivacea,* at least 84 (20%) of which showed hooks, monofilament lines, fractured phalanges, and knife cuts, indicating clear signs of interactions with different coastal fisheries other than trawls (Orrego and Arauz, 2005).

Just as the gravity of the situation with shrimp trawling was being appreciated, long line fishing began to increase along the Pacific coast of Central and South America; in part, the increase in this fishery resulted from the conversion of trawlers into long liners (see Arauz et al., 2000). Some of the only systematic information available on incidental capture of *L. olivacea* in Latin American waters is from Costa Rica. An initial study yielded catch rates of 14.4124 turtles/1,000 hooks; *L. olivacea* comprised 100% of the turtles caught, and mortality on landing was 0% (Arauz et al., 2000; with corrections from R. Arauz, in litt., June 26, 2004). Additional work involving larger sample sizes, also from within the EEZ of Costa Rica, reported that nearly 94% of the turtles were *L. olivacea,* and including fishes this was the second most commonly captured species; overall catch rate for 39,248 hooks was 6.364 turtles/1,000 hooks; all animals were landed alive (Arauz, 2005). Two additional long line sets west of the Galapagos Islands yielded a catch rate of 19.429 turtles/1,000 hooks, with 55% of the turtles being *L. olivacea* and mortality at landing 8.8% of the turtles (Arauz et al., 2000; with corrections from R. Arauz, in litt., June 26, 2004). As usual for these sorts of studies, there is no information on mortality after release, so depending on the type of hooking and trauma caused by handling, there is likely to be tremendous variation in the survivorship of the turtles released. Extrapolating these results to just the Costa Rican long-line fleet, where 553 vessels are estimated to deploy over 70,000,000 hooks a year, some 396,000 marine turtles are estimated to be caught annually (Arauz, 2005), a large proportion of which would be *L. olivacea.*

There are general comments about incidental take off the Pacific coast of Colombia and the fact that the turtles are consumed (Amorocho et al., 1992), but little quantitative information is available. A mass stranding in 1990, which included some *L. olivacea,* was attributed to incidental capture in shrimp trawls or possibly tuna and shark fisheries (Rueda-Almonacid, 1992). However, cause of death was not determined.

Large numbers of strandings also have been recorded on the mainland coast of Ecuador; in 1999 it was estimated that thousands of *L. olivacea* carcasses had washed ashore. Because the Ecuadorian fleet of shrimp trawlers is active in offshore waters, it was thought that they were responsible, although no definitive determination could be made (Alava et al., 2000). The capture of marine turtles in nets and long lines in Peruvian waters can be categorized as incidental capture, but because it is usual to land the animals, and this is legal, this fishery has been discussed above under directed take. Nonetheless, it is estimated that a relatively small proportion of the turtles caught in Peru are *L. olivacea* (J. Alfaro, in litt., June 4, 2004).

INTERACTIONS OTHER THAN CAPTURE. There are records of both juvenile and adult *L. olivacea* being entangled in marine debris around the Hawaiian Islands (Balazs, 1985). It was surmised that debris causing problems with entanglement and ingestion and boat strikes are minor problems along the west coast of the continental United States (NMFS-USFWS, 1998). There are reports of *L. olivacea* entangled in plastics, discarded sacks from tuna boats, and lost fishing gear (IATTC, 2003a). The webbing that trails from FADs used in open ocean, especially for the tuna fishery, is known to ensnare and kill turtles; there may be some 240 turtles involved annually, most of which are thought to be *L. olivacea* (IATTC, 2003b).

The fact that even major, and scientifically advanced, fisheries organizations such as the IATTC do not have even basic information on the variety of threats posed by nonfisheries anthropogenic activities to marine turtles at sea (IATTC, 2003a) emphasizes the status of knowledge overall.

CONCLUSIONS. The most remarkable at-sea exploitation of any turtles occurred on the Pacific coasts of Mexico and Ecuador. This intense, legal exploitation directed at breeding and feeding aggregations continued for more than a quarter of a century. In addition to the official figures, there are estimates of "unofficial" (black market) figures for exploitation, and in some years they are as large as the official figures for exploitation. By the 1980s, two mass nesting populations in Mexico had been decimated, and after two decades, they still have not recovered: El Playón, Jalisco, and Piedra del Talcoyunque, Guerrero (Martínez and Castellanos, 2006). The mass nesting population at Escobilla, Oaxaca, was also greatly reduced, but after the 1990 complete ban on marine turtle exploitation, it rebounded from some 55,000 nests estimated for 1988 to more than 700,000 nests estimat0ed for 1994 (Márquez-M. et al., 1996) and to over a million nests around 2000 (C. Peñaflores S., personal communication). Although much below the values from the 1960s and 1970s, illegal take of *L. olivacea* continues to this day, especially in remote costal waters of Pacific Mexico: the supply of skins and leather goods made of this species continues to be sold and confiscated in Mexico and at Mexican–U.S. border crossings (Teyeliz, 2000).

In both Pacific Mexico and Ecuador a major motivation for these intense fisheries resulted from the depleted state of crocodile stocks globally, with persistent market demands from producers, tanners, and consumers (Márquez-M. et al., 1976; Márquez-M., 1996). Hence, in both Mexico and Ecuador, the major attraction for exploitation of turtles has been to provide skins: animals weighing some 45 kg were slaughtered primarily for about 2–7 kg of skin, and at least during part of the fishery, the remainder of the carcasses was discarded, although in both Mexico and Ecuador processed meat was sold in both national and international markets. In the case of Mexico, the majority of the animals taken were reproductively active. It has been estimated that from 1965 until 1980 more than two million turtles, the vast majority *L. olivacea,* supplied these major commercial fisheries in the states of Sinaloa, Nayarit, Jalisco, Michoacan, Guerrero, and Oaxaca as well as in Ecuador (Márquez-M. et al., 1976; Cahill, 1978; Frazier, 1980c; Cliffton et al., 1982; Márquez-M. et al., 1990). Hence, in the annals of human–turtle interactions, these fisheries stand out not only for the massive numbers of turtles exploited but also for long-term damage to what had seemed to be robust populations, not to mention wanton waste and greed.

Despite the fact that at least two major massed nesting populations have been lost (Mismaloya, Jalisco, and Piedra del Talcoyunque, Guerrero), the recuperation of the numbers nesting at Escobilla has shown that at least some populations can recover after protection. This should help promote more interest in embarking on long-term conservation programs.

General Conclusions

Human–turtle interactions at sea involving both species of *Lepidochelys* have had enormous importance on both local and global scales. The history of *L. kempii,* with the catastrophic decline in numbers nesting during the last half of the twentieth century, has in many ways been the battle cry of marine turtle conservationists. As a result, this turtle has for decades held the dubious distinction of being recognized as critically endangered, and its situation has been used as a poster child to warn both the general public and policymakers about the distinct dangers of extinction as a result of irresponsible human activities. On the other hand, *L. olivacea* is considered to be the most abundant marine turtle species, and far less attention has been paid to many conservation issues involving this species. Now that there are numerous indications that even the very large populations have experienced declines in all ocean basins, and indeed some massed nesting populations have been exterminated, this species is now receiving considerable attention.

However, although threats in terrestrial habitats are relatively easy to distinguish, marine-based pressures are more difficult to identify and resolve; legislation and regulations pertaining to marine resources are also more difficult to enforce. Nonetheless, it is clear that some of the most significant threats to both species of *Lepidochelys* involve human–turtle interactions at sea. Beyond directed take, both species are known to be seriously affected by incidental capture in shrimp trawls, long lines, and other types of fishing gear. Conservation actions taken in response to these threats have had enormous ramifications on national policies, with direct impacts on international conservation activities and even wider policies affecting trade and international relations. The case of the "shrimp–turtle" dispute before the World Trade Organization is a clear example of the global ramifications that resulted from concerns about the fate of one species of ridley turtle, *L. kempii* (see Frazier, 2002; Frazier and Bache, 2002; Bache and Frazier, 2006).

The problem of fisheries by-catch and responsible fisheries is a global issue that has attracted considerable attention from many different sectors (SEAFDEC, 1997, 1999; UA-SGCP, 1997; APEC, 1998; Moore and Jennings, 2000; Bache, 2003; FAO, 2005). Because marine turtles are charismatic flagship species (Frazier, 2005b), there is considerable interest in many different countries, from different sectors of society, in resolving by-catch problems involving these reptiles.

Innovative experiments with TEDs in Pacific Costa Rica have provided basic information needed to increase the spaces between deflector bars and thereby make the TED more acceptable to shrimpers (Arauz et al., 1998b). The increase in size of the Escobilla, Mexico, nesting population following closure of the fishery also provides impetus for developing long-term conservation programs.

Although both species of *Lepidochelys* provide poignant examples of problems caused by irresponsible human–turtle interactions, they also illustrate cases for hope when concerted conservation efforts are made. The dramatic decline in nesting *L. kempii* has finally been turned around, and at last the species is on the road to recovery (Turtle Expert Working Group, 2000): at least one depressed nesting population of *L. olivacea* recently has shown dramatic signs of increase (Márquez-M. et al., 1996). With the surfeit of conservation problems involving marine turtles, it is essential that the lessons of what should be avoided and corrected be shown in a context of what is possible with adequate conservation activities, with a message that can instill hope for the future.

ACKNOWLEDGMENTS

I thank Jacqueline Castilhos (Projeto TAMAR-IBAMA, Sergipe) for sharing information about *L. olivacea* mortality in Sergipe. Leandro Bugoni, Andrés Estrades, and Alejandro Fallabrino provided information on southern Brazil and Uruguay; Damian White and Scott Whiting provided information on northern Australia; and Juan Carlos Cantú provided information from Mexico. The above geographic sections were compiled by J. Frazier, based on selected contributions provided by a team of specialists: southwestern Atlantic (eastern South America), Johan Chevalier and Matthew H. Godfrey; eastern Atlantic (western Africa), Angela Formia and Jacques Fretey; eastern Indian Ocean, Kartik Shanker; eastern Pacific, Randall Arauz; Mexico, Rene Márquez-M. Manjula Tiwari translated Jacques Fretey's contribution into English; Charles Caillouet and Wallace Jay Nichols made valuable comments on an earlier draft. J. Frazier was supported by a grant from the Department of Conservation Biolgy, Conservation and Research Center, National Zoological Park, Smithsonian Institution.

LITERATURE CITED

Abe, O. 1999. Sea turtle conservation and management in Japan. In *Report of the SEAFDEC-ASEAN Regional Workshop on Sea Turtle Conservation and Management, July 26–28, 1999, Kuala Terengganu, Malaysia*, pp. 108–117.

Alava, J. J., Chiriboga, C., Peñafiel, M., Calle, N., Jiménez, P., Aguirre, W., Amador, P., and Molina, E. 2000. Datos históricos sobre la mortalidad de tortugas marinas en varios sitios de la costa ecuatoriana y observaciones reciente acerca de una mortalidad masiva de *Lepidochelys olivacea* (Reptilia, Cheloniidae). Guayaquil, Ecuador: Fundación Natura.

Alvarado, J., and Figueroa, A. 1990. The ecological recovery of sea turtles of Michoacán, Mexico. Special attention: the black turtle, *Chelonia agassizi*. Final Report 1989–1990, U.S. Fish and Wildlife Serviced and WWF-USA.

Amorocho, D. F., Rubio T., H., and Diaz R., W. 1992. Observaciones sobre el estado actual de las tortugas marinas en el Pacífico Colombiano. In Rodríguez Mahecha, J. V., and Sánchez Paez, H. (Eds.). *Contribución al Conocimiento de las Tortugas Marinas de Colombia,* Libro 4. Santa Fe de Bogotá: Serie de Publicaciones Especiales del INDERENA, pp. 155–179.

Anderson, R. C., Zahir, H., Sakamoto, T., and Sakamoto, I. 2003. Olive ridley turtles (*Lepidochelys olivacea*) in Maldivian waters. *Rasain (Annual Fisheries Journal of the Ministry of Fisheries, Agriculture and Marine Resources)* 23:171–184.

Andrews, H. V., Krishnan, S., and Biswas, P. 2001. The status and distribution of marine turtles around the Andaman and Nicobar archipelago. GOI UNDP Sea Turtle Project Report. Tamil Nadu, India: Madras Crocodile Bank Trust.

Andrews, H. V., Krishnan, S., and Biswas, P. 2006. Distribution and status of marine turtles in the Andaman and Nicobar Islands. In Shanker, K., and Choudhury, B. C. (Eds.). *Marine Turtles of the Indian Subcontinent*. Hyderabad, India: Universities Press, pp. 33–57.

Anonymous. 2002. Experts advocate closure of Oswal Plant. *Kachhapa* 7:22.

Arauz, R. 1996. A description of the Central American shrimp fisheries with estimates of incidental capture and mortality of sea turtles. In Keinath, J., Barnard, D., Musick, J. D., and Bell, B. A. (Compilers). *Proceedings of the Fifteenth Annual Symposium on Sea Turtle Biology and Conservation*. NOAA Technical Memorandum NMFS-SEFSC-387, pp. 5–9.

Arauz, R. 1999. *Description of the Eastern Pacific High-Seas Longline and Coastal Gillnet Swordfish Fisheries of South America, Including Sea Turtle Interactions, and Management Recommendations*. Unpublished report submitted to Dr. Jim Spotila, Drexel University, Philadelphia.

Arauz, R. 2002. *Sea turtle nesting activity and conservation of leatherback turtles (*Dermochelys coriacea*) in Playa El Mogote, Río Escalante Chacocente Wildlife Refuge, Nicaragua*. Unpublished report submitted to the authorities of the Ministry of Environment and Natural Resources (MARENA), Dirección General de Areas Protegidas (DGAP), Managua, Nicaragua.

Arauz, R. 2005. Incidental capture of sea turtles by high seas longline pelagic fisheries in Costa Rica's exclusive economic zone (EEZ)—A second look. In Coyne, M., and Clark, R. D. (Compilers). *Proceedings of the Twenty-first Annual Symposium on Sea Turtle Biology and Conserzvation*. NOAA Technical Memorandum NMFS-SEFSC-528, pp. 40–42.

Arauz, R. M, Vargas, R., Naranjo, I., and Gamboa, C. 1998a. Analysis of the incidental capture and mortality of sea turtles in the shrimp fleet of Pacific Costa Rica. In Epperly, S. P., and Braun, J. (Compilers). *Proceedings of the Seventeenth Annual Symposium on Sea Turtle Biology and Conservation*. NOAA Technical Memorandum NMFS-SEFSC-415, pp. 1–5.

Arauz, R., Naranjo, I., Rojas, R., and Vargas, R. 1998b. Evaluation of the super shooter and Seymour turtle excluder devices with different deflector bar spacing in the shrimp fishery of Pacific Costa Rica. In Epperly, S. P., and Braun, J. (Compilers). *Proceedings of the Seventeenth Annual Symposium on Sea Turtle Biology and Conservation*. NOAA Technical Memorandum NMFS-SEFSC-415, pp. 114–117.

Arauz, R., Rodríguez, O., Vargas, R., and Segura, A. 2000. Incidental capture of sea turtles by Costa Rica's longline fleet. In Kalb, H., and Wibbles, T. (Compilers). *Proceedings of the Nineteenth Annual Symposium on Sea Turtle Biology and Conservation*. NOAA Technical Memorandum NMFS-SEFSC-443, pp. 62–64.

Arenas, P., and Hall, M. 1992. The association of sea turtles and other pelagic fauna with floating objects in the eastern tropical Pacific Ocean. In Salmon, M., and Wyneken, J. (Compilers). *Proceedings of the Eleventh Annual Workshop on Sea Turtle Biology and Conservation.* NOAA Technical Memorandum NMFS-SEFSC-302, pp. 7–10.

Aridjis, H. 1990. Mexico proclaims total ban on harvest of turtles and eggs. *Marine Turtle Newsletter* 50:1–3

Asia-Pacific Economic Cooperation, Marine Resources Conservation Working Group (APEC). 1998. *Proceedings of the Workshop on the Impacts of Destructive Fishing Practices on the Marine Environment, December 16–18, 1997.* Hong Kong: Agriculture and Fisheries Department.

Astuti, R. 1995. *People of the Sea: Identity and Descent among the Vezo of Madagascar.* New York: Cambridge University Press.

Aureggi, M., and Chantrapornsyl, S. 2003. Conservation project: Sea turtles at Phra Thong Island, South Thailand. *Kachhapa* 9:3–5.

Bache, S. J. 2003. *Marine Wildlife Bycatch Mitigation: Global Trends, International Action and the Challenges for Australia.* Ocean Publications, Centre for Maritime Policy; University of Wollongong, Australia.

Bache, S. J., and Frazier, J. 2006. International instruments and marine turtle conservation. In Shanker, K., and Choudhury, B. C. (Eds.). *Marine Turtles of the Indian Subcontinent.* Hyderabad, India: Universities Press, pp. 324–353.

Balazs, G. H. 1985. Impact of ocean debris on marine turtles: entanglement and ingestion. In Shomura, R. S., and Yoshida, H. O. (Eds.). *Proceedings of the Workshop on the Fate and Impact of Marine Debris.* NOAA Technical Memorandum NMFS-SWFSC-54, pp. 387–429.

Baldwin, R. M., and Al-Kiyumi, A. A. 1999. The ecology and conservation status of sea turtles of Oman. In Fisher, M., Ghazanfar, S. A., and Spalton, A. (Eds.). *The Natural History of Oman: A Festschrift for Michael Gallagher.* Leiden: Backhuys, pp. 89–98.

Barnett, L. K., Emms, C., Jallow, A., Mbenga Cham, A., and Mortimer, J. A. 2004. The distribution and conservation status of marine turtles in The Gambia, West Africa: a first assessment. *Oryx* 38:203–208.

Behera, C. R. 2006. Beyond TEDs: The TED controversy from the perspective of Orissa's trawling industry. In Shanker, K., and Choudhury, B. C. (Eds.). *Marine Turtles of the Indian Subcontinent.* Hyderabad, India: Universities Press, pp. 238–243.

Bhaskar, S. 1979. Sea turtle survey in the Andaman and Nicobars. *Hamadryad* 4:2–19.

Bhaskar, S. 1984. The distribution and status of sea turtles in India. In *Proceedings of the Workshop on Sea Turtle Conservation. CMFRI Special Publication* 18:21–35.

Bhaskar, S. 1993. *The Status and Ecology of Sea Turtles in the Andaman and Nicobar Islands.* Chennai: Centre for Herpetology, Publication ST 1/93.

Bhupathy, S., and Saravanan, S. 2002. Sea turtles along the Tamil Nadu coast, India. *Kachhapa* 7:7–13.

Bhupathy, S., and Saravanan, S. 2003. Exploitation of sea turtles along the southeast coast of Tamilnadu. *Journal of the Bombay Natural History Society* 100:628–631.

Bhupathy, S., and Saravanan, S. 2006. Marine turtles of Tamil Nadu. In Shanker, K., and Choudhury, B. C. (Eds.). *Marine Turtles of the Indian Subcontinent.* Hyderabad, India: Universities Press, pp. 58–67.

Biswas, S. 1982. A report on the olive ridley, *Lepidochelys olivacea* (Eschscholtz) (Testudines: Cheloniidae) of Bay of Bengal. *Records of the Zoological Survey of India* 79:275–302.

Bjorndal, K. A. 1997. Foraging ecology and nutrition of sea turtles. In Lutz, P. L., and Musick, J. A. (Eds.). *The Biology of Sea Turtles.* Boca Raton, FL: CRC Press, pp. 199–231.

Bowen, B. W., Clark, A. M., Abreu-Grobois, F. A., Chaves, A., Reichart, H. A., and Ferl, R. J. 1998. Global phylogeography of ridley sea turtles (*Lepidochelys* spp.) as inferred from mitochondrial DNA sequences. *Genetica* 101:179–189.

Brongersma, L. 1972. *European Atlantic Turtles.* Leiden: Zoologische Verhandelingen.

Brongersma, L. 1982. Marine turtles of the Eastern Atlantic Ocean. In Bjorndal, K. (Ed.). *Biology and Conservation of Sea Turtles* (reprinted 1995). Washington, DC: Smithsonian Institution Press, pp. 407–416.

Burchfield, P. M., Dierauf, L., and Byles, R. A. 1997. *Report on the Mexico/United States of America Population Restoration Project for* Lepidochelys kempii, *on the Coasts of Tamaulipas and Veracruz, Mexico.* Albuquerque, NM: U.S. Department of Interior, Fish and Wildlife Service.

Byles, R. A. 1988. *Satellite Telemetry of Kemp's Ridley Sea Turtle,* Lepidochelys kempii, *in the Gulf of Mexico.* Report to the National Fish and Wildlife Foundation.

Cahill, T. 1978. The shame of Escobilla. *Outside* (February):22–27,62–64.

Caillouet, C. W., Jr., Shaver, D. J., Teas, W. G., Nance, J. M., Revera, D. B., and Cannon, A. C. 1996. Relationship between sea turtle stranding rates and shrimp fishing intensities in the northwestern Gulf of Mexico: 1986–1989 versus 1990–1993. *U. S. Fishery Bulletin* 94:237–249.

Cantú, J. C., and Sánchez, M. E.. 1999. *Trade in Sea Turtle Products in Mexico.* Mexico City: Teyeliz, A.C.

Carr, A. 1955. The riddle of the ridley. *Animal Kingdom* 58(5):146–156.

Carr, A. 1977. Crisis for the Atlantic ridley. *Marine Turtle Newsletter* 4:2–3.

Chada, S., and Kar, C. S. 1999. *Bhitarkanika: Myth and Reality.* Dehra Dun, India: Natraj.

Chan, E. H., and Liew, H. C. 1988. A review on the effects of oil-based activities and oil pollution on sea turtles. In *Proceedings of the Eleventh Annual Seminar of the Malaysian Society of Marine Sciences,* pp. 159–167.

Chan, E. H., Liew, H. C., and Mazlan, A. G. 1988. The incidental capture of sea turtles in fishing gear in Terengganu, Malaysia. *Biological Conservation* 43:1–7.

Chandraratne, R. M. M. 1997. Some reptile bones from the Gedige excavation in 1985, the Citadel of Anuradhapura, Sri Lanka. *Lyriocehpalus* 3(2): 7–15.

Chantrapornsyl, S. 1994. Status of marine turtles in Thailand. In Nacu, A., Trono, R., Palma, J. A., Torres, D., and Agas, F., Jr. (Eds.). *Proceedings of the First ASEAN Symposium-Workshop on Marine Turtle Conservation.* Manila: World Wildlife Fund, pp. 123–129.

Charuchinda, M., and Chantrapornsyl, S. 1999. Status of sea turtle conservation and research in Thailand. In: *Report of the SEAFDEC-ASEAN Regional Workshop on Sea Turtle Conservation and Management, July 26–28, 1999, Kuala Terengganu, Malaysia,* pp. 160–174.

Chatto, R., Guinea, M., and Conway, S. 1995. Sea turtles killed by flotsam in northern Australia. *Marine Turtle Newsletter* 69:17–18.

Chevalier, J. 2001. Etude des captures accidentelles de tortues marines liées à la pêche au filet dérivant dans l'ouest guyanais. Unpublished report for DIREN Guyane, ONCFS.

Chevalier, J., Cazelles, B., and Girondot, M. 1998. Apports scientifiques à la stratégie de conservation des Tortues luths en Guyane français. *JATBA. Revue d'Ethnobiologie* 40:485–507.

Choudhury, B. C. 2003. TEDs in India: From conflict to consultation. *Kachhapa* 8:1–2.

Chowdhury, B. R., Das, S. K., and Ghose, P. S. 2006. Marine turtles of West Bengal. In Shanker, K., and Choudhury, B. C. (Eds.). *Marine Turtles of the Indian Subcontinent.* Hyderabad, India: Universities Press, pp. 107–116.

Cliffton, K., Cornejo, D. O., and Felger, R. S. 1982. Sea turtles of the Pacific coast of Mexico. In Bjorndal, K. (Ed.). *Biology and Conservation of Sea Turtles* (reprinted 1995). Washington, DC: Smithsonian Institution Press, pp. 199–209.

Cornelius, S. 1982. Status of sea turtles along the Pacific coast of middle America. In Bjorndal, K. (Ed.). *Biology and Conservation of Sea Turtles* (reprinted 1995). Washington, DC: Smithsonian Institution Press, pp. 211–219.

Cruz, R. D. 1999. Research, conservation and management of marine turtles in the Philippines. In *Report of the SEAFDEC-ASEAN Regional Workshop on Sea Turtle Conservation and Management, July 26–28, 1999, Kuala Terengganu, Malaysia,* pp. 146–152.

Das, I. 1985. Marine turtle drain. *Hamadryad* 10:16.

Das, I. 1989. *Sea turtles and coastal habitats in southeastern Bangladesh.* Project report to the Sea Turtle Rescue Fund/Centre for Marine Conservation, Washington, DC.

Dash, M. C., and Kar, C. S. 1990. *The Turtle Paradise Gahirmatha (An Ecological Analysis and Conservation Strategy).* New Delhi, India: Interprint.

Da Silva, A. C. C. D., de Castilhos, J. C., Rocha, D. A. S., Oliveira, F. L. C., and Weber, M. 2002. Mortalidade de tartarugas marinhas no entorno de sitios de reprodução no estado de Sergipe, Brasil. In *XXIV Congresso Brasilero de Zoologia, 17–22 Fevereiro 2002.* Santa Catarina: UNIVALI, Itajaí, Abstract 14114.

Dattatri, S., and Samarajiva, D. 1982. *The Status and Conservation of Sea Turtles in Sri Lanka.* A project of the sea turtle rescue fund, Center for Environmental Education, Washington DC.

De Grazia, T., and Smith, W. N. 1970. *The Seri Indians. A Primitive People of Tiburón Island in the Gulf of California.* Flagstaff, AZ: Northland Press.

Deraniyagala, P. E. P. 1939. *The Tetrapod Reptiles of Ceylon, vol 1. Testudinates and Crocodilians.* London: Dulau & Co.

De Silva, A. 2006. Marine turtles of Sri Lanka: A historic account. In Shanker, K., and Choudhury, B. C. (Eds.). *Marine Turtles of the Indian Subcontinent.* Hyderabad, India: Universities Press, pp. 188–199.

DF-GUM (Department of Fisheries, Ministry of Livestock and Fisheries, Government of the Union of Myanmar). 1999. Sea turtle conservation and protection activities in Myanmar. In *Report of the SEAFDEC-ASEAN Regional Workshop on Sea Turtle Conservation and Management, July 26–28, 1999, Kuala Terengganu, Malaysia,* pp. 134–141.

Dharani, S. 2003. Turtle conservation by local communities in Madras. *Kachhapa* 8:22.

Dileepkumar, N., and Jayakumar, C. 2002. *Field Study and Networking for Turtle Conservation in Kerala.* A

GOI-UNDP Sea Turtle Project Report. Trivandrum, India: THANAL Conservation Action and Information Network.

Dintheer, C., Gilly, B. J. Y., Le Gall, M., Lemoine, M., and Rosé, J. 1989. La recherche et la gestion de la pêcherie de crevettes pénéides en Guyane Française de 1958 à 1988: trente années de surf. *Equinoxe* 28:21–33.

Dobie, J. L. 1996. *Lepidochelys kempii* (Kemp's ridley turtle). Feeding on insects. *Herpetological Review* 27(4):199.

DOF (Diario Oficial de la Federación). 1990. Acuerdo que establece veda total para todas las especies y subespecies de tortugas marinas en aguas de jurisdicción nacional de los litorales del Océano Pacífico, Golfo de México y Mar Caribe. *Diario Oficial de la Federación.* México (Mayo 31): 21–22.

Dongre, S. K. M. 2003. School education to support sea turtle conservation: Experiences from Goa and Orissa. *Kachhapa* 8:20–21.

Eckert, K. L. 1993. *The Biology and Population Status of Marine Turtles in the North Pacific Ocean.* NOAA Technical Memorandum NMFS-SWFSC-186.

Epperly, S. P., Braun, J., Chester, A. J., Cross, F. A., Merriner, J. V., Tester, P. A., and Churchill, J. H. 1996. Beach strandings as an indicator of at-sea mortality of sea turtles. *Bulletin of Marine Science.* 59:289–297.

Estrades A., Domingo, A., Laporta, M., López-Mendilaharsu, M., and Fallabrino A., A. In press. Implementation and advances of the first Sea Turtle National Tagging Program in Uruguay. In: *Proceedings of the Twenty-Fourth Annual Symposium on Sea Turtle Biology and Conservation.* NOAA Technical Memorandum NMFS-SEFSC.

FAO (Food and Agriculture Organization of the United Nations). 2005. Committee on Fisheries; Twenty-sixth Session, Rome, Italy, 7–11 March 2005. Outcome of the Technical Consultation on Sea Turtles Conservation and Fisheries, Bangkok, Thailand, 29 November – 2 December 2004. Rome: FAO COFI/2005/7.

FAO (Food and Agriculture Organization of the United Nations). 2006. Report of the workshop on assessing the relative importance of sea turtle mortality due to fisheries. Rome: FAO GCP/INT.919/JPN.

Felger, R. S., and Moser, M. B. 1991. *People of the Desert and Sea: Ethnobotany of the Seri Indians.* Tucson: University of Arizona.

Firdous, F. 2000. Sea turtle conservation and education in Karachi, Pakistan. In Pilcher, N., and Ismail, G. (Eds.). *Sea Turtles of the Indo-Pacific: Research, Conservation and Management.* London: ASEAN Academic Press, pp. 45–55.

Foley, A. M., Dutton, P. H., Singel, K. E., Redlow, A. E., and Teas, W. G. 2003. The first records of olive ridleys in Florida, USA. *Marine Turtle Newsletter* 101:23–25.

Formia, A. 2002. Population and genetic structure of the green turtle (*Chelonia mydas*) in West and Central Africa; implications for management and conservation. PhD diss. Cardiff University, Cardiff, UK.

Frazier, J. G. 1980a. Marine turtles and problems in coastal management. In Edge, B. L. (Ed.). *Proceedings of the Second Symposium on Coastal and Ocean Management.* New York: American Society of Civil Engineers, Volume 3, pp. 2395–2411.

Frazier, J. 1980b. Exploitation of marine turtles in the Indian Ocean. *Human Ecology* 8:329–370.

Frazier, J. 1980c. *Marine turtle fisheries in Ecuador and Mexico: The last of the Pacific ridley?* Unpublished manuscript, Department of Zoological Research, National Zoological Park, Smithsonian Institution, Washington, DC.

Frazier, J. 1982. Subsistence hunting in the Indian Ocean. In Bjorndal, K. (Ed.). *Biology and Conservation of Sea Turtles* (reprinted 1995). Washington, DC: Smithsonian Institution Press, pp. 391–396.

Frazier, J. 1985. Misidentification of marine turtles: *Caretta caretta* and *Lepidochelys olivacea* in the East Pacific. *Journal of Herpetology* 19:1–11.

Frazier, J. G. 1991. La presencia de la tortuga marina *Lepidochelys olivacea* (Eschscholtz), en la Republica Oriental del Uruguay. *Revista de la Facultad de Humanidades y Ciencias. Serie Ciencias Biológicas, 3a época* 2(6):1–4.

Frazier, J. 1997. Sustainable development: modern elixir or sack dress? *Environmental Conservation* 24:182–193

Frazier, J. 1998. Recommendations on future CMS activities for marine turtle conservation. Convention on the Conservation of Migratory Species of Wild Animals (CMS). Eighth meeting of the CMS Scientific Council, Wageningen, The Netherlands, June 3–5, 1998.

Frazier, J. 2000. Kemp's ridley sea turtle. In Reading, R. P., and Miller, B. (Eds.). *Endangered Animals: A Reference Guide to Conflicting Issues.* Westport, CT: Greenwood Press, pp. 164–170.

Frazier, J. (Ed.). 2002. International instruments and marine turtle conservation. *Journal of International Wildlife Law and Policy* 5:1–207.

Frazier, J. 2003. Prehistoric and ancient historic interactions between humans and marine turtles. In Lutz, P. L., Musick, J. A., and Wyneken, J. (Eds.).

The Biology of Sea Turtles, Volume 2. Boca Raton, FL: CRC Press, pp. 1–38.

Frazier, J. 2004. Marine turtles of the past: a vision for the future? In Lauwerier, R. C. G. M., and Plug, I. (Eds.). *The Future from the Past: Archaeozoology in Wildlife Conservation and Heritage Management. Proceedings of the 9th Conference of the International Council for Archaeological Zoology, Durham, August 2002,* Volume 3. Oxford: Oxbow Books, pp. 103–116.

Frazier, J. 2005a. Marine Turtles: The Ultimate Tool Kit. A Review of worked bones in marine turtles. In Luik, H., Choyke, A. M., Batey, C. E., and Lõugas, L. (Eds.). *From Hooves to Horns, from Mollusc to Mammoth: Manufacture and Use of Bone Artefacts from Prehistoric Times to the Present. Proceedings of the 4th Meeting of the ICAZ Worked Bone Research Group at Tallinn, 26th–31st of August 2003. Muinasaja Teadus* 15 (Estonia), pp. 359–382.

Frazier, J. (Ed.) 2005b. Marine turtles as flagships. *MAST / Maritime Studies* Special Issue 3(2)/4(1): 1–303.

Frazier, J., and Bache, S. J. 2002. Sea turtle conservation and the "big stick": the effects of unilateral U.S. embargoes on international fishing activities. In Foley, A., and Moser, A. (Compilers). *Proceedings of the Twentieth Annual Symposium on Sea Turtle Biology and Conservation.* NOAA Technical Memorandum NMFS-SEFSC-477, pp. 118–121.

Frazier, J., and Salas, S.. 1982. Ecuador closes commercial turtle fishery. *Marine Turtle Newsletter* 20:5–6.

Frazier, J., and Salas, S. 1984. Tortugas marinas en Chile. *Boletín del Museo Nacional de Historia Natural de Santiago, Chile.* 39:63–73.

Frazier, J., Frazier, S., Hanbo, D., Zhujian, H., Ji, Z., and Ling, L. 1988. Sea turtles in Fujian and Guangdong Provinces. *Acta Herpetologica Sinica.* 6:16–46.

Fretey, J. 1989. Reproduction de la tortue olivatre (*Lepidochelys olivacea*) en Guyane Française pendant la saison 1987. *Nature Guyanaise* 1:8–13.

Fretey, J. 1999. Répartition des tortues du genre *Lepidochelys* Fitzinger, 1843. I. L'Atlantique ouest. *Biogeographica* 75:97–117.

Fretey, J. 2001. *Biogeography and Conservation of Marine Turtles of Atlantic Coast of Africa/Biogéographie et conservation des tortues marines de la côte atlantique de l'Afrique.* CMS Technical Series Publication No. 6, UNEP/CMS Secretariat, Bonn, Germany.

Fretey, J., Dontaine, J.-F., and Billes, A., 2001. Tortues marines de la façade atlantique de l'Afrique, genre *Lepidochelys.* 2. Suivi et conservation de *L. olivacea* (Eschscholtz, 1829) (Chelonii, Cheloniidae) à São Tomé et Príncipe. *Bulletin de la Société Herpétologique de France* 98:43–56.

Fretey, J., Segniagbeto, G. H., Dossou-Bodjrenou, J., and Soumah, M'M. In press. Use of marine turtles in West African pharmacopoeia and voodoo. *Proceedings of the Twenty-Fourth Annual Symposium on Sea Turtle Biology and Conservation.* NOAA Technical Memorandum NMFS-SEFSC.

Fritts, T. H., Stinson, M. L., and Márquez-M., R. 1982. Status of sea turtle nesting in southern Baja California, Mexico. *Bulletin of the Southern California Academy of Science* 81(2):51–60.

Gardner, S. C., and Nichols, W. J. 2001. Assessment of sea turtle mortality rates in the Bahía Magdalena region, Baja California Sur, México. *Chelonian Conservation and Biology* 4:197–199.

Gebremariam, T., Amer, A., Gebremariam, S., and Asfaw, M. 1998. Shrimp fishery in Eritrea: exploitation and legislation. In Wamukoya, G. M., and Salm, R. V. (Eds.). *Report of the Western Indian Ocean Turtle Excluder Device (TED) Training Workshop, Mombasa, Kenya, January 27–31, 1997.* Nairobi, Kenya: IUCN East Africa Regional Office, pp. 12–13.

Giri, V. 2001. Survey of marine turtles along the coast of Maharastra and Goa. In Shanker, K., and Choudhury, B. C. (Eds.). *Proceedings of the National Workshop for the Development of a National Sea Turtle Conservation Action Plan, Bhubaneshwar.* Dehradun, India: Wildlife Institute of India, pp. 77–79.

Giri, V., and Chaturvedi, N. 2003. Status of marine turtles in Maharashtra, India. *Kachhapa* 8: 11–15.

Giri, V., and Chaturvedi, N. 2006. Sea turtles of Maharashtra and Goa. In Shanker, K., and Choudhury, B. C. (Eds.). *Marine Turtles of the Indian Subcontinent.* Hyderabad, India: Universities Press, pp. 147–155.

Godfrey, M. H., and Chevalier, J. 2004. The status of olive ridley sea turtles in the West Atlantic. Unpublished report prepared for the IUCN/SSC Marine Turtle Specialist Group. http://members.seaturtle.org/godfreym/Godfrey2004MTSG.pdf.

Gove, D., Pacule, H., and Gonçalves, M. 2004. The impact of Sofala Bank (Central Mozambique) shallow water shrimp fishery on marine turtles and the effects of introducing TED (turtle excluder device) on shrimp fishery. Unpublished report.

Green, D., and Ortiz-Crespo, F. 1982. Status of sea turtle populations in the central eastern Pacific. In Bjorndal, K. (Ed.). *Biology and Conservation of*

Sea Turtles (reprinted 1995). Washington, DC: Smithsonian Institution Press, pp. 221–233.

Grottanelli, V. L. 1955. *Pescatori dell'Oceano indiano: saggio etnologico preliminare sui Bagiuni, Bantu cosieri dell'Oltregiuba.* Rome: Cremonese.

Guada, H. 2001. Marine turtle work in Venezuela and the Venezuelan STRAP. In Schoueten, A., Mohadin, K., Ashin, S., and McClintock, E. (Eds.). *Proceedings of the V Regional Marine Turtle Symposium for the Guianas, Paramaribo, September 25–27, 2001.* WWF technical report no. GFECP-9, pp. 41–48.

Gudger, E. W. 1919a. On the use of the sucking-fish for catching fish and turtles: Studies in *Echeneis* or *Remora,* II. *The American Naturalist* 53:446–467.

Gudger, E. W. 1919b. On the use of the sucking-fish for catching fish and turtles: Studies in *Echeneis* or *Remora,* III. *The American Naturalist* 53:515–525.

Guéguen, F. 2000. Captures accidentelles de tortues marines par la flottille crevettière de Guyane Française. *Bulletin Société Herpétologie Française* 93:27–93.

Guinea, M. L., and Chatto, R. 1992. Sea turtles killed in Australian shark fin fishery. *Marine Turtle Newsletter* 57:5–6.

Guinea, M. L., and Whiting, S. 1997. Sea turtle deaths coincide with trawling activities in northern Australia. *Marine Turtle Newsletter* 77:11–14.

Gunawardene, P. S. 1986. *National sea turtle survey progress report.* Unpublished Report for the National Aquatic Resources and Development Agency, Colombo, 1986 (cited in Kapurusinghe and Cooray, 2002, p. 6).

Hare, S. 1991. Turtles caught incidental to demersal finfish fishery in Oman. *Marine Turtle Newsletter* 53:14–16.

Haule, W. V., Kalikela, G., and Mahundu, I. 1998. Some information on the sea turtles of Tanzania. In Wamukoya, G. M., and Salm, R. V. (Eds.). *Report of the Western Indian Ocean Turtle Excluder Device (TED) Training Workshop, Mombasa, Kenya, January 27–31, 1997.* Nairobi, Kenya: IUCN East Africa Regional Office, pp. 21–22.

Hays Brown, C., and Brown, W. M. 1982. Status of sea turtles in the southeastern Pacific: Emphasis on Perú. In Bjorndal, K. (Ed.). *Biology and Conservation of Sea Turtles* (reprinted 1995). Washington, DC: Smithsonian Institution Press, pp. 235–240.

Hendrickson, J. R. 1980. The ecological strategies of sea turtles. In Symposium on the Behavioral and Reproductive Biology of Sea Turtles. *American Zoologist* 20:597–608.

Hewavisenthi, S. 1990. Exploitation of marine turtles in Sri Lanka: historic background and present status. *Marine Turtle Newsletter* 48:14–19.

Higginson, J. 1989. Sea turtles in Guatemala: threats and conservation efforts. *Marine Turtle Newsletter* 45:1–5.

Hildebrand, H. H. 1963. Hallazgo del área de anidación de la tortuga lora *Lepidochelys kempii* (Garman), en la costa occidental de Golfo de México (Rept., Chel.). *Ciencia, México* 22:105–112.

Hildebrand, H. H. 1982. A historical review of the status of sea turtle populations in the Western Gulf of Mexico. In Bjorndal, K. (Ed.). *Biology and Conservation of Sea Turtles* (reprinted 1995). Washington, DC: Smithsonian Institution Press, pp. 447–453.

Hoekert, W. E. J., Schouten, A. D., van Tienen, L. H. G., and Weijerman, M. 1996. Is the Surinam olive ridley on the eve of extinction? First census data for olive ridleys, green turtles, and leatherbacks since 1989. *Marine Turtle Newsletter* 75:1–4.

Hurtado, M. 1982. The ban on the exportation of turtle skin from Ecuador. *Marine Turtle Newsletter* 20:1–4.

Inter-American Tropical Tuna Commission (IATTC). 2003a. Review of the status of sea turtle stocks in the eastern Pacific. Working Group on Bycatch; Fourth Meeting, Kobe, Japan, January 14–16, 2004. Document BYC-4-04.

Inter-American Tropical Tuna Commission (IATTC). 2003b. Interactions of sea turtles with tuna fisheries, and other impacts on turtle populations: Interactions in the purse-seine fishery. Working Group on Bycatch; Fourth Meeting, Kobe, Japan, January 14–16, 2004. Document BYC-4-05a.

Inter-American Tropical Tuna Commission (IATTC). 2003c. Interactions of sea turtles with tuna fisheries, and other impacts on turtle populations: Interactions in longline fisheries. Working Group on Bycatch; Fourth Meeting, Kobe, Japan, January 14–16, 2004. Document BYC-4-05a.

Islam, M. Z. 2002. Threats to sea turtles in St. Martin's Island, Bangladesh. *Kachhapa* 6:6–10.

Jackson, J. B. C. 2001. What was natural in the coastal oceans? *Proceedings of the National Academy of Sciences USA* 98:5411–5418.

Jinadasa, J. 1984. The effect of fishing on turtle populations. *Loris* 16:311–314.

Jones, S., and Fernando, A. B. 1968. The present state of the turtle fishery in the Gulf of Mannar and Palk Bay. In *Symposiuum of Living Resources of the Seas Around India, Cochin,* December, 1968, pp. 712–715.

Kalim, M. H., and Dermawan, A. 1999. Marine turtle research and conservation in Indonesia. In *Report of the SEAFDEC-ASEAN Regional workshop on sea turtle conservation and management, July 26–28, 1999, Kuala Terengganu, Malaysia,* pp. 78–103.

Kamarruddin, I. 1994. The status of marine turtle conservation in Peninsular Malaysia. In Nacu, A., Trono, R., Palma, J. A., Torres, D., and Agas, F., Jr. (Eds.). *Proceedings of the First ASEAN Symposium-Workshop on marine turtle conservation.* Manila: World Wildlife Fund, pp. 87–103.

Kapurusinghe, T., and Cooray, R. 2002. *Marine turtle by-catch in Sri Lanka. Survey Report.* Panadura, Sri Lanka: Turtle Conservation Project.

Kar, C. S. 1980. The Gahirmatha turtle rookery along the coast of Orissa, India. *Marine Turtle Newsletter* 15:2–3.

Kar, C. S., and Bhaskar, S. 1982. Status of sea turtles in the eastern Indian Ocean. In Bjorndal, K. A. (Ed.). *Biology and Conservation of Sea Turtles* (reprinted 1995). Washington, DC: Smithsonian Institution Press, pp. 365–372.

Kar, C. S., and Dash, M. C. 1984. Conservation and status of sea turtles in Orissa. In Silas, E. G. (Ed.). *Proceedings of the Workshop on Sea Turtle Conservation.* Central Marine Fisheries Research Institute Special Publication 18, pp. 93–107.

Kotas, J. E., dos Santos, S., de Azevedo, V. G., Gallo, B. M. G., and Barata, P. C. R. 2004. Incidental capture of loggerhead (*Caretta caretta*) and leatherback (*Dermochelys coriacea*) sea turtles by the pelagic longline fishery off southern Brazil. *Fishery Bulletin* 102:393–399.

Kuriyan, G. K. 1950. Turtle fishing in the sea around Krusadai island. *Journal of the Bombay Natural History Society* 49:509–512.

Kutty, R. 2001. Community based conservation of sea turtle nesting sites in India. In Shanker, K., and Choudhury, B. C. (Eds.). *Proceedings of the National Workshop for the Development of a National Sea Turtle Conservation Action Plan, Bhubaneshwar.* Dehradun, India: Wildlife Institute of India, pp. 86–89.

Kutty, R. 2006. Community-based conservation of sea turtle nesting sites in India: Some case studies. In Shanker, K., and Choudhury, B. C. (Eds.). *Marine Turtles of the Indian Subcontinent.* Hyderabad, India: Universities Press, pp. 171–189.

Laurent, L. 1999a. Etude préliminaire sur les interactions entre les populations reproductrices de tortues marines du Plateau des Guyanes et les pêcheries atlantiques. Draft report for WWF France.

Laurent, L. 1999b. Sea turtle and fishery interactions in Trinidad and Tobago. Draft report for WWF Suriname, Agreement FH-13.

Laurent, L., Charles, R., and Lieveld, R. 1999. Guayana Shield Sea Turtle Conservation Regional Strategy Action Plan 2000–2005: Fishery Sector Report. Draft report for WWF Suriname, Agreement FH-13.

Lima, E. H. S. M., Brosig, C., and Ximenes, M. C. A. 2002. Tartarugas marinhas capturadas acidentalmente em Almofala, Ceará. XXIV Congresso Brasilero de Zoologia, 17–22 Fevereiro 2002, UNIVALI, Itajaí, Santa Catarina, Abstract 14022.

Limpus, C. J. 1994. The worldwide status of marine turtle conservation. In Nacu, A., Trono, R., Palma, J. A., Torres, D., and Agas, F. Jr. (Eds.). *Proceedings of the First ASEAN Symposium—Workshop on Marine Turtle Conservation.* Manila: World Wildlife Fund, pp. 43–63.

Limpus, C. J. 1995. Global overview of the status of marine turtles: A 1995 overview. In Bjorndal, K. A. (Ed.). *Biology and Conservation of Sea Turtles.* Washington, DC: Smithsonian Institution Press, pp. 605–609.

Limpus, C. J. 1997. Marine turtle populations of Southeast Asia and the western Pacific region: distribution and status. In Noor, Y. R., Lubis, I. R., Ounsted, R., Troeng, S., and Abdullas, A. (Eds.). *Proceedings of the Workshop on Marine Turtle Research and Management in Indonesia, Jember, East Java, November 1996.* Bogor, Indonesia: Wetlands International / PHPA / Environment Australia, pp. 37–[73?].

López-Mendilaharsu, M., Fallabrino, A., Estrades, A., Hernández, M., Caraccio, N., Lezama, C., Laporta, M., Calvo, V., Quirici, V., Bauza, A., and Aisenberg, A. 2003. *Proyecto Karumbé: Tortugas marinas del Uruguay. Libro de Resúmenes; 2as Jornadas de Conservación y uso Sustentable de la fauna marina; I Reunión de Investigación y Conservación de las Tortugas Marinas del Atlántico Sur Occidental.* Montevideo: UNESCO, p. 70.

Lutcavage, M. E., Plotkin, P., Witherington, B., and Lutz, P. L. 1997. *Human Impacts on Sea Turtle Survival.* In Lutz, P. L., and Musick, J. A. (Eds.). *The Biology of Sea Turtles.* Boca Raton, FL: CRC Press, pp. 387–409.

Mack, D., Duplaix, N., and Wells, S. 1982. Sea turtles, animals of divisible parts: International trade in sea turtle products. In Bjorndal, K. (Ed.). *Biology and Conservation of Sea Turtles* (reprinted 1995). Washington, DC: Smithsonian Institution Press, pp. 545–562.

Madhyastha, M. N., Sharath, B. K., and Jayaprakash Rao, I. 1986. Preliminary studies on marine turtle hatchery at Bengre beach, Mangalore. *Mahasagar Bulletin of the National Institute of Oceanography* 19:137–140.

Magane, S., Sousa, L., and Pacule, H. 1998. Summary of turtles and fisheries resources information for Mozambique. In Wamukoya, G. M., and

Salm, R. V. (Eds.). *Report of the Western Indian Ocean Turtle Excluder Device (TED) Training Workshop, Mombasa, Kenya, January 27–31, 1997*. Nairobi: IUCN East Africa Regional Office, pp. 18–20.

Man, E. H. 1883. *On the Aboriginal Inhabitants of the Andaman Islands* (reprint, 1978). New Delhi: Prakashak.

Manzella, S. A., Caillouet, C., Jr., and Fontaine, C. 1988. Kemp's ridley, *Lepidochelys kempi*, sea turtle head start tag recoveries: Distribution, habitat, and method of recovery. *Marine Fisheries Review* 50(3):24–32.

Marcano, L. A., and Alio M., J. J. 2000. Incidental capture of sea turtles by the industrial shrimping fleet off northeastern Venezuela. In Abreu-Grobois, F. A., Briseño-Dueñas, F. A., Márquez, R., and Sarti, L. (Compilers). *Proceedings of the Eighteenth International Sea Turtle Symposium.* NOAA Technical Memorandum NMFS-SEFSC-436, p. 107.

Marcovaldi, M. Â. 2001. Status and distribution of the olive ridley turtle, *Lepidochelys olivacea*, in the western Atlantic Ocean. In Eckert, K. L., and Albreu-Grobois, F. A. (Eds.). *Proceedings of the Regional Meeting "Marine Turtle Conservation in the Wider Caribbean Region: a Dialogue for Effective Regional Management." Santo Domingo, November 16–18, 1999:* WIDECAST, IUCN-MTSG, WWF, and UNEP-CEP, pp. 52–56.

Marcovaldi, M. Â., da Silva, A. C. C. D., Gallo, B. M. G., Baptistotte, C., Lima, E. P., Bellini, C., Lima, E. H. S. M., de Castilhos, J. C., Thomé, J. C. A., Moreira, L. M. P., and Sanches, T. M. 2000. Recaptures of tagged turtles from nesting and feeding grounds protected by Projeto TAMAR-IBAMA, Brasil. In Kalb, H. J., and Wibbels, T. (Compilers). *Proceedings of the Nineteenth Annual Symposium on Sea Turtle Biology and Conservation.* NOAA Technical Memorandum NMFS-SEFSC-443, pp. 164–166.

Marcovaldi, M. Â., Gallo, B. G., Lima, E. H. S. M., and Godfrey, M. H. 2001. *Nem tudo que cai na rede é peixe:* An environmental education initiative to reduce mortality of marine turtles caught in artisanal fishing nets in Brazil. In Borgese, E. M., Chricop, A., and McConnell, M. (Eds.). *Ocean Yearbook 15.* Chicago: University of Chicago Press, pp. 246–256.

Marcovaldi, M. Â., Thomé, J. C., Sales, G., Coelho, A. C., Gallo, B., and Bellini, C. 2002. Brazilian plan for reduction of incidental sea turtle capture in fisheries. *Marine Turtle Newsletter* 96: 24–25.

Margavio, A. V., and Forsyth, C. J. 1996. *Caught in the net: The Conflict between Shrimpers and Conservationists.* College Station: Texas A & M University Press.

Márquez-M., R. 1990. *Sea Turtles of the World, an Annotated and Illustrated Catalogue of Sea Turtle Species Known to Date.* FAO Species Catalogue. FAO Fisheries Synopsis No. 125, Volume 11.

Márquez-M., R. 1994a. *Sinopsis de datos biológicos sobre la tortuga lora,* Lepidochelys kempi *(Garman, 1880).* FAO synopsis sobre la Pesca, No. 152; Instituto Nacional de la Pesca.

Márquez-M., R. 1994b. *Synopsis of Biological data on the Kemp's ridley turtle,* Lepidochelys kempi *(Garman, 1880).* NOAA Technical Memorandum NMFS-SEFSC-343.

Márquez-M., R. 1996. *Las Tortugas Marinas y Nuestro Tiempo.* México: La Ciencia para Todos, Fondo Cultural Económico (2d. Ed. 2000).

Márquez-M., R., Villanueva O., A., and Peñaflores S., C. 1976. *Sinopsis de datos Biológicos sobre la Tortuga Golfina,* Lepidochelys olivacea *(Eschscholtz, 1829).* México: Instituto Nacional de la Pesca INP Sinopsis sobre la Pesca, No. 2. SAST-Tortuga Golfina.

Márquez-M., R., Vasconcelos P., J., and Peñaflores S., C. 1990. *XXV Años de Investigación, conservación y protección de la tortuga marina.* México: Secretaria de Pesca, Instituto Nacional de la Pesca.

Márquez-M., R., Peñaflores, C., and Vasconcelos, J. 1996. Olive ridley turtles (*Lepidochelys olivacea*) show signs of recovery at La Escobilla, Oaxaca. *Marine Turtle Newsletter* 73:5–7.

Martínez T., C., and Castellanos M., R. 2006. Synchronized nesting of olive ridley sea turtles (*Lepidochelys olivacea*) in Chalacatepec, Majahuas and Mismaloya beaches, Jalisco, México. In Pilcher, N. J. (Compiler). *Proceedings of the Twenty-Third Symposium on Sea Turtle Biology and Conservation.* NOAA Technical Memorandum NMFS-SEFSC-536, pp. 174–176.

Mast, R. B., Hutchinson, B. J., and Pilcher, N. J. 2004. IUCN/SSC Marine Turtle Specialist Group news, first quarter 2004. *Marine Turtle Newsletter* 104:21–22.

McGee W. J. 1898. The Seri Indians. In *Accompanying Papers to 17th Annual Report of the Bureau of American Ethnology,* Part I. *1895–1896,* Washington, DC, pp. 1–344.

Meylan, A. B., and Sadove, S. 1986. Cold stunning in Long Island Sound, New York. *Marine Turtle Newsletter* 37:7–8.

Milton, S., Lutz, P., and Shigenaka, G. 2003. Oil toxicity and impacts on sea turtles. In Shigenaka, G. (Ed.). *Oil and sea turtles: Biology, planning, and response.* U.S. Department of Commerce, National Oceanic and Atmospheric Administration, NOAA's National Ocean Service, Office of Response and Restoration, Hazardous Materials Response Division, Seattle, WA, pp. 35–47.

Moein Bartol, S., and Musick, J. A. 2003. Sensory biology of sea turtles. In Lotz, P. L., Musick, J. A., and Wyneken, J. (Eds.). *The Biology of Sea Turtles,* Volume 2. Boca Raton, FL: CRC Press, pp. 79–102.

Mohanty, B. 2002a. Effluents from Oswal Fertilisers threatens olive ridley sea turtles on the Orissa coast. *Kachhapa* 6:20.

Mohanty, B. 2002b. Comments. *Kachhapa* 7:22.

Moore, G., and Jennings, S. (Eds.). 2000. *Commercial Fishing: The Wider Ecological Impacts.* Cambridge: The British Ecological Society (Ecological Issues Series).

Murthy, T. S. N. 1981. Turtles: their natural history, economic importance and conservation. *Zoologiana* 4:57–65.

Nabhan, G. P. 2003. *Singing the Turtles to Sea: The Comcáac (Seri) Art and Science of Reptiles.* Berkeley: University of California Press.

National Marine Fisheries Service and U.S. Fish and Wildlife Service (NMFS-USFWS). 1998. *Recovery Plan for US Pacific Populations of the Olive Ridley Turtle (*Lepidochelys olivacea*).* Silver Spring, MD: National Marine Fisheries Service.

National Research Council (NRC). 1990. *Decline of the Sea Turtles: Causes and Prevention.* Washington, DC: National Academy Press.

Nichols, W. J. 2003a. Sinks, sewers, and speed bumps: the impact of marine development on sea turtles in Baja California, Mexico. In *Proceedings of the Twenty-Second Annual Symposium on Sea Turtle Biololgy and Conservation.* NOAA Technical Memorandum NMFS-SEFC-503, pp. 17–18.

Nichols, W. J. 2003b. Biology and conservation of sea turtles in Baja California, Mexico. Ph.D. diss. Department of Wildlife and Fisheries Science, University of Arizona. Tucson.

Orrego V., C. M., and Arauz, R. 2005. Mortality of sea turtles along the Pacific coast of Costa Rica. In Coyne, M., and Clark, R. D. (Compilers). *Proceedings of the Twenty-First Annual Symposium on Sea Turtle Biology and Conservation.* NOAA Technical Memorandum NMFS-SEFSC-528, pp. 265–266.

Palma, J. A. M. 1997. Marine turtle conservation in the Philippines and initiatives towards a regional management and conservation program. In Noor, Y. R., Lubis, I. R., Ounsted, R., Troeng, S., and Abdullas, A. (Eds.). *Proceedings of the Workshop on Marine Turtle Research and Management in Indonesia, Jember, East Java, November 1996.* Bogor, Indonesia: Wetlands International / PHPA / Environment Australia, pp. 121–138.

Pandav, B. 2000. Conservation and management of olive ridley sea turtles on the Orissa coast. Ph.D. diss. Utkal University, Bhubaneshwar, India.

Pandav, B. 2003. Letter to the Editor. *Kachhapa* 8:26.

Pandav, B., Choudhury, B. C., and Kar, C. S. 1994. A status survey of olive ridley sea turtle *(Lepidochelys olivacea)* and its nesting habitats along the Orissa coast, India. Dehra Dun, India: Wildlife Institute of India.

Pandav, B., Choudhury, B. C., and Kar, C. S. 1997 (revised ed.). L. olivacea *Sea Turtle (*Lepidochelys olivacea*) and Its Nesting Habitats along the Orissa coast, India. A Status Survey.* Dehra Dun, India: Wildlife Institute of India.

Pandav, B., and Choudhury, B. C. 2006. Migration and movement of olive ridley turtles along the east coast of India. In Shanker, K., and Choudhury, B. C. (Eds.). *Sea Turtles of the Indian Subcontinent.* Hyderabad, India: Universities Press, pp. 365–379.

Pandav, B., Choudhury, B. C., and Kar, C. S. 2006. Sea turtle nesting habitats on the coast of Orissa. In Shanker, K., and Choudhury, B. C. (Eds.). *Sea Turtles of the Indian Subcontinent.* Hyderabad, India: Universities Press, pp. 88–106.

Perera, L., 1986. *National Sea Turtle Summary Report.* Unpublished report submitted for National Aquatic Resources Agency (NARA), Colombo, 1986 (cited in Kapurusinghe and Cooray, 2002, p. 6).

Pinedo, M. C., and Polacheck, T. 2004. Sea turtle bycatch in pelagic longline sets off southern Brazil. *Biological Conservation* 119:335–339.

Pitman, R. L. 1990. Pelagic distribution and biology of sea turtles in the eastern tropical Pacific. In Richardson, T. H., Richardson, J. I., and Donnelly, M. (Compilers). *Proceedings of the Tenth Annual Workshop on Sea Turtle Biology and Conservation.* NOAA Technical Memorandum NMFS-SEFC-278, pp. 143–148.

Pitman, R. L. 1992. Sea turtle associations with flotsam in the eastern tropical Pacific Ocean. In Salmon, M., and Wyneken, J. (Compilers). *Proceedings of the Eleventh Annual Workshop on Sea Turtle Biology and Conservation.* NOAA Technical Memorandum NMFS-SEFSC-302, p. 94.

Plotkin, P. T., Byles, R. A., and Owens, D. W. 1993. Migratory and reproductive behavior of *Lepidochelys olivacea* in the eastern Pacific Ocean. In Schroeder, B. A., and Witherington, B. E. (Compilers). *Proceedings of the Thirteenth Annual Sea Turtle Symposium.* NOAA Technical Memorandum NMFS-SEFSC-341, p. 138.

Polunin, N. V. C., and Sumertha Nuitja, N. 1982. Sea turtle populations in Indonesia and Thailand. In

Bjorndal, K. (Ed.). *Biology and Conservation of Sea Turtles* (reprinted 1995). Washington, DC: Smithsonian Institution Press, pp. 353–362.

Pritchard, P. C. H. 1969. Sea turtles of the Guianas. *Bulletin of Florida State Museum* 13:85–140.

Pritchard, P. C. H. 1973. International migrations of South American sea turtles (Cheloniidae and Dermochelidae). *Animal Behaviour* 21:18–27.

Pritchard, P. C. H. 1997. Evolution, phylogeny, and current status. In Lutz, P. L., and Musick, J. A. (Eds.). *The Biology of Sea Turtles.* Boca Raton, FL: CRC Press, pp. 1–28.

Pritchard, P. C. H., and Plotkin, P. T. 1995. Olive ridley sea turtle, *Lepidochelys olivacea.* In Plotkin, P. T. (Ed.). *National Marine Fisheries Service and U. S. Fish and Wildlife Service Status Reviews for Sea Turtles Listed under the Endangered Species Act of 1973.* Silver Spring, MD: National Marine Fisheries Service, pp. 123–139.

PROFEPA. 2003. Aseguramientos de productos y subproductos de tortuga marina de 1995–2003. Oficio DGVP.294.03. 18 septiembre del 2003.

Qureshi, T. 2001. Information Sheet on Ramsar Wetlands: Ormara Turtle Beaches, Pakistan. Unpublished Report. Karachi: IUCN Pakistan.

Qureshi, M. T. 2006. Sea turtles in Pakistan. In Shanker, K., and Choudhury, B. C. (Eds.). *Marine Turtles of the Indian Subcontinent.* Hyderabad, India: Universities Press, pp. 217–224.

Rajagopalan, M., Vivekanandan, E., Pillai, S. K., Srinath, M., and Bastion Fernando, A. 1996. Incidental catch of sea turtles in India. *Marine Fisheries Information Service. Technical and Extension Series* 143:8–16.

Rajagopalan, M., Vivekanandan, E., Balan, K., and Narayana Kurup, K. 2001. Threats to sea turtles in India through incidental catch. In Shanker, K., and Choudhury, B. C. (Eds.). *Proceedings of the National Workshop for the Development of a National Sea Turtle Conservation Action Plan, Bhubaneshwar.* Dehradun, India: Wildlife Institute of India, pp. 12–14.

Raja Sekhar, P. S., and Subba Rao, M. V. 1988. Conservation and management of the endangered olive ridley sea turtle *Lepidochelys olivacea* (Eschscholtz) along the northern Andhra Pradesh coastline, India. *(British Chelonia Group) Testudo* 3(5):35–53.

Rakotonirina, B., and Cooke, A. 1994. Sea turtles of Madagascar—their status, exploitation and conservation. *Oryx* 28:51–61.

Ramli, M. N., and Hiew, K. W. P. 1999. Marine turtle management, conservation and protection programme in Malaysia. In *Report of the SEAFDEC-ASEAN Regional Workshop on Sea Turtle Conservation and Management, July 26–28, 1999, Kuala Terengganu, Malaysia,* pp. 122–130.

Randriamiarana, H., Rakotonirina, B., and Maharavo, J. 1998. TED experience in Madagascar. In Wamukoya, G. M., and Salm, R. V. (Eds.). *Report of the Western Indian Ocean Turtle Excluder Device (TED) Training Workshop, Mombasa, Kenya, January 27–31, 1997.* Nairobi: IUCN East Africa Regional Office, pp. 16–17.

Rashid, S. M. A., and Islam, M. Z. 2006. Status and conservation of marine turtles in Bangladesh. In Shanker, K., and Choudhury, B. C. (Eds.). *Marine Turtles of the Indian Subcontinent.* Hyderabad, India: Universities Press, pp. 200–216.

Raut, S. K., and Nandi, N. C. 1988. Present status of marine turtle conservation and management in West Bengal. In Silas, E. G. (Ed.) *Proceedings of the Symposium on Endangered Marine Animals and Marine Parks.* Cochin, India: Marine Biological Association of India, pp. 255–259.

Reichart, H. A. 1993. Synopsis of Biological Data on the Olive Ridley Sea Turtle *Lepidochelys olivacea* (Eschscholtz, 1839) in the Western Atlantic. NOAA Technical Memorandum NMFS-SEFSC-336.

Reichart, H. A., and Fretey, J. 1993. *WIDECAST Sea Turtle Recovery Action Plan for Suriname.* In Eckert, K. L. (Ed.). CEP Technical Report No. 2. Kingston, Jamaica: UNEP Caribbean Environment Programme.

Riding, A. 1985. *Distant Neighbors. A Portrait of the Mexicans.* New York: Alfred A. Knopf.

Roeger, S. 2004. *Entanglement of Marine Turtles in Netting: Northeast Arnhem Land, Northern Territory, Australia.* Report to: Alcan Gove Pty Limited; World Wide Fund for Nature (Australia); Humane Society International; Northern Land Council. Dhimurru Land Management Corporation; Nhulunbuy Northern Territory, Australia.

Ross, J. P., and Barwani, M. A.. 1982. Review of sea turtles in the Arabian area. In Bjorndal, K. A. (Ed.). *Biology and Conservation of Sea Turtles* (reprinted 1995). Washington, DC: Smithsonian Institution Press, pp. 373–383.

Roychoudhury, B. 2001. Survey of sea-turtles in the coasts of West Bengal. In Shanker, K., and Choudhury, B. C. (Eds.). *Proceedings of the National Workshop for the Development of a National Sea Turtle Conservation Action Plan, Bhubaneshwar.* Dehradun, India: Wildlife Institute of India, pp. 50–52.

Rueda-Almonacid, J. V. 1992. Anotaciones sobre un caso de mortalidad masiva de tortugas marinas en la costa Pacífica de Colombia. In Rodríguez Mahecha, J. V., and Sánchez Paez, H. (Eds.). *Con-*

tribución al Conocimiento de las Tortugas Marinas de Colombia, Libro 4. Serie de Publicaciones Especiales del INDERENA, Santa Fe de Bogotá, pp. 181–190.

Sankar, O. B., and Raju, M. A. 2003. Implementation of the Turtle Excluder Device in Andhra Pradesh. *Kachhapa* 8:2–5.

Sankar, O. B., and Raju, M. A. 2006. Implementation of the TED in Andhra Pradesh. In Shanker, K., and Choudhury, B. C. (Eds.). *Marine Turtles of the Indian Subcontinent.* Hyderabad, India: Universities Press, pp. 262–267.

Schulz, J. P. 1975. Sea turtles nesting in Surinam. *Zoologische Verhandelingen* 143:1–143.

SEAFDEC (Southeast Asian Fisheries Development Center). 1997. *Proceeding of the Regional Workshop on Responsible Fishing, Bangkok, Thailand, June 24–27, 1997.* Samut Prakarn, Thailand: SEAFDEC.

SEAFDEC (Southeast Asian Fisheries Development Center). 1999. *Regional Guidelines for Responsible Fisheries in Southeast Asia.* Bangkok, Thailand: SEAFDEC.

Sekhsaria, P. 2004a. Caught in a corporate web. *The Hindu Sunday Magazine* March 28, 2004.

Sekhsaria, P. 2004b. Reliance vs. the olive ridley turtle. *InfoChange News & Features,* www.infochangeindia.org/features177.jsp, June 14, 2004.

Serafina, T. Z., Soto, J. M. R., and Celini, A. A. O. S. 2002. Registro da captura de tartaruga-olivácea, *Lepidochelys olivacea* (Eschscholtz, 1829) (Testudinata, Cheloniidae), por espinhel-pelágico no Rio Grande do Sul, Brasil. XXIV Congresso Brasileriro de Zoologia, INIVALI, Itajal Santa Catarina.

Shanker, K. 2003. Thirty years of sea turtle conservation on the Madras coast: A review. *Kachhapa* 8:16–19.

Shanker, K., and Kutty, R. 2005. Sailing the flagship fantastic: Different approaches to sea turtle conservation in India. In Frazier, J. (Ed.). *MAST/Maritime Studies* Special Issue: Marine turtles as flagships 3(2)/4(1): 213–240.

Shanker, K., and Pilcher, N. J. 2003. Marine turtle conservation in South and Southeast Asia: Hopeless cause of cause for hope? *Marine Turtle Newsletter* 100:43–51.

Shanker, K., Pandav, B., and Choudhury, B. C. 2003. An assessment of the olive ridley turtle (*Lepidochelys olivacea*) nesting population in Orissa, India. *Biological Conservation* 115:149–160.

Shanker, K., Ramadevi, J., Choudhury, B. C., Sing, L., and Aggarwal, R. K. 2004. Phylogeography of olive ridley turtles (*Lepidochelys olivacea*) on the east coast of India: implications for conservation theory. *Molecular Ecology* 13:1899–1909.

Sharath, B. K. 2002. Status survey of sea turtles along the Karnataka coast, India. A GOI UNDP Project Report. Department of Biosciences, University of Mysore, Karnataka.

Sharath, B. K. 2006. Sea turtles along the Karnataka coast. In Shanker, K., and Choudhury, B. C. (Eds.). *Marine Turtles of the Indian Subcontinent.* Hyderabad, India: Universities Press, pp. 141–146.

Shigenaka, G. (Ed.). 2003. *Oil and Sea Turtles: Biology, Planning, and Response.* U.S. Department of Commerce, National Oceanic and Atmospheric Administration, NOAA's National Ocean Service, Office of Response and Restoration, Hazardous Materials Response Division, Seattle, WA.

Silas, E. G., Rajagopalan, M., Fernando, A. B., and Dan, S. S. 1983a. Marine turtle conservation and management: A survey of the situation in Orissa 1981/82 and 1982/83. Central Marine Fisheries Research Institute, Cochin, India. *Marine Fisheries Information Service* 50:13–23.

Silas, E. G., Rajagopalan, M., and Dan, S. S. 1983b. Marine turtle conservation and management: A survey of the situation in West Bental 1981/82 and 1982/83. Central Marine Fisheries Research Institute, Cochin, India. *Marine Fisheries Information Service* 50:24–32.

Silas, E. G., Rajagopalan, M., Dan, S. S., and Bastian Fernando, A. 1984. Observations on the mass nesting and immediate postmass nesting influxes of the olive ridley *Lepidochelys olivacea* at Gahirmatha, Orissa—1984 season. Central Marine Fisheries Research Institute, Cochin, India. *CMFRI Bulletin* 35:76–82, Plates. I–IV.

Singh, A., Andrews, H., and Shanker, K. 2003. Report on the GOI-UNDP Sea turtle workshop, Andaman & Nicobar Island, India. *Kachhapa* 9:19–20.

Siraimeetan, P. 1988. Observations on the green turtle *Chelonia mydas* along the Gujarat Coast. In Silas, E. G. (Ed.). *Proceedings of the Symposium on Endangered Marine Animals and Marine Parks.* Cochin, India: Marine Biological Association of India, pp. 290–297.

Smith, W. N. 1974. The Seri Indians and the sea turtles. *Journal of Arizona History* 15:139–158.

Somander, K. 1963. Jaffna's turtle trials. *Loris* 9: 312–314.

Soto, J. M. R., and Beheregaray, R. C. P. 1997. New records of *Lepidochelys olivacea* (Eschscholtz, 1829) and *Eretmochelys imbricata* (Linnaeus, 1766) in the Southwest Atlantic. *Marine Turtle Newsletter* 77:8–9.

Stevens, S. W. L. 1998. *Present Status of Shrimp-Marine Turtle Interaction in Pakistan: a Literature Overview.* Unpublished Report, IUCN Pakistan.

Sunderraj, S. F. W., Vijay Kumar, V., Joshua, J., Serebiah, S., Patel, I. L., and Saravana Kumar, A. 2001. Status of the breeding population of sea turtles along the Gujarat coast. In Shanker, K., and Choudhury, B. C. (Eds.). *Proceedings of the National Workshop for the Development of a National Sea Turtle Conservation Action Plan, Bhubaneshwar.* Dehradun, India: Wildlife Institute of India, pp. 65–69.

Sunderraj, S. F. W., Joshua, J., and Vijay Kumar, V. 2006. Sea turtles and their nesting habitats in Gujarat. In Shanker, K., and Choudhury, B. C. (Eds.). *Marine Turtles of the Indian Subcontinent.* Hyderabad, India: Universities Press, pp. 156–169.

Taha, S. H. M. 1999. The management and conservation of marine turtles in Brunei Darussalam: Country report. In *Report of the SEAFDEC-ASEAN Regional Workshop on Sea Turtle Conservation and Management, July 26–28, 1999, Kuala Terengganu, Malaysia,* pp. 64–68.

Tambiah, C. R. 1994. Saving sea turtles or killing them: The case of U.S. regulated TEDs in Guyana and Suriname. In Bjorndal, K. A., Bolten, A. B., Johnson, D. A., and Eliazar, P. J. (Compilers). *Proceedings of the Fourteenth Annual Symposium on Sea Turtle Biology and Conservation.* NOAA Technical Memorandum NMFS-SEFSC-351, pp. 149–151.

Teas, W. G. 1993. Species composition and size class distribution of marine turtle strandings on the Gulf of Mexico and Southeast United States coasts, 1985–1991. NOAA Technical Memorandum NMFS-SEFSC-315.

Teyeliz, A. C. 2000. Historia del Tráfico de Tortugas Marinas en México antes de 1990. Unpublished report. Mexico City.

Thomé, J. C. A., Marcovaldi, M. A., Marcovaldi, G. G. D., Bellini, C., Gallo, B. M. G., Lima, E. H. S. M., da Silva, A. C. D. D., Sales, G., and Barata, P. C. R. 2003. An overview of Projeto TAMAR-IBAMA's activities in relation to the incidental capture of sea turtles in Brazilian fisheries. In Seminoff, J. A. (Compiler). *Proceedings of the Twenty-Second Annual Symposium on Sea Turtle Biology and Conservation.* NOAA Technical Memorandum NMFS-SEFSC-503, pp. 119–120.

Thorbjarnarson, J. B., Platt, S. G., and Khaing, S. T. 2000. Sea turtles in Myanmar: Past and present. *Marine Turtle Newsletter* 88:10–11.

Tow, S. K., and Moll, E. O. 1982. Status and conservation of estuarine and sea turtles in West Malaysian waters. In Bjorndal, K. (Ed.). *Biology and Conservation of Sea Turtles* (reprinted 1995). Washington, DC: Smithsonian Institution Press, pp. 339–347.

TRAFFIC. 2004. The trade in marine turtle products in Viet Nam. A report prepared for the Marine Turtle Conservation and Management Team, Viet Nam. Hanoi: Traffic Southeast Asia–Indochina.

Tripathy, B. 2002. Is Gahirmatha the world's largest sea turtle rookery? *Current Science* 83(11): 1299.

Tripathy, B., Choudhury, B. C., and Shanker, K. 2002. Marine turtles of Lakshadweep Islands, India. *Kachhapa* 7:3–7.

Tripathy, B., Shanker, K., and Choudhury, B. C. 2003. Important nesting habitats of olive ridley turtles (*Lepidochelys olivacea*) along the Andhra Pradesh coast of eastern India. *Oryx* 37:454–463.

Tripathy, B., Shanker, K., and Choudhury, B.C. In press. The status of sea turtles and their habitats in the Lakshadweep Archipelago, India. *Journal of the Bombay Natural History Society.*

Tripathy, B., Shanker, K., and Choudhury, B.C. 2006. Sea turtles and their habitats in the Lakshadweep Islands. In Shanker, K., and Choudhury, B. C. (Eds.). *Marine Turtles of the Indian Subcontinent.* Hyderabad, India: Universities Press, pp. 119–135.

Try, I. 1999. Country report on status of sea turtle in Cambodia. In *Report of the SEAFDEC-ASEAN Regional Workshop on Sea Turtle Conservation and Management, July 26–28, 1999, Kuala Terengganu, Malaysia,* pp. 72–74.

Turtle Expert Working Group (TEWG). 2000. Assessment Update for the Kemp's Ridley and Loggerhead Sea Turtle Populations in the Western North Atlantic. NOAA Technical Memorandum NMFS-SEFSC-444.

Ulaiwi, W. 1997. Marine turtle research and management in Papua New Guinea. In Noor, Y. R., Lubis, I. R., Ounsted, R., Troeng, S., and Abdullas, A. (Eds.). *Proceedings of the Workshop on Marine Turtle Research and Management in Indonesia, Jember, East Java, November 1996.* Bogor, Indonesia: Wetlands International/PHPA/Environment Australia, pp. 111–120.

Univeristy of Alaska Sea Grant College Program (UA-SGCP). 1997. *Fisheries Bycatch: Consequences and Management.* Alaska Sea Grant College Program Report No. 97-02, University of Alaska Fairbanks.

U.S. Fish and Wildlife Service and National Marine Fisheries Service (USFWS-NMFS). 1992. Recovery Plan for the Kemp's Ridley Sea Turtle (*Lepidochelys kempii*). St. Petersburg, FL: National Marine Fisheries Service.

Valliapan, S., and Pushparaj, S. 1973. Sea turtles in Indian waters. *Cheetal* 16:26–30.

Vinh, C. T., and Thuoc, P. 1999. Research, Conservation and Management of Marine Turtles in Vietnam. In *Report of the SEAFDEC-ASEAN Regional Workshop on Sea Turtle Conservation and Management, July 26–28, 1999, Kuala Terengganu, Malaysia,* pp. 178–187.

Wamukoya, G. M., Mbendo, J. R., and Eria, J. 1998. Bycatch in shrimp trawls in Kenya with specific reference to sea turtles. In Wamukoya, G. M., and Salm, R. V. (Eds.). *Report of the Western Indian Ocean Turtle Excluder Device (TED) Training Workshop, Mombasa, Kenya, January 27–31, 1997.* airobi: IUCN East Africa Regional Office, pp. 14–15.

Weber, M. 1995. Kemp's ridley sea turtle. In Plotkin, P. T. (Ed.). *National Marine Fisheries Service and U. S. Fish and Wildlife Service Status Reviews of Sea Turtles Listed under the Endangered Species Act of 1973.* Silver Spring, MD: National Marine Fisheries Service; Washington, DC: U. S. Fish and Wildlife Service, pp. 109–122.

Weber, M. 1996. Book Review: Caught in the net: The conflict between shrimpers and conservationists. *Marine Turtle Newsletter* 75:31–32.

Weber, M., Crouse, D., Irvin, R., and Iudicello, S. 1995. *Delay and Denial: A Political History of Sea Turtles and Shrimp Fishing.* Washington, DC: Center for Marine Conservation.

Wickremasinghe, S. 1981. Turtles and their conservation. *Loris* 15:313–315.

Witzell, W. N. 1994a. *The U.S. Commercial Sea Turtle Landings.* NOAA Technical Memorandum NMFS-SEFSC-350.

Witzell, W. N. 1994b. The origin, evolution, and demise of the U.S. Sea turtle fisheries. *Marine Fisheries Bulletin* 56(4):8–23.

WMS-EA (Wildlife Management Section, Biodiversity Group, Environment Australia). 1998. *Draft Recovery Plan for Marine Turtles in Australia.* Canberra: Environment Australia.

WPSI (Wildlife Protection Society of India). 2003. Operation Kachhapa News. *Kachhapa* 8:26–27.

Wright, B., and Mohanty, B. 2002. Olive ridley mortality in gill nets in Orissa. *Kachhapa* 6:18.

Wright, B., and Mohanty, B. 2006. Operation Kachhapa: An NGO Initiative for Sea Turtle Conservation in Orissa. In Shanker, K., and Choudhury, B. C. (Eds.). *Sea Turtles of the Indian Subcontinent.* Hyderabad, India: Universities Press, pp. 290–302.

DONNA J. SHAVER
THANE WIBBELS

14

Head-Starting the Kemp's Ridley Sea Turtle

HEAD-STARTING is the term used to describe a procedure whereby hatchling sea turtles are reared in captivity for periods ranging from a few days to a few years before they are released into the ocean (Pritchard et al., 1983; Magnuson et al., 1990; U.S. Fish and Wildlife Service and National Marine Fisheries Service, 1992). Some scientists define head-started turtles as hatchlings reared in captivity for any duration. However, Caillouet et al. (1997) used a more specific definition based on the amount of time in captivity. For example, they defined head-started turtles as turtles reared in captivity and released after approximately one year. Head-starting generally has been considered an "experimental" management technique (Klima and McVey, 1982; Magnuson et al., 1990; U.S. Fish and Wildlife Service and National Marine Fisheries Service, 1992) and has been controversial because it is unproven, it removes turtles from their natural environment, and it does not reduce the threats that cause population declines (Pritchard et al., 1983; Magnuson et al., 1990). The rationale behind head-starting is that hatchlings held in captivity will avoid the high level of predation associated with the posthatchling life stage in the wild and will be larger and susceptible to fewer predators on their release. Thus, survival chances of the turtles should increase, thereby helping the recovery of sea turtle populations (Klima and McVey, 1982; Anonymous, 1993).

Historically, hatchling sea turtles have been reared in captivity in a variety of projects. This includes long-term projects with green turtles (*Chelonia mydas*) (Huff, 1989) and hawksbill turtles (*Eretmochelys imbricata*) (Sato and Madriasau, 1991). For example, the Florida Department of Natural Resources began head-starting green turtles as early as 1959, and during a 30-year period, over 18,000 green turtles were head-started for 6–12 months before their release (Huff, 1989). Head-starting also was undertaken on a more limited scale with olive ridley turtles (*Lepidochelys olivacea*) in India and Thailand (Phasuk and Rongmuangsart, 1973; Subba Rao and Raja Sekhar, 1998; Subba Rao and Ratna Kumari, 2002). Many head-start projects were discontinued because of a lack of data needed to evaluate their success (Huff, 1989; Sato and Madriasau, 1991).

The Kemp's Ridley Sea Turtle (*Lepidochelys kempii*) Head Start Experiment began in 1978 as part of a larger Mexico–United States Kemp's Ridley Recovery Program to save the species. The Kemp's ridley had experienced a precipitous decline in the number of nesting turtles at its primary nesting beach in Rancho Nuevo, Mexico (U.S. Fish and Wildlife Service and National Marine Fisheries Service, 1992). As part of the Head Start Experiment, from 1978 to 2000, more than 23,000 Kemp's ridley hatchlings were raised in captivity for approximately one year at the National Marine Fisheries Service (NMFS) Laboratory in Galveston, Texas (Mrosovsky, 1978; Balazs, 1979; Woody, 1981; Klima and McVey, 1982; Manzella et al., 1988; Magnuson et al., 1990; Anonymous, 1993; Caillouet et al., 1993). In addition, a few hundred head-started Kemp's ridleys were retained in captivity beyond a year at the Cayman Turtle Farm (CTF) and at several aquaria to preserve a gene pool for the species in case of extinction in the wild (Mrosovsky, 1979; Brongersma et al., 1979; Balazs, 1979; Wood, 1982; Caillouet and Revera, 1985).

Head-starting Kemp's ridleys has been the subject of intense discussion and controversy for many reasons but primarily because the project was a large-scale experiment using the most endangered sea turtle in the world. Its critically endangered status drew attention to and scrutiny of all facets of the Kemp's Ridley Recovery Program, particularly head-starting. In the following sections, we provide a broad overview of the history, accomplishments, and controversy regarding the Kemp's Ridley Head Start Experiment.

Head-Starting the Kemp's Ridley Sea Turtle

Circumstances Leading to the Development of the Kemp's Ridley Head Start Experiment

The Kemp's ridley experienced a devastating decline between the late 1940s and the mid-1980s. Estimations from the historic Herrera film of 1947 suggested as many as 40,000 turtles nested in a single *arribada* (i.e., the mass nesting behavior characteristic of ridley sea turtles) at the primary nesting beach near Rancho Nuevo, Tamaulipas, Mexico (Carr, 1963; Hildebrand, 1963). By the time scientists discovered the beach in the early 1960s, the population had declined significantly, and in 1966, agencies in Mexico began conservation and research activities to protect the primary nesting beach (Chavez, 1968a, 1968b). The largest arribadas recorded from 1966 to 1968 ranged from approximately 1,500 turtles to 5,000 turtles (Chavez, 1968b; Pritchard, 1969). The population continued to decline during the 1970s, and by the late 1970s, the maximum size of an arribada had declined to approximately several hundred turtles (U.S. Fish and Wildlife Service and National Marine Fisheries Service, 1992).

Concerns that the Kemp's ridley would soon become extinct inspired a group of state and federal agencies from Mexico and the United States to meet in 1977 to develop a binational recovery program for the Kemp's ridley (Woody, 1989). As a result, the Kemp's Ridley Recovery Program was initiated in 1978 and included a variety of recovery actions. These actions were supported and carried out by a number of U.S. and Mexican agencies. During the initial years of the program, the primary contributors included Instituto Nacional de la Pesca, U.S. Fish and Wildlife Service, NMFS, National Park Service, Texas Parks and Wildlife Department, Florida Audubon Society, and the Gladys Porter Zoo. The program's main objective was to intensify efforts on the primary nesting beach, including utilizing the expertise of a binational team of biologists that attempted to protect all nesting females, eggs, and hatchlings (Woody, 1989). The program also included an experiment to establish a nesting beach for Kemp's ridleys on Padre Island National Seashore (PAIS) along the Texas coast near Corpus Christi (Fig. 14.1). This was a logical location for establishing a nesting beach because it was a protected area, and historically there were reports of scattered nesting along the south Texas coast (Hildebrand, 1982; Woody, 1989). Further, the idea of establishing a nesting beach on the south Texas coast was not new and had been attempted on a smaller scale by private citizens back in the 1960s (Woody, 1989). The experiment to establish a nesting beach at PAIS included the translocation of a small percentage of the total eggs from Rancho Nuevo (from 1978–1988 an average of 2.8% of the total eggs) to PAIS for incubation, hatching, and experimental imprinting (Caillouet, 1995a). It has been hypothesized that hatchlings "imprint" on their natal beach and then females return to that beach to nest after they reach sexual maturity (Meylan et al., 1990a). The rationale for this experiment was that establishing a second nesting colony of Kemp's ridleys at PAIS would enhance the survival of the species if a disaster (e.g., hurricane, oil spill) occurred at the primary nesting beach (Woody, 1989).

Because mortality is presumably high during the early life history of sea turtles, the imprinting experiment was coupled with the "Head Start Experiment," also initiated in 1978 at the NMFS Galveston Laboratory (Klima and McVey, 1982). The main objective of the Head Start Experiment was to increase survival of post-hatchling Kemp's ridleys by raising them in captivity for 9–11 months. According to Klima and McVey (1982), head-starting was an important component of the Kemp's Ridley Recovery Program because the Kemp's ridley population had declined to a level that might not recover naturally, and head-starting the experimentally imprinted hatchlings might be necessary for the establishment of a nesting beach at PAIS. Further, they indicate that the Head Start Experiment provided a

Fig. 14.1. Release locations of head-started Kemp's ridleys. The number of turtles released at each location and the year are listed in Table 14.1.

means of evaluating head-starting as a management technique for sea turtle recovery.

Experimental Imprinting and the Source of Hatchlings for Kemp's Ridley Head Start Experiment

The turtles used in the Kemp's Ridley Head Start Experiment were hatched at Rancho Nuevo or at PAIS. From 1989 through 2000, all hatchlings reared in the Head Start Experiment were obtained directly from Rancho Nuevo. Additionally, several hundred hatchlings from the 1978, 1979, 1980, and 1983 year classes were also from Rancho Nuevo, but the majority of the hatchlings from the earlier years of the Head Start Experiment (i.e., 1978 through 1988) were obtained from eggs that had been transferred from Rancho Nuevo to PAIS for experimental imprinting at PAIS (Shaver, 1989, 1990a; Shaver and Miller, 1999).

PAIS is the longest stretch of undeveloped barrier island beach in the United States and is administered and protected by the National Park Service. Sporadic Kemp's ridley nesting has been documented at PAIS (Werler, 1951; Hildebrand, 1963; Carr, 1967; Francis, 1978), although the historic number of nesting ridleys is unknown. Kemp's ridley eggs were collected at Rancho Nuevo from 1978 to 1988 (total of 22,507 eggs) and transferred to PAIS for incubation and experimental imprinting. These eggs were incubated in PAIS sand (see Shaver, 2005, for detailed explanation of incubation). Hatchlings were released on the beach at PAIS and allowed to crawl down the beach and allowed to swim briefly in the Gulf of Mexico. After hatchlings swam approximately 5–10 m, they were captured using aquarium dip nets and shipped to the NMFS Galveston Laboratory for head-starting. Overall, 77.1% of the eggs incubated at PAIS produced hatchlings (total of 15,857 hatchlings) that were transferred to the Head Start Experiment from 1978 to 1988 (Shaver, 2005).

Sex Ratios of Kemp's Ridleys in the Head Start Experiment

As with all sea turtles, the Kemp's ridley possesses temperature-dependent sex determination (TSD) in which the incubation temperature of the egg determines the sex of the hatchling (Shaver et al., 1988; Wibbels et al., 1989b; reviewed by Wibbels, 2003). In sea turtles, warmer temperatures produce females, and cooler temperatures produce males (reviewed by Wibbels, 2003). Sex ratios resulting from TSD can vary widely (e.g., the sex ratio from a given nest can vary from all male to all female depending on the incubation temperature) and thus can significantly affect the reproductive ecology and recovery of an endangered species. For example, a female bias in a population might increase egg production if males are not a limiting factor in reproduction. Monitoring sex ratios of hatchlings used in the Head Start Experiment was of particular importance because the eggs were artificially incubated in Styrofoam boxes. However, because of the endangered status of Kemp's ridleys, they could not be sacrificed to determine gender, so a variety of techniques (i.e., necropsy and gonadal histology, laparoscopy, serum testosterone assay, tail length evaluation of adults) were used to determine the sex of dead embryos, hatchlings, posthatchings, and older captive turtles that died during captive rearing in the Head Start Experiment (Shaver et al., 1988; Wibbels et al., 1989b; Shaver and Fletcher, 1992; Caillouet, 1995b; Caillouet et al., 1995b). These data were used to estimate sex ratios of individual year classes. Males predominated in the 1978, 1979, 1981, and 1983 year classes. The egg hatcheries at Rancho Nuevo and PAIS were subsequently modified to increase incubation temperatures to increase the proportion of females. This was successful, and 77.5% of the turtles examined from the 1985–1988 year classes were females (Shaver et al., 1988). Overall, 59.6% of the 1978–1988 year classes collectively were females, for an overall sex ratio of 1.5F:1M (Shaver, 2005). Because over 90% of the head-start turtles from the 1989–2000 year classes were females (Caillouet et al., 1995b; B. Higgins, personal communication), sex ratios could lead to a higher representation of these year classes among the group of head-start Kemp's ridley turtles that could be eventually recorded nesting. However, the earlier years were male dominated, which certainly reduced the probability of those year classes producing nesters.

Head Start Methodology

During the 23 years of the Head Start Experiment (1978–2000), the methods utilized for rearing turtles were periodically enhanced, but many general aspects of the methodology were retained throughout the program. The methods have been described in detail previously (Klima and McVey, 1982; Caillouet et al., 1989; Fontaine et al., 1989b; Caillouet, 2000), so only a brief description is provided in the current review. Turtles were reared in a building that had the capacity to hold approximately 2,000 turtles. The building was metal framed, with a synthetic top that allowed ambient sunlight to illuminate the facility. Inside the building were long holding tanks (i.e., raceways) that each held approximately 100 smaller containers (buckets, baskets, crates) for isolating individual turtles. Isolation of individual turtles prevented them from biting one another, thus preventing wounds and potential subsequent infections. During the early years of the program, plastic buckets with holes in the bottom were used to separate turtles within the raceways. In later years, custom-built plastic crates with flow-through bottoms were utilized. The buckets and crates were suspended in seawater, which filled the raceways. Seawater was pumped directly from the Gulf of Mexico and was initially held and heated (if necessary) in large holding tanks located in the turtle building. The water in the raceways was changed three times weekly. Mean water temperature in the raceways varied from 23°C to 31°C with an average of approximately 27°C (Fontaine and Shaver, 2005). Mean salinity was approximately 31 parts per thousand (Fontaine and Shaver, 2005). Turtles were fed a dry, pelleted, floating food that was developed and produced commercially. They were fed twice daily for a total daily ration of approximately 1–2% of their body weight. Turtles were reared for approximately 9–10 months, tagged (the tagging techniques are discussed in detail below), and released into the wild.

Tagging Methodology

Evaluating the effectiveness of the Head Start Experiment was contingent on recapturing and identifying turtles after their release (Klima and McVey, 1982; Caillouet et al., 1989). Therefore, a variety of methods for marking turtles were evaluated and utilized during the Head Start Experiment (Fontaine et al., 1993). Throughout the program, external metal tags were attached to the trailing edge of front flippers (Hasco Type 681 tags made of monel or inconel alloys) (Manzella et al., 1988). All turtles received at least one metal flipper tag, and most were tagged on the right front flipper (Fontaine et al., 1993). Because of concerns that metal tags might eventually fall off, several permanent tags were evaluated and implemented. Starting with the 1984 year class, all head-start turtles were marked with "living tags" in which a small disk of lighter color tissue from the plastron was transplanted to the darker carapace (Hendrickson and Hendrickson, 1981; Caillouet et al., 1989; Fontaine et al., 1993). Each year, the disk of tissue from the plastron was transplanted to a different location on the carapace to facilitate identification of specific year classes of turtles (Caillouet et al., 1986). Living tags were used on a small subset of the head-start turtles during the early 1980s. Virtually all turtles were marked with living tags from 1984 through 2000 (Fontaine et al., 1993; Fontaine and Shaver, 2005). Some turtles also received internal permanent tags. Starting with the 1984 year class, magnetized wire tags were inserted into the right or left front flipper of all turtles. The wire tags identified the year class but did not identify individual turtles. Detecting the wire tags requires the use of a magnetometer or x-ray machine. Starting with the 1990 year class, all turtles were marked with passive integrated transponder tags (i.e., PIT tags), which were injected under the skin near the base of the left front flipper, (Fontaine et al., 1989b, 1993). Additionally, a few turtles in the early year classes received PIT tags. A portable electronic PIT tag reader is required to detect and read PIT tags, each of which has a unique alphanumeric code that identifies individual turtles.

Summary of Turtles Reared and Released by the Kemp's Ridley Head Start Experiment

During each year of the Kemp's Ridley Head Start Experiment, approximately 1,000–2,000 turtles were raised in captivity annually. The only exceptions were in 1983, when only 230 hatchlings were reared because of the low hatch rate, and 1993–2000, when 178–200 hatchlings were received each year for potential use in TED testing. Over 23 years, a total of 27,137 Kemp's ridley hatchlings were transported to the NMFS Galveston Laboratory from either PAIS or directly from Rancho Nuevo (see details regarding egg incubation and experimental imprinting above in the Experimental Imprinting section). Of those hatchlings, 23,987 (88.4%) were successfully reared, tagged, and released into the Gulf of Mexico (Fontaine and Shaver, 2005) at sizes comparable to late pelagic stage or early postpelagic stage wild Kemp's ridleys (Ogren, 1989). In the later years of the program, more than 90% of hatchlings were successfully reared for 9–11 months (Fontaine and Shaver, 2005).

Under the controlled conditions of head-starting, Kemp's ridleys grew to an average weight of 1.25 kg and an average SCL of 19.5 cm within a year (Caillouet et al., 1989). Because most turtles were released after 9–10 months of captive rearing, they typically weighed approximately 1 kg at the time of release (Fontaine et al., 1989b).

Head-started turtles were released in a variety of locations (Fig. 14.1; Table 14.1) (Klima and McVey, 1982; Manzella et al., 1988; Fontaine et al., 1989a; Caillouet et al., 1995a). Most release sites were chosen because they were believed to represent appropriate habitats for late pelagic or postpelagic Kemp's ridleys (Klima and McVey, 1982). The majority of turtles in the first two-year classes (1978 and 1979) were released off the Gulf coast of Florida. Based on the recovery of turtles released there, there was concern that many turtles were dispersing into the Atlantic Ocean and might not return to the Gulf of Mexico for breeding (Manzella et al., 1988; Fontaine et al., 1989a; Caillouet et al., 1995a). Therefore, from 1980 through 2000, turtles were released in the western Gulf of Mexico, primarily off the Texas coast, with the majority of the releases occurring in the waters off Padre Island.

Movements, Growth, and Behavior of Head-Started Turtles Following Their Release into the Wild

The effectiveness of head-starting is dependent on the survival, adaptation, and eventual breed-

Table 14.1 Yearly release locations of Kemp's ridley sea turtles from the Head Start Program

Year class	Release site	Number of turtles released
1978	Sandy Key, FL	307
	East Cape, FL	219
	Homosassa, FL	1,380
	Padre Island, TX	113
1979	Homosassa, FL	1,339
	Key Largo, FL	24
	Padre Island, TX	5
	Galveston, TX	1
1980	Padre Island, TX	1,526
	Bay of Campeche, Mexico	197
1981	Padre Island, TX	1,521
	Sabine Pass, TX	118
1982	Padre Island, TX	1,159
	Nueces Bay, TX	96
	Sabine Pass, TX	69
	Mustang Island, TX	1
1983	Mustang Island, TX	190
1984	Padre Island, TX	1,017
1985	Copano Bay, TX	448
	Indian Bend, TX	22
	Port Bay, TX	49
	Padre Island, TX	961
	Galveston, TX	54
1986	Mustang Island, TX	1,630
1987	Padre Island, TX	1,230
1988	Padre Island, TX	808
1989	Galveston, TX	1,894
1990	Galveston, TX	1,979
1991	Galveston, TX	1,943
1992	Galveston, TX	1,963
1993	Mustang Island, TX	158
	Panama City, FL	29
	High Island, TX	1
1994	Galveston, TX	170
1995	Galveston, TX	168
1996	Galveston, TX	174
1997	Galveston, TX	178
1998	Galveston, TX	162
1999	Galveston, TX	176
2000	Galveston, TX	170

Note: Turtles were released offshore from these locations. Based on data from Manzella et al. (1988); Fontaine et al. (1989a); Caillouet et al. (1995b); Fontaine and Shaver (2005); B. Higgins (personal communication).

ing of the turtles following their release. Therefore, it is paramount to assess the turtles' movements, habitat utilization, and behavior after release. Information on these topics can be gathered from a variety of sources, including tag returns, tracking studies, and dietary studies. Here we assess the effectiveness of the Head Start Experiment based on empirical data on the movements, growth, and behavior of head-started Kemp's ridleys following their release in the Gulf of Mexico.

Methods of Recapture

Head-started Kemp's ridleys were recaptured by a wide variety of methods (reviewed by Fontaine and Shaver, 2005). A relatively large percentage of the recaptures were found stranded on land dead (30.4%) or alive (13.3%) (Fontaine and Shaver, 2005). Many were also captured by various fishing methods including shrimp trawl (16.8%), hook and line (12.1%), entanglement net (3.5%), gill net (2.9%), cast net (0.4%), beach seine (0.2%), oyster dredge (0.2%), pound net (0.1%), and trammel net (0.1%). Some turtles also were captured by hand (5.4%) or by dip net (1.6%), and the recapture method for some turtles was not reported (11.7%).

Many of the methods listed above are also methods by which wild Kemp's ridleys have also been captured (e.g., shrimp trawl, hook and line, entanglement nets, gill nets, pound nets). In contrast, wild Kemp's ridleys are not noted for being captured by hand or by dip net, and such methods might reflect behavior resulting from the captive rearing. Certain individual turtles appeared susceptible to such capture techniques (Manzella et al., 1988). However, such behavior does not appear to be the norm for all head-started turtles because some turtles showed evasive behaviors soon after their release (Klima and McVey, 1982; Wibbels, 1984).

Recapture Rate

The average number of recaptures for all year classes of head-started turtles was reported to be approximately 3.9% of the total number of turtles released, with most year classes having approximately 2–3% (Fontaine and Shaver, 2005). Many of the recaptures occurred soon after release. For example, Manzella et al. (1988) analyzed data from the 1978 through 1986 year classes and found that 36.6% of the recaptures occurred during the first 2 months following release.

The percentage of head-started turtles that have been recaptured has varied among year classes, ranging from approximately 1.6% to 11.1% of the total number of turtles released in each year (Fontaine and Shaver, 2005). Some of the variation in the number of recaptures appears to be related to the release location. The highest recapture rate (11.1%) was for the 1982 year class, when most of the turtles were released into sargassum patches in the nearshore waters off Padre Island, Texas (Manzella et al., 1988).The sargassum washed ashore within 2 weeks, and with it came many head-start turtles. The majority of recaptures from that year class occurred within 2 weeks of release, and many turtles were coated with oil that may have accumulated in the sargassum patches (Manzella et al., 1988). Turtles released into Texas bays also showed relatively high recapture rates. A subset of turtles from the 1982 year class were released into Nueces Bay, Texas ($n = 96$), and a subset of turtles from the 1985 year class were released into Nueces Bay and Copano Bay, Texas ($n = 519$). Approximately 14.0% of those turtles were recaptured, and over 90% of those recaptures occurred in the bay in which they were released (Manzella et al., 1988). However, the majority of those recaptures occurred well after the release, averaging 129 days from the release date for the Nueces Bay release and 105 days for the Copano Bay release.

Movements of Head-Started Kemp's Ridleys Following Release

The recapture locations of turtles have varied widely throughout the history of the Head Start Experiment. Some of this variation is related to release locations, and some is a result of variation among individual turtles. During the first two years, the majority of turtles were released in the eastern Gulf of Mexico (Fig. 14.1; Table 14.1). Tag return data from those releases revealed a variety of movement patterns (Klima and McVey, 1982; McVey and Wibbels, 1984; Manzella et al., 1988; Fontaine et al., 1989a). Many of the turtles released off southwest Florida were recaptured along the Atlantic coast, ranging from the Florida Keys to as far north as

Long Island, New York (Klima and McVey, 1982; McVey and Wibbels, 1984; Manzella et al., 1988). Turtles released on the central west coast of Florida near Homosassa appeared to move north or south of the release location, with some traveling around the southern portion of Florida and moving into the Atlantic, and others moving north and west along the Gulf coast with some traveling as far west as Texas (Klima and McVey, 1982; McVey and Wibbels, 1984). The long-distance recaptures typically occurred approximately one to two years after release. One turtle from the 1979 year class that had been released near Homosassa, Florida, was recovered 568 days later along the coast of France (Wibbels, 1983a; Manzella et al., 1988). A second turtle from that same year class was recovered along the coast of Morocco 898 days after its release (Manzella et al., 1988). Additionally, some turtles were recaptured relatively close to the Homosassa release site, but most of those recaptures occurred within the first few months after release. However, one turtle was recaptured approximately 80 km from the release site 434 days after its release. Overall, turtles released in the eastern Gulf of Mexico during the first two years of the Head Start Experiment dispersed widely from the release areas, with the majority of recaptures (66% as reported by Manzella et al., 1988) occurring along the Atlantic coast.

The release locations were changed to the western Gulf of Mexico for the 1980 through 2000 year classes of turtles (Manzella et al., 1988; Fontaine et al., 1989a; Caillouet et al., 1995a; B. Higgins, personal communication). The majority of these locations were along the Texas Gulf coast, ranging from Padre Island to Sabine Pass (Fig. 14.1). However, 197 turtles from the 1980 year class were released from a NOAA research ship in the Bay of Campeche located in the southeastern Gulf of Mexico. Additionally, two subsets of turtles were released into Texas bays. Manzella et al. (1988) summarized recapture data for the 1978 through 1986 year classes and reported that 96.3% of the recaptures from the Texas releases occurred in the coastal waters of the Gulf of Mexico ranging from the Bay of Campeche to the Gulf coast of central Florida, with the majority of these recaptures occurring along the coasts of Texas and Louisiana. A small percentage of the recaptures (3.6%) from the Texas releases occurred along the Atlantic coast of the United States. One turtle from the 1981 year class, released off Padre Island, Texas, was recovered on the coast of France 1,394 days after its release. Another turtle released along the Texas coast was recovered off the coast of Nicaragua (Manzella et al., 1991).

Radio Tracking of Head-Started Kemp's Ridleys Following Release

To evaluate the movement of turtles after release, a small number of head-started turtles from the 1978 ($n = 2$) and 1979 ($n = 10$) year classes were tracked using radio transmitters for approximately 3 weeks after their release near Homosassa (Klima and McVey, 1982; Wibbels, 1984). Results from the 1979 year class suggest that turtles moved randomly relative to geographic direction and were being affected by currents. The final locations of the turtles during the study ranged from as far south as Clearwater, Florida, to as far north as Cedar Keys, Florida. However, many of the turtles remained relatively close to the release area (Wibbels, 1984).

Growth of Head-Started Kemp's Ridleys Following Release

The growth of head-started turtles following release is one indicator of how well the turtles adapt to the wild. Growth data for head-started Kemp's ridleys have been reported in several previous studies (McVey and Wibbels, 1984; Fontaine et al., 1989a; Caillouet et al., 1995b). Growth data suggest that many head-started turtles adapted successfully to the wild. For example, 13 turtles from the 1979 year class were weighed after recapture, and 11 had at least doubled or tripled their weight within approximately one to two years following their release (McVey and Wibbels, 1984). Fontaine et al. (1989a) indicated that turtles recaptured in the Gulf of Mexico exhibited higher growth rates than those recaptured in the Atlantic. Growth rates of head-started turtles in the wild appear to fit a von Bertalanffy growth model (Caillouet et al., 1995b). Based on growth data from turtles recaptured in the Gulf of Mexico, that model suggests that head-started turtles would reach the size of sexually mature Kemp's ridleys (i.e., approximately 60 cm SCL) in approximately 10 years (Caillouet et al., 1995b). It would take longer than 10 years for turtles in the Atlantic (Caillouet et al., 1995b). These estimates are consistent with previous predictions for wild Kemp's ridleys, which range from 5.5 years to over 15 years to reach sexual maturity (Márquez-M., 1994; Zug et al., 1997). The average growth rates of head-started Kemp's ridleys in the wild were lower than those of Kemp's ridleys that were permanently reared in captivity at Galveston Sea Arama and Miami Seaquarium (McVey and Wibbels, 1984: Fontaine et al., 1989a). This might be expected because captive turtles are fed high-protein diets and probably expend less energy (i.e., less exercise) than wild turtles. Sexual maturity may be attained at a younger age in captivity because Kemp's ridleys permanently reared in captivity have reached sexually maturity in as little as 5–7 years at the CTF (Caillouet et al., 1995b).

Diet of Head-Started Kemp's Ridleys Following Release

The diet of head-started turtles after release is also an indicator of their ability to successfully adapt to the wild. Two studies examined post-release feeding behavior of head-started Kemp's ridley turtles (Shaver, 1991a; Werner, 1994; Werner and Landry, 1994). In the first study, digestive tract contents were examined from 101 dead Kemp's ridleys stranded on south Texas beaches from 1983 to 1989 (Shaver, 1991a). Of the 101 turtles, 55 were wild and 51 were head-started turtles from various year classes and were recovered 10–2,100 days after their release. Many of the head-started turtles were recovered from bay shorelines (n = 46), and five were recovered on Gulf beaches. In contrast, 45 of the wild turtles were stranded on Gulf beaches and 5 on bay shores. However, some of the head-started turtles were released in the bays, so the difference in stranding location might be expected. Curved carapace lengths of wild turtles (5.2–71.0 cm) were significantly greater than those of head-started turtles (14.6–48.2 cm). The study by Shaver (1991a) concluded that head-started turtles had adapted to feeding in the wild, and there were similarities between the diets of both groups. However, the diets of wild and head-started turtles varied significantly in the amount and percentage frequency of seven food groups (crabs, fish, mollusks, vegetation, shrimp, marine debris, and other materials). The most noteworthy differences were that wild turtles consumed greater amounts of crabs, and head-started turtles ate greater amounts of mollusks, fish, and other materials. Size and habitat differences were potential causes of dietary divergences of wild and head-started Kemp's ridley turtles. Larger Kemp's ridleys tended to consume more crabs and less of other food types, and turtles found stranded in the bays tended to consume a narrower range of crab species than those found stranded on Gulf beaches. Alternatively, it is possible that the dietary differences resulted from behavioral differences between head-started and wild turtles.

In the second study, the diets of similar sized wild and head-started Kemp's ridleys inhabiting the same habitat were compared. Fecal material was collected from 86 live Kemp's ridleys (74 wild and 12 head start) captured during entanglement netting at Sabine Pass, Texas, located on the coast near the Texas–Louisiana border, from April to November 1993 (Werner, 1994; Werner and Landry, 1994). All head-started ridleys were from the 1991 year class and had been in the wild for 343–451 days before capture. All 12 of the head-started turtles measured less than 40 cm straight-line carapace length (center of the nuchal notch to the posterior tip of the postcentral scute) (SCL), and the 11 used in the study were 26.0–35.9 cm; most wild turtles were less than 40 cm. The fecal material from the 11 head-started turtles was compared to the fecal material from the 27 similar-size wild turtles. No significant difference in frequency of occurrence or dry mass of prey items or species was detected between wild and head-started turtles. This study indicated that head-started turtles had recently consumed prey items similar to their wild counterparts, indicating that they had adapted to feeding in the wild. Thus, these results indicate that head-started Kemp's ridleys can transition from the artificial diets fed during captivity to natural prey after their release in the wild (Werner, 1994).

Movements, Growth, and Adaptation of Head-Started Kemp's Ridleys in the Wild

A few general conclusions can be derived from the recapture and radio-tracking data from head-started turtles. The site of release obviously had an effect on the movements and distribution of head-started turtles. These data indicate that turtles released in the western Gulf of Mexico had a greater probability of remaining in the Gulf of Mexico than those released in the eastern Gulf of Mexico. Many recaptures occurred during the first few months following release. The turtles released at offshore locations had a general tendency to disperse widely from their release sites. In contrast, many of the turtles released in Texas bays were recaptured in the same general areas, even after relatively long time periods. Overall, head-started Kemp's ridleys dispersed widely from release areas and were reported throughout the natural range of the species and in habitats where wild individuals occur (Carr, 1952, 1957; Carr and Caldwell, 1958; Pritchard and Márquez-M., 1973; Hildebrand, 1982; McVey and Wibbels, 1984; Ogren, 1989; Shaver, 1991a; Werner, 1994; Werner and Landry, 1994; Coyne, 2000). In general, turtles that were at large for more than 60 days appeared to be distributed in the same areas as wild turtles, with notable exceptions being turtles recovered from Morocco and Nicaragua (Manzella et al., 1988; Fontaine and Shaver, 2005). Growth and dietary data from recaptured and stranded turtles indicate that head-started turtles can adapt to feeding in the wild. The distribution and growth of recaptured head-started Kemp's ridley turtles indicate that (at least) some of these turtles successfully adapt to wild conditions (Klima and McVey, 1982; McVey and Wibbels, 1984; Manzella et al., 1988; Caillouet et al., 1995a).

Although some turtles appear to adapt readily to the wild, others may exhibit behaviors that are atypical of wild turtles. Some head-started Kemp's ridleys have been reported displaying aberrant behavior in the water days to months after release (Fontaine et al., 1989a; Taubes, 1992; Meylan and Ehrenfeld, 2000). Some head-started turtles released at Panama City, Florida, did not exhibit the escape and avoidance response to humans typical of wild ridleys; for up to five months after release, they approached humans in shallow water and were easy to capture by hand (L. Ogren, personal communication). In South Texas, some Kemp's ridleys that had been held for extended periods of captive rearing (several years) and were at large for a few days to weeks were reported bumping into surf fishermen and swimming up to small fishing boats, enabling people to pick them up and read their tag numbers (P. Plotkin, personal communication). However, the percentage of turtles exhibiting aberrant behaviors was relatively low (i.e., approximately 6%; Fontaine and Shaver, 2005).

Survival of Head-Started Kemp's Ridleys in the Wild

The majority of recaptures (91%) occur during the first two years after release, and very few recaptures occur after four years or more. Although the decrease in recaptures could reflect a number of factors (e.g., tag loss, behavioral changes over time that affect recapture rate, sampling variability), turtle mortality could be a major factor contributing to this trend. Caillouet et al. (1995a) used recapture data to estimate survival rates of head-started turtles following release and attempted to adjust for possible tag loss. They estimated that survival rates varied among year classes and increased over time after release (Caillouet et al., 1995a). Mortality rates were estimated to be relatively high in the wild, such that few turtles would be expected to reach maturity (i.e., at least 10 years of age). However, because of the lack of an available control group (i.e., wild Kemp's ridleys with tags), it was not possible to compare mortality rates of head-started turtles to those of wild Kemp's ridleys.

In view of the continued decline of this species and the relatively low numbers of nesting females throughout the 1980s and early 1990s, rates likely were high for both wild and head-started turtles. This is particularly applicable before mandatory use of turtle excluder devices (TEDs) starting in 1989 (Shaver and Fletcher, 1992; Wibbels, 1992; Werner, 1994; Shaver and Caillouet, 1998; Fontaine and Shaver, 2005; Shaver, 2005). Mortality of Kemp's ridleys was so high during the 1980s that survival to adulthood was unlikely for most turtles (Wibbels, 1992; Caillouet et al., 1995a). From 1986 to 2003, more wild adult Kemp's ridley turtles were found stranded dead in Texas than in any other state in the United States. Moreover, strandings became increasingly concentrated on south Texas Gulf of Mexico beaches. The greatest numbers of strandings were found after 1994 at the same time when more ridley nests were found there (Shaver, 2005; W. Teas, personal communication). From 1995 to 2003, 152 of the 268 wild stranded adult Kemp's ridleys documented by the Sea Turtle Stranding and Salvage Network in the United States were located on south Texas Gulf beaches (Shaver, 2005). Only three confirmed head-started adult Kemp's ridleys were reported stranded through 2003, but some of those classified as wild could have been head-started turtles that were not recognizable because their tags were shed or their decomposed condition prohibited thorough examination for tags (Shaver and Caillouet, 1998; Shaver, 2005).

There are a variety of natural and anthropogenic causes for mortality of Kemp's ridley turtles in the waters of the United States and Mexico (Turtle Expert Working Group, 1998). However, Magnuson et al. (1990) concluded that the most important human-associated source of sea turtle mortality for benthic immatures and adults in the coastal waters is incidental capture in shrimp trawls, which account for more deaths than all other human activities combined. It should be noted that most release locations for head-started turtles were areas that were known to be frequented by shrimp trawlers, and "incidental capture by trawl" was one of the reported methods of capture. The mandatory use of TEDs began in U.S. offshore waters in 1990, and most trawlers operating in southeastern U.S. waters were required to use TEDs year-round after December 1992 (Caillouet et al., 1996; Turtle Expert Working Group, 1998). Despite mandatory TED regulations, a correlation between shrimp fishing and sea turtle strandings continued through 2000 (Caillouet et al., 1991, 1996; Turtle Expert Working Group, 1998; Lewison et al., 2003; Shaver, 2005). Of the 152 dead adult ridleys found on South Texas Gulf beaches from 1995 to 2003, 142 (93.4%) stranded during times when Gulf waters off Texas were open to shrimp trawling (mid-July through mid-May) (Shaver, 2005).

In an attempt to decrease mortality of the Kemp's ridley in Texas waters, several environmental groups and biologists suggested the creation of a marine reserve or area closed to commercial fishing (Plotkin, 1999; McDaniel et al., 2000; Shore, 2000). They noted the large number of dead adults found stranded on South Texas Gulf beaches (the site of the experimental imprinting), and they noted that most Kemp's ridley nests in the United States were at that same location. Texas Parks and Wildlife Department (TPWD) regulations, passed in August 2000 to prevent overfishing in the shrimping industry in Texas, established a new annual closure of Gulf waters to shrimp trawling off North Padre Island, South Padre Island, and Boca Chica Beach out to 8 km from shore, from December 1 through mid-May each year, preceding the existing annual Texas Closure which typically extends from mid-May through mid-July. This regulation went into effect on December 1, 2000, and the annual closure helps protect adult Kemp's ridley turtles in South Texas.

Nesting of Head-Started Kemp's Ridleys

Nesting of Head-Started Kemp's Ridleys in Captivity

Assimilation of head-started turtles into the breeding population is the primary objective of head-starting programs (Klima and McVey, 1982; Caillouet et al., 1995b). Head-started Kemp's ridleys have been documented nesting both in the wild and in captivity. Head-started Kemp's ridley turtles held in captivity until sexual maturity have nested, but hatching and emergence success of their offspring generally have been lower than those of head-started and wild Kemp's ridleys nesting in the wild. Head-started Kemp's ridleys were transferred from NMFS Galveston Laboratory to numerous aquaria during the late 1970s to mid-1980s for captive breeding, to serve as a breeding stock in case the wild population continued to decline despite conservation efforts (Caillouet, 2000). CTF received most of the turtles and was the first to breed Kemp's ridleys successfully. At the CTF, ridleys reached sexual maturity at a minimum age of 5 years, and the first viable hatchlings were produced from 7-year-old turtles in 1986 (Wood, 1982; Wood and Wood, 1984, 1988, 1989; Caillouet, 2000). Of the ridleys held at the CTF, 30% had begun nesting by 7 years of age at 56–57 cm curved carapace length (Wood and Wood, 1988), slightly younger and smaller than documented for the head-started Kemp's ridleys that nested in the wild (Table 14.2). Clutch size for these captive ridleys (n = 10 clutches laid between 1984 and 1986) averaged 60 eggs (range 7–103 eggs) (Wood and Wood, 1988), which is less than the average clutch size of 94 eggs for head-started Kemp's ridleys that nested in the wild (Table 14.2) and 100 eggs for wild Kemp's ridleys at Rancho Nuevo (U.S. Fish and Wildlife Service and National Marine Fisheries Service, 1992). However, nesters at Rancho Nuevo comprise multiple age groups, likely including some old, large Kemp's ridleys with generally higher fecundity than neophyte nesters. Captive green turtles (*Chelonia mydas*) held at the CTF also attained sexual maturity and laid smaller clutches than typically recorded for wild individuals. Only 4 of the 10 clutches laid by CTF ridleys hatched, with hatching success ranging from 5% to 45% (Wood and Wood, 1988). This is lower than the hatching success recorded for head-started Kemp's ridleys that nested in the wild (Table 14.2) and wild Kemp's ridleys. From 1990 to 1992, Kemp's ridleys laid 11 clutches of eggs in captivity at Texas A&M University and Sea World of Texas. These were head-started turtles from the NMFS Galveston Laboratory that were transferred to the Clearwater Aquarium (Clearwater, Florida) for extended rearing and later transported back to Texas (Shaver, 1992). These clutches were placed into Styrofoam boxes and transferred to PAIS for incubation (Shaver, 1992). The mean clutch size was 59 eggs (range 24–107 eggs). Only one clutch hatched, and hatching and emergence success was 11%. Fewer than half of these ridleys had been observed mating before nesting, and embryonic development was not observed in most of the unhatched eggs (D. Owens, personal communication; Shaver, 1990b, 1991b, 1992). Similar to the clutches laid at CTF, clutch size was smaller, and hatching success was lower, for these head-started turtles in comparison to wild turtles and to head-started Kemp's ridleys nesting in the wild.

The captive-breeding program was terminated in 1988, and most of the surviving turtles and their offspring produced in captivity were released into the wild (Caillouet, 2000). On May 20, 1994, 13 adult Kemp's ridleys (7 males and 6 females) were released at PAIS (Shaver, 1994). These turtles had been experimentally imprinted to PAIS between 1978 and 1992, head-started, and held in captivity at various facilities throughout their lives. Following their release, two females and one male emerged on the beach at PAIS within 48 hours. One female emerged once but did not lay eggs. The other female was seen on three occasions; during her first emergence, visitors frightened her back into the water, during her second, they picked her up and placed her back into the water thinking that they were aiding her, and during her third, she was seen reentering the water, and it was impossible to determine whether she had nested or not (Shaver, 1994).

Efforts to Detect Nesting Head-Started Kemp's Ridleys in the Wild

Thorough monitoring of beaches for nesting head-started ridleys is necessary to evaluate the effectiveness of the Head Start Experiment. Nesting Kemp's ridleys have been monitored for several decades near Rancho Nuevo, Mexico, and more recently at several other beaches in the states of Tamaulipas and Veracruz, Mexico (Márquez-M. et al., 1999, 2001; Burchfield, 2004). Efforts to detect and protect nesting Kemp's ridleys and their eggs on PAIS and adjacent beaches began in 1986 (Shaver, 1990a). These beaches were monitored at least once per day, 2–5 days each week, from 1986 to 1994 and at least once per day 7 days each week from 1995 to 1997. From 1998 to 2004, North Padre Island was repeatedly traversed each day, as has occurred in Rancho Nuevo for several years. Such repeated coverage greatly increases the likelihood of observing nesting females and locating their eggs. From 1986 to 1998, North Padre Island was the only area on the Texas coast specifically patrolled to detect nesting sea turtles. However, repeated daily patrols also have been conducted on Boca Chica Beach since 1999 and on South Padre Island since 2000. In conjunction with the beach monitoring, educational programs alert-

Table 14.2 Confirmed nesting outside captivity by head-started Kemp's ridley turtles from 1996 to 2002

Turtle no.	Year-class	Location imprinted[a]	Date nested	Location nested[a]	Method identified[b]	Carapace length (cm)[c]	No. of eggs laid	Hatching success (%)/ hatching emergence success (%)[d]
1	1983	PAIS	May 27, 1996	PAIS	L	NA	89	95.5/95.5
2	1986	PAIS	May 29, 1996	PAIS	L, C	67.5 CCL-US	88	32.6/29.5
3	1984	PAIS	April 25, 1998	MI	L, C	62.9 SCL-US	97	77.3/77.3
	1984	PAIS	May 21, 1998	MI	L, C	62.9 SCL-US	101	41.6/41.6
	1984	PAIS	April 24, 2001	NPI	L, C	63.0 SCL-US	113	93.8/93.8
4	1987	PAIS	May 3, 1998	MX	L, C	65.0 CCL-MX	88	63.6/59.1
5	1986	PAIS	May 22, 1998	PAIS	L, C	61.0 SCL-US	99[e]	32.7/31.6
6	1987	PAIS	June 2, 1998	PAIS	L, C	59.7 SCL-US	93	97.8/97.8
	1987	PAIS	April 19, 2000	NPI	L, C	60.5 SCL-US	85	82.4/82.4
	1987	PAIS	April 26, 2002	PAIS	L, C	60.9 SCL-US	95	77.9/74.7
	1987	PAIS	May 16, 2002	PAIS	L, C	60.9 SCL-US	105	59.0/54.3
7	1984	PAIS	April 8, 1999	PAIS	L, C	64.1 SCL-US	101	100.0/100.0
	1984	PAIS	May 1, 1999	PAIS	L, C	64.1 SCL-US	98[e]	94.8/94.8
8	1989	MX	April 15, 1999	MX	L, C, E	64.5 CCL-MX	97	NA
	1989	MS	May 4, 1999	MX	L, C, E	64.0 CCL-MX	71	NA
9	1986	PAIS	April 20, 1999	PAIS	L, C	59.9 SCL-US	103	95.1/95.1
	1986	PAIS	May 16, 1999	PAIS	L, C	59.9 SCL-US	97	99.0/99.0
10	1988	PAIS	May 6, 1999	MI	L	NA	94	0.0/0.0
	1988	PAIS	May 26, 1999	MI	L	NA	101	0.0/0.0
	1988	PAIS	May 18, 2002	NPI	L	NA	103	0.0/0.0
	1988	PAIS	May 21, 2004	MI	L, C	59.4 SCL-US	105	95.2/95.2
11	1986	PAIS	May 10, 1999	PAIS	L, C, E, P	62.2 SCL-US	98	96.9/96.9
12	1987	PAIS	May 20, 2001	PAIS	L, C	63.1 SCL-US	96	61.4/58.3
13	1991	MX	May 18, 2002	PAIS	L, C	58.1 SCL-US	82	51.2/51.2
14	1989	MX	May 23, 2002	PAIS	L, C, P	58.2 SCL-US	70	0.0/0.0
15	1992	MX	June 9, 2002	G	L	NA	95	87.4/87.4
16	1993	MX	April 9, 2003	PAIS	L, C, E	60.1 SCL-US	93	98.9/98.9
	1993	MS	May 4, 2003	PAIS	L, C, E	60.1 SCL-US	70	96.8/95.2
17	1992	MX	May 5, 2003	PAIS	L, C	61.4 SCL-US	102	98.0/98.0
18	1987	PAIS	May 13, 2003	PAIS	L, C	61.7 SCL-US	95	96.8/96.8
19	1989	MX	April 5, 2004	G	L, C	58.5 SCL-US	77	70.1/70.1
20	1991	MX	April 17, 2004	G	P	62.5 SCL-US	83	96.4/96.4
21	1987	PAIS	May 1, 2004	PAIS	L, C	64.2 SCL-US	105	96.2/96.2
	1987	PAIS	May 30, 2004	PAIS	L, C	64.2 SCL-US	97	88.7/88.7
22	1992	MX	May 11, 2004	BP	L, C	63.0 SCL-US	99	97.0/97.0
23	1986	PAIS	May 22, 2004	N	L, C	61.2 SCL-US	102	97.1/97.1

Source: Data from Shaver (1997b, 1999b, 2000, 2001, 2002, 2004), D. J. Shaver (unpublished data), Shaver and Caillouet (1998), and M. A. Carrasco (personal communication).

[a]Locations that the head-started nesting females had been exposed to as eggs and hatchlings: PAIS, experimentally imprinted to Padre Island National Seashore, Texas; MX, hatched in Mexico. Locations where the head-started females nested: PAIS, Padre Island National Seashore; MI, Mustang Island, Texas; N, North Padre Island, Texas, north of PAIS; MX, Tamaulipas, Mexico; G, Galveston Island, Texas; BP, Bolivar Peninsula.

[b]Tag type used to identify the turtles as being head-started: L, living tag; C, internal coded wire tag; E, external flipper tag; P, passive integrated transponder (PIT).

[c]Carapace length of the head-started female at nesting: NA, data not available because the turtle was not measured; CCL-US, curved carapace length as measured in the United States from the center of the nuchal notch to the posterior tip of the postcentral scute; SCL-US, straight-line carapace length as measured in the United States from the center of the nuchal notch to the posterior tip of the postcentral scute; CCL-MX, curved carapace length as measured in Mexico from the anterior tip of the nuchal notch to the posterior tip of the postcentral scute on the opposite side of the carapace.

[d]Hatching success (number of live and dead hatchlings/number of eggs incubated) and emergence success (number of live hatchlings that emerged from the nest/number of eggs incubated) for clutches laid by the head-started females; NA, data not available.

[e]One of these eggs was broken and is included here in the number of eggs laid but excluded from the number of eggs incubated during calculations of hatching and emergence success.

ing South Texas beach visitors to report nesting Kemp's ridleys were implemented in the mid-1980s and have expanded since that time (Shaver, 1990a; Shaver and Miller, 1999). These programs were highly successful, and a large proportion of the nesting Kemp's ridleys were found as a result of visitor reports (Shaver, 2005). Whenever possible, nesting Kemp's ridleys have been examined for tags used to mark head-started turtles. Unfortunately, not all nesting Kemp's ridleys have been examined because many returned to the water before biologists arrived.

Documented Nesting of Head-Started Kemp's Ridleys in the Wild

Some head-started Kemp's ridleys have assimilated into the wild population and nested in the wild, meeting a primary objective of the Head Start Experiment. Through 2004, 23 different head-started Kemp's ridleys have nested in the wild ($n = 36$ nests) (Table 14.2; Shaver and Caillouet, 1998; Shaver, 2005; D. J. Shaver, unpublished data; Fontaine and Shaver, 2005; M. Carrasco, personal communication). These represent the first confirmed head-started sea turtles of any species nesting in the wild. Moreover, this is the first time that a sea turtle nested on a nonnatal beach it was imprinted to (Shaver, 1996a, 1996b, 1997a; Shaver and Caillouet, 1998; Fontaine and Shaver, 2005; Shaver, 2005). They are also the first nests in the wild by known-aged Kemp's ridley turtles (Shaver, 2005; Fontaine and Shaver, 2005).

Nesting head-started turtles from various year classes were identified by their marking tags. The head-started turtles were from nine different year classes (1983, 1984, 1986, 1987, 1988, 1989, 1991, 1992, 1993) (Table 14.2). They were 10–18 years of age when first detected nesting, although it is unknown if they were nesting for the first time at that time. All but one turtle were identified by their living tags; one lacking a living tag was identified by its PIT tag. All were examined for metal tags, but only two possessed them, including one that nested on the Texas coast that had been reared in captivity for three years before its release (i.e., it was "super-head-started"). Most of the turtles were also examined for magnetic and PIT tags.

Head-started turtles were documented nesting in Texas and Mexico. Of the 21 head-started turtles that nested on the Texas coast ($n = 33$ nests), 13 had been experimentally imprinted to PAIS, and the other 8 had been taken directly from Mexico and subsequently reared at the NMFS Laboratory (Table 14.2). The two turtles that nested in Mexico ($n = 3$ nests) included one that had been experimentally imprinted to PAIS and one had been taken from Mexico and reared at the NMFS Laboratory.

The sizes at maturity for these head-started turtles were larger than the minimum size at maturity reported for wild Kemp's ridleys nesting in Mexico (Márquez-M., 1990): they ranged in size from 58.1 to 64.3 cm SCL (Table 14.2).

The age of the head-started turtles (10–18 years) was similar to model predictions for age at maturity for wild and head-started Kemp's ridley turtles. Growth models predicted that wild Kemp's ridleys mature at 11–16 years of age, based on a mature size of 65.0 cm SCL (Zug et al., 1997). Based on a mature size of 60.0 cm SCL, Caillouet et al. (1995b) estimated maturity at 10 years for head-started Kemp's ridleys, and Zug et al. (1997) predicted 9–13 years.

The internesting and remigration intervals of head-started turtles were equivalent to those for wild turtles. Of the 23 head-started turtles, 9 nested more than once either within a nesting season or during different nesting seasons (Table 14.2). The internesting intervals of turtles that had nested more than once within a season ranged from 20 to 29 days, similar to the 20–28 days reported for ridleys in Mexico (Miller, 1997). The remigration interval of turtles that nested during different nesting seasons was two to three years, which is comparable to the two years estimated in Mexico (Turtle Expert Working Group, 1998).

Fecundity of head-started Kemp's ridleys was similar to that of wild individuals in Mexico from 1996 to 2004. Clutch size of head starts ranged from 70 to 113 eggs (mean = 94.1, SD = 10.3, $n = 36$) (Table 14.2), nearly identical to clutch size in Mexico (mean = 93.0, $n = 45{,}577$ nests) (Burchfield, 2004). The 33 clutches laid in Texas were incubated in Styrofoam boxes at PAIS (Shaver, 1997b, 1999, 2000, 2001, 2002, 2004; D. J. Shaver, unpublished data), and the three laid in Mexico were incubated in a protective corral there. Hatching success ranged from 0 to 100% (mean = 72.7, SD = 33.3, $n = 34$) and emergence success from 0 to 100% (mean = 72.1, SD = 33.6,

$n = 34$) in the clutches for which data were available (Table 14.2). This is comparable to the hatching success for the nests incubated in corrals at Mexico (mean = 63.1%, $n = 45{,}577$ nests) (Burchfield, 2004).

Possible Factors Limiting the Detection and Documentation of Nesting in the Wild

Several factors have possibly limited the detection and documentation of nesting in the wild by head-started Kemp's ridleys. Numbers of Kemp's ridley nests have increased in Mexico since 1986 (Burchfield, 2004) and in the United States since 1994 (Shaver and Caillouet, 1998; Shaver, 2005). The increase in Mexico is likely a result of enhanced monitoring of beaches, expansion of the area monitored, and growth of the Kemp's ridley population. In Texas, the increase likely reflects an increase in the number of ridleys nesting, improved monitoring of beaches, elevated awareness of nesting turtles and reporting by the public, or a combination of all (Shaver, 2005). The return of head-started turtles to nest on Texas beaches has contributed to the increased number of nests found there. Of the 170 Kemp's ridley nests found on the Texas coast from 1988 to 2004, 33 were from head-started turtles (D. J. Shaver, unpublished data). Head-start nests represent 19.4% (33 of 170) of all Kemp's ridley nests documented in Texas, 37.9% (33 of 87) of all nests in Texas at which the nesting females were seen and examined for tags (Shaver, 2005; D. J. Shaver, unpublished data), and fewer than 1% of all nests worldwide (36 of 57,798), from 1988 to 2004.

A variety of factors likely have limited observations of head-started ridleys, including insufficient monitoring and tag loss. Thus, the 23 head-started Kemp's ridleys ($n = 36$ nests) is a minimum estimate of reproduction in the wild in the United States and Mexico through 2004.

Variations and limitations in beach monitoring in the United States and Mexico (influenced largely by funding and logistical constraints) likely limited the number of head-started Kemp's ridleys observed nesting in the wild (Shaver and Fletcher, 1992; Shaver, 2005). More head-started turtles might have been detected if monitoring had been more frequent and covered more geographic area beginning in 1988 rather than nearly a decade later. Additional monitoring might have increased opportunities to examine nesting turtles for tags and perhaps locate other nesting that went undetected. Because it takes less than an hour for a Kemp's ridley to nest, nests are often located without the turtle being seen. Of the 170 Kemp's ridley nests laid in Texas from 1988 to 2004, only 87 of the nesting turtles were observed and examined for tags (Shaver, 2005; D. J. Shaver, unpublished data). Some Kemp's ridley nests likely went unnoticed and unrecorded in Texas, even after monitoring increased, particularly on stretches of beach that were difficult to travel and sparsely visited or monitored (Shaver, 2005). Nests go undetected at egg laying when the nesting turtles and tracks are not found, as evidenced by emergence of hatchlings from three nests discovered during 2002 and one during 2004 (Shaver, 2004; D. J. Shaver, unpublished data).

Tag loss also likely limited observations of head-started ridleys nesting in the wild (Shaver and Fletcher, 1992; Shaver, 2005; Fontaine and Shaver, 2005). All of the head-started ridleys were tagged with external flipper tags, but these tags were still attached to only 3 of the 23 head-started turtles that were found nesting (Table 14.2). More head-started turtles might have been detected nesting if the turtles from the early year-classes had received more than just a flipper tag (Shaver and Fletcher, 1992; Wibbels, 1992; Shaver, 1998, 2005). Some of the unmarked ridleys found nesting in the United States and Mexico could have been from one of the early year classes. Furthermore, in Mexico, the large number of turtles nesting there, coupled with the limited number of magnetic and PIT tag readers available, may have limited observation of head-started turtles.

It is also possible that head-started ridleys may have nested on other beaches, but this hypothesis is not supported by tag returns. Kemp's ridley nests have been documented in Florida ($n = 15$), South Carolina ($n = 1$), North Carolina ($n = 2$), and Alabama ($n = 2$) (Meylan et al., 1990b; Anonymous, 1992; Palmatier, 1993; Godfrey, 1996; Libert, 1998; Johnson et al., 1999, 2000; Foote and Mueller, 2002; Nicholas et al., 2004; G. Harman, personal communication; S. MacPherson, personal communication; M. Nicholas, personal communication; M. Rickard, personal communication; C. South, personal communication; J. Steiner, personal communication). Examination of some of the nesting ridleys revealed no external tags. Bowen et al. (1994) speculated that these ridleys could have been head-started turtles because there were no previous records of Kemp's ridleys nesting on these beaches. Johnson et al. (1999) suggested that the ridleys that nested in Florida might be colonists from the wild population and did not discount the possibility that scattered nesting by wild Kemp's ridleys occurred in Florida historically and went undetected or misidentified. Kemp's ridley turtles may have nested in Louisiana historically (Hildebrand, 1982), but there are no confirmed records. Hildebrand (1963) suggested that scattered nesting of Kemp's ridley in south Texas and beaches adjacent to Rancho Nuevo might be remnants of nesting colonies that existed before the species declined. In fact, during the last few years more nesting has occurred in these areas as the Kemp's ridley nesting population has increased (Shaver and Caillouet, 1998; Turtle Expert Working Group, 1998; Márquez-M. et al., 1999, 2001; Shaver, 2005).

Satellite Tracking of Postnesting Head-Started Kemp's Ridleys

The movements of postnesting head-started ridleys in comparison to movements of wild individuals are indicators of the ability of the head-started individuals to successfully adapt to the wild. Satellite telemetry was used to monitor the movements of head-started ($n = 11$) and wild ($n = 14$) adult female Kemp's ridley turtles after they nested on North Padre and Mustang Islands, Texas, between 1997 and 2004 (Shaver, 1999, 2000, 2001, 2002, 2004). Two of the turtles (one head-started and one wild) were recaptured after nesting two to three years later, and a second transmitter was attached. One wild turtle was recaptured three times at 2- to 3-year remigration intervals and received three transmitters. The 11 head-started turtles were 10–17 years of age when tracked; 10 had been captive for approximately one year, and 1 for three years.

Most of the head-started and wild ridleys monitored exhibited similar movement patterns. Swimming speeds were similar for head-started and wild turtles, and most locations identified for all turtles monitored were in 20 fathoms water depth or less. After completion of the nesting season, all of the 14 wild ridleys left waters off North Padre Island; some traveled southward to Mexico, but most traveled directly northward, and all were last located where they established residency off the coastal United States in the northern or eastern Gulf of Mexico. Nine of the 11 head-started individuals (including the individual that had been head-started for nearly three years) followed a pattern similar to the majority of the wild turtles. However, two of the head-started ridleys tracked beginning in April 1999 established residency and remained off the South Texas coast until transmissions ceased at the beginning of October and end of November 1999, well after the nesting season was completed. Minor differences noted for the two could have been caused by a year-specific habitat variation because both were tracked during 1999.

Thus, the movement data suggest that many of the nesting head-started Kemp's ridleys were exhibiting behaviors similar to wild turtles. Similar studies of other sea turtle species are limited to immature green and loggerhead turtles. Captive-reared immature green turtles traveled thousands of kilometers after release and were later found at foraging sites used by wild green turtles (Pelletier et al., 2003). However, diving behavior of immature head-started loggerheads was different from that of wild loggerheads, with head starts making more and shorter dives and spending less time submerged (Keinath, 1993).

Mating and Nesting by Other Species of Head-Started Sea Turtles in the Wild

In addition to Kemp's ridleys, other head-started sea turtles have survived after release and have successfully bred in the wild. Subsequent to the first detection of head-started Kemp's ridleys nesting in the wild in 1996 (Shaver, 1996a, 1996b, 1997a, 1997b), some head-started green turtles were recorded mating and nesting in the wild. A 1-year-old head-started green turtle that had been released in 1981 nested four times in 2000 and seven times in 2002 on a beach at Maui, Hawaii, about 125 miles from the release site (Balazs et al., 2004; G. Balazs, personal communication). For the nests found in 2000, the first contained 94 eggs, of which only 14% resulted in hatchlings, possibly because of dry sand conditions, and the last three contained 76, 76, and 88 eggs, of which 63%, 57%, and 55% (respectively) resulted in live hatchlings (Balazs et al., 2004). A 15-year-old head-started green turtle was observed nesting on Melbourne Beach, Florida, about 60 miles north of her natal beach (R. Ernest, personal communication, cited in Bell and Parsons, 2002).

Adult green turtles from the head-start program at CTF have been found mating and/or nesting in the wild, including a male released as a hatchling in 1983 that was recaptured in 2002 while mating just offshore from CTF, a female released as a yearling in 1988 that nested at Grand Cayman in 2002, an adult female released as a hatchling in 1985 that was observed nesting at Grand Cayman in 2002, and a male released in 2001 and recaptured mating with a wild female at Bastimentos Island National Marine Park, Bocas del Toro Province, Panama, a known green turtle courtship and mating ground (Bell and Parsons, 2002; P. Meylan, personal communication).

To our knowledge, no other head-started sea turtles have been documented and reported nesting in the wild. However, factors similar to those described previously that might have limited the recaptures of head-started Kemp's ridleys could also have limited recaptures of other head-started turtles.

Panel Reviews of the Kemp's Ridley Head Start Experiment

The results from the Kemp's Ridley Head Start Experiment were reviewed at annual meetings of the Kemp's Ridley Working Group, which consisted of scientists from the various Mexican and U.S. agencies involved in the Kemp's ridley recovery program, as well as other invited scientists. In addition to these meetings, during 1989 and 1992, two separate panels of scientists were convened at the request of the NMFS to review the effectiveness the Kemp's Ridley Head Start Experiment. The first review occurred during August 1989 (Wibbels et al., 1989a). Data presented by the NMFS Galveston Laboratory to the initial review panel indicated that many head-started ridleys had survived and grown in the wild following their release. However, there was no evidence at that time that any ridleys had reached sexual maturity and integrated into the breeding population. The panel indicated that the criterion for evaluating the success of head-starting would be a comparison of the number of head-started turtles recruited into the breeding population versus the number of wild ridleys recruited into the breeding population, with the number of hatchlings used in the Head Start Experiment relative to the number of hatchlings released at Rancho Nuevo taken into account. The panel concluded that it was difficult if not impossible to evaluate the effectiveness of the program at that time because ridley mortality in the wild was extremely high. Although tens of thousands of hatchlings had been released yearly at Rancho Nuevo, and 1,000–2,000 turtles had been head-started yearly, the nesting population was still critically small. Therefore, it appeared that very few ridleys (wild or head-start turtles) were reaching sexual maturity in the wild. Fortunately, the ridley mortality in the wild was expected to decrease because of the impending implementation of TED regulations in the U.S. Gulf of Mexico and Atlantic Ocean. Further, if any head-started turtles had reached maturity, their recruitment into the breeding population would have been difficult to assess at that time because internal permanent tags (coded wire tags) had been used only since 1984, and PIT tags had not yet been used. Therefore, to evaluate the effectiveness of head-starting, the panel recommended that the Head Start Experiment continue after TED use became mandatory, and with the continued use of permanent tags on all head-started turtles. The panel recommended that the project be continued for a period of 10 additional years because it would enable head-started turtles with permanent tags to reach maturity.

The 1989 panel made several other recommendations. They suggested that "provisional criteria" might be useful to evaluate the effectiveness of the program in the short term. These criteria related to survival, growth, and behavior of juvenile head-started turtles after release compared to juvenile wild Kemp's ridleys (while taking into account possible sampling bias resulting from the presence of tags on turtles). The panel also indicated that the Head Start Experiment should clearly be considered an experiment. The panel was concerned that placing too much emphasis on the Head Start Experiment as a method for saving the ridley could be detrimental to the overall recovery program because it could decrease the emphasis on the primary element in the recovery program, the protection of turtles in their natural habitat.

During September 1992, a second panel of scientists was convened by the NMFS to review the Head Start Experiment (Eckert et al., 1994; Caillouet, 1998). They concluded that a major obstacle preventing the evaluation of the program was inadequate monitoring of nesting beaches for head-started turtles. Furthermore, they concluded that enough head-started turtles already had been released with permanent tags and that it was unnecessary to continue to release more head-started turtles to evaluate the program. Instead, they recommended that efforts be concentrated on detecting head-started turtles on nesting beaches. The panel also suggested establishing a tagged control group. The panel indicated that increased use of TEDs during the previous year had decreased mortality rates in the wild. Therefore, calculating survival rates of head-started versus wild ridleys would have to be based on new tagging and recapture studies. The panel suggested developing a rigorous mark–recapture program that would allow evaluation of the Head Start Experiment as well as provide valuable life history data on the growth and survival of Kemp's ridleys in the wild. This tagging program was to include in-water netting/capture studies around the Gulf of Mexico and the large-scale permanent internal wire tagging of hatchlings from Rancho Nuevo. Caillouet (1998) suggested that the only way to conduct a direct comparison was to tag wild and head-started ridleys of similar age.

Controversy Regarding the Kemp's Ridley Head Start Experiment

The Kemp's Ridley Head Start Experiment was started at a time when there was serious concern that this species would soon be extinct. Although it was an unproven technique, head-starting was implemented as one of several facets of the recovery program to prevent extinction of the Kemp's ridley (Woody, 1981, 1989). By head-starting a turtle through its initial year of life, it was hoped that the high mortality rates of post-hatchling turtles could be reduced (Klima and McVey, 1982). Because only a small percentage of the ridleys collected from Rancho Nuevo were used in the Head Start Experiment, the project was endorsed by many during its early years. In fact, the concept of the Head Start Experiment helped facilitate development of the Kemp's Ridley Recovery Program between Mexico and the United States that included increased protection of the nesting beach by a binational team of biologists (Woody, 1989). However, the Head Start Experiment gradually became the subject of increased controversy. Finally, in 1993, the U.S. Fish and Wildlife Service did not issue permits to allow the import of 2,000 hatchlings, and NMFS did not protest this action (Byles, 1993; Williams, 1993). This represented the official termination of the "large-scale" Kemp's Ridley Head Start Experiment (Mrosovsky, 1983; Woody, 1991; Taubes, 1992). However, the Head Start Experiment, in effect, continued at the NMFS Galveston Laboratory through 2000, with 178–200 turtles per year from 1993 to 2000. These turtles were reared 9–11 months and then used in TED testing before or during their release. The Kemp's Ridley Head Start Experiment was terminated for a variety of factors, some of which were not related to its effectiveness as a management tool.

The criticisms of head-starting ranged from scientific to political in nature (Mrosovsky, 1983; Woody, 1991; Taubes, 1992; Wibbels, 1992). The criticisms focused on the experimental and unproven effectiveness of head-starting. Many of these criticisms also were controversial and, thus, were not universally endorsed. A study supported by the National Academy of Sciences indicated that head-starting was an experimental program and the actual benefits had yet to be evaluated (Magnuson et al., 1990). To evaluate head-starting, the survival and breeding rates of head-started turtles following release needed to be determined and compared to those of wild turtles. Unfortunately, this is a very difficult task because it requires permanent tagging of head-started turtles and a control group of wild turtles for comparison. However, if all head-started turtles had permanent tags, then wild untagged ridleys could act as the control group (Caillouet, 1998).

Although there have been head-start projects before the Kemp's Ridley Head Start Experiment, their effectiveness as a management tool was never determined. For example, in 1989 the State of Florida terminated its 30-year-old green turtle and loggerhead (*Caretta caretta*) head-start programs because of uncertainty regarding their success (Huff, 1989). They specifically cited potential problems such as "possible interference with imprinting mechanisms that guide turtles to the nesting beach, imbalance in sex ratios from artificial incubation of eggs, nutritional deficiency from captive rearing of hatchlings, and behavioral modifications" (Huff, 1989). However, it is not clear if, or to what extent, these potential problems affected the survival and potential breeding of those head-started turtles. Thus, there were no empirical data to evaluate the effectiveness of those programs.

Another criticism of head-starting was the potential for abnormal behavior after a turtle was released into the wild (Mrosovsky, 1983; Taubes, 1992). Critics suggested that head-started turtles (at least initially after release) might have reduced fitness compared to wild turtles because they had not been exposed to natural prey, predators, currents, seasonal temperature fluctuations, etc. (Werner, 1994). Additionally, turtles head-started in relatively small containers such as buckets or crates could have reduced swimming performance (Stabenau et al., 1992). Further, it is possible that head-started turtles might become habituated to pelleted food. However, feeding experiments with 9- to 11-month-old head-started Kemp's ridleys indicated that some turtles quickly adapted and showed no hesitation to begin feeding on natural prey such as crabs, shrimp, and fish (Klima and McVey, 1982). Additionally, recapture data and diet data (reviewed above) suggest that many head-started turtles survive and grow in the natural environment, and some data indicate evasive behaviors soon after release (McVey and Wibbels, 1984; Wibbels, 1984). However, there also are many examples of head-started turtles exhibiting atypical evasive behaviors (see section on Overview of Movements, Growth, and Adaptation above). Some aberrant behavior is indicated by atypical capture techniques, such as "captured by hand or by dip net" (see Methods of Recapture section above; also see Meylan and Ehrenfeld, 2000). Aberrant behavior of some of the newly released turtles might contribute to the increased rate of recapture during the first few months after release (see Percentage and Rate of Recaptures section above). The percentage of head-started turtles that exhibited aberrant behaviors is unknown, and it is unknown if these turtles eventually adopt natural behavior patterns. Unusual locations of some head-started ridleys could also be related to aberrant behavior. Although most recaptures of head-started ridleys have occurred within the normal distribution or range of Kemp's ridleys, turtles recaptured in Nicaragua could be indicative of altered behavior of some head-started turtles (Woody, 1991).

Results from an educational program that included the temporary captive rearing of Hawaiian green turtles suggested that many of the captive-reared turtles adapted to the wild, whereas others exhibited atypical behaviors. Of 103 captive-reared turtles released from 1990 to 1999 (SCLs of 25.5–68.0 cm), 18 turtles were recaptured. Of these 18 turtles, 6 appeared to have successfully adapted to the wild. The growth rates of these turtles compared favorably with those of naturally occurring Hawaiian green turtles. However, six other turtles were found stranded (four of which were alive), and three other turtles were fed pelleted fish food and lettuce by tourists while in the wild (Balazs et al., 2002). Thus, it would appear that the behavior varied in these captive-reared turtles, with many adapting to the natural environment, and others, at least initially, exhibiting behaviors atypical of wild turtles.

Another criticism of the Kemp's Ridley Head Start Experiment was that the Kemp's ridley was not an appropriate species for a head-start experiment (Woody, 1991). Because of the experimental nature of head-starting, it may not have been prudent to use a critically endangered species such as the Kemp's ridley. Woody (1991) suggested that it might have been better for the survival of the species to release the hatchlings at Rancho Nuevo from the approximate 25 clutches of eggs per year that were used in the Head Start Experiment rather than rear them in captivity for 9–11 months. At that time, 25 clutches of eggs represented approximately 3% of the total annual hatchling production (Woody, 1991). However, others disagreed and strongly supported the use of the Kemp's ridley in the Head Start Experiment (Allen, 1990, 1992). In fact, even Woody (1989) provided a rationale for using the Kemp's ridley in the Head Start Experiment.

Another criticism of the Kemp's Ridley Head Start Experiment related to potential problems associated with "imprinting." The subject of imprinting in sea turtles is controversial. Although data support the concept that some sea turtles return to their natal beaches to nest (Owens et al., 1982; Meylan et al., 1990a), the mechanisms are unknown. During the early years of the Head Start Experiment (1978–1988), most of the hatchlings were "experimentally" imprinted to PAIS by incubating them in PAIS sand and letting hatchlings crawl down the beach and into the surf (see Experimental Imprinting section above). During the later years of the project (1989–2000), all hatchlings came from eggs incubated and hatched in corrals at Rancho Nuevo. It is of interest that the majority of documented nestings of head-started Kemp's ridleys have been from Texas beaches. Of the 23 head-started ridleys recorded nesting (as of 2004), 21 nested in Texas. Of those 21, 13 had been imprinted to PAIS, and 8 to Rancho Nuevo. These are the first reports that a sea turtle imprinted to an area returned to nest there (Shaver, 1996a, 1996b, 1997a; Shaver and Caillouet, 1998; Fontaine and Shaver, 2005; Shaver, 2005). These results do not validate "experimental imprinting" as a management tool nor verify that imprinting occurs in sea turtles, but some of these data are consistent with the imprinting hypothesis (Owens et al 1982; Meylan et al., 1990a).

There was also criticism regarding the cost-effectiveness of the Head Start Experiment. For example, Woody (1991) estimated the cost of head-starting a Kemp's ridley to be conservatively at least $125 per turtle compared to $0.65 per hatchling released at Rancho Nuevo. Taubes (1992) reported that the total annual budget for imprinting and head-starting Kemp's ridleys was approximately $250,000 or more. Woody (1991) suggested that reallocating the resources from the Head Start Experiment into proven conservation techniques could enhance the recovery of the Kemp's ridley. Alternatively, if head-starting significantly enhanced a hatchling's chances of survival to sexual maturity and breeding, a price of $125 might be considered reasonable by some conservationists.

There were also political concerns regarding the Kemp's Ridley Head Start Experiment. Although head-starting was an experimental technique, it was a highly visible aspect of the recovery program (Mrosovsky, 1983; Woody, 1991). The release of healthy head-started turtles into the wild had great public and political appeal and seemed to justify the program to the public as well as to many conservationists (Mrosovsky, 1983; Woody, 1991). The Kemp's Ridley Recovery Plan (U.S. Fish and Wildlife Service and National Marine Fisheries Service, 1992) indicated that, although the Kemp' Ridley Head Start Experiment was the focus of public interest, it was an experiment and not a recovery action. As such, there was concern that the Head Start Experiment might not be viewed as experimental and might be considered more important than, or a substitute for, protection of turtles in their natural habitat. The National Academy of Sciences Report (Magnuson et al., 1990) specifically indicated that head-starting should not substitute for other essential conservation measures.

Finally, there is the basic criticism that, if head-starting worked, then there should now be relatively large numbers of head-started Kemp's ridleys nesting in Texas and/or Mexico (Woody, 1991). As previously mentioned, as of 2004, 23 head-started ridleys have been documented nesting. This verifies that head-started turtles can survive to adulthood and breed, but it is not clear if this number of nesting turtles is indicative of the success or failure of the Kemp's Ridley Head Start Program. There are many factors that confound the accurate evaluation of the effectiveness of head-starting. For example, many factors likely limited the detection of nesting head-started turtles (see Possible Factors Limiting the Detection and Documentation of Nesting section above). Further, most of the Head Start Experiment occurred before TEDs became mandatory in U.S. waters. Thus, the survival of head-started ridleys was likely very low. The high mortality of Kemp's ridleys in the wild is illustrated by the decrease in the number of ridleys nesting during the late 1970s and early to mid-1980s, despite intense conservation efforts on the nesting beach. Also, sex ratios in the early year classes in the Head Start Experiment were male biased. Further, accurate identification of head start turtles was likely problematic. Although all head-started turtles were marked with metal flipper tags, permanent tags were not used until the seventh year (i.e., 1984) of the program (see Tagging Methodology section above).

Many of the head-started turtles from the early years of the program probably are no longer identifiable because of tag loss. Further, depending on the type of permanent tag used, a trained biologist and specialized equipment are required to detect a tag. One literally has to be at the right place at the right time (i.e., the one hour or less that the turtle is on the beach) with the proper training and equipment to verify whether or not a nesting turtle is a head-started ridley. Although difficult, this may not be an insurmountable task because the beaches near Rancho Nuevo and those in south Texas are regularly monitored by biologists looking for nesting ridleys. Some critics would argue that documented nesting of 23 head-started turtles (as of 2004) seems low, and the significance of these data is debatable.

Summary and Comments

The Kemp's Ridley Head Start Experiment was initiated in 1978 as an experimental facet of the recovery program for this critically endangered sea turtle. It initially included an experimental imprinting project at PAIS, and during 23 years of the Head Start Experiment, over 23,000 turtles were released into the Gulf of Mexico. Tag return data indicate that many of the head-started turtles survived and grew in the wild. More recent data indicate that some turtles have survived to maturity and are nesting. Although these findings appear encouraging, opponents of head-starting have enumerated a variety of criticisms suggesting that head-starting was not an effective or an appropriate management tool for the Kemp's ridley (see Controversy Regarding the Kemp's Ridley Head Start Experiment section above). Both critics and supporters of head-starting have valid points, but neither can comprehensively support their case based on the data available. That is the basic problem that has confounded head-starting: accurate validation of the effectiveness or ineffectiveness of the technique (Mrosovsky, 1983).

The ultimate goal of head-starting was to decrease the early-life-stage mortality of turtles so that a greater percentage of turtles would reach sexual maturity and become integrated into the breeding population. Unfortunately, it is very difficult to evaluate such a program when working with a slowly maturing species that lives in the ocean and can experience relatively high mortality before attaining sexual maturity. To evaluate head-starting, the survival rates and eventual reproductive success of head-started turtles must be compared to wild counterparts (i.e., what is the average reproductive success of hatchlings that are head-started versus hatchlings released into the Gulf of Mexico at Rancho Nuevo?). This type of analysis ideally includes the tagging (preferably permanent tags) of experimental and control groups (i.e., head-started turtles versus wild turtles). It also requires comprehensive surveys of nesting beaches and evaluation of nesting turtles for all possible tags. Although this is logistically difficult, the Kemp's Ridley Head Start Experiment may represent an unprecedented opportunity to evaluate head-starting. Turtles from the latter years of the Head Start Experiment all received permanent PIT tags, and turtles from many of the year classes of head-started turtles received living tags. Additionally, the number of hatchlings entering the Gulf of Mexico at Rancho Nuevo is known for the past three decades, and relatively large numbers of hatchlings at Rancho Nuevo have been marked with permanent internal wire tags in several recent years. Biologists routinely monitor ridley nesting beaches in Mexico and Texas, and they have the equipment to scan nesting turtles for PIT tags. A major concern would be the percentage of nesting turtles that are actually examined. It is logistically impossible to monitor all nesting turtles. Further, as the size of arribadas increases in Mexico, it is possible that less emphasis will be placed on evaluation of nesting turtles for tags, and more emphasis will be placed on relocating the large and increasing numbers of nests to protected corrals. Thus, there is still a potential window of opportunity for evaluating the effectiveness of the Head Start Experiment, but it is not clear if the opportunity will be utilized.

The question still remains if an experimental management technique should have been used in the recovery program for a critically endangered turtle. Some considered the Kemp's Ridley Head Start Experiment a "conservation gamble" (Mrosovsky, 1983). It was part of a multifaceted recovery program that was initiated because there was serious concern that the Kemp's ridley was headed for extinction. During the late 1980s and early 1990s, the Kemp's ridley population began to increase slowly. At that point, some felt that it was no longer necessary to continue with an experimental technique (Woody, 1991; Taubes, 1992). The more conventional techniques of protecting turtles on the nesting beaches and in their natural habitat appeared to be working. Therefore, the program was officially terminated in 1992 (although 178–200 turtles from the 1993–2000 year classes were reared in captivity for use in TED testing), and all emphasis was placed on protecting the Kemp's ridley in their natural habitat. But the question still remains: Did the Kemp's ridley win or lose on the gamble? This question applies not only to the Kemp's ridley experiment but to other head-start projects such as the 30 years of head-starting green and loggerhead turtles in Florida (Witham and Futch, 1977). It is quite possible that these programs have contributed to the recovery of these populations. Unfortunately, as with previous head-start projects, the data are currently inconclusive. For example, some head-started Kemp's ridleys have returned to nest, but it is currently unclear how many documented nestings of head-started turtles are required to indicate the success of the program. Other head-started Kemp's ridleys may have nested without being detected by biologists, but there is no way of verifying that speculation. It is also possible that the number of nesting females in South Texas may continue to increase as hatchlings produced from those beaches mature and reach sexual maturity. However, as indicated above, the ultimate measure will be to determine the percentage of head-started turtles that survive and reproduce in comparison to the percentage survival and reproduction of wild ridleys. Fortunately, it may still be possible to evaluate the effectiveness of the Kemp's Ridley Head Start Experiment. Twenty-three year classes of head-started Kemp's ridleys have been released into the Gulf of Mexico, many with permanent tags. Many have had time to mature. We now need to closely monitor the nesting beaches in Mexico and Texas in order to accurately determine what percentage of head-started turtles have survived to maturity and have integrated into the breeding population. Such information would provide the most comprehensive database thus far for evaluating head-starting. In fact, this is the best opportunity to date, and in the foreseeable future, to accurately evaluate head-starting.

ACKNOWLEDGMENTS

We thank Marc Zelickson and Alyssa Geis for comments on this manuscript. Rene Márquez-M., the Instituto Nacional de la Pesca, the government of Mexico, and their partners are acknowledged for providing eggs for the imprinting project, hatchlings for head-starting, and information regarding detection of head-started turtles that nested in Mexico.

We thank numerous individuals who aided with imprinting and head-starting activities in the United States. Peter Bohls, Darrell Echols, Cynthia Rubio, Jock Whitworth, and many others are acknowledged for their support and assistance with activities at Padre Island National Seashore. Charles Caillouet, Jr., Tim Fontaine, Ben Higgins, Dickie Revera, Jo Williams, and Ted Williams provided information on the marking and release of head-started turtles. We also thank the many people in the United States who assisted with the detection and investigation of nesting and stranded adult Kemp's ridley turtles and Kemp's ridley nests. Wendy Teas and the Sea Turtle Stranding and Salvage Network provided information on stranded turtles. We also acknowledge HEART for their support of Kemp's ridley conservation.

LITERATURE CITED

Allen, C. 1990. Give "headstarting" a chance. *Marine Turtle Newsletter* 51:12–16.

Allen, C. 1992. It's time to give Kemp's ridley headstarting a fair and scientific evaluation. *Marine Turtle Newsletter* 56:21–24.

Anonymous. 1992. First Kemp's ridley nesting in South Carolina. *Marine Turtle Newsletter* 59:23.

Anonymous. 1993. U.S. terminated "head-start" for Kemp's ridley turtles. *Marine Turtle Newsletter* 63:27–28.

Balazs, G. H. 1979. An additional strategy for possibly preventing the extinction of Kemp's ridley, *Lepidochelys kempi*. *Marine Turtle Newsletter* 12:3–4.

Balazs, G. H., Murakawa, S. K. K., Parker, D. M, and Rice, M. R. 2002. Adaptation of captive-reared green turtles released into Hawaiian coastal foraging habitats, 1990–99. In Mosier, A., Foley, A., and Brost, B. (Compilers). *Proceedings of the Twentieth Annual Symposium on Sea Turtle Biology and Conservation.* NOAA Technical Memorandum NMFS-SEFSC-477, pp. 187–189.

Balazs, G. H., Nakai, G. L., Hau, S., Grady, M. J., and Gilmartin, W. G. 2004. Year 2000 nesting of a captive-reared Hawaiian green turtle tagged and released as a yearling. In Coyne, M. S., and Clark, R. D. (Compilers). *Proceedings of the Twenty-First Annual Symposium on Sea Turtle Biology and Conservation.* NOAA Technical Memorandum NMFS-SEFSC-528, pp. 100–102.

Bell, C. D. L., and Parsons, J. 2002. Cayman Turtle Farm Head-starting Project yields tangible success. *Marine Turtle Newsletter* 98:5–6.

Bowen, B. W., Conant, T. A., and Hopkins-Murphy, S. R. 1994. Where are they now? The Kemp's Ridley Head-start Project. *Conservation Biology* 8:853–856.

Brongersma, L. D., Pritchard, P. C. H., Ehrhart, L., Mrosovsky, N., Mittag, J., Márquez-M., R., Hughes, G. H., Witham, R., Hendrickson, J. R., Wood, J. R., and Mittag, H. 1979. Statement of intent. *Marine Turtle Newsletter* 12:2–3.

Burchfield, P. M. 2004. Report on the Mexico/United States of America Population Restoration Project for the Kemp's Ridley Sea Turtle, *Lepidochelys kempii,* on the coasts of Tamaulipas and Veracruz, Mexico 2004. Unpublished report, U.S. Fish and Wildlife Service, Albuquerque, NM.

Byles, 1993. The headstart experiment no longer rearing Kemp's ridleys. *Marine Turtle Newsletter* 63:1–3.

Caillouet, C. W., Jr. 1995a. Egg and hatchling take for Kemp's Ridley Headstart Experiment. *Marine Turtle Newsletter* 68:13–15.

Caillouet, C. W., Jr. 1995b. An update of sample sex ratio composition data for head started Kemp's ridley sea turtles. *Marine Turtle Newsletter* 69: 11–14.

Caillouet, C. W., Jr. 1998. Testing hypotheses of the Kemp's Ridley Headstart Experiment. *Marine Turtle Newsletter* 79:16–18.

Caillouet, C. W., Jr. 2000. Sea turtle culture: Kemp's ridley and loggerhead turtles. In Stickney, R. R. (Ed.). *Encyclopedia of Aquaculture.* New York: John Wiley & Sons, pp. 786–798.

Caillouet, C. W., Jr., and Revera, D. B. 1985. Brood stock of captive-reared Kemp's ridley to be listed in international species inventory system. *Marine Turtle Newsletter* 34:3–6.

Caillouet, C. W., Jr., Fontaine, C. T., Manzella, S. A., and Williams, T. D. 1986. Scutes reserved for living tags. *Marine Turtle Newsletter* 36:5–6.

Caillouet, C. W., Jr., Manzella, S. A., Fontaine, C. T., Williams, T. D., Tyree, M. G., and Koi, D. B. 1989. Feeding, growth rate and survival of the 1984 year-class of Kemp's ridley sea turtles (*Lepidochelys kempi*) reared in captivity. In Caillouet, C. W., Jr., and Landry, A. M., Jr. (Eds.). *Proceedings of the First International Symposium on Kemp's Ridley Sea Turtle Biology, Conservation, and Management.* Texas A&M University, Sea Grant College Program, TAMU-SG-89-105, pp. 165–177.

Caillouet, C. W., Jr., Duronslet, M. J., Landry, A. M., Jr., and Shaver, D. J. 1991. Sea turtle strandings and shrimp fishing effort in the northwestern Gulf of Mexico, 1986–1989. *Fishery Bulletin* 89:712–718.

Caillouet, C. W., Jr., Fontaine, C. T., and Flanagan, J. P. 1993. Captive rearing of sea turtles: head-starting Kemp's ridley, *Lepidochelys kempi.* In Jung, R. E. (Ed.). *Proceedings of the American Association of Zoo Veterinarians. St. Louis, MO,* October 10–15, 1993, pp. 8–12.

Caillouet, C. W., Jr., Fontaine, C. T., Manzella-Tirpak, S. A., and Shaver, D. J. 1995a. Survival of head-started Kemp's ridley sea turtles (*Lepidochelys kempii*) released into the Gulf of Mexico or adjacent bays. *Chelonian Conservation and Biology* 1:285–292.

Caillouet, C. W., Jr., Fontaine, C. T, Manzella-Tirpak, S. A., and Williams, T. D. 1995b. Growth of head-started Kemp's ridley sea turtles (*Lepidochelys kempii*) following release. *Chelonian Conservation and Biology* 1:231–234.

Caillouet, C. W., Jr., Shaver, D. J., Teas, W. G., Nance, J. N., Revera, D. B., and Cannon, A. C. 1996. Relationship between sea turtle strandings and shrimp fishing effort in the Northwestern Gulf of Mexico: 1986–1989 versus 1990–1993. *Fishery Bulletin* 94:237–249.

Caillouet, C. W, Jr., Robertson, B. A., Fontaine, C. T., Williams, T. D., Higgins, B. M., and Revera, D. B. 1997. Distinguishing captive-reared from wild Kemp's ridleys. *Marine Turtle Newsletter* 77:1–6.

Carr, A. F. 1952. *Handbook of Turtles of the United States, Canada, and Baja California.* Ithaca, NY: Cornell University Press.

Carr, A. F. 1957. Notes on the zoogeography of the Atlantic sea turtles of the genus *Lepidochelys. Revista Biologica Tropical* 5:45–61.

Carr, A. F. 1963. Panspecific reproductive convergence of the Atlantic sea turtles of the genus *Lepidochelys. Ergebnisse der Biologie* 26:298–303.

Carr, A. F. 1967. *So Excellent a Fishe: A Natural History of Sea Turtles.* New York: Scribners, 1984 revised edition.

Carr, A. F., and Caldwell, D. 1958. The problem of

the Atlantic ridley turtle (*Lepidochelys kempi*) in 1958. *Revista Biologica Tropical* 6:245–262.

Chaloupka, M., and Zug, G. R. 1997. A polyphasic growth function for the endangered Kemp's ridley sea turtle, *Lepidochelys kempii. Fishery Bulletin* 95:849–856.

Chavez, H. 1968a. On the coast of Tamaulipas, part one. *International Turtle and Tortoise Society Journal* 4:20–29, 37.

Chavez, H. 1968b. On the coast of Tamaulipas, part two. *International Turtle and Tortoise Society Journal* 5:16–19, 27–34.

Coyne, M. S. 2000. Population sex ratio of the Kemp's ridley sea turtle (*Lepidochelys kempii*): Problems in population modeling. Ph.D. diss. Texas A&M University, College Station.

Eckert, S. A., Crouse, D., Crowder, L. B., Maceina, M., and Shah, A. 1994. Review of the Kemp's Ridley Sea Turtle Head-start Program. NOAA Technical Memorandum NMFS-OPR-3.

Fontaine, C. T., and Shaver, D. J. 2005. Headstarting the Kemp's ridley sea turtle, *Lepidochelys kempii*, at the NMFS Galveston Laboratory, 1978 to 1992: A review. *Chelonian Conservation and Biology* 4:838–845.

Fontaine, C. T., Manzella, S. A., Williams, T. D., Harris, R. M., and Browning, W. J. 1989a. Distribution, growth and survival of head-started, tagged, and released Kemp's ridley sea turtles (*Lepidochelys kempi*) from year-classes 1978–1983. In Caillouet, C. W., Jr., and Landry, A. M., Jr. (Eds.). *Proceedings of the First International Symposium on Kemp's Ridley Sea Turtle Biology, Conservation and Management.* Texas A&M University, Sea Grant Program, TAMU-SG-89-105. College Station, TX, pp. 124–144.

Fontaine, C. T., Williams, T. D., Manzella, S. A., and Caillouet, C. W., Jr. 1989b. Kemp's ridley sea turtle head-start operations of the NMFS SEFC Galveston Laboratory. In Caillouet, C. W., Jr., and Landry, A. M., Jr. (Eds.). *Proceedings of the First International Symposium on Kemp's Ridley Sea Turtle Biology, Conservation and Management.* Texas A&M University, Sea Grant Program, TAMU-SG-89-105. College Station, TX, pp. 96–110.

Fontaine, C. T., Revera, D. B., Williams, T. D., and Caillouet, C. W., Jr. 1993. *Detection, Verification and Decoding of Tags and Marks in Head-started Kemp's Ridley Sea Turtles,* Lepidochelys kempii. NOAA Technical Memorandum NMFS-SEFC-334.

Foote, J. J., and Mueller, T. L. 2002. Two Kemp's ridley (*Lepidochelys kempii*) nests on the central Gulf coast of Sarasota County, Florida, USA. In Mosier, A., Foley, A., and Brost, B. (Compilers). *Proceedings of the Twentieth Annual Symposium on Sea Turtle Biology and Conservation.* NOAA Technical Memorandum. NMFS-SEFSC-477, pp. 252–253.

Francis, K. 1978. Kemp's ridley sea turtle conservation programs at South Padre Island, Texas, and Rancho Nuevo, Tamaulipas, Mexico. In Henderson, G. E. (Ed.). *Proceedings of the Florida and Interregional Conference on Sea Turtles.* Florida Marine Research Publications No. 33, pp. 51–52.

Godfrey, D. 1996. New riddles about Kemp's ridley. *Velador* 1996:1–5.

Hendrickson, J. R., and Hendrickson, L. P. 1981. A new method for marking sea turtles. *Marine Turtle Newsletter* 19:6–7.

Hildebrand, H. H. 1963. Hallazgo del area de anidacion de la tortuga marina "lora," *Lepidochelys kempi* (Garman), en la costa occidental del Golfo de Mexico (Rept., Chel.). *Sobretiro de Ciencia, Mexico* 22:105–112.

Hildebrand, H. H. 1982. A historical review of the status of sea turtle populations in the western Gulf of Mexico. In Bjorndal, K. A. (Ed.). *Biology and Conservation of Sea Turtles.* Washington, DC: Smithsonian Institution Press, pp. 447–453.

Huff, J. A. 1989. Florida (USA) terminates "headstart" program. *Marine Turtle Newsletter* 46:1–2.

Johnson, S. A., Bass, A. L., Libert, B., Marshall, M., and Fulk, D. 1999. Kemp's ridley (*Lepidochelys kempii*) nesting in Florida. *Florida Scientist* 62:194–204.

Johnson, S. A., Bass, A. L., Libert, B., Marshall, M., and Fulk, D. 2000. Kemp's ridley (*Lepidochelys kempii*) nesting in Florida, USA. In Kalb, H., and Wibbels, T. (Compilers). *Proceedings of the Nineteenth Annual Symposium on Sea Turtle Conservation and Biology.* NOAA Technical Memorandum NMFS-SEFC-443, p. 283.

Keinath, J. A. 1993. Movements and behavior of wild and head-started sea turtles. Ph.D. diss. College of William & Mary, Williamsburg, VA.

Klima, E. F., and McVey, J. P. 1982. Head-starting the Kemp's ridley turtle, *Lepidochelys kempii.* In Bjorndal, K. A. (Ed.). *Biology and Conservation of Sea Turtles.* Washington, DC: Smithsonian Institution Press, pp. 481–487.

Lewison, R. L., Crowder, L. B., and Shaver, D. J. 2003. The impact of Turtle Excluder Devices and fisheries closures on loggerhead and Kemp's ridley strandings in the western Gulf of Mexico. *Conservation Biology* 17:1089–1097.

Libert, B. 1998. Kemp's ridley nesting in Volusia County. In Epperly, S. P., and Braun, J. (Compilers). *Proceedings of the Seventeenth Annual Symposium on Sea Turtle Biology and Conservation.* NOAA Technical Memorandum NMFS-SEFC-415, p. 219.

Magnuson, J. J., Bjorndal, K. A., DuPaul, W. D.,

Graham, G. L., Owens, D. W., Peterson, C. H., Pritchard, P. C. H., Richardson, J. I., Saul, G. E., and West, C. W. 1990. *Decline of the Sea Turtles: Causes and Prevention.* Washington, DC: National Research Council, National Academy Press.

Manzella, S. A., Caillouet, C. W., Jr., and Fontaine, C. T. 1988. Kemp's ridley, *Lepidochelys kempi,* sea turtle head-start tag recoveries: distribution, habitat and method of recovery. *Marine Fisheries Review* 50(3):24–32.

Manzella, S., Bjorndal, K., and Lagueux, C. 1991. Head-started Kemp's ridley captured in Caribbean. *Marine Turtle Newsletter* 54:13–14.

Márquez-M., R. 1990. *FAO Species Catalogue,* Volume 11: *Sea Turtles of the World, An Annotated and Illustrated Catalogue of Sea Turtle Species Known to Date.* Food and Agriculture Organization of the United Nations, FAO Species Synopsis No. 125, Volume 11, FIR/S125.

Márquez-M., R. 1994. *Synopsis of Biological Data on the Kemp's Ridley Turtle,* Lepidochelys kempi *(Garman, 1880).* NOAA Technical memorandum NMFS-SEFSC-343.

Márquez-M., R., Díaz, J., Sánchez, M., Burchfield, P., Leo, A., Carrasco, M., Peña, J., Jiménez, C., and Bravo, R. 1999. Results of the Kemp's ridley nesting beach conservation efforts in México. *Marine Turtle Newsletter* 85:2–4.

Márquez-M., R., Burchfield, P., Carrasco, M. A., Jiménez, C., Díaz, J., Garduño, M., Leo, A., Peña, J., Bravo, R., and González, E. 2001. Update on the Kemp's ridley turtle nesting in Mexico. *Marine Turtle Newsletter* 92:2–4.

McDaniel, C. J., Crowder, L. B., and Priddy, J. A. 2000. Spatial dynamics of sea turtle abundance and shrimping intensity in the U.S. Gulf of Mexico. *Conservation Ecology* 4(1):15 [online]. www.consecol.org/vol4/iss1/art15/.

McVey, J. P., and Wibbels, T. 1984. *The Growth and Movements of Captive-Reared Kemp's Ridley Sea Turtles,* Lepidochelys kempi, *Following Their Release in the Gulf of Mexico.* NOAA Technical Memorandum NMFS-SEFC-145.

Meylan, A. B., and Ehrenfeld, D. 2000. Conservation of marine turtles. In Klemens, M. W. (Ed.). *Turtle Conservation.* Washington, DC: Smithsonian Institution Press, pp. 96–125.

Meylan, A., Bowen, B. W., and Avise, J. C. 1990a. A genetic test of the natal homing versus the social facilitation models for green turtle migration. *Science* 248:724–727.

Meylan, A., Castaneda, P., Coogan, C., Lozon, T., and Fletemeyer, J. 1990b. Kemp's ridley sea turtle reproduction. *Herpetological Review* 21:19–20.

Miller, J. D. 1997. Reproduction in sea turtles. In Lutz, P. L., and Musick, J. A. (Eds.). *The Biology of Sea Turtles.* Boca Raton, FL: CRC Press, pp. 51–81.

Mrosovsky, N. 1978. Editorial. *Marine Turtle Newsletter* 7:1.

Mrosovsky, N. 1979. Editorial. *Marine Turtle Newsletter* 12:1–2.

Mrosovsky, N. 1983. *Conserving Sea Turtles.* London: The British Herpetological Society.

Nicholas, M. A., Davis, T. L., Berry, K. A., Russell, R. R., and Diller, A. P. 2004. Two additional species of marine turtles nest within Gulf Islands National Seashore's Florida District. In Coyne, M. S., and Clark, R. D. (Compilers). *Proceedings of the Twenty-First Annual Symposium on Sea Turtle Biology and Conservation.* NOAA Technical Memorandum NMFS-SEFSC-528, p. 264.

Ogren, L. 1989. Distribution of juvenile and subadult Kemp's ridley turtles: Preliminary results from 1984–1987 surveys. In Caillouet, C. W., Jr., and Landry, A. M., Jr. (Eds.). *Proceedings of the First International Symposium on Kemp's Ridley Sea Turtle Biology, Conservation and Management,* Texas A&M University, Sea Grant College Program, TAMU-SG-89-105, pp. 116–123.

Owens, D. W., Grassman, M. A., and Hendrickson, J. R. 1982. The imprinting hypothesis and sea turtle reproduction. *Herpetologica* 38:124–135.

Palmatier, R. 1993. Kemp's ridley nesting. *Herpetological Review* 24:149–150.

Pelletier, D., Roos, D., and Ciccione, S. 2003. Oceanic survival and movements of wild and captive-reared immature green turtles (*Chelonia mydas*) in the Indian Ocean. *Aquatic Living Resources* 16:35–41.

Phasuk, B., and Rongmuangsart, S. 1973. Growth studies on the ridley turtle, *Lepidochelys olivacea olivacea* Eschscholtz, in captivity and the effect of food preference on growth. Phuket, Thailand: Phuket Marine Biological Center, *Research Bulletin* No. 1.

Plotkin, P. T. 1999. Resolutions of the participants at the 19th Annual Symposium on Sea Turtle Biology and Conservation. *Marine Turtle Newsletter* 85:20–24.

Pritchard, P. C. H. 1969. The survival status of ridley sea turtles in American waters. *Biological Conservation* 2:13–17.

Pritchard, P., Bacon, P., Berry, F., Carr, A., Fletemeyer, J., Gallagher, R., Hopkins, S., Lankford, R., Márquez-M., R., Ogren, L., Pringle, W., Jr., Reichart, H., and Witham, R. 1983. *Manual of Sea Turtle Research and Conservation Techniques, Second Edition.* Bjorndal, K. A., and Balazs, G. H. (Eds.). Washington, DC: Center for Environmental Education.

Pritchard, P. C. H., and Márquez-M., R. 1973. Kemp's ridley sea turtles or Atlantic ridley *Lepidochelys kempi*. IUCN Monograph 2: Marine Turtle Series k.

Sato, F., and Madriasau, B. B. 1991. Preliminary report on natural reproduction of hawksbill sea turtles in Palau. *Marine Turtle Newsletter* 55:12–14.

Shaver, D. J. 1989. Results from eleven years of incubating Kemp's ridley sea turtle eggs at Padre Island National Seashore. In Eckert, S. A., and Eckert, K. L. (Compilers). *Proceedings of the Ninth Annual Workshop on Sea Turtle Conservation and Biology*. NOAA Technical Memorandum NMFS-SEFC-232, pp. 163–165.

Shaver, D. J. 1990a. Kemp's Ridley Project at Padre Island enters a new phase. *Park Science* 10(1):12–13.

Shaver, D. J. 1990b. *Padre Island National Seashore Kemp's Ridley Sea Turtle Project 1990 Report*. Unpublished report, National Park Service, Corpus Christi, TX.

Shaver, D. J. 1991a. Feeding ecology of wild and head-started Kemp's ridley sea turtles in south Texas waters. *Journal of Herpetology* 25:327–334.

Shaver, D. J. 1991b. *Padre Island National Seashore Kemp's Ridley Sea Turtle Project 1991 Report*. Unpublished report, National Park Service, Corpus Christi, TX.

Shaver, D. J. 1992. *Padre Island National Seashore Kemp's Ridley Sea Turtle Project 1992 Report*. Unpublished report, National Park Service, Corpus Christi, TX.

Shaver, D. J. 1994. *Padre Island National Seashore Kemp's Ridley Sea Turtle Project 1994 Report*. Unpublished report, National Park Service, Corpus Christi, TX.

Shaver, D. J. 1996a. Head-started Kemp's ridley turtles nest in Texas. *Marine Turtle Newsletter* 74:5–7.

Shaver, D. J. 1996b. A note about Kemp's ridleys nesting in Texas. *Marine Turtle Newsletter* 75:25.

Shaver, D. J. 1997a. Kemp's ridley turtles from an international project return to Texas to nest. In University of New Orleans (Compiler). *Proceedings from the Sixteenth Annual Gulf of Mexico Information Transfer Meeting*. Minerals Management Service, Gulf of Mexico OCS Region, MMS 97-0038, pp. 38–40.

Shaver, D. J. 1997b. *Padre Island National Seashore Kemp's Ridley Sea Turtle Project 1996 Report*. Unpublished report, Department of the Interior, U.S. Geological Survey.

Shaver, D. J. 1998. Kemp's ridley sea turtle nesting on the Texas coast, 1979–1996. In Epperly, S. P., and Braun, J. (Compilers). *Proceedings of the Seventeenth Annual Symposium on Sea Turtle Biology and Conservation*. NOAA Technical Memorandum NMFS-SEFSC-415, pp. 91–94.

Shaver, D. J. 1999. *Padre Island National Seashore Kemp's Ridley Sea Turtle Project and Texas Sea Turtle Strandings 1998 Report*. Unpublished report, Department of the Interior, U.S. Geological Survey.

Shaver, D. J. 2000. *Padre Island National Seashore Kemp's Ridley Sea Turtle Project and Texas Sea Turtle Nesting and Stranding 1999 Report*. Unpublished report, Department of the Interior, U.S. Geological Survey.

Shaver, D. J. 2001. *Padre Island National Seashore Kemp's Ridley Sea Turtle Project and Texas Sea Turtle Nesting and Stranding 2000 Report*. Unpublished report, Department of the Interior, U.S. Geological Survey.

Shaver, D. J. 2002. *Kemp's Ridley Sea Turtle Project at Padre Island National Seashore and Texas Sea Turtle Nesting and Stranding 2001 Report*. Unpublished report, Department of the Interior, U.S. Geological Survey.

Shaver, D. J. 2004. *Kemp's Ridley Sea Turtle Project at Padre Island National Seashore and Texas Sea Turtle Nesting and Stranding 2002 Report*. Unpublished report, Department of the Interior, U.S. National Park Service.

Shaver, D. J. 2005. Analysis of the Kemp's ridley imprinting and headstart project at Padre Island National Seashore, Texas, 1978–1988, with subsequent Kemp's ridley nesting and stranding records on the Texas coast. *Chelonian Conservation and Biology* 4:846–859.

Shaver, D. J., and Caillouet, C. W., Jr. 1998. More Kemp's ridley turtles return to south Texas to nest. *Marine Turtle Newsletter* 82:1–5.

Shaver, D. J., and Fletcher, M. R. 1992. Kemp's ridley sea turtles. *Science* 257:465–466.

Shaver, D. J., and Miller, J. E. 1999. Kemp's ridley sea turtles return to Padre Island National Seashore. *Park Science* 19(2):16–17,39.

Shaver, D. J., Owens, D. W., Chaney, A. H., Caillouet, C. W., Jr., Burchfield, P., and Márquez-M., R. 1988. Styrofoam box and beach temperatures in relation to incubation and sex ratios of Kemp's ridley sea turtles. In Schroeder, B. A. (Compiler). *Proceedings of the Eighth Annual Workshop on Sea Turtle Conservation and Biology*. NOAA Technical Memorandum NMFS-SEFC-214, pp. 103–108.

Shore, T. 2000. Creating a Kemp's ridley marine reserve in Texas: The missing link is a proven protection strategy. *Endangered Species Update* 17:35–39.

Stabenau, E. K., Landry, A. M., Jr., and Caillouet, C. W., Jr. 1992. Swimming performance of captive reared Kemp's ridley sea turtles, *Lepidochelys kempi* (Garman). *Journal of Experimental Marine Biology and Ecology* 162:213–222.

Subba Rao, M. V., and Raja Sekhar, P. S. 1998. Eggs and hatchlings of the endangered olive ridley sea turtle, *Lepidochelys olivacea* (Eschscholtz). In Epperly, S. P., and Braun, J. (Compilers). *Proceedings of the Seventeenth Annual Sea Turtle Symposium.* NOAA Technical Memorandum NMFS-SEFSC-415, pp. 79–83.

Subba Rao, M. V., and Ratna Kumari, M. 2002. Captive breeding of olive ridley turtle hatchlings *Lepidochelys olivacea* (Eschscholtz) for conservation purposes. In Mosier, A., Foley, A., and Brost, B. (Compilers). *Proceedings of the Twentieth Annual Symposium on Sea Turtle Biology and Conservation.* NOAA Technical Memorandum NMFS-SEFSC-477, pp. 128–131.

Taubes, G. 1992. A dubious battle to save the Kemp's ridley sea turtle. *Science* 256:614–616.

Turtle Expert Working Group (TEWG). 1998. *An Assessment of the Kemp's Ridley (Lepidochelys kempii) and Loggerhead (Caretta caretta) Sea Turtle Populations in the Western North Atlantic.* NOAA Technical Memorandum NMFS-SEFSC-409.

U.S. Fish and Wildlife Service and National Marine Fisheries Service. 1992. *Recovery Plan for the Kemp's Ridley Sea Turtle (*Lepidochelys kempii*).* St. Petersburg, FL: National Marine Fisheries Service.

Werler, J. E. 1951. Miscellaneous notes on the eggs and young of Texas and Mexican reptiles. *Zoologica* 36:37–48.

Werner, S. A. 1994. Feeding ecology of wild and head-started Kemp's ridley sea turtles. M.S. Thesis, Texas A&M University, College Station.

Werner, S. A., and Landry, A. M., Jr. 1994. Feeding ecology of wild and head-started Kemp's ridley turtles (*Lepidochelys kempii*). In Bjorndal, K. A., Bolten, A. B., Johnson, D. A., and Eliazar, P. J. (Compilers). *Proceedings of the Fourteenth Annual Symposium on Sea Turtle Biology and Conservation.* NOAA Technical Memorandum NMFS-SEFC-351, p. 163.

Wibbels, T. R. 1983a. A transatlantic movement of a head-started Kemp's ridley. *Marine Turtle Newsletter* 24:15–16.

Wibbels, T. R. 1983b. Recapture of a "living-tagged" Kemp's ridley. *Marine Turtle Newsletter* 24:16–17.

Wibbels, T. R. 1984. *Orientation characteristics of immature Kemp's ridley sea turtles,* Lepidochelys kempii. NOAA Technical Memorandum NMFS-SEFC-131.

Wibbels, T. 1992. Kemp's ridley sea turtles. *Science* 257:465.

Wibbels, T. 2003. Critical approaches to sex determination in sea turtles. In Lutz, P. L., Musick, J. A., and Wyneken, J. (Eds.). *The Biology of Sea Turtles,* Volume 2. Boca Raton, FL: CRC Press, pp. 103–134.

Wibbels, T., Frazer, N., Grassman, M., Hendrickson, J., and Pritchard, P. 1989a. *Blue Ribbon Panel Review of the National Marine Fisheries Service Kemp's Ridley Head-start Program.* Unpublished report to the National Marine Fisheries Service.

Wibbels, T. R., Morris, Y. A., Owens, D. W., Dienberg, G. A., Noell, J., Leong, J. K., King, R. E., and Márquez-M., R. 1989b. Predicted sex ratios from the international Kemp's ridley sea turtle head-start research project. In Caillouet, C. W., Jr., and Landry, A. M., Jr. (Eds.). *Proceedings of the First International Symposium on Kemp's Ridley Sea Turtle Biology, Conservation and Management.* Texas A&M University, Sea Grant Program, TAMU-SG-89-105, pp. 77–81.

Williams, P. 1993. NMFS to concentrate on measuring survivorship, fecundity of head-started Kemp's ridleys in the wild. *Marine Turtle Newsletter* 63:3–4.

Witham, R., and Futch, C. R. 1977. Early growth and oceanic survival of pen-reared sea turtles. *Herpetologica* 33:404–409.

Wood, J. R. 1982. Captive rearing of Atlantic ridleys at Cayman Turtle Farm Ltd. *Marine Turtle Newsletter* 20:7–9.

Wood, J. R., and Wood, F. E. 1984. Captive breeding of the Kemp's ridley. *Marine Turtle Newsletter* 29:12.

Wood, J. R., and Wood, F. E. 1988. Captive reproduction of Kemp's ridley *Lepidochelys kempii. Herpetology Journal* 1:247–249.

Wood, J. R., and Wood, F. E. 1989. Captive rearing and breeding Kemp's ridley sea turtles at Cayman Turtle Farm (1983) Ltd. In Caillouet, C. W., Jr., and Landry, A. M., Jr. (Eds.). *Proceedings of the First International Symposium on Kemp's Ridley Sea Turtle Biology, Conservation and Management.* Texas A&M University Sea Grant College Program, TAMU-SG-89-105, pp. 237–240.

Woody, J. B. 1981. Head-starting of Kemp's ridley. *Marine Turtle Newsletter* 19:5–6.

Woody, J. B. 1989. International efforts in the conservation and management of Kemp's ridley sea turtle (*Lepidochelys kempii*). In Caillouet, C. W., Jr., and Landry, A. M., Jr. (Eds.). *Proceedings of the First International Symposium on Kemp's Ridley Sea Turtle Biology, Conservation and Management.* Texas A&M University, Sea Grant College Program, TAMU-SG-89-105, pp. 1–3.

Woody, J. B. 1991. Guest Editorial: It's time to stop head-starting Kemp's ridley. *Marine Turtle Newsletter* 55:7–8.

Zug, G. R., Kalb, H. J., and Luzar, S. J. 1997. Age and growth in wild Kemp's ridley sea turtles *Lepidochelys kempii* from skeletochronological data. *Biological Conservation* 80:261–268.

SELINA S. HEPPELL
PATRICK M. BURCHFIELD
LUIS JAIME PEÑA

15

Kemp's Ridley Recovery

How Far Have We Come, and Where Are We Headed?

THE KEMP'S RIDLEY SEA TURTLE (*Lepidochelys kempii*), once hovering on the brink of extinction, is making a remarkable comeback. Over 35 years of dedicated efforts by Mexican and U.S. biologists, wildlife managers, and volunteers have reduced anthropogenic mortality and increased the productivity of the population through a variety of interventions. Funding for these efforts has come from a myriad of public and private sources. The result is an average population growth rate that exceeds any other estimate for a marine turtle. However, our estimate of the number of nesting females for the entire species is still at least an order of magnitude below figures from the 1940s (Hildebrand, 1963). How long will the population increase exponentially? Can we hope to witness a Kemp's ridley *arribada* that rivals those images from 1947? At this early stage in population recovery, it is impossible to tell. Carrying capacity has likely changed and may prevent the species from returning to its original levels. However, we can use our limited knowledge of Kemp's ridley life history and the time series of nests and hatchlings from the turtle camps in Mexico to model changes in Kemp's ridley population dynamics. These analyses can then identify how conservation efforts may have benefited the species and how recovery may respond to changes in those efforts.

Although the small population size and limited nesting distribution of Kemp's ridley are among the main causes of its critical endangerment, they also provide us a remarkable time series of information on an entire species

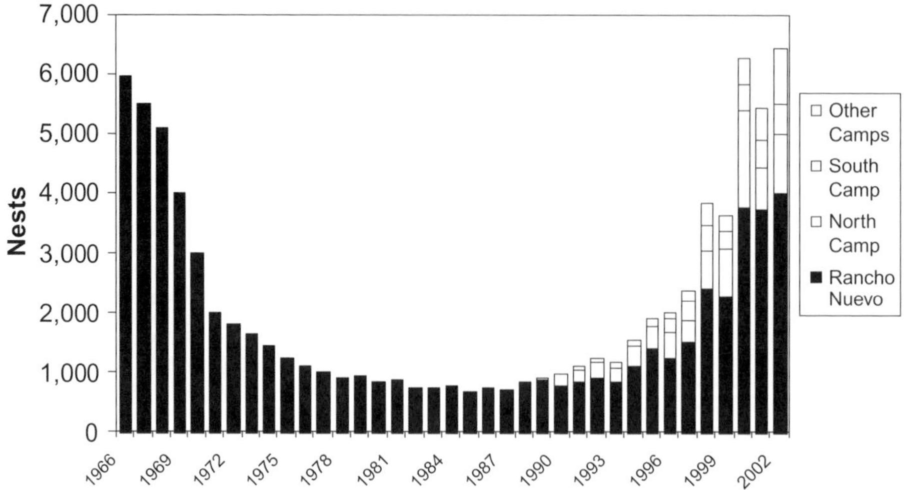

Fig. 15.1. Time series of Kemp's ridley (*Lepidochelys kempii*) nests, 1966–2002. Records from 1966–1977 are from the Turtle Expert Working Group report (Turtle Expert Working Group, 1998), originally from Instituto Nacional de la Pesca records, and 1978–2002 figures are from U.S. Fish and Wildlife Service and Gladys Porter Zoo reports. Note that population estimates from a film shot at Rancho Nuevo in 1947 suggest that an order of magnitude more turtles nested there in a single day (Hildebrand, 1963).

of sea turtle (Fig. 15.1). This greatly reduces uncertainties that are common in sea turtle population assessment because extrapolation from individual nesting beaches to entire populations or species is always problematic. Likewise, the hatchling production time series offers the only estimate we have of cohort strength for a species of sea turtle. We have a 35-year record of "input" for Kemp's ridley with relatively little uncertainty, particularly after 1978, when nest protection methods became standardized and nearly all nests were recorded and moved to protective corrals (Márquez-M. et al., 1999). However, since the mid-1990s, the level of uncertainty in the true population size has increased because of spatial expansion of the population and an increase in protection efforts north and south of the primary nesting site at Rancho Nuevo (Fig. 15.2). Nevertheless, thorough monitoring by cooperating U.S. and Mexican biologists has been the key to this unique time series of the near extinction and subsequent recovery of an endangered species.

History of Recovery Efforts

Efforts to prevent extinction of Kemp's ridley began in 1966, when Mexico's Instituto Nacional de Investigaciones Biologico-Pesqueras sent biologists Humberto Chávez, Martin Contreras, and Eduardo Hernandez to Rancho Nuevo to survey, study, and begin to protect the Kemp's ridley population, which had suffered from massive egg exploitation (Table 15.1). The Kemp's Ridley Working Group, an informal, multiagency, binational team, first met in January 1977 to develop a recovery plan for the rapidly declining species. In 1978, at Rancho Nuevo, Peter C. H. Pritchard led a U.S. contingent to join Rene Márquez-M. and other Mexican biologists in an effort to prevent the species' extinction. It was decided first that the protection of the few remaining nesting females, their eggs, and subsequent hatchlings at Rancho Nuevo, Tamaulipas, was the number-one priority. Fenced "corrals" served as hatcheries, with the goal of 100% nest relocation to assure maximum protection from predators and egg collectors. Second, it was clear that something needed to be done to reduce incidental mortality among the immature and adult life stages of Kemp's ridley in various fishing operations, primarily shrimp trawling. This led to efforts to eliminate fishing operations in the vicinity of Rancho Nuevo during the nesting season and to the development and implementation of turtle excluder devices (TEDs). Third, because of the extremely grave status of the population, an experimental imprinting and head-start effort

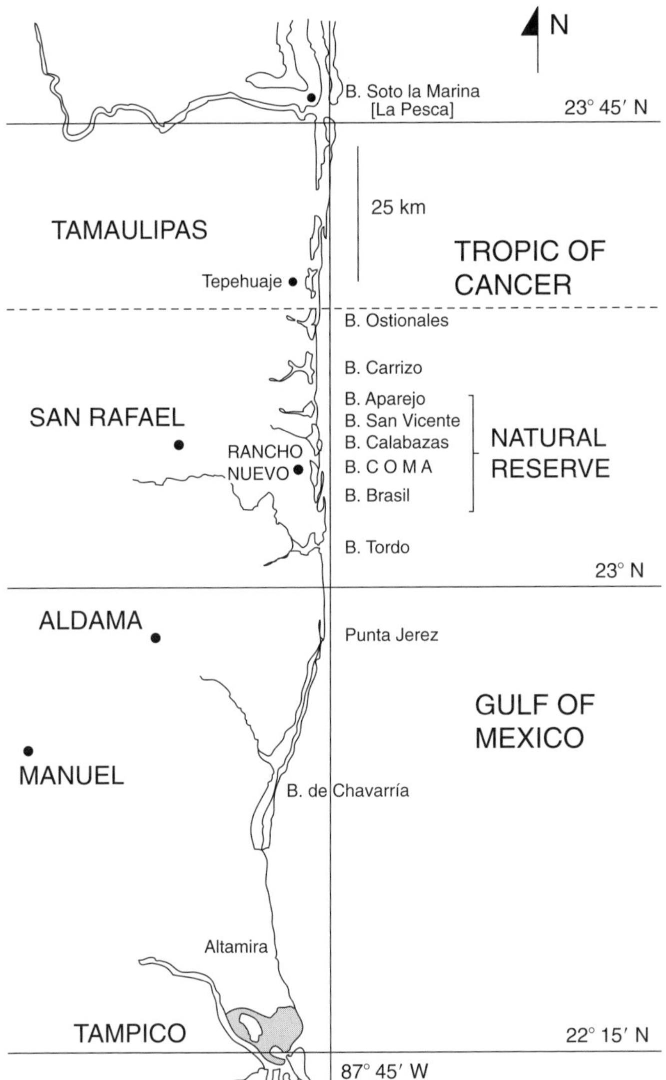

Fig. 15.2. Map of the egg relocation camps and survey areas for Kemp's ridley nests in Tamaulipas.

geared toward establishing a second nesting population at the Padre Island National Seashore in Texas was undertaken. Details of this experiment are given in Chapter 14 of this volume.

Nest protection continues as the fundamental conservation task in Tamaulipas. Each nesting season, biologists, students, and volunteers at the Mexican camps relocate as many egg clutches as possible to artificial nests in fenced corrals near the base camps. On completion of incubation, researchers quantify the success of each clutch of eggs and release the viable hatchlings onto the beach and into the surf. This procedure effectively eliminates land-based predation from the life cycle. Annual hatch success is high, up to 75%, which is considerably higher than estimates for other sea turtle populations, and millions of hatchlings have been released since 1966. Over time, nest protection efforts have expanded north and south of Rancho Nuevo (Fig. 15.2). In 2002, researchers began a study to determine the feasibility and logistics of leaving nests in situ. The main aspect of this study has been to determine the logistics of when, where, and how many nests can be left in situ, both protected and unprotected. It has been determined that temporal stochastic factors (duration and strength of the rainy season) and location (predation levels) have a tremendous impact on nest survivorship, which has exhibited a dramatically lower mean

Table 15.1 Chronology of events in the conservation effort to save the Kemp's ridley sea turtle

Year	Event	Demographic or conservation benefit
1966	Egg protection begins at Rancho Nuevo through the National Program for Research and Conservation	Egg survival increased from near 0 to 5–15%
1976	Expanded efforts at Rancho Nuevo	Egg survival increased to 30%
1977	Rancho Nuevo designated as the first National Reserve for the Management and Conservation of Sea Turtles in Mexico	Long-term protection of primary nesting beach and adjacent waters
1978	United States joins cooperative conservation efforts at Rancho Nuevo; Head Start Program begins	Egg survival increased to 60–75%. Increase in first-year survival for 2–3% of hatchlings through head-starting. Reintroduction of Kemp's to south Texas nesting beaches.
1980	U.S. shrimp trawlers excluded from Mexican waters	Benthic juvenile and adult survival possibly increased
1981	U.S. Fish and Wildlife Service designates Gladys Porter Zoo, Brownsville, Texas, as primary administrator of U.S. efforts to increase production of hatchlings in Mexico	Improved coordination of research activities and data collection, increased public awareness in United States
1990	Turtle excluder devices (TEDs) required in the United States.	Benthic juvenile and adult survival increased
1990	Ban on sea turtle product trade in Mexico	Adult survival possibly increased
1995	TEDs required in Mexico, overall reduction in fishing effort off the primary nesting beaches	Adult and subadult survival increase
1999	Mexican shrimping season closure coincides with primary nesting period	Nesting female survival increase
2002	Protected area designated off South Padre Island, Texas	Protection of nesting females and benthic feeding turtles in nearshore waters off Padre Island

Source: Márquez-M. et al. (1998, 1999); Turtle Expert Working Group (1998).

survival rate (<30%). This will be taken into consideration as this study continues in 2004.

The 26-year Mexico–United States binational effort to recover the Kemp's ridley sea turtle population to self-sustaining numbers has evolved into a holistic effort involving many federal, state, and local government agencies along with NGOs, schools, and the private sector (Table 15.1). Since the program's inception, there has been an emphasis on community involvement to promote public awareness and concern for conservation of the Kemp's ridley in the United States and in Mexico. In Texas, the HEART (Help Endangered Animals—Ridley Turtle) program raised public awareness and critical research dollars through advertising campaigns and school programs. In Mexico, operations at the second most productive hatchery are largely supported by the Mexican and U.S. fishing industries and other private donors. Once the majority of nesting is complete, members of the binational conservation team either travel to local communities or bring school children and family members to the turtle camps to teach them about sea turtles, the marine environment, and why it is so important to protect them. Over the years, there has been a dramatic change in attitude toward the turtles by the local inhabitants. This change in attitude resulted in large part from the long-term education process. Over 20,000 bilingual sea turtle conservation books provided by Conoco-Phillips Petroleum Company have been distributed to area children over the past several years. In recent seasons, the State of Tamaulipas has mounted an even more aggressive education program at the state-run and -operated camp and ecoeducation center at La Pesca. Several events geared toward educating large numbers of individuals are undertaken each nesting season. School groups visiting the camp are given educational programs on sea turtle and Gulf of Mexico ecology. Personnel from the State of Tamaulipas, in cooperation with Sea Turtle Inc. of South Padre Island, Texas, also do an active ex situ education program, visiting many sites and area schools.

All of the aforementioned interventions, activities, and programs are only a temporary solution if a critical underlying problem is not addressed: local economic need. In that regard, a small artisanal ceramics factory has been established with the help of the State of Tamaulipas and the Darden Restaurant Foundation to try to supply income and an alternative to taking turtles or their eggs. This is the pilot program of an intended six programs that are planned, one at each of the major Kemp's ridley turtle nesting beaches. In each instance, local artisans will handcraft objets d'art with a sea turtle theme.

In addition to expanding protection of nests and eggs, conservation measures in the 1990s have focused on decreasing mortality of turtles at sea (Table 15.1). The Mexican ban on the sale of sea turtle products is thought to have greatly benefited olive ridley turtles (*Lepidochelys olivacea*) and may have had a positive influence on survival rates of Kemp's ridleys as well (Márquez-M. et al., 1998). Turtle excluder devices have been shown to decrease mortality of individuals caught in shrimp trawls, and TED regulations in the United States and Mexico are thought to have decreased incidental mortality rates at the population level (Turtle Expert Working Group, 1998). Seasonal and spatial closures of the shrimp fishery reduce mortality of nesting females and other nearshore turtles and have been directly linked to decreased strandings (Caillouet et al., 1991, 1996; Lewison et al., 2003). Coupled with large cohorts produced on the primary nesting beaches, these efforts have led to the rapid population growth rates we see today.

Current Status

The status and prospectus for the Kemp's ridley were reported in 1998 and 2000 by the Turtle Expert Working Group (TEWG), a team of biologists, managers, and stakeholder representatives assembled by the National Marine Fisheries Service, Southeast Fisheries Science Center. Here, we include some analyses from those reports with updates from population sizes reported since 1998–1999.

We can use the data from nesting beaches to estimate rates of decline and recovery over various time intervals (Table 15.2). Clearly, the population is increasing, although the true rate of increase is somewhat obscured by a lack of information on the coastwide distribution of nesting through time. Nests have been reported in considerable numbers since the early 1970s north of the main camp located in Barra Coma (mostly in the area between Barra Carrizo and Barra Ostionales), but because of limited resources, it was many years before an actual nest protection effort was undertaken (Fig. 15.2). It is inappropriate to assume that the increase in nests counted is entirely biological because efforts to locate, report, and protect nests have increased in recent years. The TEWG debated this problem at great length (Turtle Expert Working Group, 1998, 2000) and finally settled on Rancho Nuevo, Tepehuajes (North Camp), and Barra del Tordo (South Camp) as the best indicators of true population growth. However, it is our opinion that the only possible clear indicator of population growth is the area comprised of approximately 19 km north of Barra Coma and the 13.5 km south to Barra del Tordo, designated as "Rancho Nuevo" in published records. Continued efforts to identify tagged turtles and determine nest site fidelity may improve the estimate of population growth rate in the future.

Projecting population growth with appropriate confidence intervals provides a range of estimated recovery dates, assuming that the observed annual rate of increase continues (Table 15.2). The "momentum" of population increase, fueled by high egg survival rates in the corrals and reduced at-sea mortality, may allow rapid exponential growth for several more years (Keyfitz, 1966; Turtle Expert Working Group, 2000). However, it is probably overly optimistic to assume that the observed exponential increase of 12–19% per year will persist for many more years, as survival rates in various life stages are likely to decrease as the population becomes larger. Rates of increase calculated for hatchling production, rather than nests, are already slowing, indicating a gradual reduction in average nest survival or fecundity (average eggs per nest per female).

Changes in population trajectories at Rancho Nuevo since 1966 have dramatically altered the

Table 15.2 Results of exponential trend regression fits to Kemp's ridley nests surveyed and hatchlings released over a range of time intervals

					Rancho Nuevo only		Rancho Nuevo, N and S camps		All camps	
	1947–1966	1966–1978	1966–1985	1978–1985	1985–2002	1990–2002	1985–2002	1990–2002	1985–2003	1990–2003
Nests										
Slope of ln-transformed data	−0.1024	−0.169	−0.116	−0.041	0.112	0.154	0.135	0.160	0.142	0.171
SE of slope		0.010	0.009	0.009	0.010	0.011	0.007	0.009	0.008	0.009
R-square		0.964	0.898	0.788	0.881	0.944	0.953	0.960	0.952	0.965
Population growth rate in percent per year: $(\exp^{slope} - 1) \times 100$	−9.7	−15.5	−10.9	−4.0	11.9	16.7	14.5	17.4	15.3	18.6
Predicted nests in 2020 (95% confidence limits)					12,400–56,400	33,800–127,100	35,700–106,700	59,500–190,800	43,100–138,400	80,100–255,500
Hatchlings										
Slope of ln-transformed data		0.026	0.057	−0.006	0.101	0.135	0.124	0.147	0.129	0.150
SE of slope		0.033	0.016	0.037	0.009	0.011	0.008	0.011	0.008	0.011
R-square		0.057	0.427	0.005	0.877	0.924	0.935	0.936	0.945	0.941
Population growth rate in percent per year: $(\exp^{slope} - 1) \times 100$		2.6	5.87	−0.6	10.63	14.45	13.20	15.84	13.77	16.18

Note: Projected population sizes are based on the 95% confidence intervals of the slopes.

probability of extinction, as measured by the diffusion approximation method (Dennis et al., 1991; Holmes, 2001, 2003) (Fig. 15.3). This viability analysis treats population growth as a diffusion process in which the probability of extinction is based purely on the mean trajectory of the population and its year-to-year variance. To account for the remigration interval and to reduce sampling error bias (Holmes, 2001), we used a 2-year running sum and a variance modifier $\tau = 3$ (see the "slope method" described in Holmes, 2001). Extinction risk is arbitrarily based on a quasiextinction threshold of eight nests over 2 years. The purpose of this exercise is to show how the trajectories for Kemp's ridleys, over different time periods, have changed the potential for extinction via stochastic processes. For example, if the trend and nest number variance measured from 1966 to 1978 had continued, there would have been a 50% chance of extinction in 35 years and a 100% chance of extinction within 50 years. Because the variance and decline in nest numbers were less dramatic in the early 1980s, the probability of extinction decreased to about a 15% chance in 35 years for the trend calculated over the 20-year time period from 1966 to 1985. Over the 1978–1987 time period, extinction risk dropped to a 5% chance in 150 years. If current rapid population growth and moderate variance levels continue, the species has little chance of extinction in the next 250 years, according to this simple analysis.

Reasons for Recovery

What factors have contributed to Kemp's ridley recovery? Clearly, the answer is that all conservation efforts have contributed in some way. Critical events in the conservation of Kemp's ridley have resulted in increased survival rates for various life stages (Table 15.1). A precise, quantitative assessment of the relative impacts of all of these events is impossible because (1) the events overlap in time and are not always independent of one another and (2) the change in survival rate was not quantified in most cases. There also may be contributions from natural factors such as cli-

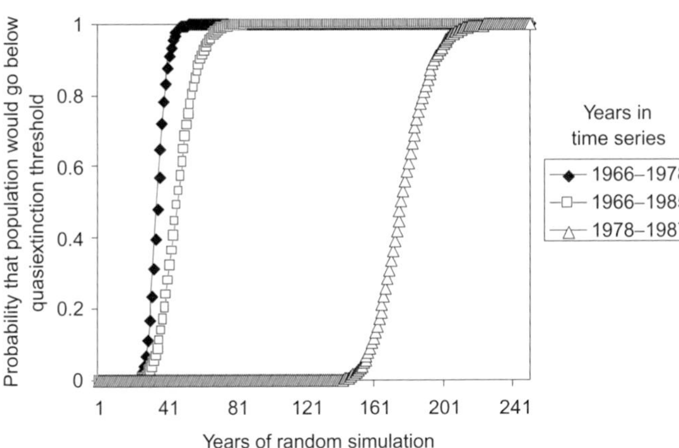

Fig. 15.3. Extinction risk curves generated from various time series of Kemp's ridley sea turtle nest abundance, using diffusion approximation methods of Dennis et al. (1991) and Holmes (2001). Extinction risk is the probability of a population trajectory dropping below a threshold of eight nests over two years (roughly equivalent to one turtle), based on the trajectory and variance in nest numbers given by the particular time series. To account for remigration interval, the analysis is done on the two-year running sum total of nests counted. Improved population trajectories, a result of intense conservation efforts, have decreased the species' probability of extinction from stochastic events.

mate change or indirect benefits through shifts in predator or prey populations. Simply stated, population growth occurs when births exceed deaths and/or immigration exceeds emigration. We can ignore the latter for Kemp's because we have data for the entire species. Over the past 35 years, the difference between births and deaths has increased because of a decrease in mortality for various, perhaps all, life stages.

Egg protection is the primary conservation tool for sea turtles and can dramatically increase cohort sizes. We know that egg survival for Kemp's ridley was extremely low before 1966 as a result of high harvest and predation (Márquez-M. et al., 1999). Current hatchling emergence rates in the corrals average around 65%, high for any sea turtle nesting beach and certainly higher than the rate that would occur without nest protection.

Increased survival rate in other life stages is more difficult to document. Although we do not have mark–recapture information to directly estimate a change in survivorship to maturity, the relationship between hatchling production (cohort size) and nests counted after an appropriate time lag (10 years, 12 years, or 14 years, representing age at maturation) can serve as indirect evidence (Turtle Expert Working Group, 2000). This relationship between hatchlings and nests is valid only if the proportion of first-time nesters represented in a given nest count is relatively constant. Regardless of the lag to maturity used, the graphs showed no relationship between the cohort size and the number of nests counted after the lag for cohorts produced in the 1960s and 1970s. Larger cohorts did not result in more nests. Cohorts produced in the mid-1980s to late 1980s, on the other hand, show a strong positive relationship between cohort size and the number of nests counted after a lag, suggesting that more turtles in these cohorts survived to adulthood. Although this apparent increase in survivorship cannot be attributed to a particular management effort, the fact that TEDs became widely used in 1990 is probably not coincidental (Turtle Expert Working Group, 2000). However, 1990 also marks the cessation of legal trade of sea turtle products in Mexico, which may have had a positive effect on subadult or adult survival rates (Márquez-M. et al., 1998). Regardless of the cause, the fact that nest numbers have been increasing more rapidly than hatchlings for many years, longer than the suggested generation time (Table 15.2), suggests that survival rates at sea have improved. The size distributions of stranded Kemp's ridleys have changed, with a greater proportion of larger turtles appearing in the samples (Turtle Expert Working Group, 2000; Fig. 15.4). This may indicate that turtles are living longer, leading to an accumulation of animals in larger size classes.

TEWG also found evidence for increased survivorship for Kemp's ridley in the 1990s through an analysis of a simple age-structured model (Turtle Expert Working Group, 2000; Heppell et al., 2005). The models suffer from a lack of empirical data to estimate survival parameters but can serve as a starting point for analysis. The Group ran a catch curve analysis, based on size distributions of stranded turtles, to estimate the annual survival rate of Kemp's ridleys between 2 and 6 years of age (30–60 cm SCL; age defined by a von Bertalanffy growth curve). Survival parameters for pelagic juveniles, large benthic juveniles (age 6 years to maturity), and adults were fit using the known cohort sizes from hatchling production to achieve a model time series of nests that matched observations of nests at Rancho Nuevo, North and South camps (1966–1996). The models could not match the observed increase in nest abundance unless they included a decrease in mortality in the late 1980s or early 1990s (Turtle Expert Working Group, 2000). The Group decided to fit a fourth parameter, a multiplier for mortality of benthic-stage turtles that affected the survival rate after 1990. This multiplier ranged between 0.4 and 0.7 for the different models, representing a 30–60% decrease in instantaneous mortality or a 5–35% increase in annual survival for benthic juveniles and adults. The models were used to estimate recovery times and population sizes for the species under a range of management scenarios and parameter uncertainty.

The time series of olive ridley nests in Oaxaca, Mexico also suggests that reductions in the mortality rates of adult or juvenile turtles are needed for population recovery (Cornelius et al., Chapter 12). As with Kemp's ridley, the olive ridley population stopped declining when egg protec-

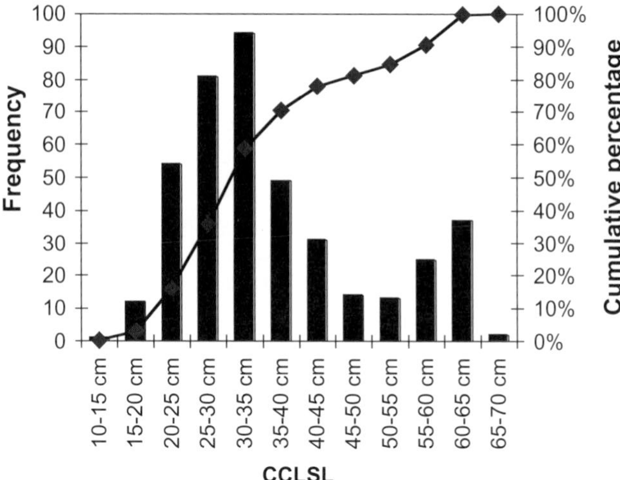

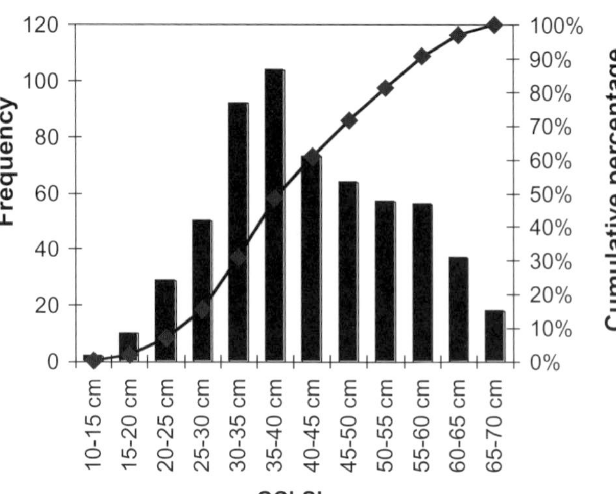

Fig. 15.4. Shifts in the size distribution of stranded Kemp's ridley sea turtles, reported by the Turtle Expert Working Group (Turtle Expert Working Group, 2000). CCLSL is the curved carapace length converted to straight carapace length when no straight measurement was made.

tion measures were adopted, but rapid population growth occurred after subadult and adult harvest mortality decreased. The extremely high growth rate of this population of olive ridleys was probably the result of a combination of large cohorts produced through nest protection, a strongly female-biased sex ratio, and higher survival rates of adult females after harvest for the leather trade was halted in 1990 (Márquez-M. et al., 1998; Peñaflores Salazar et al., 2000). Olive ridleys were likely harvested at much higher rates than Kemp's ridley, but incidental mortality in fisheries was probably lower.

Like most sea turtle populations, Kemp's ridley requires more research on mortality and growth rates. Much of our limited information on demographic parameters comes from analysis of captive-reared animals that were part of the Kemp's Ridley Head Start Experiment (Caillouet et al., 1995; Manzella et al., 1988; Caillouet, 2000). Mark–recapture studies in foraging grounds are contributing to our estimates of demographic parameters for wild turtles (Schmid, 1998; Schmid et al., 2002), but the highly mobile nature of juvenile and adult ridleys make large-scale studies nearly impossible. Many nesting females have been tagged at Rancho Nuevo (Márquez-M. et al., 1999); although flipper tag loss is high, and PIT tagging and tag reading have been variable from year to year, the data represent a real opportunity to assess adult female survival because the time series is long. Researchers at Rancho Nuevo also have collected information on the size and condition of nesting females that allows a rough estimation of the proportion of "new" ("neophyte") nesters, which can tell us something about recruitment and adult survival (Leo Paredo et al., 2000). New aging methodologies are contributing to our estimates of growth variability, time of settlement, and age at maturity (Chaloupka and Zug, 1997; Zug et al., 1997; Snover et al., 2002; Snover et al., Chapter 5). Analysis of humerus growth layers may also provide an opportunity to estimate adult survival through changes in the mean "age" at death (in years postmaturation) before and after TED implementation (Snover, 2002).

What Does the Future Hold?

Kemp's ridley nests have exceeded levels observed in the 1960s, when around 6,000 nests were counted at Rancho Nuevo. If current exponential rates of increase continue, Kemp's ridleys will reach preexploitation levels in a few decades. However, we cannot assume that the current growth rate will continue indefinitely because of changes in egg survival rates as a result of decreased per capita protection on nesting beaches, the potential for increased natural or anthropogenic threats, and possible density-dependent changes in survival and growth. These changes in vital rates may not occur until the species has surpassed desired recovery levels and may only occur locally. Continued monitoring of the nesting population is essential, not only because it can give us information about the status of the breeding population but because it is our most reliable indicator of population change.

As the number of nests increases in Tamaulipas (Márquez-M. et al., 2001), it will be impossible to move all of the eggs to the corrals, and the overall egg survival rate will go down. Currently, many in situ nests receive some protection, and new protection methods for these nests are under development. Fortunately, models predict that modest decreases in egg survival may slow, but not prevent, species recovery if at-sea survival rates remain high (Turtle Expert Working Group, 2000). More severe reductions in egg survival or a decrease in fecundity (eggs per nest, nests per female, or frequency of nesting) could eventually stagnate population growth. The exact definition of recovery is currently under debate by the Kemp's Ridley Recovery Team, appointed by NOAA Fisheries and the U.S. Fish and Wildlife Service.

In addition to population spread north and south of Rancho Nuevo, the number of Kemp's ridley nests is increasing at South Padre Island, Texas, and elsewhere in the United States and Mexico. Population spread and establishment of new nesting beaches (or reestablishment of historical nesting beaches) will provide insurance against catastrophes in Tamaulipas and create the potential for enhanced genetic variance through local adaptation. New nesting beaches may also improve the primary sex ratio, which is currently biased toward females (T. Wibbels, personal communication). Identification, monitoring, and protection of Kemp's ridley nests outside of Tamaulipas will depend on the cooperation of sea turtle biologists and wildlife managers as well as increased public awareness.

We know very little about the survival rates of pelagic and nearshore juveniles and nothing about the survival rates of males. Although we think that TEDs have decreased mortality, enforcement is the key to continuing this benefit (Lewison et al., 2003). Careful monitoring of other fisheries that may affect survival of sea turtles is also a necessity in all habitats used by the species. In particular, gill nets and nearshore longlines in the United States and Mexico may take substantial numbers of turtles, but the impacts of these fisheries are rarely assessed (Turtle Expert Working Group, 1998; Frazier et al., Chapter 13). Continued research and testing of fishing gear to reduce incidental catch and mortality are essential. Nearshore areas frequented by foraging juvenile and adult Kemp's ridleys may require additional protections, through time/area closures (McDaniel et al., 2000; Shore, 2000; Lewison et al., 2003). It will be important to evaluate the benefits of gear improvements and closures through careful monitoring of strandings and mark–recapture research.

Kemp's ridley abundance is still far below historical levels, and conservation efforts should result in continued population growth for the next several years. However, reductions in habitat and prey availability may have decreased the maximum population size that the species can hope to attain. Reductions in the carrying capacity of the environment can be obvious, such as a restriction in the amount of available nesting habitat, or they can be more subtle, as in a reduced prey base through competition with fisheries or interspecific competition with recovering loggerhead turtles. More beaches in Tamaulipas, Vera Cruz, and southern Texas should be designated as Kemp's ridley conservation zones to ensure that an adequate amount of high-quality nesting habitat is available to the recovering population. Loss of high-quality foraging habitat is a more difficult problem to assess, but we have learned much in recent years about the prey and preferred habitat of Kemp's ridley in the Gulf of Mexico (Morreale et al., Chapter 11). A large proportion of the most productive area of the northern Gulf of Mexico has been lost to the "Dead Zone," a recurring region of anoxia caused by high-nutrient runoff from the Mississippi River. As we learn more about the foraging ecology of Kemp's ridley, the impacts of this loss of foraging habitat may become more apparent.

The future of Kemp's ridley is improving, but it is not yet secure. We are still extremely uncertain about critical aspects of the ecology of ridley turtles, which hampers our ability to predict changes in population size in response to natural and anthropogenic perturbations. Recovery will require protection efforts on land and at sea and should be monitored on both nesting beaches and foraging grounds. With continued conservation efforts, international cooperation, and community involvement and public support, the Kemp's ridley may again be the most abundant sea turtle in the Gulf of Mexico.

ACKNOWLEDGMENTS

We thank the editor for her infinite patience and two anonymous reviewers who provided thoughtful commentary and helpful editorial suggestions. All of the models and speculations on Kemp's ridley population biology stem from the efforts and discussions of the Turtle Expert Working Group. In particular, Wendy Gabriel was a key contributor to the formulation of the models, and Sheryan Epperly and Terry Henwood did most of the work on catch curve analyses. This study was supported in part by the Oregon Agricultural Experiment Station under project ORE00102.

LITERATURE CITED

Caillouet, C. W., Jr. 2000. Sea turtle culture: Kemp's ridley and loggerhead turtles. In Stickney, R. (Ed.). *Encyclopedia of Aquaculture.* New York: John Wiley & Sons, pp. 786–798.

Caillouet, C. W., Jr., Duronslet, M. J., Landry, A. M., Jr., Revera, D. B., Shaver, D. J., Stanley, K. M., Heinly, R. W., and Stabenau, E. K. 1991. Sea turtle strandings and shrimp fishing effort in the northwestern Gulf of Mexico, 1986–89. *Fishery Bulletin* 89:712–718.

Caillouet, C. W., Jr., Fontaine, C. T., Manzella-Tirpak, S. A., and Shaver, D. J. 1995. Survival of head-started Kemp's ridley sea turtles (*Lepidochelys kempii*) released into the Gulf of Mexico or adjacent bays. *Chelonian Conservation and Biology* 1:285–292.

Caillouet, C. W., Jr., Shaver, D. J., Teas, W. G., Nance, J. M., Revera, D. B., and Cannon, A. C. 1996. Relationship between sea turtle stranding rates and shrimp fishing intensities in the northwestern Gulf of Mexico: 1986–1989 versus 1990–1993. *Fishery Bulletin* 94:237–249.

Chaloupka, M., and Zug, G. R. 1997. A polyphasic growth-function for the endangered Kemp's ridley sea-turtle, *Lepidochelys kempii. Fishery Bulletin* 95:849–856.

Dennis, B., Munholland, P. L., and Scott, J. M. 1991. Estimation of growth and extinction parameters for endangered species. *Ecological Monographs* 61:115–143.

Heppell, S. S., Crouse, D., Crowder, L., Epperly, S., Gabriel, W., Henwood, T., and Márquez-M., R. 2005. A population model to estimate recovery time, population size and management impacts on Kemp's ridley sea turtles. *Chelonian Conservation and Biology* 4:767–773.

Hildebrand, H. H. 1963. Hallazgo del área de anidación de la tortuga marina "lora" *Lepidochelys kempi* (Garman), en la costa occidental del Golfo de México. *Sobretiro de Ciencia, México* 22:105–112.

Holmes, E. E. 2001. Estimating risks in declining populations with poor data. *Proceedings of the National Academy of Sciences USA* 98:5072–5077.

Holmes, E. E. 2003. Beyond theory to application and evaluation: diffusion approximations for population viability analysis. *Ecological Applications* 14:1272–1293.

Keyfitz, N. 1966. Sampling variance of standardised mortality rates. *Human Biology* 38:309–317.

Leo Peredo, A. S., Castro Melendez, R. G., Cruz Flores, A. L., Gonzalez Cruz, A., and Conde Galaviz, E. 2000. Distribution and abundance of the Kemp's ridley (*Lepidochelys kempii*) neophytes at the Rancho Nuevo nesting beach, Tamaulipas, Mexico, during 1996–1998. In Kalb, H. J., and Wibbels, T. (Eds.). *Proceedings of the Nineteenth Annual Symposium on Sea Turtle Biology and Conservation.* NOAA Technical Memorandum NMFS-SEFSC-443, p. 272.

Lewison, R. L., Crowder, L. B., and Shaver, D. L. 2003. The impact of turtle excluder devices and fisheries closures on loggerhead and Kemp's ridley strandings in the Western Gulf of Mexico. *Conservation Biology* 17:1089–1097.

Manzella, S. A., Caillouet, C. W., Jr., and Fontaine, C. T. 1988. Kemp's ridley, *Lepidochelys kempii,* sea turtle head start tag recoveries: distribution, habitat and method of recovery. *Marine Fishery Review* 50:24–32.

Márquez-M., R., Jiménez, M. D., Carrasco, M. A., and Villanueva, N. A. 1998. Comments on the population trends of sea turtles of the *Lepidochelys* genus, after total ban of 1990. *Oceanides* 13:41–62.

Márquez-M., R., Díaz, J., Sánchez, M., Burchfield, P., Leo, A., Carrasco, M., Peña, J., Jiménez, C., and Bravo, R. 1999. Results of the Kemp's ridley nesting beach conservation efforts in México. *Marine Turtle Newsletter* 85:2–4.

Márquez-M., R., Burchfield, P., Carrasco, M. A., Jiménez, C., Díaz, J., Garduño, M., Leo, A., Peña, J., Bravo, R., and González, E. 2001. Update on the Kemp's ridley turtle nesting in México. *Marine Turtle Newsletter* 92:2–4.

McDaniel, C. J., Crowder, L. B., and Priddy, J. A. 2000. Spatial dynamics of sea turtle abundance and shrimping intensity in the U.S. Gulf of Mexico. *Conservation Ecology* 4:15.

Peñaflores Salazar, C., Vasconcelos Perez, J., Albavera Padilla, E., and Márquez-M., R. 2000. Twenty-five years nesting of olive ridley sea turtle *Lepidochelys olivacea* in Escobilla Beach, Oaxaca, Mexico. In Abreu-Grobois, F. A., Briseño-Dueñas, R., Márquez-M., R., Sarti, L. (Compilers). *Proceedings of the Eighteenth International Sea Turtle Symposium.* NOAA Technical Memorandum NMFS-SEFSC-436, pp. 27–29.

Schmid, J. R. 1998. Marine turtle populations on the west-central coast of Florida: results of tagging studies at the Cedar Keys, Florida, 1986–1995. *Fishery Bulletin* 96:589–602.

Schmid, J. R., Bolten, A. B., Bjorndal, K. A., and Lindberg, W. J. 2002. Activity patterns of Kemp's ridley turtles, *Lepidochelys kempii,* in the coastal waters of the Cedar Keys, Florida. *Marine Biology* 140:215–228.

Shore, T. 2000. Creating a Kemp's ridley marine reserve in Texas: the missing link in a proven protection strategy. *Endangered Species Update* 17:35–39.

Snover, M. L. 2002. Growth and ontogeny of sea turtles using skeletochronology: methods, validation and application to conservation. Ph.D. diss. Duke University, Durham, NC.

Snover, M. L., Hohn, A. A., and Macko, S. A. 2002. Skeletochronology in juvenile Kemp's ridleys: validation, settlement and growth. In Mosier, A., Foley, A., and Brost, B. (Eds.). *Proceedings of the Twentieth Annual Symposium on Sea Turtle Biology and Conservation.* NOAA Technical Memorandum NMFS-SEFSC-477, p. 76.

Turtle Expert Working Group (TEWG). 1998. *An assessment of the Kemp's ridley (*Lepidochelys kempii*) and loggerhead (*Caretta caretta*) sea turtle populations in the Western North Atlantic.* NOAA Technical Memorandum NMFS-SEFSC-409.

Turtle Expert Working Group (TEWG). 2000. *Assessment update for the Kemp's ridley and loggerhead sea turtle populations in the Western North Atlantic.* NOAA Technical Memorandum NMFS-SEFSC-444.

Zug, G. R., Kalb, H. J., and Luzar, S. J. 1997. Age and growth in wild Kemp's ridley sea turtles *Lepidochelys kempii* from skeletochronological data. *Biological Conservation* 80:261–268.

PAMELA T. PLOTKIN

Near Extinction and Recovery

MY FIRST RIDLEY EXPERIENCE occurred 25 years ago, when I volunteered to assist an ambitious, experimental last-ditch effort to increase the number of Kemp's ridleys. At the time, the species had hit rock bottom. Only a few hundred reproductively active females remained in the world, and the survival outlook for the species was grim. A lot has changed in 25 years. Kemp's ridley is increasing, and if population growth continues on its current trajectory, recovery might well be attainable.

Underlying this very optimistic forecast for Kemp's ridley lie several very controversial issues regarding past and future recovery efforts. High on this list is the Head Start Experiment (Shaver and Wibbels, Chapter 14). This experiment was in the public spotlight for many years, inspiring simultaneous praise and criticism from around the world as well as an extremely political and polarizing debate. To some it was a feel-good program that distracted attention and critical resources from other issues more deserving of support. To others, it was a program that might increase the number of turtles, establish a second nesting colony, and generate much needed biological information. Undoubtedly there were many other perceptions and feelings that fell between these two ends of the spectrum. Politics and emotions aside, the experiment yielded a lot of biological information and one very interesting result that deserves much more attention than it has yet to receive: some of the Kemp's ridley eggs that were taken from Mexico to Texas to hatch and then grow in captivity for one year before their release into the Gulf of Mexico did survive to adulthood after their release and returned to lay their eggs on the same Texas beach where they were "experimentally imprinted." Scientifically that is pretty exciting because it provides evidence of an alternative mechanism to the natal homing mechanism, which has been broadly and inappropriately applied to explain the remigration of all female sea turtles to specific nesting beaches.

I have to be honest and admit that I didn't think anyone would ever find a head-started turtle nesting anywhere. Not because it was a biological impossibility but because the logistics of finding such a turtle seemed so improbable. There was no well-established plan to evaluate the Head Start Experiment before it began (i.e., to systematically survey Texas beaches for nesting turtles), and once an evaluation plan did take shape, there was very little funding to support the effort. This lack of support remains a critical issue.

Another critical issue regarding past and future recovery efforts for this species is the need for a quantitative evaluation of the impact that the various recovery efforts had toward reversing the population decline (Heppell et al., Chapter 15). There were three main recovery actions: protection of eggs at the nesting beach, the mandatory use of turtle excluder devices (TEDs) by the shrimp fishery, and the Head Start Experiment. We know the Head Start Experiment did little to contribute to the current population increase. But what about egg protection and TED use? What were the relative contributions of reducing egg mortality and reducing at-sea mortality to the current population increase? The answers to these questions are important to Kemp's ridley recovery planning, where recovery activities are determined and prioritized. Also critical to recovery planning for this species remains the ever-elusive question "What is a recovered Kemp's ridley population?" What is the criterion we will use to judge the species recovered? This is another important question that remains unresolved.

The future for Kemp's ridley is much brighter today than it was 25 years ago, but many challenges remain. Increasing our biological and ecological knowledge of all life stages is necessary. Protecting important nesting sites is critical. Continuing to survey the coast for stranded turtles and determining the cause of their mortality are essential to mitigating threats at sea and identifying new threats as they emerge in an ever-changing world. And protecting critical marine habitats is vital.

In striking contrast, the world's most abundant sea turtle is perhaps the least understood biologically. Much of our understanding of the signature behavior of *Lepidochelys*, the *arribada*, is speculative. How these aggregations form and why this behavior evolved continue to be a mystery and likely will remain so for some time (Bernardo and Plotkin, Chapter 4). The arribada phenomenon has captured the attention of the world and has obscured what is really most newsworthy: olive ridleys exhibit three nesting strategies within and among populations (solitary nesting, arribada nesting, and a mixed strategy). We know virtually nothing about this reproductive polymorphism: how it evolved, how it is maintained, and the relative contribution of each phenotype to the existence of populations and the species.

Because so much attention has been drawn to the large olive ridley nesting aggregations, comparatively little is known about solitary nesting. In fact, we still do not know the entire range of solitary nesting in many parts of the world. One would expect that these large charismatic vertebrates that haul themselves out on beaches to lay eggs and leave well-formed tracks in their wake would be highly visible and therefore well documented throughout their range, but this is not the case. In places such as West Africa, we are still acquiring new distribution information, and I suspect there are still many undocumented beaches in the world where ridleys nest. Gathering comprehensive distribution data and monitoring key or select solitary nesting beaches worldwide should be considered as important as are the efforts to monitor turtles at arribada beaches.

Large olive ridley nesting aggregations also attracted inauspicious attention. This large and fairly predictable pulse of resource facilitated large-scale use of the turtles' meat and skin (Campbell, Chapter 2; Cornelius et al., Chapter 12). Olive ridleys were at one time the most commercially important sea turtle in the world.

Their eggs were also commercialized and remain an economically important resource in some areas of Central America where controlled use is legal. We know that the large-scale take of adult turtles resulted in significant nesting population declines at several Mexican beaches. We know much less about the impact of controlled egg use.

Since the olive ridley slaughterhouses shut down over 15 years ago, the number of ridleys nesting on some beaches in Mexico has increased dramatically and now exceeds the number that nested before commercialization. Scientifically this is noteworthy because, similar to what we have seen with Kemp's ridley, the olive ridley also responds favorably to protection. Moreover, it appears to respond quickly, something very uncharacteristic of a long-lived animal.

Although the olive ridley is the most abundant sea turtle in the world and is very broadly distributed, its future remains uncertain, and the challenges ahead are more numerous and complicated than those of Kemp's ridley. Our dearth of biological and ecological information is one of the most serious threats to the olive ridley today. Why do arribadas occur at specific beaches? How important are the solitary nesters versus the arribada nesters? Where do the juveniles live? What do they eat? At what age do they attain sexual maturity? How long do they live? What role do they play in our world's oceans? The dearth of information is not because no one wants to study olive ridleys. There are many biologists studying these magnificent turtles around the world, but the results of much of this research remain in the unpublished literature, as abstracts in conference proceedings where full details are absent, or as government reports. Making research findings widely available is critical to the future of this species because these results have the potential to greatly influence conservation planning and policy.

Another significant threat for the olive ridley is the false sense of security associated with the arribada phenomenon. When one sees thousands of turtles emerging onto a beach to lay eggs, it is difficult to imagine that a resource as bountiful as this might one day vanish. But current abundance is hardly a guarantee. We know from past mistakes that even plentiful resources can vanish within a short time frame. Organisms such as the passenger pigeon, the buffalo, and the Carolina parakeet were once abundant and are now extinct or critically endangered. These organisms were not cryptic species that were out of sight and thus out of mind. They were highly visible yet still slipped away from existence.

Finally, the future for the olive ridley is as precarious today as it was a few decades ago. The challenges ahead that were identified for Kemp's ridley are also applicable to the olive ridley: increasing our biological and ecological knowledge of all life stages, identifying and protecting important nesting sites, surveying the coast for stranded turtles and determining the cause of their mortality, identifying new threats, and protecting critical marine habitats. But unlike with Kemp's ridley, the use issue further complicates matters for olive ridleys, and it will remain an issue for as long as the species gathers in great numbers.

CONTRIBUTORS

Randall Arauz
PRETOMA
1203-1100 Tibás
San José, Costa Rica

Joseph Bernardo
Department of Biology
College of Charleston
Charleston, South Carolina

Patrick M. Burchfield
Gladys Porter Zoo
Brownsville, Texas

Lisa Campbell
Nicholas School of the Environment and Earth Sciences
Duke University Marine Laboratory
Beaufort, North Carolina

Johan Chevalier
Office National de la Chasse et de la Faune Sauvage Simarouba
Kourou, French Guiana

Stephen E. Cornelius
Sonoran Institute
Tucson, Arizona

Michael Coyne
Nicholas School of the Environment and Earth Sciences
Duke University
Durham, North Carolina

Larry B. Crowder
Nicholas School of the Environment and Earth Sciences
Duke University Marine Laboratory
Beaufort, North Carolina

Angela Formia
School of Biosciences
Cardiff University
Cardiff, United Kingdom

Jack Frazier
Conservation and Research Center
Smithsonian Institution
Front Royal, Virginia

Jacques Fretey
Muséum National d'Histoire Naturelle
Paris, France

Matthew H. Godfrey
North Carolina Wildlife Resources Commission
Beaufort, North Carolina

Selina S. Heppell
Department of Fisheries and Wildlife
Oregon State University
Corvallis, Oregon

Aleta A. Hohn
Center for Coastal Habitat and Fisheries Research
NOAA Fisheries, Southeast Fisheries Science Center
Beaufort, North Carolina

Kristina Kichler Holder
Tallahassee, Florida

Mark T. Holder
School of Computational Science and Information Technology
Florida State University
Tallahassee, Florida

Heather J. Kalb
Biology Department
University of Evansville
Evansville, Indiana

Andre M. Landry, Jr.
Sea Turtle and Fisheries Ecology Research Laboratory
Texas A&M University at Galveston
Galveston, Texas

Duncan S. MacKenzie
Department of Biology
Texas A&M University
College Station, Texas

Rene Márquez-M.
Manzanillo, Colima, Mexico

Stephen J. Morreale
Department of Natural Resources
Cornell University
Ithaca, New York

Bivash Pandav
Wildlife Institute of India
Dehradun, Uttaranchal, India

Luis Jaime Peña
Gladys Porter Zoo
Brownsville, Texas

Pamela T. Plotkin
Office of Research and Sponsored Programs
East Tennessee State University
Johnson City, Tennessee

Peter C. H. Pritchard
Chelonian Research Institute
Oviedo, Florida

David C. Rostal
Georgia Southern University
Statesboro, Georgia

Kartik Shanker
c/o Madras Consultancy Group
Madras, India

Donna J. Shaver
Padre Island National Seashore
National Park Service
Corpus Christi, Texas

Melissa L. Snover
Pacific Fisheries Environmental Laboratory
NOAA Fisheries, Southwest Fisheries Science Center
Pacific Grove, California

Erich K. Stabenau
Biology Department
Bradley University
Peoria, Illinois

Roldán A. Valverde
Department of Biological Sciences
Southeastern Louisiana University
Hammond, Louisiana

Thane Wibbels
Department of Biology
University of Alabama at Birmingham
Birmingham, Alabama

INDEX

Page numbers followed by the letters *f* and *t* indicate figures and tables, respectively.

abundance, primordial, management for, 17–19

acaparadores, 26

acid–base balance, 120–26, 122t

ACTH. *See* adrenocorticotropic hormone

adder (*Vipera berus*), multiple mating in, 195

ADIO. *See* Asociación de Desarrollo Integral de Ostional

adrenal physiology, 128–38, 128f; salinity and, 136–37; stress and, 128–36

adrenocorticotropic hormone (ACTH), in stress response, 129, 131, 132, 133

Africa: arribadas in, 11–12, 15, 54; land-based harvest in, 233, 237–38; population stability in, 243; sea-based interactions in, 261–65

Agardy, T., 233

age at sexual maturation (ASM), 89; in captivity, 307–8; with mark–recapture data, 91, 91t, 92, 103; in population models, 195; with skeletochronology, 96–97, 103

age estimates, for Kemp's ridleys, 89–103; future research needs on, 102–3; with mark–recapture data, 89–92; with skeletochronology, 90, 92–103, 95t, 96f, 97f, 97t, 98t, 99f, 100f, 101f, 102f

Aguilar, H. R., 172

albatrosses, migration of, 215

albumin glands, 154–55, 159

albumin layer, 154–55, 158, 159, 159f

aldosterone, salinity and, 136–37

Amazon River, nesting on, 68

ammonia, in sand, 78

amphibians, reproductive synchrony in, 69

Andaman Islands: land-based harvest in, 234–35, 236f; sea-based interactions in, 269, 271, 272

Andhra Pradesh (India), sea-based interactions in, 267, 268, 270, 271–72

anesthesia, 125–26

Angola: land-based harvest in, 233; nesting in, 11

Annual Symposium on Sea Turtle Biology and Conservation, Nineteenth (1999), 4

annuli, 92, 94–95

anoxia, 334

aquaria, nesting at, 307, 308

Arauz, R., 34–36

Arauz Almengor, M., 39, 41

arginine vasotocin (AVT), 127, 130–31, 160
arribadas, 7–20, 59–79; changes in location of, 12–13; cues for, 62–67; definitions of, 3, 60–61, 119; density-dependent nest mortality in, 3, 76–78, 77f, 236; differences among, by location, 14; discovery of, 4, 7–8, 60–61; endocrine regulation of, 127, 128, 132, 141, 158; evolutionary perspective on, 12, 67–79; failure of, 14; film and photos of, 6f, 8, 10, 14, 58f, 60f, 237, 298; and fitness, 70–72, 75–79, 75t, 77f; gaps in understanding of, 3–4, 59, 338; geographic extent of, 15, 61–62; historical accounts of, 11; internesting intervals in, 151, 152, 160, 310; of Kemp's versus olive ridleys, 7–8, 10, 14, 15; management goals for, 13–14; management history of, 15–19; migration to, 221–24; number of individuals in, 61, 119; and predator satiation, 72–73; prehistoric, 254; remote locations of, 4, 11; reproductive physiology and, 160–62; shifting baseline for, 10–12; significance of, 12–13; social facilitation of, 66–67; versus solitary nesting, 73–79, 75t, 76t; stress response during, 130–35; synchronicity of, 61, 67–79; timing of, 61; zoogeography of, 52. *See also* egg collection; *specific nesting locations*
artisanal fishing, in Atlantic Ocean, 262
Asian trawlers, in Atlantic Ocean, 263
ASM. *See* age at sexual maturation
Asociación de Desarrollo Integral de Ostional (ADIO), 37, 38, 39
Atlantic Ocean: arribada research in, 9–11; arribada sites in, 15, 61, 62t, 64f, 65f; growth rates in, 91–92, 93, 94, 97, 98, 98t, 102–3; head-started turtles recaptured in, 303–4; land-based harvest along, 232–33, 237–38; in phylogeography, 109–11, 110f, 112f; population stability in, 237–38; sea-based interactions in, 255, 258–64; solitary nesting sites in, 61, 63t, 64f, 65f; species distribution in, 52–55
atresia, 159–60
Australia, sea-based interactions in, 273–76
AVT. *See* arginine vasotocin

Bajun, 264
Baluchistan (Pakistan): land-based harvest in, 235; sea-based interactions in, 264–65
Band 3 protein, 124
Bangladesh: land-based harvest in, 234, 239; sea-based interactions in, 268, 269, 271, 272
baselines, shifting, 10–12
basking turtles, stress response in, 130–33, 131f, 133f
Bastardschildkröte, 7, 45
bats, migration of, 215
Baur, G., 12, 46
beaches, arribada: characteristics of, 3; deterioration over time, 78; discovery of, 4, 11; locations of, 4, 11, 61–62, 62t, 63t, 64f, 65f, 74; migration to, 221–24; passive versus active selection of, 72. *See also specific beaches*
beaches, solitary nesting: location of, 61, 63t, 64f, 65f; migration to, 221–24. *See also specific beaches*
Bengal, Bay of: migration in, 219; sea-based interactions in, 267
Benin: land-based harvest in, 233; sea-based interactions in, 262
Bentley, T. B., 121, 125
Bernardo, J., 76
Bhaskar, S., 235
Bhupathy, S., 239
biogeography, 108–11
Bioko Island: land-based harvest on, 233; nesting on, 11
birds: migration of, 215; reproductive synchrony in, 69; thyroid hormones in, 141
Bjorndal, K. A., 98, 100
black turtle (*Chelonia agassizi*), 51, 52, 108
Bleakney, S., 52
blood: oxygen in, 125; in respiratory physiology, 120–26; in sex determination, 194, 196; in submergence studies, 120–24; thyroid hormones in, 139–40, 139t
body proportional hypothesis (BPH), 94–95
body size: bone dimensions and, 92; sex ratios and, 196–97, 197f, 200; shifts in distribution of, 331, 332f. *See also* growth
body temperature, and thyroid hormones, 142
bones, 48–51; in skeletochronology, 92–93
Booth, J., 158
bottom trawling, in Atlantic Ocean, 262
Bowen, B. W., 47, 55, 108, 109, 111, 312
BPH. *See* body proportional hypothesis
brain: in sex determination, 175; in stress system, 129
Brazil: arribadas in, 15, 54; land-based harvest in, 232, 237; population genetics in, 111; population size in, 240, 259; population stability in, 237; sea-based interactions in, 258–61
breeding: captive, 307–8; migrations for, 221–23; nearshore, 222–23; offshore, 223. *See also* reproduction
Brongersma, L. D., 52
Burke, V. J., 72
Butler, P. J., 125

Cahill, Tim, 26
Caillouet, C. W., Jr., 4, 55, 297, 306, 310, 314
calcium, in female reproductive physiology, 156
Cambodia, sea-based interactions in, 274
Cameroon: land-based harvest in, 233; sea-based interactions in, 261, 262, 263
Campbell, L. M., 37–39, 41, 42
Caporaso, Fred, 9
captive breeding, 307–8
captive nesting, 307–8
captive-reared turtles: hormone samples from, 126–27; respiratory physiology of, 120–26; submergence studies of, 120–24. *See also* head-started turtles
carapace, 48–49; of arribada versus solitary nesters, 76; functional interpretation of, 49; human use of, 233, 262; length of, in skeletochronology, 93–98, 96f, 97f; morphology of, 48–49; scratches on, in sex determination, 193; sex ratios and, 197, 197f
Caretta: evolutionary relationships of, 47; versus *Lepidochelys,* 12, 47; neural bones of, 49; plastra of, 49
Caretta caretta. See loggerhead sea turtle
Carettinae, 47
Carettini, 47
Carib. *See* Kali'na
Caribbean, 54
caribou, migration of, 215
Carr, A. F., Jr., 7–8, 12, 19, 29, 46, 47, 52–53, 60–61, 75, 76, 217, 240
Carr, David, 8
Carr, J. L., 54
carrying capacity: in arribada management, 19; changes to, 325, 334
case studies, strengths and weaknesses of, 24, 41
Casichelydia, 47
catecholamine hormones, 137
"cat-eye" appearance, of follicles, 159f, 160
Cayman Turtle Farm (CTF): endocrinology studies at, 127, 141; head-starting at, 298, 307–8, 313; management approach at, 19; reproductive studies at, 113–14, 152
CCL. *See* curved carapace length
central nervous system (CNS), and stress, 128–29
cervical vertebrae, 51
Chacocente (Nicaragua): arribadas in, 12, 15, 30; politics of egg collection at, 29–34, 36–37; productive conservation at, 30, 32; sea-based interactions in, 280

Chacocente–Rio Escalente Wildlife Refuge, 30, 34–35
Chada, S., 268
Chaloupka, M. Y., 90, 92, 93, 95, 100–101, 195
Chamorro, Violetta, 33
Chan, E. H., 275
Chaves, Anny, 38
Chávez, Humberto, 52, 326
Chelonia (genus): evolutionary relationships of, 47; plastra of, 49
Chelonia (order), 46
Chelonia agassizi. See black turtle
Chelonia (Natator) depressa, 46
Chelonia mydas. See green sea turtle
Chelonian Conservation and Biology (journal), 4
Chelonia olivacea, 12
Cheloniidae, 47–48
Cheloniinae, 47
Chelonini, 47
Chelonioida, 47
Chelydra serpentina. See snapping turtle
Chesapeake Bay, 52
Chesser, R. K., 195, 205
Chile, sea-based interactions in, 278
Chin, C. C. Q., 67
China, sea-based interactions in, 276
Choudhury, B. C., 70, 239
Christianity, and turtle meat, 234–35, 265
chromaffin cells, 137
Chrysemys picta. See painted turtle
CITES, 278
Cliffton, K., 25
climate, in migration, 216
Clusella, S., 240
clutch size: of arribada versus solitary nesters, 76; in captivity, 308; of head-started turtles, 308, 310–11
CNS. *See* central nervous system
Cocibolca, Fundación, 34, 35, 42
coded wire tag (CWT) experiment, age and growth data from, 90–91, 93, 102
Colombia, sea-based interactions in, 278, 280
colonial nesting: rarity of, 73; among turtle species, 60, 68. *See also* arribadas
Colpochelys, 45, 46
Colpochelys kempii, 45
commerce. *See* market
commercial fishing, versus industrial fishing, 25
compensatory growth, 98–100
Concho water snake (*Nerodia harteri paucimaculata*), 198
Conoco-Phillips Petroleum Company, 328
conservation efforts: accessibility in, 191; age and growth data from, 90–91, 102; changes in arribadas and, 13; cultural issues in, 40; through egg collection, controlled, 241; through egg purchasing, 10, 40; future of, 333–34; hatcheries in, 241–42; history of, 326–29, 328t; IUCN definition of, 23–24; for Kemp's versus olive ridleys, 24; land-based harvest and, 242–45; politics of, 30–36, 316; population genetics and, 108, 111; productive, 30, 32; reproductive physiology and, 160–62; respiratory physiology and, 143; sex determination and, 167, 170–71, 191–93; sex ratio manipulation in, 170–71, 193, 195, 202, 204–6; sex ratios influencing, 167, 168, 170–71, 176, 192–93; shifting baseline and, 10–12; at solitary nesting sites, 75; status of, 329–30; success of, 330–33; thyroid physiology and, 138, 143
consumptive use, sustainability of, 24
Contras (Nicaragua), 30
Contreras, Martin, 326
cooperatives: egg, in Nicaragua, 32–33; fishing, in Mexico, 25–26, 27
Cooray, R., 270
Cornelius, S. E., 4–5, 14, 15, 61
corruption, government, 27
cortex, of ovary, 174, 174f
corticosteroids, in stress response, 130, 131
corticosterone: basal values for, 130; in oviposition, 160; salinity and, 136–37; in stress response, 129–37, 131f, 133f, 134f, 135f
corticotropin-releasing hormone (CRH): in stress response, 129, 131, 133, 133f, 135; and thyroid hormones, 138
cortisol, in stress response, 129
costal bones, 48–49
Costa Rica: arribadas in, 12, 14, 15; ecotourism in, 38–39; egg collection in, 37–39, 41, 233; migration near, 219, 220f; sea-based interactions in, 279–80, 282. *See also specific nesting locations*
courtship, 157–58
Coyne, M. S., 197, 199, 200
crabs, 305
CRH. *See* corticotropin-releasing hormone
cryptic female choice, 71
Cryptodira, 46, 47
CTF. *See* Cayman Turtle Farm
cues: for arribadas, 62–67; biology of processing, 67; for reproductive synchrony, 69
Cuixmala Beach (Mexico), 176
cultural aspects: of egg collection, 39–40; of turtle meat, 233, 234, 264, 265
currais, 260
currents. *See* ocean currents
curved carapace length (CCL): sex ratios and, 197; in skeletochronology, 94
Customs Department, U.S., 256
CWT. *See* coded wire tag
Daiocryptodira, 47
Danton, C., 197
Darden Restaurant Foundation, 329
Das, I., 235, 268
Dash, M. C., 4, 268
Dattatri, S., 235, 267
Dead Zone, 334
deiodinases, 139–40
density, of Pacific Ocean arribadas, 62t, 69f
density-dependent nest mortality, 3, 76–78, 77f, 236
Deraniyagala, P. E. P., 12, 53, 254
Dermochelyidae, 47
Dermochelys coriacea. See leatherback turtle
De Silva, A., 268
Devi (India), sea-based interactions in, 266–67, 268
diamondback terrapin (*Malaclemys terrapin*), 68
diet, of head-started turtles, 305, 315. *See also* feeding
differential selection, in arribada versus solitary nesters, 76–78
directed take: in Atlantic Ocean, 259, 261–62; in Gulf of Mexico, 255–56; in Indian Ocean, 264–65, 267–69; of Kemp's ridleys, 255–56; of olive ridleys, 259–79; in Pacific Ocean, 276, 277–79; in Sunda Sea, 274. *See also* fishing
dispersal, versus migration, 215
distribution. *See* geographic distribution
DNA analysis: mitochondrial, 108, 109, 111, 115; nuclear, 111, 113, 115; and zoogeography, 55
dorsal vertebrae, 48
Dutton, P. H., 108

E_2. *See* estradiol
Eckrich, C. E., 72–73, 76
ecology: of migration, 215–16; sex ratios and, 170
economics: in Costa Rica, 37–39; of ecotourism, 38; of egg collection, 24–39, 241; of fishing, 24–28; in Honduras, 28–29; in Mexico, 24–28; in Nicaragua, 29–37; in recovery efforts, 329
ecotourism: arribadas in, 20, 27; in Costa Rica, 38–39; economics of, 38; in Mexico, 27; in Nicaragua, 35
Ecuador: fishing in, 238, 278, 281; sea-based interactions in, 278, 280, 281
education, in recovery efforts, 328–29
egg(s): development of, 155, 159–60, 159f; thyroid hormones in, 141
egg collection: on Atlantic Ocean, 232–33, 237–38; in conservation strategies, 10, 24, 32; controlled, 241; in Costa Rica, 37–39, 41, 233; cultural aspects of, 39–40; economics of, 24–39, 241; hatcheries and, 241–42; history of, 231–36; in Honduras, 28–29, 233;

egg collection (*continued*)
maximum sustainable yield of, 16–17; in Mexico, 24–28, 233–34, 238; nature of, 231–36; in Nicaragua, 29–37, 41, 238, 241; on Pacific Ocean, 233–34, 238; politics of, 29–37; and population stability, 13, 236–40, 243, 244t; versus purchasing eggs for conservation, 10, 40; and significance of arribadas, 12; social aspects of, 37–39; in Suriname, 39–40, 232, 237
egg laying. *See* oviposition
egg mortality. *See* nest mortality
egg predation, in predator satiation hypothesis, 72–73
egg traders/sellers: in Mexico, 26, 28–29; in Nicaragua, 31–33, 36
Eilanti Beach (Suriname): arribadas on, 9–10, 13, 15; egg collection on, 40; land-based harvest at, 232; management approach at, 19; winds on, 14
El Salvador: hatcheries in, 242; land-based harvest in, 233, 238; population stability in, 238; sea-based interactions in, 279–80; sex ratios in, 176
Emys orbicularis. See European pond turtle
endangered species: Kemp's ridley as, 151; mating systems of, 113; temperature-dependent sex determination in, 167
endocrine physiology, 126–43; adrenal, 128–38; in oviposition, 127, 128, 131, 160; in reproductive behavior, 127, 128, 132, 141, 158; in sex determination, 175; techniques for studying, 126–28; thyroid, 138–43
energy storage, stress and, 136, 137
enforcement, law, 25, 27, 241, 264
entanglement net capture stress: and endocrine physiology, 131–32, 137; and respiratory physiology, 126
entoplastra, 49–50
environmentalism, in Nicaragua, politics of, 30
epididymis, 152
epinephrine, in stress response, 137
epiplastra, 50
"epithecal" bone, 49
Eretmochelys, plastra of, 49
Eretmochelys imbricata. See hawksbill turtle
Eritrea, sea-based interactions in, 265
erythrocytes, 124
Eschscholtz, J. F., 12
estradiol (E_2): in ovarian cycle, 156–57, 157f; in oviposition, 160; in vitellogenesis, 155
estrogen, in sex determination, 175, 192
Eucyptodira, 47
European pond turtle (*Emys orbicularis*), 194
evolution: and arribadas, 12, 67–79; of reproductive polymorphism, 73–78; of reproductive synchrony, 69–73; and temperature-dependent sex determination, 169–70
evolutionary relationships, 12, 46–48, 108
exploitation, connotations of term, 254
extinction, of Kemp's ridley: near, research and recovery influenced by, 4, 151, 337–39; probability of, 330, 331f
extremities, osteology of, 50–51

Faber, D., 30, 31, 36
fecundity: definition of, 162; of head-started turtles, 310–11; in population models, 201, 203–4, 204f, 205
fecundity index, 161–62
feeding: adrenal physiology and, 136; and evolutionary relationships, 48; by head-started turtles, 305, 315; and recovery efforts, 334; thyroid physiology and, 142
female(s): gonadal differentiation in, 174–75, 174f; migration by, 216–19, 218f, 220f, 221–24; multiple mating's effects on, 70–71; reproductive physiology of, 154–63; reproductive polymorphisms in, 74; reproductive strategies of, 70–71; secondary sex characteristics of, 193; stress response in, 130–34, 131f, 134f, 135f; thyroid physiology of, 140–41
female choice, cryptic, 71
Female:Male:Female (FMF) pattern, 168, 169f
femur, 50, 51
fertilization, 154
fertilizers, 272
Figler, R. A., 127, 160
film, arribada, 8, 10, 14, 237, 298
Fish and Wildlife Service, U.S. (USFWS), 18, 314
fishing, turtle: in arribada management strategies, 16; in Atlantic Ocean, 259, 261–62; in conservation strategies, 16, 24; economics of, 24–28; government management of, 25–26, 27, 277; for green turtles, 232, 261, 265, 268, 269; in Gulf of Mexico, 255–56; in Indian Ocean, 235–36, 264–65, 267–69; industrial versus commercial, 25; for Kemp's ridleys, 255–56; in Mexico, 16, 24–28, 234, 255–56, 277–78, 281; for olive ridleys, 259–79; in Pacific Ocean, 276, 277–79; and population stability, 238, 240; in Sunda Sea, 274
fitness: of arribada versus solitary nesters, 73, 75–79, 75t, 77f; of head-started turtles, 315; reproductive polymorphism and, 73, 75–78; reproductive synchrony and, 69–72, 75–79; and sex ratios, 170
Fitzinger, L. J., 12
flatback turtle (*Natator depressus*): evolutionary relationships of, 46, 47; migration by, 214, 221
Florida: head-starting in, 298, 315; in zoogeography, 54–55
Florida Department of Natural Resources, 298
FMF. *See* Female:Male:Female
follicles. *See* ovarian follicles
follicle-stimulating hormone (FSH), 127
Fonseca, Golfo de: arribadas in, 12, 28; egg collection in, 28–29
Fontaine, C. T., 92
fontanelles: intercostal, 48, 49; lateral, 50
foraging migrations, 215, 217–19, 221–24. *See also* feeding
forelimbs, 50
forensic science, 115
French Guiana: arribadas in, 15, 259; land-based harvest in, 232, 236, 237; population stability in, 236, 237, 259; sea-based interactions in, 258–61
Fretey, J., 11, 48, 51, 54, 260
FSH. *See* follicle-stimulating hormone

Gadow, H., 46
Gaffney, E. S., 47
Gahirmatha (India): arribadas on, 11, 15; land-based harvest at, 235, 239; migration to, 222; population stability at, 239; research on, 4; sea-based interactions in, 266–67, 268, 270; sex determination in, 171, 172; sex ratios at, 176
Galápagos Islands, 7, 53, 280
Galveston (Texas). *See* National Marine Fisheries Service Laboratory
GAM. *See* generalized additive models
Garman, S., 45, 46, 51
Geijskes (1945), 40, 232
generalized additive models (GAM), 91
genetics: of arribada versus solitary nesters, 76; of multiple mating, 70–71, 113–15, 114t, 195; in sex determination, 175, 183. *See also* population genetics
genotypes, 109, 113–14
geographic distribution: in Atlantic Ocean, 52–55; and gaps in research, 4; in Gulf of Mexico, 52, 54; in Indian Ocean, 53; in Pacific Ocean, 53; and sea-based interactions, 253, 255
GH. *See* growth hormone
Ghana, land-based harvest in, 233
gillnets, mortality from: in Atlantic Ocean, 260, 262; in Gulf of Mexico, 257; in Indian Ocean, 271, 272; in Pacific Ocean, 279, 280; in Sunda Sea, 275
Girondot, M., 76–77

Gist, D. H., 195
glucocorticoids: functions of, 129; regulation of, 129; in stress response, 129, 130, 136
glucose, stress and, 135–36, 137, 138
Glyptochelone suyckerbuyki, 48
Goa (India), sea-based interactions in, 265, 266
Godfrey, M. H., 195
gonads: differentiation in, 174–75, 174f; in sexing of turtles, 192, 194, 196, 197–98. *See also* ovaries; testes
gopher tortoise (*Gopherus polyphemus*), 130
Grande, Playa, 13
granulosa cells, 155
Greater Sunda Sea, sea-based interactions in, 273–76
green sea turtle (*Chelonia mydas*): adrenal physiology of, 130, 132, 135; anesthesia in, 125; in captivity, 308; density-dependent nest mortality in, 76; fishing of, 232, 261, 265, 268, 269; head-starting of, 298, 312–13; historical accounts of, 11; in Indian Ocean, 265, 267, 268, 269; intraspecific divergence in, 108; migration by, 217, 221, 224; multiple mating in, 70–71, 114; nesting behavior of, 18, 68, 308; phylogenetics of, 45, 108; reproductive physiology of, 152, 153, 158, 162; respiratory physiology of, 125; sexing of, 192, 194; stress response in, 135; submergence studies on, 121; thyroid physiology of, 138; zoogeography of, 51
Green Turtle and Man, The (Parsons), 11
Gregory, L. F., 197
growth, pituitary hormones in, 127–28
growth hormone (GH), 127–28
growth rates, for Kemp's ridleys, 89–103; future research needs on, 102–3; head-started, 89, 90–91, 92, 93, 102, 304–5; with mark–recapture data, 89–92; patterns in, 98–102; with skeletochronology, 90, 92–103, 95t, 96f, 97f, 97t, 98t, 99f, 100f, 101f, 102f
Guam, sea-based interactions in, 276
Guanacaste Conservation Area (Costa Rica), 5, 15
Guatemala: hatcheries in, 242; land-based harvest in, 233, 238; sea-based interactions in, 278, 279–80
Guéguen, F., 259–60
Guerrero (Mexico), 16
Guianas. *See* French Guiana; Guyana; Suriname
Guinea, Gulf of, sea-based interactions in, 262, 263
Guinea-Bissau: land-based harvest in, 233; nesting in, 11; sea-based interactions in, 263
Gujarat (India): hatcheries in, 242; sea-based interactions in, 265–66
Guyana: arribadas in, 9–10, 15, 53; land-based harvest in, 232, 237; population stability in, 237; sea-based interactions in, 259

habitat utilization: in migrations, 213–24; and sea-based interactions, 253–54
halothane, 125
Hamann, M., 137, 152, 155, 156
haplotypes, 108–13, 110f
harvest, connotations of term, 254. *See also* fishing; land-based harvest
hatcheries, 241–42; effectiveness of, 241–42; at Rancho Nuevo, 176–79, 177f, 242, 326; sex ratios in, 176–79
hatchlings: and adult migration constraints, 216–17; of arribada versus solitary nesters, 76; gonadal differentiation in, 174–75, 174f; ovarian cycle and, 157; in predator satiation hypothesis, 72–73; sex ratios of, 176–79, 183
Hawaiian Islands, sea-based interactions in, 279, 281
hawksbill turtle (*Eretmochelys imbricata*): adrenal physiology of, 132; challenges in conservation of, 13; head-starting of, 298; intraspecific divergence in, 108; migration by, 214; phylogenetics of, 47, 108; reproductive physiology of, 152; sexing of, 194
Haynes, S. P., 139
Hays, G. C., 71
HDT. *See* high disturbance threshold
head-started turtles, 297–318; adaptation of, 306; age and growth data from, 89, 90–91, 92, 93, 102, 304–5; in arribada management strategy, 18–19; controversy, 297, 298, 314–17, 337; cost of, 316; definition of, 297; diet of, 305, 315; duration of captivity, 297; imprinting in, 299, 300, 316, 338; methodology for, 301; movements of, 303–4, 306, 312; nesting of, 307–13, 309t; non-ridley species of, 298, 312–13; origins of program, 298–300, 326–27; outcome of, 18, 313–14, 337–38; rationale for, 297; recapture locations of, 303–4; recapture methods for, 303; recapture rate of, 303, 306; release locations for, 299f, 302–4, 302t; reviews of program, 313–14; sex ratios of, 179–81, 196, 197, 300; source of hatchlings for, 298, 300; survival of, 306–7; and zoogeography, 53, 55
HEART program, 328
Heck, J., 141
Heming, T. A., 124, 125
Henderson–Hasselbalch constants, 124–25
Hendrickson, J. R., 54
Hernandez, Eduardo, 326
Herrera, Andres, 8, 10, 14
high disturbance threshold (HDT), endocrinology of, 135
Hildebrand, H. H., 7–8, 17, 46, 60, 162, 237, 255, 312
Hill, R. L., 54
hindlimbs, 50–51
Hinduism, 234, 264, 265, 268
Hochachka, P. W., 121
Hoekert, W. E. J., 113, 114, 115
Hohn, A. A., 93, 94
homeostasis, disruption of. *See* stress
Honduras: economics of human use in, 28–29; egg collection in, 28–29, 233; hatcheries in, 242; land-based harvest in, 233; sea-based interactions in, 279–80
Hoopes, L. A., 126, 137
Hope, R. A., 30, 31, 34–36, 37–39, 41
hormones: research on, 126–27; techniques for measuring, 127–28. *See also* endocrine physiology
HPA. *See* hypothalamo-pituitary-adrenal
Huaves, 233
hueveros, 12
Hughes, D. A., 72
Hughes, G. R., 194
human use, 23–42; in arribada management strategies, 16–17; consumptive versus non-consumptive, 24; in definitions of conservation, 23–24; economics of, 24–39; enforcement of laws against, 25, 27, 241; history of, 231–36; in Honduras, 28–29; of Kemp's versus olive ridleys, 24; in Mexico, 24–28; nature of, 231–36; in Nicaragua, 29–37; politics of, 29–37; sustainability of, 23–24. *See also* egg collection; land-based harvest; sea-based interactions
humerus, 50; in skeletochronology, 92, 93, 93f, 94–95
H-Y antigen assay, in sexing of turtles, 194
hybrid theory, 45–46, 254
hyoplastra, 49–50
hypophagia, 136, 138
hypoplastra, 49–50
hypothalamo-pituitary-adrenal (HPA) axis: organization of, 128f, 129; salinity and, 136; stress and, 129–35, 136, 137
hypothalamo-pituitary-thyroid axis, 128f, 138, 140

IAC. *See* Inter-American Convention for the Protection and Conservation of Sea Turtles
IATTC. *See* Inter-American Tropical Tuna Commission
immature turtles: sexing of, 181, 193–94; sex ratios of, 179–82, 180t, 183–84. *See also* juveniles

imprinting, 299, 300, 316, 338
inbreeding, 71
incentives, economic, for illegal human use, 27
incidental capture: in Atlantic Ocean, 259–61, 262–63; in Gulf of Mexico, 256–57; in Indian Ocean, 265–66, 269–73; of Kemp's ridleys, 255, 256–57; of olive ridleys, 259–81; in Pacific Ocean, 276, 279–81; in Sunda Sea, 274–75. *See also* trawler mortality
Inconel tags, 90
incubation, thermosensitive period of, 168, 171
India: arribadas in, 4, 11, 14, 15; hatcheries in, 242; head-starting in, 298; land-based harvest in, 234–36, 236f, 239; migration to and from, 219, 222; population genetics in, 111; population stability in, 238–39; sea-based interactions in, 264–73. *See also specific nesting locations*
Indian Ocean: arribada sites in, 15, 61, 62t, 64f; fishing in, 235–36, 264–65, 267–69; land-based harvest along, 234–36, 238–39; in phylogeography, 109–11, 110f, 112f; population stability in, 238–39; sea-based interactions in, 264–73; solitary nesting sites in, 61, 63t, 64f; species distribution in, 53
Indian Ocean and South-East Asia (IOSEA), Memorandum of Understanding on the Conservation and Management of Marine Turtles and Their Habitats of the, 258
Indonesia, sea-based interactions in, 274
industrial fishing: in Africa, 262; in Mexico, 25–26
inner ring deiodinases (IRD), 139–40
Instituto Nicaragüense de Recursos Naturales y del Ambiente (IRENA), 30, 31–33
insurance hypothesis, 71
interactions, human–turtle: connotations of term, 254; gaps in knowledge of, 254–55; prehistoric, 254–55. *See also* land-based harvest; sea-based interactions
Inter-American Convention for the Protection and Conservation of Sea Turtles (IAC), 258
Inter-American Tropical Tuna Commission (IATTC), 279, 281
international agreements, 257–58
International Commission for Zoological Nomenclature, 46
International Symposium on Kemp's Ridley Sea Turtle Biology, Conservation and Management, First (1985), 4
internesting intervals, 151, 152, 160, 310
ion transport, 124–26, 128
IOSEA. *See* Indian Ocean and South-East Asia
IRD. *See* inner ring deiodinases
IRENA. *See* Instituto Nicaragüense de Recursos Naturales y del Ambiente
Islam, 234, 264, 265, 269
isoflurane, 125
IUCN. *See* World Conservation Union
Ixtapilla, Playa (Mexico), 17

Jaffna (Sri Lanka), sea-based interactions in, 267
Jaguars Ate My Flesh (Cahill), 26
Jalisco (Mexico): arribadas in, 16; hatcheries in, 242
Japan, sea-based interactions in, 276
jaws, 51
Jenkins, Jorge, 30
Jiménez Gómez, G., 38
Jiménez-Quiroz, M. C., 62–65
Johnson, S. A., 312
Jones, J. M., 195
Jumbo Roto tags, 90
juveniles: growth rates for, 89–103; osteology of, 48–51; sexing of, 181, 193–94; sex ratios of, 179–82, 180t, 183–84; stress response in, 131–32, 135–36; zoogeography of, 52–53

Kalb, H. J., 76, 222
Kali'na peoples, 39–40
Kapurusinghe, T., 270
Kar, C. S., 4, 268
Karl, S. A., 47, 55
Karnataka (India), 266
Kaufmann, R., 52
Kemp, Richard M., 45
Kemp's Ridley Recovery Program: establishment of, 298; growth data from, 90; head-starting in, 298; objectives of, 298–99
Kemp's ridley sea turtle (*Lepidochelys kempii*): gaps in understanding of, 4–5, 337–39; phylogenetics of, 12, 45–48, 108
Kemp's Ridley Working Group, 313–14, 326
Kenya, sea-based interactions in, 264, 265
Kerala (India), 266
Kichler, K., 113, 114, 115
kin recognition, 71
Klima, E. F., 299
Kloos, P., 39, 40

lactate, blood, 120–21, 126
La Escobilla (Mexico): adrenal physiology studies at, 130; economics of egg collection at, 26–27; hatcheries at, 242; illegal human use at, 24–25, 26–27; land-based harvest at, 238, 240; management of, 16, 240, 242; population stability at, 238, 240, 243
La Flor (Nicaragua), 12, 15; egg collection at, 29–30, 34–37, 241; law enforcement at, 241; politics of, 29–30, 34–37
LAG. *See* lines of arrested growth
Lagueux, C., 12, 28–29
land-based harvest, 231–45; on Atlantic Ocean, 232–33, 237–38; history of, 231–36, 254–55; impact on global populations, 236–45, 244t; on Indian Ocean, 234–36, 238–39; nature of, 231–36; on Pacific Ocean, 233–34, 238; and sea-based interactions, 254. *See also* egg collection
land-based stressors, 239
Landry, A. M., Jr., 4
laparoscopy, sex determination with, 192, 194, 196
laws. *See* legislation
Lazell, J. D., Jr., 53
leather, olive ridley, 25
leatherback turtle (*Dermochelys coriacea*): density-dependent nest mortality in, 76–77; intraspecific divergence in, 108; migration by, 213–14; multiple paternity in, 195; nesting sites of, 11–12; olive ridley confused with, 12; reproductive behavior of, 72, 78; reproductive physiology of, 152, 156; respiratory physiology of, 125; sex determination in, 168
Lee, P. L., 71
legislation, on human use: in Costa Rica, 37–38; enforcement of, 25, 27, 241, 264; in Mexico, 24–28, 256, 277; in Suriname, 40, 232; in United States, 256, 257, 260
length-at-age, 96, 97, 97t
Lepidochelys, evolutionary relationships of, 46–48
Lepidochelys kempii. See Kemp's ridley sea turtle
Lepidochelys olivacea. See olive ridley sea turtle
LH. *See* luteinizing hormone
Licht, P., 158
Liew, H. C., 275
life histories: of arribada versus solitary nesters, 76; gaps in understanding of, 191, 194–95; in population models, 191, 194–95
limbs, 50–51
Limpus, C. J., 47, 195, 221, 242, 275
Limpus, D. J., 221
lines of arrested growth (LAG), 92, 93, 93f, 94–95
living tags, 301
loggerhead sea turtle (*Caretta caretta*): anesthesia in, 125; growth patterns in, 98, 100; head-starting of, 312; intraspecific divergence in, 108; migration by, 221, 224; multiple paternity in, 195; phylogenetics of, 45, 46, 108; population models for, 195; reproductive physiology of, 152, 153, 155, 160, 162; respiratory physiology of, 121, 123–24, 125; ridleys confused with, 46, 254; sex determination in, 168; sexing of, 192, 194; stress response in, 134–35,

135f; submergence studies on, 121, 123–24; winter torpor in, 52; zoogeography of, 52, 53
Long Island Sound (New York), 53
long line fishing, 280
lunar cycles, and arribadas, 65–66
lungs, 125
luteinizing hormone (LH), 127; in oviposition, 160; in ovulation, 158
Lutz, P. L., 121, 125

Madagascar, sea-based interactions in, 264, 265
Madras (India): hatcheries in, 242; land-based harvest in, 235
Madsen, T., 195
Magnuson, J. J., 307
Maharashtra (India), sea-based interactions in, 266
Maigret, J., 53
Malaclemys terrapin. *See* diamondback terrapin
Malaysia: land-based harvest in, 239; sea-based interactions in, 273–74
Maldives, sea-based interactions in, 266
Male:Female (MF) pattern, 168, 169f
males: gonadal differentiation in, 174–75, 174f; migration by, 218f, 219–21, 220f, 222–24; reproductive physiology of, 152–53, 162; reproductive polymorphisms in, 74; reproductive strategies of, 70; secondary sex characteristics of, 152, 193; stress response in, 130, 132, 134, 134f; thyroid physiology of, 140–41
Mannar, Gulf of, sea-based interactions in, 267, 268
Manzella, S. A., 303, 304
MARENA. *See* Ministerio de Agricultura y Reforma Agraria
Marine Turtle Specialist Group, IUCN, 24
market: development of, 232; for eggs, in Central America, 28–29, 31, 39; for meat, in Africa, 233
mark–recapture data: age and growth estimates from, 89–92; fecundity estimates from, 162; for head-started turtles, 303; methodology of, 90–91; problems with, 90, 162; recapture interval in, 90, 91
Márquez-M., R., 4, 233, 326
Massachusetts, 53
Mast, R. B., 54
mate finding, 70, 71
maternal care, 72
mating: female reproductive physiology and, 157–58; migrations for, 221–24
mating systems: and population genetics, 113–15; reproductive synchrony and, 69–72. *See also* multiple mating; reproduction
Mauritania, sea-based interactions in, 262
maxillae, 51
maximum sustainable yield, management for, 16–17
McCoy, C. J., 174
McPherson, R. J., 153
McVey, J. P., 299
measurement errors, 90
meat, turtle: in Africa, 233, 262; in arribada management strategies, 16; export from Suriname, 39–40; religion and, 233, 234, 264, 265, 268, 269
medullary cords, of ovary, 174, 174f
Memorandum of Understanding on the Conservation and Management of Marine Turtles and Their Habitats of the Indian Ocean and South-East Asia (IOSEA), 258
Mendonça, Mary, 8
Merchant-Larios, H., 174
mesoplastra, 50
Metachelydia, 47
meteorological cues, for arribadas, 14, 62–65
Mexico: arribadas in, 8–9, 13, 15; economics of human use in, 24–28; egg collection in, 24–28, 233–34, 238; fishing in, 16, 24–28, 234, 255–56, 277–78, 281; hatcheries in, 242; head-started turtles nesting in, 310–11; history of human use in, 233; in Kemp's Ridley Recovery Program, 298; land-based harvest in, 24–28, 233–34, 238; legislation in, 24–28, 256, 277; population stability in, 238, 243; sea-based interactions in, 255–56, 276–78, 279, 281; social facilitation in, 66. *See also specific nesting locations*
Mexico, Gulf of: arribadas in, 8; Dead Zone in, 334; growth rates in, 91–92, 93, 102–3; head-started turtles released in, 302–4, 306; migration in, 214, 217–19, 218f, 221, 222; sea-based interactions in, 255–58; species distribution in, 52, 54
MF. *See* Male:Female
migration(s), 213–24; to breeding grounds, 221–24; definition of, 215; ecological influences on, 215–16; of head-started turtles, 310, 312; nesting constraints on, 216–17; of other turtle species, 213–14; in phylogeography, 109–11; and population genetics, 111, 115; postreproductive, 217–21, 218f, 220f; and sex ratios, 198–99; social facilitation of, 66; testosterone in, 198–99, 199f; variation in, 213–14, 215, 223
Ministerio de Agricultura y Reforma Agraria (MARENA, Nicaragua), 33, 34–35
mitochondrial DNA (mtDNA), 108, 109, 111, 115
Mohanty-Hejmadi, P., 11
Monge Artavia, K., 38
monopoly, 26
monopsony, 26
moon, and arribadas, 65–66
Moon, D.-Y., 140, 142
Moon, P. F., 125
Moore, M. K., 115
Mora, J. M., 67
morphology, 48–51; of arribada versus solitary nesters, 75; of Kemp's versus olive ridleys, 46, 108; of loggerheads, 46; sex determination through, 174–75
Morris, Y. A., 136, 137, 199
Morro Ayuta (Mexico), 16
mortality: mass, and arribada management, 16; and sex ratios, 198. *See also* nest mortality; trawler mortality
Mozambique, sea-based interactions in, 264, 265
Mrosovsky, N., 195
mtDNA. *See* mitochondrial DNA
Muccio, C., 242
multiple mating: and fitness, 70–72, 78–79; and population genetics, 70–71, 113–15, 114t, 195; in population models, 203–4
muscles, oxygen in, 125
Musick, J. A., 52, 90, 92, 95
Myanmar: land-based harvest in, 235, 239; sea-based interactions in, 267

Nancite Beach (Costa Rica), *60;* adrenal physiology studies at, 130, 132; beaches adjacent to, 74, 74f; decline in arribadas at, 15; hatching success at, 12, 15; internesting intervals in, 160; lunar cycles and, 65; management of, 15–16, 17; mate finding at, 70; migration to, 222; versus Ostional arribadas, 14, 65; population stability at, 236, 243; predation at, 72–73; Rathke's glands and, 67; research on, 5, 9; sex determination at, 171, 172; social facilitation at, 66; versus solitary nesting, 76; stress response in, 132
Nandi, N. C., 269
Naranjo, Playa (Costa Rica), 74, 74f
natal homing: and inbreeding, 71; and population genetics, 111–13
Natator, evolutionary relationships of, 46, 47
Natator depressus. *See* flatback turtle
National Academy of Sciences, 256, 314, 316
National Marine Fisheries Service (NMFS), U.S., 18, 115
National Marine Fisheries Service Laboratory (Galveston): endocrinology studies at, 130, 140–41; head-starting at, 298, 299, 300; respiratory studies at, 120
National Research Council, 121
natural selection, and sex ratios, 170
Navy, U.S., 257

necropsy, sex determination with, 194, 196, 198
Nerodia harteri paucimaculata. See Concho water snake
nesting behavior: in captivity, 307–8; carapace in, 49; discovery of, 4; endocrine regulation of, 127, 128; in head-started turtles, 307–13, 309t; migration constrained by, 216–17; migration for, 221–24; of other turtle species, 60, 68; polymorphism in, 4, 60, 73–78; reproductive physiology and, 160–62; stress response during, 130–35, 133f, 135f; zoogeography of, 52. *See also* arribadas; solitary nesting
nesting sites. *See* beaches; *specific sites*
nest (egg) mortality: density-dependent, 3, 76–78, 77f, 236; in predator satiation hypothesis, 72–73
nets: capture stress with, 126, 131–32, 137; mesh turtle, 261
neural bones, 48, 49
neurohypophysis, 130–31
neurophysin (NP), 127, 160
New York, 53
Nicaragua: arribadas in, 12, 15; economics and politics of human use in, 29–37; ecotourism in, 35; egg collection in, 29–37, 41, 238, 241; population stability in, 238; sea-based interactions in, 279–80. *See also* Chacocente; La Flor
Nicobar Islands: land-based harvest in, 234–35, 236f; sea-based interactions in, 269, 271
Nigeria, sea-based interactions in, 263
Ningo ethnic group, 233
Nixtayolero, 35
NMFS. *See* National Marine Fisheries Service
norepinephrine, in stress response, 137
North Carolina, 53, 54–55
NP. *See* neurophysin
nuchal bone, 49
nuclear genes, 111, 113, 115

Oaxaca (Mexico), 9; arribadas in, 15, 16, 24–25; fishing in, 16, 25; human use in, 16, 24–25; management approach in, 16. *See also* La Escobilla
ocean-based interactions. *See* sea-based interactions
ocean currents: and adult migration, 216–17; passive displacement by, 72; and zoogeography, 54, 55
offspring: in predator satiation hypothesis, 72–73; of solitary versus arribada nesters, 72–73, 75–76. *See also* hatchlings
oil industry: in Atlantic Ocean, 263; in Indian Ocean, 272; in Sunda Sea, 275
oil spills, 257, 263
olfaction, in arribadas, 67
olive ridley sea turtle (*Lepidochelys olivacea*): gaps in understanding of, 4–5, 12, 46, 338–39; leatherback turtle confused with, 12; loggerhead turtle confused with, 46, 254; phylogenetics of, 12, 46–48, 108
Oman, sea-based interactions in, 265
orbit, 51
ORD. *See* outer ring deiodinases
Orissa (India): arribadas in, 11, 14, 15, 16; land-based harvest in, 235–36, 239; management approach in, 16; mate finding in, 70; sea-based interactions in, 266–69, 270–73
Ortiz, R. M., 136–37
osmoregulation, 128, 136–37
osteology, 48–51
Osteopygis, carapace of, 49
Ostional (Costa Rica): ecotourism in, 38–39; egg collection at, 17, 37–39, 41, 233, 234f, 241, 243; hatching success at, 12, 17; history of arribadas at, 13; lunar cycles and, 65; management of, 17, 241, 243; versus Nancite arribadas, 14, 65; population stability at, 240, 243; sea-based interactions in, 280
Ostional Wildlife Refuge, 37
outer ring deiodinases (ORD), 139–40
Outside Magazine, 26
ova: development of, 159–60, 159f; in ovarian cycle, 157; in reproductive system, 155. *See also* egg
ovarian cycle, 156–57, 157f, 162
ovarian follicles: anatomy of, 154–55, 155f; in egg development, 159–60, 159f; in nesting patterns, 161, 161f; in ovarian cycle, 156–57; in ovulation, 158; vitellogenesis in, 155–56
ovaries: anatomy of, 154–55, 155f; in ovarian cycle, 156–57; sexual differentiation in, 174–75, 174f; vitellogenesis in, 155–56
oviducts: anatomy of, 154–55, 155f; egg development in, 155, 159–60, 159f; in ovarian cycle, 156–57
oviposition, 160; dangers of, 160; endocrine regulation of, 127, 128, 131, 160. *See also* nesting behavior
ovulation, 154, 158, 161
Owens, D. W., 72–73, 76, 122, 153, 194, 196, 197–98, 199
oxygen, in blood and muscles, 125

Pacific Ocean: arribada density in, 62t, 69f; arribada research on, 8–9; arribada sites in, 15, 60–61, 62t, 64f; land-based harvest along, 233–34, 238; lunar cycles and, 65; migration in, 219, 220f, 221, 223; in phylogeography, 109–11, 110f, 112f; population stability in, 238; sea-based interactions in, 276–81; solitary nesting sites in, 61, 63t, 64f; species distribution in, 53
Padre Island National Seashore (PAIS/PINS): head-starting at, 179, 299, 300; sex determination at, 192–93
painted turtle (*Chrysemys picta*): migration by, 198; nesting behavior of, 68
PAIS. *See* Padre Island National Seashore
Pakistan: hatcheries in, 242; land-based harvest in, 235, 239; sea-based interactions in, 264–65
palate, 51
Palmer, K. S., 195
Panama: arribadas in, 15; hatcheries in, 242; land-based harvest in, 233, 238; population stability in, 238; sea-based interactions in, 279
Panama, Isthmus of, 54, 109
Pandav, B., 70, 239
Papua New Guinea, sea-based interactions in, 274
Paris, B., 233
Parsons, J. J., 11
passive integrated transponder (PIT) tags, 90, 162, 301
paternity, multiple, 113–15, 114t, 195, 203–4
penis, 152
peripheral bones, 48–49
Peru, sea-based interactions in, 278, 280–81
Pesca, Departamento de (Mexico), 18
Pesquera Industria de Oaxaca (PIOSA), 16, 25–26
Peters, J. A., 158
phenotypic variation, in arribada versus solitary nesters, 76
Philippines, sea-based interactions in, 274, 276
philopatry: and population genetics, 111–13; and significance of arribadas, 12
phylogenetics, 46–48, 108, 110f
phylogeography, 107–15
physiology: endocrine, 126–43; reproductive, 151–63; respiratory, 119–26; of sex determination, 175–76
PINS. *See* Padre Island National Seashore
PIOSA. *See* Pesquera Industria de Oaxaca
PIT. *See* passive integrated transponder
pituitary hormones: in stress response, 133; techniques for measuring, 127–28
pivotal temperatures: definition of, 168, 169f, 192; evaluation of, 172–73, 193; evolution of, 170, 172; for Kemp's versus olive ridleys, 171–73, 173t; of other turtle

species, 172–73, 173t; variation in, 168, 172, 183, 192, 193
plasma T: in sex determination, 196; and sex ratios, 198–200
plastra, 49–50, 152, 193
Plotkin, P. T., 66, 67, 70, 76, 153, 260
poaching, connotations of term, 254
Podocnemis expansa, nesting behavior of, 68
politics: of egg collection, 29–37; of head-starting, 316
pollution: in Atlantic Ocean, 261, 263; forms of, 257; in Gulf of Mexico, 257; in Indian Ocean, 266, 271–72; Kemp's ridleys affected by, 257; in Pacific Ocean, 276, 281; in Sunda Sea, 275
polyandry. *See* multiple mating
Polycryptodira, 47
polygyny. *See* multiple mating
polymorphism, in nesting behavior, 4, 60, 73–78
Polynesia, sea-based interactions in, 278–79
Pomaros, 233
population genetics, 107–15; multiple mating and, 70–71, 113–15, 114t, 195; philopatry and, 111–13
population growth: cases of, 240, 329–30; future of, 333–34; management for, 19; in recovery, 325, 329–34; techniques for estimating, 240
population models, 191–207; assumptions of, 201; conceptual representation of, 203f; definition of variables in, 202t; development of, 201; future research needed on, 206–7; life histories in, 191, 194–95; limits on data used in, 191, 194–95; in management decisions, 191, 194; and recovery, 325–26, 329–34; sex ratios in, 191–92, 200–206, 203f, 204f
population stability: cases of, 240; egg collection and, 13, 236–40, 243, 244t; hatcheries in, 241–42; land-based harvest's impact on, 236–45, 244t
precautionary principle, 255
predator satiation hypothesis (PSH), 72–73; density-dependent nest mortality and, 76, 77f; and reproductive polymorphism, 73, 75
prehistoric human–turtle interactions, 254–55
Prescott, R., 197
preservation, management for, 15–16
primordial abundance, management for, 17–19
Pritchard, P. C. H., 29, 46, 48, 54, 61, 67, 76, 109, 162, 260, 326
PRL. *See* prolactin
Proceedings of the First International Symposium on Kemp's Ridley Sea Turtle Biology, Conservation and Management (Caillouet and Landry), 4
Procoelocryptodira, 47
Procolpochelys grandaeva, 48
Procuraduria Federal de Protección al Ambiente (PROFEPA, Mexico), 25, 27
productive conservation: definition of, 30; in Nicaragua, 30, 32
Productos Pesqueros Mexicanos (PROPEMEX), 26, 27
PROFEPA. *See* Procuraduria Federal de Protección al Ambiente
progesterone: in ovarian cycle, 156–57, 157f; in oviposition, 160; in ovulation, 158
Project TAMAR, 259
prolactin (PRL), 127, 128
PROPEMEX. *See* Productos Pesqueros Mexicanos
protein hormones, techniques for measuring, 128
proteins, thyroid hormones binding to, 139–40, 139t, 142
Protostegidae, 47
Pseudemys scripta. See slider turtle
PSH. *See* predator satiation hypothesis
public awareness, in recovery efforts, 328–29
Punta Ratón (Honduras): economics of egg collection in, 28–29; hatcheries in, 242
Puppigerus camperi, 49
pygal bone, 49

radioimmunoassay (RIA), 127, 138, 192, 194, 196
radio tracking, of head-started turtles, 304
Rajagopalan, M., 272
Rajasekhar, P. S., 235, 236
Rancho Nuevo (Mexico): arribadas in, 8, 14, 15, 52, 61; hatcheries at, 176–79, 177f, 242, 326; head-starting at, 298, 300, 326–27; history of recovery efforts at, 326–27, 327f; land-based harvest at, 237; migration to and from, 217–19, 218f, 221, 222; nesting patterns in, 160–62, 161f; population genetics in, 113–14; population stability at, 237, 240, 329–30; and sea-based interactions, 255, 256, 326; sex determination at, 173, 192–93; sex ratios at, 176–79, 177f, 178f, 182
Rathke's glands: and arribadas, 14, 67; functions of, 50, 67; and plastra, 50; secretions of, 67
Ratones, Isla, 12
Raut, S. K., 269
recapture. *See* mark–recapture
recovery, 325–34, 337–39; controversies, 337–38; current status of, 329–30, 330t; future of, 333–34, 337–39; history of, 326–29, 328t, 337–39; reasons for, 330–33. *See also* conservation
red cell ion transport, 124–26
refuges. *See* reserves and refuges
Reichart, H. A., 259, 260
religion, and turtle meat, 233, 234, 264, 265, 268, 269
reproduction: in captivity, 307–8; endocrine regulation of, 127, 128, 132, 141, 158; migration after, 217–21, 218f, 220f; migration to sites for, 221–23; nearshore, 222–23; offshore, 223. *See also* arribadas; nesting
reproductive convergence, panspecific, 52
reproductive physiology, 151–63; female, 154–63; gaps in understanding of, 152, 160–61; male, 152–53, 162; model for studying, 152; unique aspects of, 151
reproductive polymorphism, 4, 60, 73–78; evolution of, 73–78; fitness and, 72, 75–78; in other organisms, 73–74
reproductive synchrony, 67–73; in definition of arribadas, 61; effectiveness of, 12–13; and fitness, 69–72, 75–79; mechanisms of, 68–69; in other organisms, 68–70; and predator satiation, 72–73
reproductive system: female, 154–55, 155f; male, 152–53
reserves and refuges: in Costa Rica, 37; in Nicaragua, 30, 34–35; in Suriname, 40
resource tracking, in migration, 215–16
respiratory physiology, 119–26; anesthesia and, 125–26; and conservation, 143; submergence studies on, 120–24, 122t
rhamphothecae, 51
RIA. *See* radioimmunoassay
ribs, 48–49
Richard, J. D., 72
Robinson, D. C., 12, 15, 17, 67
Rostal, D. C., 141, 157, 162, 163, 199
rotation system, in egg collection, 34
Ruckdeschel, C., 52
Ruiz, G. J., 122
Rushikulya (India), sea-based interactions in, 266–67

salinity, and endocrine physiology, 128, 136–37
Salm, R. V., 235
salt balance, endocrine regulation of, 128, 136–37
salt glands, 128, 136–37
Samarajiva, D., 235, 267
San Agustinillo (Mexico), slaughterhouse in, 16, 25, 26, 130
sand composition, 78
Sandinistas (Nicaragua), 30–33, 36
Santa Rosa National Park (Costa Rica), 4–5

São Thomé (island): meat trafficking on, 233, 262; nesting on, 11; sea-based interactions in, 262
Saravanan, S., 239
satellite telemetry: migration studies with, 214, 217–24; of postnesting head-started turtles, 312
scales: carapace and, 49; and evolutionary relationships, 46, 47
scapular fork, 50
scapular neck, 50
Schevill, W., 53
Schmid, J. R., 90, 91, 92, 93, 103, 197
Schulz, J. P., 39–40
Schwantes, N., 130
SCL. *See* straight carapace length
scutes: carapace and, 49; and evolutionary relationships, 46, 47, 54; of Kemp's versus olive ridleys, 54; plastra and, 50
sea-based interactions, 253–82; and arribada management strategies, 16; in Atlantic Ocean, 255, 258–64; geographic distribution and, 253, 255; in Gulf of Mexico, 255–58; habitat preference and, 253–54; history of, 254–55; in Indian Ocean, 264–73; with Kemp's ridleys, 255–58, 282; with olive ridleys, 258–82; in Pacific Ocean, 276–81; in Sunda Sea, 273–76. *See also* fishing
seasonality: of ovarian cycle, 156–57, 157f, 162; of testicular cycle, 153, 153f, 154f, 162; of thyroid physiology, 140–41, 142; of vitellogenesis, 155–56, 156f
Sea Turtle Inc., 329
Sea Turtle Restoration Project, 34
Sea Turtles of Santa Rosa, The (Cornelius), 4–5
Secretaría del Medio Ambiente Recursos Naturales y Pesca (SEMARNAP), 27
selection: differential, in arribada versus solitary nesters, 76–78; natural, and sex ratios, 170
SEMARNAP. *See* Secretaría del Medio Ambiente Recursos Naturales y Pesca
seminiferous tubules, 153, 174
Senegal, sea-based interactions in, 262
sensory biology, of cue processing, 67
Sergipe (Brazil): land-based harvest in, 232, 237; population size in, 240, 259; sea-based interactions in, 258–61
Seris, 233, 277
sex: and growth rates, 101f, 102, 102f, 103; and migration, 216–24, 218f, 220f; and stress response, 130, 132, 134, 134f; techniques for determining, 181, 192, 193–94, 196; and thyroid hormones, 140–41
sex characteristics, secondary: female, 193; male, 152, 193; in sexing turtles, 192, 193
sex determination, temperature-dependent (TSD), 167–76, 183; characteristics of, 168, 192; and conservation efforts, 167, 170–71, 191–93; ecological implications of, 170; estrogen in, 175, 192; evolutionary theory and, 169–70; gonadal differentiation in, 174–75, 174f; in Kemp's versus olive ridleys, 171–73; migration constraints and, 216; patterns of, 168, 169f; physiology of, 175–76; pivotal temperatures in, 168, 170–73, 192, 193; temperature variation in, 168, 193
sexing, techniques for, 181, 192, 193–94, 196
sex ratios, 167–84, 191–207; of adults, 179–82, 180t; body size and, 196–97, 197f, 200; conservation implications of, 167, 168, 170–71, 176, 192–93; difficulty of assessing, 191; ecological implications of, 170; evolutionary theory and, 169–70; future research needed on, 206–7; of hatchlings, 176–79; in head-start programs, 179–81, 196, 197, 300; human manipulation of, 170–71, 193, 195, 202, 204–6; of juveniles, 179–82, 180t; methodology for characterizing, 195–97; in population models, 191–92, 200–206, 203f, 204f; review of literature on, 197–200, 198t; with temperature-dependent sex determination, 167, 170; testosterone and, 181, 198–99, 199f, 200; in transitional range of temperatures, 168
sexual dimorphism, 193–94
sexual maturation, age at. *See* age at sexual maturation
Shanker, K., 109, 111–13
Shaver, D. J., 171, 197, 200, 305
Shaw, S. L., 125
shell formation, 155, 158, 159, 159f
shoulder girdles, 50
shrimp trawling. *See* incidental capture; trawler mortality
size. *See* body size
size-at-age data: from head-starting, 91; in skeletochronology, 93, 95, 96–98
skeletochronology, 92–103; age and growth estimates with, 90, 92–95, 95t, 96f, 97f, 97t, 98t, 99f, 100f, 101f, 102f; application of, 93–94; future research with, 102–3; methodology of, 92–93, 94–95; problems with, 92; results of, 96–102; validation of, 93
skull, 51
slaughter. *See* land-based harvest
slider turtle (*Pseudemys scripta*): migration by, 198–99; respiratory physiology of, 124; thyroid physiology of, 138–39, 140, 141, 142
snapping turtle (*Chelydra serpentina*), 195
Snover, M. L., 93, 94
social aspects, of egg collection, in Costa Rica, 37–39
social facilitation: of arribadas, 66–67; definition of, 66
solitary nesting, 60, 73–79; and fitness, 75–79, 75t, 77f; frequency of, 73; geographic extent of, 61, 63t, 64f, 65f; lack of literature on, 75, 338; migration for, 222; reproductive characteristics of, 76, 76t; stress response during, 132
Somoza, Anastasio, 30, 31
sonic pollution, 257, 263
South Carolina, 54–55
Southeast Asia: land-based harvest in, 234–35; sea-based interactions in, 273–76
SOX9 gene, 175
species divergences, 108
sperm, storage of, 154, 195, 204
spermatids, 153
spermatocytes, 153
spermatogenesis, seasonality of, 153, 153f, 154f
Sri Lanka: arribadas in, 15, 53; hatcheries in, 242; land-based harvest in, 235, 236f, 239; migration near, 219; population genetics in, 111; population stability in, 239; sea-based interactions in, 267–73
Stabenau, E. K., 123, 124, 125, 197, 200
steroid hormones: salinity and, 136–37; techniques for measuring, 128
Stewart, A. Y., 30–34, 36, 42
straight carapace length (SCL): maximum, 89; sex ratios and, 197, 197f; in skeletochronology, 94, 96–98, 96f, 97f, 97t, 98t
strandings: in Atlantic Ocean, 52, 263; in Gulf of Mexico, 256–57; of head-started turtles, 307; in Indian Ocean, 265–66, 270–71; of Kemp's ridleys, 256–57; in Pacific Ocean, 280; and sex ratios, 179–81, 180t
stress, 128–37; and adrenal physiology, 128–36; catecholamine hormones and, 137; definition of, 129; glucose and, 135–36, 137, 138; and respiratory physiology, 126; salinity and, 136–37
stressors: definition of, 129; land-based, 239
stress response, 128–37
stress system, concept of, 129
Suárez, Antonio, 8–9, 16, 26
Subba Rao, M. V., 235, 236

submergence studies, 120–24, 122t, 126
subsistence use, history of, 232, 233, 234, 238
Sugg, D. W., 195, 205
Sunda Sea, sea-based interactions in, 273–76
supraoccipital process, 51
suprapygals, 49
Suriname: arribadas in, 9–10, 14, 53; cultural aspects of use in, 39–40, 232; decline in arribadas in, 113, 259; egg collection in, 39–40, 232, 237; history of human use in, 39–40, 232; land-based harvest in, 232, 236, 237; mating systems in, 113, 114; population genetics in, 111, 113, 114; population stability in, 236, 237, 259; sea-based interactions in, 258–61. *See also* Eilanti Beach
survivorship: of head-started turtles, 306–7; in population models, 195; in recovery, 330–34
sustainable use: connotations of term, 254; in definition of conservation, 23–24
synchronous reproduction. *See* reproductive synchrony

T_3. *See* triiodothyronine
T_4. *See* thyroxine
tagging: fecundity estimates from, 162; in head-starting, methodology for, 301; in migration studies, 214–15. *See also* mark–recapture data
take, neutrality of term, 254. *See also* directed take
TAMAR, Project, 259
Tamaulipas (Mexico), 8, 15; decline in arribadas in, 13; egg harvesting in, 17–18; history of recovery efforts at, 326–29, 327f; management of, 17–18; population genetics in, 113–14; in zoogeography, 52
Tamil Nadu (India): land-based harvest in, 235, 239; sea-based interactions in, 267, 268, 270
Tamils, 267, 268
Tangley, Laura, 8
Tanzania, sea-based interactions in, 265
tar, 263
Taubes, G., 316
taxonomy, 12, 45–46
TED. *See* turtle excluder device
TED wars, 257–58, 282
Tehuantepec, Isthmus of, 54
temperature, body, and thyroid hormones, 142
temperature, water: in migration, 216; and testicular cycle, 153; and thyroid hormones, 140–41
temperature-dependent sex determination (TSD). *See* sex determination
testes: anatomy of, 152–53, 154f; sexual differentiation in, 174–75, 174f
testicular cycle, 153, 153f, 154f, 162
testosterone, 133; in female ovarian cycle, 153f, 156–57; in male testicular cycle, 153, 153f; in mating, 158; in migration, 198–99, 199f; in nesting patterns, 161, 161f; in oviposition, 160; reproductive status determined from, 162–63; and sex ratios, 181, 198–99, 199f, 200; thyroid hormones and, 141
Testudines, 46, 47
Testudo mydas minor, 46
tetraiodothyronine. *See* thyroxine
TEWG. *See* Turtle Expert Working Group
Texas: arribadas in, 15; hatcheries in, 242; head-started turtles nesting in, 308–11, 316; history of recovery efforts in, 327–28
Thailand: head-starting in, 298; land-based harvest in, 239; sea-based interactions in, 274, 275
Thalassochelys kempii, 45
Thalassochelys olivacea, 46
Theobald, W., 235
thermosensitive period, of incubation, 168, 171
Thinochelys, 49
Thorbjarnarson, J. B., 239
thyroid hormones, 138–43; binding to proteins, 139–40, 139t, 142; and conservation, 138, 143; functions of, 138; regulation of, 138; techniques for measuring, 128, 140
thyroid physiology, 128f, 138–43
thyrotropin (TSH), 127, 138
thyrotropin-releasing hormone (TRH), 138
thyroxine (T_4), 138–43
tidal cycles, and arribadas, 65–66
time series, 325–26, 326f
Tlacoyunque, Piedra de (Mexico), 9
Togo: land-based harvest in, 233; sea-based interactions in, 262
torpor, winter, 52
total protection policy, 15–16
tourism. *See* ecotourism
Toxochelyidae, 47
Toxochelys, 49
transitional range of temperatures (TRT): definition of, 168; in Kemp's versus olive ridleys, 171, 172
trawler mortality: annual estimates of, 120; and arribada management, 16; in Atlantic Ocean, 10, 259–60; in Gulf of Mexico, 256–57; of head-started turtles, 307; in Indian Ocean, 265–66, 269–73; of Kemp's ridleys, 256–57; in Pacific Ocean, 276, 279–81; and population stability, 13, 237; reduction of, 329, 334; submergence times and, 120–24; in Sunda Sea, 274–75. *See also* incidental capture
Trebbau, P., 46
TRH. *See* thyrotropin-releasing hormone
triiodothyronine (T_3), 138–43
Trinidad, H., 25–27, 28, 41
Tripathy, B., 235
trochanters, 51
TRT. *See* transitional range of temperatures
TSD. *See* sex determination, temperature-dependent
TSH. *See* thyrotropin
tubercles, vertebral, 49
tuna fishery, 281
tuna migration, 215
turning stress, 132, 133–35, 133f, 134f, 136
turtle excluder device (TED) programs: certification trials for, 121–24; development of, 121–24, 256, 257; effectiveness of, 329; in management strategies, 18; and respiratory physiology, 121–24; wars over, 257–58, 282; in West Africa, 264
Turtle Expert Working Group (TEWG), 329, 331–32
Turtle Paradise Gahirmatha, The (Dash and Kar), 4

Ubina, Magali, 31–32
ultrasonography, 159, 159f, 163
University of Costa Rica, 37
Uruguay, sea-based interactions in, 260–61
USFWS. *See* Fish and Wildlife Service, U.S.

Valverde, R. A., 132, 196
Vargas, P., 156
VBGF. *See* von Bertalanffy growth function
Venezuela, sea-based interactions in, 259–60
ventral prong, of scapular fork, 50
vertebrae, 51; cervical, 51; dorsal, 48
vertebral tubercles, 49
vertebrates, stress in, 128–29, 135
Vezo, 264
Vietnam, sea-based interactions in, 274–75
Vietti, K. R. N., 123, 124
Vipera berus. See adder
Virginia, 52
Vishnu (god), 264
vision, in arribadas, 50, 67
vitamin D, 139
vitellogenesis, 141, 155–56, 156f
vitellogenin, 141
von Bertalanffy growth function (VBGF), 91–92, 91t, 95, 96–98

Waku Waku Animal World, 9
water balance, endocrine regulation of, 128

weather, and arribadas, 14, 62–65
Werler, J. E., 8
Wermuth, H., 46
West Africa: arribadas in, 11, 15, 54; land-based harvest in, 233, 237–38; population stability in, 243; sea-based interactions in, 261–64
West Bengal, sea-based interactions in, 267–69, 271
Wibbels, T. R., 153, 155, 156, 174, 175, 198
wildebeest, 73
Wildlife Conservation Law 6919 (Costa Rica), 37
Williams, E. E., 51
Wilson, J., 25–27, 28, 41
winds, and arribadas, 14, 62–65
Windward Road, The (Carr), 8, 46
winter: growth rates in, 90; torpor in, 52
Witzell, W. N., 91, 93
Woody, J. B., 315, 316
World Bank, 27
World Conservation Union (IUCN), conservation defined by, 23–24
World Trade Organization, 258, 282
World Wildlife Fund, 10, 40

xiphiplastra, 49–50

yolk: in ovarian cycle, 157; during oviductal egg development, 159, 159f; vitellogenesis and, 155, 156

Zaborski, P., 194
Zangerl, R., 46–47, 50
zoogeography, 51–55
Zug, G. R., 92, 93, 100–101